METHODS IN MOLECULAR BIOLOGY™

Series Editor
John M. Walker
School of Life Sciences
University of Hertfordshire
Hatfield, Hertfordshire, AL10 9AB, UK

For further volumes:
http://www.springer.com/series/7651

Neurodegeneration

Methods and Protocols

Edited by

Giovanni Manfredi

Department of Neurology and Neuroscience, Weill Medical College, Cornell University, New York, NY, USA

Hibiki Kawamata

Weill Medical College of Cornell University, New York, NY, USA

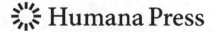 Humana Press

Editors
Giovanni Manfredi
Department of Neurology and Neuroscience
Weill Medical College
Cornell University
New York, NY, USA
gim2004@med.cornell.edu

Hibiki Kawamata
Weill Medical College of Cornell University
New York, NY, USA
hik2004@med.cornell.edu

ISSN 1064-3745   e-ISSN 1940-6029
ISBN 978-1-61779-327-1   e-ISBN 978-1-61779-328-8
DOI 10.1007/978-1-61779-328-8
Springer New York Dordrecht Heidelberg London

Library of Congress Control Number: 2011935542

Printed on acid-free paper

Humana Press is part of Springer Science+Business Media (www.springer.com)

Preface

Neurodegenerative disorders, such as Alzheimer's and Parkinson's disease, are becoming a growing burden on our society, in parallel with the aging of the world's population. Understanding and treating these common diseases is one of the major current challenges for the biomedical field.

In *Neurodegeneration: Methods and Protocols*, experts in the field tackle this challenge, and present cutting edge analytical and technological approaches to study the mechanisms underlying neurodegenerative processes.

Chapters include protocols for isolating and culturing cells from the nervous system, production and use of animal models, characterization of cell death, analytical tools to study disease mechanisms, and therapeutic approaches for neurodegeneration.

Each chapter composed in the highly successful *Methods in Molecular Biology* series format contains introduction of the topic, list of necessary reagents, step-by-step methods, and notes on troubleshooting and avoiding known pitfalls, for reproducible laboratory results.

Target audience: Cell biologists, molecular biologists, biochemists, pathologists, and researchers in the field of neurodegeneration.

New York, NY, USA   *Giovanni Manfredi*
New York, NY, USA   *Hibiki Kawamata*

Contents

PART I   NEURODEGENERATION

PART II   NEURAL CELL CULTURE TECHNIQUES

PART III   MODELS OF NEURODEGENERATION

Contributors

ANDREY Y. ABRAMOV • *Department of Molecular Neuroscience, UCL Institute of Neurology, London, UK*

BRIAN J. BACSKAIQ • *MassGeneral Institute for Neurodegenerative Diseases, Massachusetts General Hospital, Charlestown, MA, USA*

NIGEL S. BAMFORD • *Department of Neurology, University of Washington and Seattle Children's Hospital, Seattle, WA, USA; Department of Pediatrics and Psychology, University of Washington and Seattle Children's Hospital, Seattle, WA, USA; The Graduate Program in Neurobiology and Behavior, University of Washington and Seattle Children's Hospital, Seattle, WA, USA; Center on Human Development and Disability, University of Washington and Seattle Children's Hospital, Seattle, WA, USA*

ANTONI BARRIENTOS • *Department of Neurology, University of Miami, Miller School of Medicine, Miami, FL, USA; Department of Biochemistry and Molecular Biology, University of Miami, Miller School of Medicine, Miami, FL, USA*

M. FLINT BEAL • *Department of Neurology and Neuroscience, Weill Cornell Medical College, New York, NY, USA*

EILEEN H. BIGIO • *Department of Pathology, Northwestern Alzheimer Disease Center, Chicago, IL, USA*

GUY A. CALDWELL • *Department of Biological Sciences, The University of Alabama, Tuscaloosa, AL, USA; Department of Neurology, University of Alabama at Birmingham, Birmingham, AL, USA; Department of Neurobiology, Center for Neurodegeneration and Experimental Therapeutics, University of Alabama at Birmingham, Birmingham, AL, USA*

KIM A. CALDWELL • *Department of Biological Sciences, The University of Alabama, Tuscaloosa, AL, USA; Department of Neurology, University of Alabama at Birmingham, Birmingham, AL, USA; Department of Neurobiology, Center for Neurodegeneration and Experimental Therapeutics, University of Alabama at Birmingham, Birmingham, AL, USA*

STUART M. CHAMBERS • *Developmental Biology Program, Sloan-Kettering Institute, New York, NY, USA*

FUJUN CHEN • *Department of Neuroscience, UT Southwestern Medical Center, Dallas, TX, USA*

CHRISTOS CHINOPOULOS • *Department of Medical Biochemistry, Semmelweis University, Budapest, Hungary*

CHARLEEN T. CHU • *Department of Pathology, University of Pittsburgh School of Medicine, Pittsburgh, PA, USA*

ANDREW COOPER • *Repligen Corporation, Waltham, MA, USA*

GIOVANNI COPPOLA • *Program in Neurogenetics, Department of Neurology, David Geffen School of Medicine, #1524 Gonda Neuroscience and Genetics Research Center, University of California Los Angeles, Los Angeles, CA, USA*

RUBEN K. DAGDA • *Department of Pathology, University of Pittsburgh School of Medicine, Pittsburgh, PA, USA*

HAN-XIANG DENG • *Division of Neuromuscular Medicine, Davee Department of Neurology and Clinical Neurosciences, Northwestern University Feinberg School of Medicine, Chicago, IL, USA*

MARK DENHAM • *Centre for Neuroscience, University of Melbourne, Parkville, VIC, Australia*

MIRELLA DOTTORI • *Centre for Neuroscience, University of Melbourne, Parkville, VIC, Australia; Department of Pharmacology, University of Melbourne, Parkville, VIC, Australia*

MICHAEL R. DUCHEN • *Department of Cell and Developmental Biology and Consortium for Mitochondrial Research, University College London, London, UK*

MAGALI DUMONT • *Department of Neurology and Neuroscience, Weill Cornell Medical College, New York, NY, USA*

Maria Figueiredo-Pereira • *Department of Biological Sciences, Hunter College of City University of New York, New York, NY, USA*

KINDIYA GEGHMAN • *Department of Neurology and Neuroscience, Weill Cornell Medical College, New York, NY, USA*

BERNARDINO GHETTI • *Department of Pathology and Laboratory Medicine, Indiana Alzheimer Disease Center, Indiana University School of Medicine, Indianapolis, IN, USA*

DAVID GIFONDORWA • *Department of Neurobiology and Anatomy, Wake Forest University School of Medicine, Winston-Salem, NC, USA*

JOEL M. GOTTESFELD • *Department of Molecular Biology, The Scripps Research Institute, La Jolla, CA, USA*

GUNNAR K. GOURAS • *Experimental Dementia Research Unit, Department of Experimental Medical Science, Wallenberg Neuroscience Center, Lund University, Lund, Sweden, NY, USA*

ADAM J. HARRINGTON • *Department of Biological Sciences, The University of Alabama, Tuscaloosa, AL, USA*

QIAN HUANG • *Department of Biological Sciences, Hunter College of City University of New York, New York, NY, USA*

BRADLEY T. HYMAN • *MassGeneral Institute for Neurodegenerative Diseases, Massachusetts General Hospital, Charlestown, MA, USA*

COSTANTINO IADECOLA • *Division of Neurobiology, Department of Neurology and Neuroscience, Weill Cornell Medical College, New York, NY, USA*

KATHERINE JACKMAN • *Division of Neurobiology, Department of Neurology and Neuroscience, Weill Cornell Medical College, New York, NY, USA*

MARLON JANSEN • *Department of Biological Sciences, Hunter College of City University of New York, New York, NY, USA*

MICHAEL G. KAPLITT • *Laboratory of Molecular Neurosurgery, Department of Neurological Surgery, Weill Cornell Medical College, New York, NY, USA*

HIBIKI KAWAMATA • *Department of Neurology and Neuroscience, Weill Cornell Medical College, New York, NY, USA*

HYUN JEONG KIM • *Department of Neurology and Neuroscience, Weill Cornell Medical College, New York, NY, USA*

Doo Yeon Kim • *Neurobiology of Disease Laboratory, Genetics and Aging Research Unit, MassGeneral Institute for Neurodegenerative Disease, Massachusetts General Hospital, Harvard Medical School, Charlestown, MA, USA*

Peter Klivenyi • *Department of Neurology, University of Szeged, Szeged, Hungary*

Dora M. Kovacs • *Neurobiology of Disease Laboratory, Genetics and Aging Research Unit, MassGeneral Institute for Neurodegenerative Disease, Massachusetts General Hospital, Harvard Medical School, Charlestown, MA, USA*

Kishore V. Kuchibholta • *MassGeneral Institute for Neurodegenerative Disease, Massachusetts General Hospital, Harvard Medical School, Charlestown, MA, USA*

Alexander Kunz • *Division of Neurobiology, Department of Neurology and Neuroscience, Weill Cornell Medical College, New York, NY, USA*

Angela S. Laird • *Laboratory of Neurobiology, Department of Neurology, K.U. Leuven, Leuven, Belgium; Vesalius Research Center, VIB, Leuven, Belgium*

Carli Lattarulo • *MassGeneral Institute for Neurodegenerative Disease, Massachusetts General Hospital, Harvard Medical School, Charlestown, MA, USA*

Angelo C. Lepore • *Department of Neurology, The Johns Hopkins University School of Medicine, Baltimore, MD, USA*

Chenjian Li • *Department of Neurology, Friedman Brain Institute, Mount Sinai School of Medicine, New York, NY, USA*

Weichun Lin • *Department of Neuroscience, UT Southwestern Medical Center, Dallas, TX, USA*

Yun Liu • *Department of Neuroscience, UT Southwestern Medical Center, Dallas, TX, USA*

Jordi Magrané • *Department of Neurology and Neuroscience, Weill Cornell Medical College, New York, NY, USA*

Giovanni Manfredi • *Department of Neurology and Neuroscience, Weill Cornell Medical College, New York, NY, USA*

Nicholas J. Maragakis • *Department of Neurology, The Johns Hopkins University School of Medicine, Baltimore, MD, USA*

Roberta Marongiu • *Laboratory of Molecular Neurosurgery, Department of Neurological Surgery, Weill Cornell Medical College, New York, NY, USA*

Maria Jose Metcalfe • *Department of Biological Sciences, Hunter College of City University of New York, New York, NY, USA*

Yvonne Mica • *Gerstner Sloan-Kettering Graduate School of Biomedical Sciences, Sloan-Kettering Institute, New York, NY, USA*

Carol Milligan • *Department of Neurobiology and Anatomy, Wake Forest University School of Medicine, Winston-Salem, NC, USA*

Teresa A. Milner • *Division of Neurobiology, Department of Neurology and Neuroscience, Weill Cornell Medical College, New York, NY, USA*

Peter F. Morgenstern • *Laboratory of Molecular Neurosurgery, Department of Neurological Surgery, Weill Cornell Medical College, New York, NY, USA*

Sergei A. Musatov • *Neurologix, Inc., Ft. Lee, NJ, USA*

Natura Myeku • *Department of Biological Sciences, Hunter College of City University of New York, New York, NY, USA*

NICOLE NANICHE • *Frances and Joseph Weinberg Unit for ALS research, Farber Institute for Neuroscience, Thomas Jefferson University, Philadelphia, PA, USA; Department of Neuroscience, Thomas Jefferson University, Philadelphia, PA, USA*

ALEJANDRO OCAMPO • *Department of Biochemistry and Molecular Biology, University of Miami, Miller School of Medicine, Miami, FL, USA*

PIERA PASINELLI • *Frances and Joseph Weinberg Unit for ALS research, Farber Institute for Neuroscience, Thomas Jefferson University, Philadelphia, PA, USA; Department of Neuroscience, Thomas Jefferson University, Philadelphia, PA, USA*

JOSEPH P. PIERCE • *Division of Neurobiology, Department of Neurology and Neuroscience, Weill Cornell Medical College, New York, NY, USA*

HEATHER L. PLASTERER • *Repligen Corporation, Waltham, MA, USA*

LINGHUA QIU • *Department of Biochemistry and Molecular Pharmacology, University of Massachusetts Medical School, Worcester, MA, USA*

WIM ROBBERECHT • *Laboratory of Neurobiology, Department of Experimental Neurology, K.U. Leuven, Leuven, Belgium; Vesalius Research Center, VIB, Leuven, Belgium; Department of Neurology, K.U. Leuven, Leuven, Belgium*

DANIELLE C. ROBINSON • *Harold and Margaret Milliken Hatch Laboratory of Neuroendocrinology, The Rockefeller University, New York, NY, USA*

JAMES R. RUSCHE • *Repligen Corporation, Waltham, MA, USA*

THOMAS SCHMIDT-GLENEWINKEL • *Department of Biological Sciences, Hunter College of City University of New York, New York, NY, USA*

TEEPU SIDDIQUE • *Division of Neuromuscular Medicine, Davee Department of Neurology and Clinical Neurosciences, Northwestern University Feinberg School of Medicine, Chicago, IL, USA*

ELISABETTA SORAGNI • *Department of Molecular Biology, The Scripps Research Institute, La Jolla, CA, USA*

ANATOLY A. STARKOV • *Department of Neurology and Neuroscience, Weill Cornell Medical College, New York, NY, USA*

LORENZ STUDER • *Developmental Biology Program, Department of Neurosurgery, Sloan-Kettering Institute, New York, NY, USA*

DANIELA SAU • *Istituto di endocrinologia e Centro di eccellenza sulle Malattie Neurodegenrative, Univerista' di Milano, Milan, Italy*

YOSHIE SUGIURA • *Department of Neuroscience, UT Southwestern Medical Center, Dallas, TX, USA*

DAVID SULZER • *Department of Neurology, Columbia University, New York, NY, USA; Department of Pharmacology, Columbia University, New York, NY, USA*

DAVIDE TAMPELLINI • *Experimental Dementia Research Unit, Department of Experimental Medical Science, Wallenberg Neuroscience Center, Lund University, Lund, Sweden, NY, USA*

VADIM TEN • *Department of Pediatrics, Columbia University, New York, NY, USA*

BOBBY THOMAS • *Department of Neurology and Neuroscience, Weill Cornell Medical College, New York, NY, USA*

DIANA THYSSEN • *MassGeneral Institute for Neurodegenerative Diseases, Massachusetts General Hospital, Charlestown, MA, USA*

MARK J. TOMISHIMA • *SKI Stem Cell Research Facility, Sloan-Kettering Institute, New York, NY, USA*

MICHELLE L. TUCCI • *Department of Biological Sciences, The University of Alabama, Tuscaloosa, AL, USA*

LASZLO VECSEI • *Department of Neurology, University of Szeged, Szeged, Hungary*

RUBEN VIDAL • *Department of Pathology and Laboratory Medicine, Indiana Alzheimer Disease Center, Indiana University School of Medicine, Indianapolis, IN, USA*

ELIZABETH M. WATERS • *Division of Neurobiology, Department of Neurology and Neuroscience, Weill Cornell Medical College, New York, NY, USA*

CHUNPING XU • *Department of Molecular Biology, The Scripps Research Institute, La Jolla, CA, USA*

ZUOSHANG XU • *Department of Biochemistry and Molecular Pharmacology, Cell Biology, Neuroscience Program, University of Massachusetts Medical School, Worcester, MA, USA*

CHUNXING YANG • *Department of Biochemistry and Molecular Pharmacology, University of Massachusetts Medical School, Worcester, MA, USA*

LICHUAN YANG • *Department of Neurology and Neuroscience, Weill Cornell Medical College, New York, NY, USA*

CHUN-HUNG YEH • *Department of Biological Sciences, Hunter College of City University of New York, New York, NY, USA*

MINERVA Y. WONG • *Department of Neurology, Columbia University, New York, NY, USA*

STEVEN F. ZHANG • *Department of Neurology and Neuroscience, Weill Cornell Medical College, New York, NY, USA*

JIANHUI ZHU • *Department of Pathology, University of Pittsburgh School of Medicine, Pittsburgh, PA, USA*

Part I

Neurodegeneration

Chapter 1

Introduction to Neurodegenerative Diseases and Related Techniques

Hibiki Kawamata and Giovanni Manfredi

Abstract

Neurodegenerative disorders are common diseases that afflict our society with tremendous medical and financial burdens. As a whole, neurodegeneration strikes individuals of all ages, but becomes increasingly frequent with age, coming to affect a very large share of our elderly population. Due to the very complex nature of these diseases, which often result from combined genetic and environmental pathogenic factors, the scientific community that researches the causes and the therapy of neurodegeneration faces remarkable challenges, requiring constant technological advancements. This book describes a collection of experimental protocols used to study various aspects of neurodegeneration. In this introduction, we provide an overview of the field and the comprehensive technological approaches required to advance the knowledge in the pathogenesis and cure of neurodegeneration. Then, we introduce the general topics addressed herein with a brief description of each chapter.

1. Neurodegeneration and the Challenges of the Field

The term neurodegeneration encompasses a large number of disease conditions, which affect various parts of the nervous system, individually or in combination. Under this very general, and vague, definition are included diseases of the central and the peripheral nervous systems and diseases of infancy, adulthood, and old age. In certain instances, the diseases have clear genetic causes, but in most cases they arise from a complex interplay of predisposing genetic and environmental factors.

When we think of neurodegenerative diseases, our attention focuses on the most common and renowned disorders that affect large shares of the elderly population, such as Alzheimer's (AD) and Parkinson's (PD) diseases. It could be argued that chronic, progressive degeneration of certain neuronal cells and neural functions is

Giovanni Manfredi and Hibiki Kawamata (eds.), *Neurodegeneration: Methods and Protocols*,
Methods in Molecular Biology, vol. 793, DOI 10.1007/978-1-61779-328-8_1, © Springer Science+Business Media, LLC 2011

part of the normal aging process in populations that have experienced a rapid increase in average life expectation in the past decades. However, not every elderly individual presents symptoms of memory loss or movement disorder, posing the question of what are the triggering factors that promote neurodegeneration in certain individuals. Furthermore, it is becoming increasingly clear that several of the most common neurodegenerative disorders are in fact multisystemic disorders that affect different organs and tissues, despite the fact that degeneration of specific neuronal cells may be the most pervasive symptom of the disease. The mechanisms that cause degeneration and death of specific types of neurons, thereby shaping the manifestations of the disease and defining the characteristics of each neurodegenerative disease, are probably the most important targets of discovery in the field.

Within each common neurodegenerative disease condition, there is often a mixture of hereditary and "sporadic" forms. This distinction obviously underscores our ignorance on the true pathogenic causes of these conditions. While, for example, the identity of several genes that are mutated in familial forms of AD, PD, and amyotrophic lateral sclerosis (ALS) is known, we often do not understand precisely the function of these genes and how their mutations cause neuronal degeneration. The converse is also true, where in the majority of sporadic forms we claim the existence of genetic and epigenetic factors that affect the manifestation of the disease, but we have not been able to pinpoint many of them. Nevertheless, the discovery of genetic forms of neurodegeneration is extremely important because it often allows the development of valuable models of disease, utilizing a vast array of living systems, from yeast to mammals, which can then be used to investigate the biology and pathophysiology of the disease processes and to test therapeutic approaches.

The concept of disease modeling is not limited to genetics. In some diseases, environmental toxins have been shown to cause degeneration of particular neuronal populations and to cause diseases that strongly resemble or are phenocopies of the sporadic or genetic forms. Similarly, common neurological conditions, such as cerebral ischemia, are caused by changes in the supplies of fundamental metabolic components, like oxygen and glucose, to tissues and cells. The controlled use of toxic compounds or changing the availability of metabolic supplies allows for the generation of useful experimental models of neurodegeneration in cultured cells and animals.

Studying the causes of neurodegeneration at the pathological, cellular, molecular, and biochemical levels presents many procedural challenges often making it a daunting task. Many neurodegenerative diseases are not cell-autonomous, meaning that multiple non-neuronal cell types, such as astrocytes, microglia, oligodendroglia, Schwann cells, etc., participate directly or indirectly to the disease process, ultimately causing neurons to degenerate. The cellular complexity of the disease process renders the generation of relevant model systems often very difficult, expensive, and time consuming. In addition, the primary neuronal cells that we would like to study often exist in limited numbers and are virtually impossible to derive from adult patients or even from adult animal models. Moreover, in many diseases, the specific populations of neuronal cells affected by the disease are very difficult to access for in situ studies, such as live imaging of cells and tissues, but also for bioptic purposes. For this reason, investigators often have to resort to studying postmortem tissues, which have the obvious limitations of representing a snapshot in time of a terminally degenerated cells and are prone to artifacts associated with cellular death.

Therapeutic attempts aimed at treating or curing neurodegenerative diseases are clearly very challenging. In many diseases, the diagnosis is made only when the neurons involved are already largely dysfunctional or dead. Most neurons are incapable of replication and therefore, even if a therapy can slow down or arrest the disease process, a full recuperation of the lost functions is difficult. On the other hand, the nervous system has great plasticity, which may support residual function for a long time. This capacity may be exploited as a target for therapy. Aside of current pharmacological therapies that seek to compensate for the loss of neurons by providing the lacking neurotransmitters, such as dopamine in PD, recent developments in cell replacement and viral delivery of genetically lacking proteins hold great promises for the future. These approaches have straightforward rationales, but are also often difficult to perform.

2. Contents of This Book

Investigators who approach the field of neurodegeneration have to tackle arduous methodological issues. This book provides step-by-step descriptions of specific protocols that scientists involved in the field have developed in their laboratories with the intent of facilitating the setting up of these protocols. In addition, the book includes some overview chapters that provide the reader with a critical knowledge and reference tools to evaluate the suitability of specific techniques to their research needs.

2.1. Part I: Neurodegeneration

This section of the book describes methods to evaluate the degenerative processes in neuronal cells. Chapter 2 discusses the procedures to investigate apoptotic cell death in cultured neurons. Chapter 3 deals with protocols to image the neurodegenerative processes affecting the nervous system directly at the ultrastructural level using immunoelectron microscopy.

2.2. Part II: Neural Cell Culture Techniques

This part comprises chapters devoted to describing detailed methods on how to generate primary cultured neural cells from animal models and neuronal cells from human-derived pluripotent stem cells. Chapter 4 describes the procedures to dissect and grow primary neurons from mouse embryos and astrocytes from early postnatal mouse brain. Chapter 5 illustrates an unprecedented methodological approach to derive adult motor neurons from mouse spinal cord. Chapters 6 and 7 are complementary to each other and describe novel approaches to handle and differentiate human pluripotent cells into neurons.

2.3. Part III: Models of Neurodegeneration

This section provides a comprehensive view of the many possibilities to model and study neurodegeneration in living systems. It covers a wide range of models from yeast to mice that recapitulate various aspects of genetically or toxin-induced neurodegeneration. Chapter 8 describes methods to exploit the powerful means of yeast genetics to obtain models of inherited forms of neurodegeneration in a convenient setting that offers valuable biochemical tools to investigate disease pathogenesis. Chapter 9 describes the detailed procedures to generate genetically modified worm, *Caenorhabditis elegans*, for the study of neurodegenerative disease, with special focus on degeneration of the dopaminergic neurons that are affected in PD. Chapter 10 deals with the methodology to investigate the impairment of basic cellular functions in another powerful genetic system, the fruit fly, where mutants of virtually any gene can be studied. In Chapter 11, the methods for the production and analyses of genetically modified zebra fish embryo models are illustrated. This is a relatively new model system to investigate genetic modifications and their phenotypic effects in a vertebrate animal, where the ability to silence specific genes allows for a rapid and yet highly efficient way to study gene function and disease pathogenesis. Chapter 12 provides a critical overview of the various approaches to produce and study genetically modified mouse models, illustrating the general principles of standard transgenics, new-generation BAC transgenics, knock-in, and knock-out approaches. Reference tools, as well as advantages and disadvantages for each approach, are described. Chapter 13 describes the steps to set up a unique model of focal cerebral ischemia in mice and to investigate its effects on the brain tissues. Chapter 14 illustrates the methods used to produce and investigate toxicological models of PD in mice while Chapter 15 describes on the precise methodology to measure phenotypic parameters in mouse models of neurodegeneration from PD to ALS to AD.

2.4. Part IV: Neurodegenerative Disease Mechanisms – Proteinopathies

This part of the volume includes three chapters devoted to the methods to identify and characterize dysfunction in cellular protein-handling processes that can lead to neuronal degeneration. Proteinopathies are thought to be at the basis of many common neurodegenerative disorders, such as AD, PD, and ALS. Chapter 16 describes the approaches to detect and characterize amyloid protein pathology in autoptic brain samples. Chapter 17 provides protocols to reveal the presence of pathogenic protein aggregates in degenerating nervous tissues, when the very nature of the aggregated material renders the recognition of the protein epitopes difficult. Chapter 18 illustrates the methods to investigate dysfunctions of the proteasome machinery, which cells depend on to turn over proteins that are no longer needed. When protein turnover fails for a variety of reasons, protein accumulation and aggregation may ensue, leading to neuronal degeneration.

2.5. Part V: Neurodegenerative Disease Mechanisms – Mitochondria

Mitochondria are primarily involved in energy metabolism in most cell types. Many cells, including neurons, are strongly dependent on their mitochondria not only for energy conversion, but also for the biosynthesis and catabolism of cellular building blocks and for the maintenance of calcium homeostasis. When mitochondria fail, neurons often become exposed to severe damage, including energy depletion, oxidative stress, calcium imbalance, and apoptosis. Chapter 19 explains how to image and measure the occurrence of calcium-dependent mitochondrial permeability transition in living cells. Chapter 20 illustrates the protocols for the isolation and measurements of enzymatically active mitochondria from small amounts of brain tissue for the characterization of mitochondrial dysfunction in animal models of neurodegeneration. Chapter 21 addresses the protocol for assaying in neurons mitochondrial autophagy (mitophagy), a crucial step in mitochondrial quality control maintenance, which appears to be abnormal in a number of neurodegenerative diseases, especially familial PD.

2.6. Part VI: Neurodegenerative Disease Mechanisms – Biological Functions

This section of the book contains chapters that describe methods used to study a number of fundamental neuronal functions in health and disease, ranging from organellar trafficking, endo- and exocytosis, membrane depolarization, and neurotransmitter synthesis to gene expression. Chapter 22 describes the procedures to investigate vesicular transport, which is of great importance for the synaptic functions of living neurons. Chapter 23 deals with the protocols to estimate the trafficking of anion channels to the surface of neurons, in culture and in organotypic hippocampal slices. Chapter 24 describes technology to image and measure exocytosis in brain slices. Chapter 25 illustrates the procedures to image intracellular calcium dynamics in living neurons, at the single cell level. Chapter 26 explains how to assess synaptic transmission at the neuromuscular junction of mouse muscle using electrophysiological approaches. Chapter 27 describes protocols to measure catecholamine levels in the brain by HPLC. Chapter 28 provides a detailed overview of the application, protocols, interpretation, and pitfalls of techniques of microarray studies of gene expression.

2.7. Part VII: Therapeutic Approaches in Neurodegeneration

The therapy of neurodegeneration faces formidable challenges, as discussed above, largely due to the lack of information on disease causes and mechanisms, especially in the sporadic forms of the diseases. However, significant inroads have been made in the development of approaches to treat genetic forms of neurodegenerative diseases. Gene silencing is used to remove toxic mutant proteins, gene replacement to supplement the lack of functional proteins, cell replacement to replace dying or dysfunctional cells, and gene expression modulation to modify the expression of disease genes directly in tissues. Chapter 29 describes the technology for viral gene delivery directly to the brain that can be used to replace mutant genes, but also to express mutant genes and generate animal models of disease. Chapter 30 illustrates the protocol for designing and utilizing efficient silencing of toxic genes, such as mutant SOD1 in ALS. Chapter 31 describes the approaches to generate and implant astrocytic cell progenitors in the nervous tissue to supplement the loss of function caused by genetic diseases. Chapter 32 explains the design and use of selective small molecules with histone deacetylase inhibitor function to enhance gene expression in diseases, where the gene of interest is naturally downregulated, as for example, in the case of intronic expansions in Friedreich ataxia.

3. Conclusions

The field of neurodegeneration faces great challenges. A comprehensive set of approaches is needed to accomplish the identification of disease causes and mechanisms through integration of a vast array of genetic, epigenetic, epidemiological, molecular, and biochemical information and the generation of meaningful disease models to be used for studying pathogenesis and testing novel therapies. Exciting developments in the fields of stem cell replacement and gene therapy have to be integrated with emerging information on disease processes. This volume contains a compendium of experimental protocols, which provide the scientists involved in the field of neurodegeneration with a set of tools that are extremely useful for their research.

Chapter 2

In Vivo and In Vitro Determination of Cell Death Markers in Neurons

Nicole Naniche, Daniela Sau, and Piera Pasinelli

Abstract

Mitochondria are key regulators of cellular death. The mitochondrial membranes contain essential enzyme complexes for maintaining metabolic homeostasis and meeting the energy requirements of the cell (Tait and Green, Nat Rev Mol Cell Biol 11:621–632, 2010 and Galluzzi et al., Apoptosis 12:803–813, 2007). Thus, any perturbation of outer or inner mitochondrial membranes can lead to disruptions in the normal fluxes of key ions and metabolic proteins (i.e., ADP/ATP exchange), leading to eventual cellular death. In addition to maintaining cellular viability, mitochondria play a critical role in the initiation of programmed cell death. As initiators of the cell death process, key mitochondrial proteins [Cytochrome C (Cyt C) one of the most well-studied among them] are released from the intermembrane space during early cell death events eventually leading to caspase activation.

Release of Cyt C is a crucial step during cellular death (Tait and Green, Nat Rev Mol Cell Biol 11:621–632, 2010). Therefore, the measurement of Cyt C release can give vital information about cell death signaling. Immunolabeling against Cyt C can give an easy readout of mitochondrial integrity as well, allowing for simultaneous identification of mitochondrial viability (and/or damage) and initiation of intracellular death processes.

In this chapter, we use Cyt C as a dual marker of mitochondrial integrity and cell death and review several protocols to measure Cyt C localization into intact mitochondria and its release into the cytosol. The goal is to offer an array of assays that, combined, provide both qualitative and quantitative analysis of the relationship between mitochondrial viability and activation of an intracellular cell death process. Immunofluorescence, Western blot, and ELISA measurements of Cyt C as are discussed in detail.

Key words: Mitochondria, Cytochrome C, Immunofluorescence, Western blot, ELISA

1. Introduction

Mitochondria are essential for both life and death processes, including many metabolic pathways, energy production, calcium storage, production/regulation of reactive oxygen species (ROS),

Giovanni Manfredi and Hibiki Kawamata (eds.), *Neurodegeneration: Methods and Protocols*,
Methods in Molecular Biology, vol. 793, DOI 10.1007/978-1-61779-328-8_2, © Springer Science+Business Media, LLC 2011

and regulation of apoptosis. Intact mitochondria are required for healthy cellular metabolism and for regulation of the mitochondrial transmembrane potential ($\Delta\Psi m$). Once the outer membrane of the mitochondria loses its integrity, the mitochondrial ionic and metabolic homeostasis is disrupted, leading to a series of deathly consequences for the cell (1, 2). Therefore, maintaining integrity of the mitochondrial membranes is extremely important to maintain the metabolic viability of the cell. Mitochondria can also directly trigger cell death via the release of apoptogenic proteins into the cytosol. This is accomplished by permeabilization of the mitochondrial outer membrane (MOM) – which leads to decreased ATP production, change in mitochondrial membrane conductances, increases in cytosolic calcium levels, increase in ROS production and eventually cell death. Cyt C, which normally resides within the intermembrane space (IMS) of the mitochondria, binds to the apoptotic protease-activating factor 1 (APAF1) upon cytosolic localization. This leads to oligomerization of APAF1 and formation of the so-called apoptosome (3). The apoptosome serves as a platform for dimerization and activation of initiator caspase 9, which in turn cleaves and activates executioner caspases 3 and 7, finally leading to cell death (1, 4, 5).

Neurons are highly dependent on energy provided by the mitochondria and are therefore particularly sensitive to mitochondrial dysfunction. For fully differentiated cells like neurons, cell death signals released from damaged mitochondria represent a point of no return. Therefore, when developing means to measure death markers in neurons, the release of Cyt C from the mitochondria can be a doubly useful tool to define the time window between a decrease in energy supply and commitment to death (6–8). Cyt C also represents a valuable marker for mitochondrial integrity (6): it is normally retained in intact mitochondria and morphologically it appears in a distinct punctate staining typical of healthy mitochondria. It is released into the cytosol from damaged mitochondria (which lose their characteristic punctate structure) and rapidly diffuses into the cells. At this point, it serves as a key initiator of the cell death cascade and a powerful marker to define the timing in which the cell is committed to die (1, 6, 9).

In the following sections, we discuss several methods to detect retention vs. release of Cyt C from the mitochondria. We discuss using both neuronal NSC34 cells and primary cortical neurons. We originally optimized the protocols described below in NSC34 cells and used these protocols extensively in other neuronal lines like SH-SY5Y and H4 cells (6). The NSC34 cells are a hybrid neuronal line formed by the fusion of primary mouse motor neurons with neuroblastoma (10). As such, they retain many characteristics of primary neurons with the advantage of a cell line easy to maintain in culture. All the protocols have been tested and confirmed in primary cortical neurons as described below (6).

2. Materials

2.1. Cell Culture and Transfection

1. Dulbecco's modified Eagle's medium (DMEM) supplemented with 10% fetal bovine serum (FBS), 1× penicillin/streptomycin (Pen/Strep).
2. 0.05% trypsin/0.53 mM EDTA in HBSS.
3. Lipofectamine/Plus transfection reagents (Invitrogen, Carlsbad, CA).

2.2. Cell and Spinal Cord Mitochondrial Isolation

1. Buffer A: 70 mM sucrose, 190 mM mannitol, 20 mM Hepes, pH 7.5, complete protease inhibitor cocktail (Roche, Basel, Schweiz). Note: this should be kept at 4°C for short-term use and stored at −20°C for long-term storage.
2. Buffer B: 150 mM KCl, 1 mM EGTA, 20 mM HEPES, 2 mM EDTA, pH 7.2 complete protease inhibitor cocktail. Note: this should be kept at 4°C for short-term use and stored at −20°C for long-term storage.
3. Bovine serum albumin (BSA). Note: this should be kept at 4°C.

2.3. Cell and Spinal Cord Lysis

1. Buffer A: 250 mM sucrose, 20 mM Hepes-KOH pH 7.5, 10 mM KCl, 1.5 mM $MgCl_2$, 1 mM EDTA, 1 mM EGTA, 1 mM DTT, complete protease inhibitor cocktail. Note: this should be kept at 4°C for short-term use and stored at −20°C for long-term storage.
2. Buffer B: 70 mM sucrose, 190 mM mannitol, 20 mM Hepes, pH 7.5, complete protease inhibitor cocktail. Note: this should be kept at 4°C for short-term use and stored at −20°C for long-term storage.
3. Digitonin extraction buffer: 0.025% digitonin, 250 mM sucrose, 10 mM KCl, 1.5 mM $MgCl_2$, 1 mM EDTA, 1 mM EGTA, 20 mM HEPES at pH 7.5, supplemented with complete protease inhibitor. Note: this should be kept at 4°C for short-term use and stored at −20°C for long-term storage.
4. RIPA buffer: 150 mM NaCl, 1% nonidet p40, 12 mM deoxycolic acid Na salt, 0.1% SDS, 50 mM Tris–HCl (pH 8).

2.4. ELISA

1. Cytochrome c Quantikine ELISA Kit (R&D Systems, Minneapolis, MN).

2.5. SDS–Polyacrylamide Gel Electrophoresis

1. 15% precast linear gradient polyacrylamide gel.
2. Sample buffer: 60 mM Tris–HCl pH 6.8, 25% glycerol, 2% SDS, 5% 2-mercaptoethanol, 0.1% bromophenol blue, store in aliquots at −20°C.

3. Precision plus kaleidoscope protein standard (BioRad, Hercules, CA).

4. Running buffer: 25 mM Tris base, 192 mM glycine, 0.1% SDS, pH 8.3 following dilution to 1× with water.

2.6. Western Blotting

1. Nitrocellulose membrane, 0.2 μm.

2. Blot absorbent filter paper.

3. Transfer buffer: Tris/glycine buffer.

4. Tris-buffered saline with Tween (TBS-T): Prepare 10× stock with 1.37 M NaCl, 27 mM KCl, 250 mM Tris–HCl, pH 7.4, 1% Tween-20. Dilute 100 mL with 900 mL water for use.

5. Blocking buffer: 5% (w/v) nonfat dry milk in 1× TBS-T.

6. Primary antibodies: Rabbit anti-Cytochrome C (Cell Signaling, Danvers, MA, USA) Mouse anti-map 2 (Sigma Aldrich, 1:1,000) for neuronal staining.

7. Secondary antibody: Anti-rabbit HRP conjugate (GE Healthcare, Buckinghamshire, UK), anti-mouse HRP conjugate (GE Healthcare, Buckinghamshire, UK).

8. ECL advanced Western blotting detection system (GE Healthcare, Buckinghamshire, UK).

2.7. Immuno-fluorescence

1. Fixative solution: Prepare a fresh solution of 2% paraformaldehyde/2% sucrose in PBS.

2. Ice-cold methanol.

3. PBS 1×: 37 mM NaCl, 2.7 mM KCl, 4.3 mM Na_2HPO_4, 1.47 mM KH_2PO_4, pH 7.4, store at room temperature.

4. Permeabilization solution: 0.5% (v/v) Triton X-100 in PBS.

5. Blocking solution: PBS/5% FBS.

6. Primary fluorescent-conjugated mouse-anti-Cytochrome C (1:50, Alexa Fluor 546, Invitrogen), mouse anti-map 2 (Sigma Aldrich, 1:200–1:500) for neuronal staining.

7. Secondary antibody: Fluorescent conjugated anti-mouse (i.e., AlexaFluor from Invitrogen).

8. Mounting media: ProLong AntiFade, with or without DAPI (Invitrogen).

2.8. Cortical Neuron Isolation

1. Neurobasal media (Invitrogen) 4°C.

2. 10% penicillin streptomycin solution (Fisher).

3. 1× primiocen (InvivoGen).

4. 2% B27 (Invitrogen).

5. 5% FBS (Cellgro).

6. 0.3 mM l-glutamine (CellGro).

3. Methods

Cyt C is a highly soluble protein, physiologically associated with the inner mitochondrial membrane, where it acts as a component of the electron transport chain (11). Cyt C also plays a crucial role in cell death mechanisms and, in particular, it is released from mitochondria in response to toxic stimuli (12, 13). Cyt C release from mitochondria is, therefore, a good marker of cell death, especially as associated with mitochondrial toxicity (4–6). Ideally, one would like to define timing and mode of Cyt C release from damaged mitochondria both qualitatively and quantitatively (5). We have described methods for a qualitative immunofluorescence analysis of Cyt C localization and mitochondrial morphology as well as semiquantitative and quantitative analyses of the amount of Cyt C either retained or released from mitochondria by using Western blot and ELISA techniques. Our approach has been to use, whenever possible, all three techniques for a comprehensive analysis of the relationship between mitochondria morphology and activation of the cell death pathway (6).

Each of the above methods has its own values: (a) immuno-fluorescence staining followed by confocal analysis of Cyt C localization allows for visualization of mitochondrial morphology and initiation of cell death cascade, (b) Western blot allows for a semiquantitative evaluation of Cyt C retention in damaged mito-chondria, and (c) ELISA allows a definite quantitative estimate of Cyt C in the cytosol.

When staining cells for Cyt C, there are three possible outcomes depending on the viability of the mitochondria and eventually of the cell: (a) in healthy mitochondria, Cyt C has a punctate staining (see Fig. 1a), which nicely correlates with the typical punctate-like structure of healthy mitochondria and displays a similar staining to a classical mitochondrial markers (i.e., MitoTracker) (see Fig. 1a); (b) in damaged, dysfunctional mitochondria Cyt C appears fuzzy staining still within the imme-diate outer membrane; and (c) finally, in cells with fully damaged mitochondria, Cyt C staining is diffused across the cytosol. Only at this late point, there is activation of executioner caspases in the cytosol (4, 5). Upon toxic insults, there is a gradual change in staining pattern: it progressively becomes fuzzy, losing the defined punctate structure, until it completely diffuses across the cell (see Fig. 1b).

If a semiquantitative approach is desired, mitochondrial and cytosolic fractions can both be analyzed by Western blot analysis for Cyt C content (14). An advantage of this system is that Cyt C can be semiquantitatively measured from in vitro isolated mito-chondria and: (1) treated with toxic insult of choice or (2) it can be measured by purifying mitochondria from cells and/or tissues

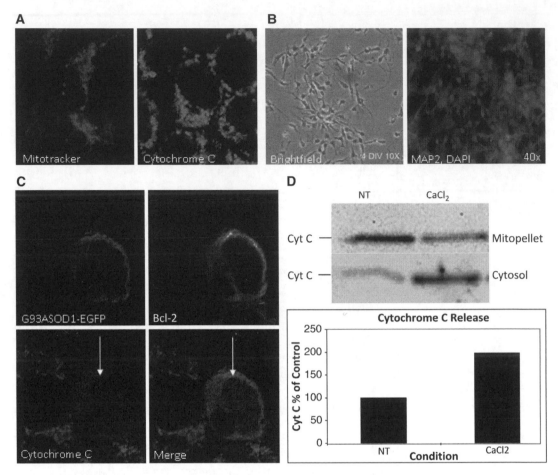

Fig. 1. Determining Cytochrome C (Cyt C) release in cell lines. (**a**) Nontreated NSC34 cells were stained with MitoTracker (*left panel*) or probed for Cyt C (*right panel*) for observation of intact, healthy mitochondria. Comparison of Cyt C with known mitochondrial stain gives same intact, punctate expression pattern across the cell. (**b**) NSC34 cells were transfected with a toxic protein (amyotrophic lateral sclerosis-linked mutant SOD1-EGFP) to induce release of Cyt C. In the transfected (*cell with arrow*), Cyt C displays a diffuse staining across the cell, indicating release. Conversely, nontransfected cells retain punctate staining, indicating intact mitochondria and no release into the cytosol. (**c**) Cortical neurons were stained for neuronal marker MAP2 and nuclear marker DAPI. (**d**) Mitopellet and cytosolic fractions of nontransfected (NT) and positive control (CaCl$_2$) NSC34 cells were run on Western blot and probed for Cyt C to determine amount retained in mitochondria and released into cytosol. Cytosolic fraction was also analyzed by ELISA; representative blot shown in *lower panel*.

previously exposed to toxic insults. This method follows a classical Western blot readout and measures the intensity of the Cyt C band in the "mitopellet" fraction (see Fig. 1d, upper panel); a potential caveat with Western blot analysis is that cytoplasmic proteins can be too diffuse to detect. However, ELISA allows for a quantitative analysis of cytoplasmic proteins and provides a more sensitive output than the Western blot (see Fig. 1d, lower panel). The Cyt C ELISA assay is a colorimetric assay – higher Cyt C cytosolic concentrations result in an intense coloration of the cytosolic sample with increased absorbance at 450 nm (6).

Oftentimes, the nature of the experimental systems used and the technical difficulties associated with them limit the analysis to one technique over the other. For instance, when using Cyt C as a cell death marker, one relies on cell culture systems. Toxic proteins are often used to induce cell toxicity and transient transfection and/or expression systems are commonly used tools to assay toxicity of proteins in neuronal cells. Experimental limitations typical of transiently transfected cellular models are linked to the fact that, in these systems, the cell population is not completely homogeneous and, at any given time, only a percentage of cells express the toxic protein(s), and it is therefore poised to undergo cell death. In this case, a qualitative – single cell – analysis of Cyt C localization by immunofluorescence may be the only possible experimental approach. Moreover, at least in transiently transfected cells, the time window to observe changes in mitochondrial function and subsequent Cyt C distribution is usually short (6–72 h) making it difficult to evaluate cumulative toxic effects or end point phenomena. Defining the precise time course of toxicity is also crucial and may also represent a limiting factor in choosing more quantitative analysis than immunofluorescence/confocal microscopy analysis of Cyt C localizations.

While single cell analysis of Cyt C localization following a toxic stimulus, represents a good starting point to define Cyt C as a marker of cell (and mitochondrial) toxicity, more quantitative measures of its translocation from the mitochondria to the cytosol are needed to define (a) its role in triggering cell death, (b) the threshold of Cyt C release needed to get to the point-of-no return into the death pathway, and (c) to quantify the extent of toxicity of the toxic insult used in the experiments studied. Therefore, upon optimization of the experimental conditions used (e.g., high transfection efficiency of a toxic protein, correct dose of toxin), it is recommended to combine the immunofluorescence data with Western blot and ELISA analyses of Cyt C content in isolated mitochondria and cytosol, respectively.

The ELISA method described above is ultimately the most accurate and quantitative assay to establish toxicity and the extent of cell death in any given sample.

3.1. Isolating Mitochondria from NSC34 Cells (see Materials in Subheading 2.2)

Protocol adapted from Pallotti and Lenaz (15, 16).

Before Starting, Make Up Solutions

1.   1 mL BSA + 4 mL buffer A.

2.   2 mL solution (1) + 2 mL buffer B.

3.   2 mL buffer A + 2 mL buffer B.

1. Use ~6 confluent dishes of NSC34 cells.

2. Collect cells, spin at $600 \times g$, 4°C, 10 min at 4°C (see Note 1).

3. Remove supernatant, resuspend pellet in 4 mL cold PBS (from now on, keep on ice).

4. Spin resuspended pellet down at $600 \times g$, 10 min at 4°C.

5. Remove supernatant, resuspend pellet gently in 800 µL solution (2); once completely resuspended transfer solution to 2-mL glass homogenization tube.

6. Homogenize resuspended pellet at 4°C.

7. Transfer homogenized solution to a clean Eppendorf tube.

8. Spin down homogenate at $1,000 \times g$, 10 min at 4°C.

9. Discard pellet, transfer supernatant into new, clean Eppendorf tube.

10. Centrifuge at $10,000 \times g$ for 10 min at 4°C.

11. Gently pipette off supernatant.

12. Resuspend pellet in 100 µL solution (3).

13. Centrifuge at $10,000 \times g$ for 10 min at 4°C.

14. Resuspend pellet in 100 µL buffer A.

15. Centrifuge at $100,000 \times g$ for 1 h at 4°C.

16. Resulting pellet is mitochondrial fraction, supernatant is cytosolic fraction.

3.2. Isolating Mitochondria from Spinal Cord (Use Materials in Subheading 2.2)

Protocol adapted from Fieni et al. (17).

Before Obtaining Spinal Cord, Prepare Solutions

1. 1 mg/mL trypsin + 2 mg/mL collagenase in PBS w/out Ca^{2+} or Mg^{2+}.

2. Pre-incubate Solution to 37°C.

3. 3 mL BSA + 22 mL buffer A.

4. 2 mL solution (1) + 2 mL buffer B.

5. 2 mL buffer A + 2 mL buffer B.

1. Once spinal cord is obtained, cut into small pieces and let it sit in 10-mL beaker in 2 mL of solution (1) at 37°C for 15 min.

2. After 15 min, add 2 mL of solution (2) to cord, place in homogenization tube.

3. Homogenize cord with Potter homogenizer (Teflon/glass; 4°C, $600 \times g$).

4. Transfer cord homogenate to ultracentrifuge tube. Spin at $800 \times g$ for 10 min at 4°C.

5. Resuspend the first nuclear pellet in 4 mL solution (2) and rehomogenize. Spin both supernatants at $800 \times g$, 10 min, 4°C (see Note 2).

6. Combine the two supernatants obtained (8 mL), rehomogenize the bigger pellet in one of the tubes after resuspending in 8 mL solution (2).

7. Spin everything (two tubes, 8 mL) at $800 \times g$, 10 min, 4°C.

8. Combine the supernatants (16 mL), get rid of the pellets, and spin at $10,000 \times g$ for 20 min, 4°C.

9. Resuspend this pellet in 2 mL solution (3). Spin at $10,000 \times g$ for 20 min, 4°C.

10. Resuspend pellet in 2 mL solution (4), spin at $100,000 \times g$ for 1 h, 4°C. Resulting pellet is mitochondrial fraction, supernatant is cytosolic fraction.

3.3. Preparing the Samples to Assay CytC Release in ELISA and WB, from Mouse Spinal Cord Extracts or Transiently Transfected Cells (Use Materials in Subheading 2.3)

1. Homogenize mouse spinal cords and/or lysates of NSC34 cells in 800 μL of buffer A, and spin at $750 \times g$ at 4°C for 10 min.

2. Rehomogenize the pellet in 400 μL of buffer A, spin at $750 \times g$ at 4°C for an additional 10 min.

3. Combine the supernatants from these two homogenization steps in one solution; spin combined supernatant at $750 \times g$ at 4°C for 10 min. Repeat spin with resulting supernatant.

4. Spin supernatant from step 3 once at $10,000 \times g$ at 4°C for 15 min.

5. Resuspend each pellet in 80 μL of buffer B.

6. Incubate 25 μL of mitochondria with 20 μL of buffer B. Spin samples at $100,000 \times g$ for 1 h at 4°C.

7. Analyze the resulting supernatant for Cyt C by ELISA following the manufacturer's instruction (see Notes 3 and 4), whereas the resulting mitopellet will be resuspended in buffer B and analyzed by WB.

8. Mitopellet for WB analysis of Cyt C release obtained as follows: collect 6–7 confluent Petri dishes (10 mm) of NSC34 cells.

9. After 24 h, detach and wash cells with 1 mL of PBS. Spin down at $500 \times g$ and lyse in 100 mL of digitonin extraction buffer supplemented with complete protease inhibitor.

10. Spin down lysates at $15,000 \times g$ and lyse the resulting pellet containing mitochondria in 100 mL of RIPA buffer.

11. Prepare 20 μg of proteins for WB by adding 5× sample buffer and boil for 5 min.

3.4. SDS–PAGE and WB

1. Assemble the gel unit using 15% polyacrylamide gels, add running buffer and connect to a power supply (see Note 5).

2. Load the samples, and run the gels at 100 V.

3. The samples that have been separated by SDS–polyacrylamide gel electrophoresis (SDS–PAGE) are transferred to supported nitrocellulose membranes electrophoretically.

4. Assemble the transfer unit to allow protein transfer from the gel to the nitrocellulose membrane, add transfer buffer, connect the unit to a power supply and apply 100 V for 1 h.

5. Once the transfer is complete the cassette is taken out of the tank and carefully disassembled.

6. The nitrocellulose is then incubated in blocking buffer for 1 h at room temperature on a rocking platform.

7. The blocking buffer is discarded and the Cyt C primary antibody is added on the membrane (1:1,000 dilution in blocking buffer) for 1 h at room temperature on a rocking platform.

8. The primary antibody is removed and the membrane is washed three times for 10 min each, in TBS-T.

9. The anti-rabbit secondary antibody is freshly prepared for each experiment in a 1:10,000 dilution in blocking buffer and added to the membrane for 1 h at room temperature on a rocking platform.

10. The secondary antibody is removed and the membrane is washed three times for 10 min each, in TBS-T.

11. During the final wash, 2 mL aliquots of each portion of the ECL reagent are warmed separately to room temperature and the remaining steps are done in a dark room under safe light conditions. Once the final wash is removed from the blot, the ECL reagents are mixed together and then immediately added to the blot, which is then rotated by hand for 1 min to ensure even coverage; chemiluminescence has been evaluated using a ChemiDoc acquiring system.

3.5. Immunofluorescence Analysis of Cyt C Release

If Transfecting a Toxic Protein

1. Plate NSC34 cells in two-chamber slides with 500,000 cells/chamber.

2. 24 h after plating transfect cells with 200 ng of plasmid; lipofectamine PLUS is recommended (follow manufacturer's protocol).

If not Transfecting, Begin Here

3. 24 h after transfection fix cells for 20 min with a solution of 2% paraformaldehyde/2% sucrose at room temperature.

4. Remove fixing solution and treat cells with ice-cold methanol, for 10 min, at room temperature.

5. Wash slides once in PBS, 5 min at room temperature, gently shaking.

6. Add blocking buffer (PBS/5% FBS) for 1 h at room temperature, rotating.

7. Incubate slides were then incubated with fluorescent-conjugated mouse anti-Cyt C (1:50, AlexaFluor 546) for 1 h in blocking buffer.

8. Rinse slides three times in PBS.

9. Mount coverslips on slides using ProLong Antifade, dry for 2 h, nailpolish and analyze (see Notes 6 and 7).

3.6. Cortical Neuron Isolation

Isolate cortical neurons from nontransgenic E15 mice.

1. Dissect embryos and place cortices in dissection solution.

2. Spin cortices at $80 \times g$ for 5 min and placed in trypsin/EDTA-HBSS for 20 min at 37°C.

3. Spin sample three times at $80 \times g$ for 5 min to remove trypsin.

4. Triturate sample with a fire-polished Pasteur pipette until the suspension is homogeneous before an additional spin at $80 \times g$ for 5 min.

5. Discard supernatant, resuspend cells, and if transfecting or running IF, count them before plating.

6. Once plated, cortical neurons can be assayed for Cyt C as described.

4. Notes

1. For Cyt C measurements by Western blot and/or ELISA, after collecting the cells all steps should be performed on ice or at 4°C. Mitochondria are highly temperature sensitive and Cyt C retention into mitochondria and its detection are also negatively affected by prolonged incubation at room temperature.

2. When isolating spinal cord mitochondria, the spin at $800 \times g$ should be repeated at least three times. This spin is performed to clean the final mitochondrial pellet as much as possible of nuclei, debris, etc. Therefore, performing this spin several times results in a "cleaner" final pellet.

3. When running ELISA the first time, run three dilutions of your sample with a tenfold change in dilution factor against the standard curve. This will roughly give you an idea of what protein concentration your samples are and whether they will fit within the linear range of the standard curve.

4. Avoid bubbles as much as possible when running ELISA – bubbles/foam can affect the readings.

5. These instructions assume the use of a BioRad Mini Protean System gels but they are easily adaptable to other formats.

6. When analyzing Cyt C release by immunofluorescence, it is best to visualize the cells under the microscope when cells are not directly on top of each other. Therefore, plating conditions should be optimized in order not to have an overconfluent filed of cells. This is particularly important for dividing neuronal cells (e.g., NSC-34). For primary neurons, it is crucial to disperse the cells at plating to avoid overlapping. If cells are overconfluent, the Cyt C staining in the juxtaposing cell may interfere with the signal derived from the cell that is being visualized.

7. When staining for Cyt C release in immunofluorescence, use a secondary antibody/cellular marker (i.e., DAPI staining of nuclei) to mark the cells – when Cyt C becomes diffuse, it can be very difficult to distinguish cellular outlines by mitochondrial staining alone.

References

1. Tait, S. W., and Green, D. R. (2010) Mitochondria and cell death: outer membrane permeabilization and beyond, *Nat Rev Mol Cell Biol 11*, 621–632.

2. Galluzzi, L., Zamzami, N., de La Motte Rouge, T., Lemaire, C., Brenner, C., and Kroemer, G. (2007) Methods for the assessment of mitochondrial membrane permeabilization in apoptosis, *Apoptosis 12*, 803–813.

3. Wei, M. C., Lindsten, T., Mootha, V. K., Weiler, S., Gross, A., Ashiya, M., Thompson, C. B., and Korsmeyer, S. J. (2000) tBID, a membrane-targeted death ligand, oligomerizes BAK to release cytochrome c, *Genes Dev 14*, 2060–2071.

4. Pasinelli, P., Borchelt, D. R., Houseweart, M. K., Cleveland, D. W., and Brown, R. H., Jr. (1998) Caspase-1 is activated in neural cells and tissue with amyotrophic lateral sclerosis-associated mutations in copper-zinc superoxide dismutase, *Proc Natl Acad Sci USA 95*, 15763–15768.

5. Pasinelli, P., Houseweart, M. K., Brown, R. H., Jr., and Cleveland, D. W. (2000) Caspase-1 and -3 are sequentially activated in motor neuron death in Cu,Zn superoxide dismutase-mediated familial amyotrophic lateral sclerosis, *Proc Natl Acad Sci USA 97*, 13901–13906.

6. Pedrini, S., Sau, D., Guareschi, S., Bogush, M., Brown, R. H., Jr., Naniche, N., Kia, A., Trotti, D., and Pasinelli, P. (2010) ALS-linked mutant SOD1 damages mitochondria by promoting conformational changes in Bcl-2, *Hum Mol Genet 19*, 2974–2986.

7. Bobba, A., Atlante, A., Giannattasio, S., Sgaramella, G., Calissano, P., and Marra, E. (1999) Early release and subsequent caspase-mediated degradation of cytochrome c in apoptotic cerebellar granule cells, *FEBS Lett 457*, 126–130.

8. Bobba, A., Atlante, A., de Bari, L., Passarella, S., and Marra, E. (2004) Apoptosis and cytochrome c release in cerebellar granule cells, *In Vivo 18*, 335–344.

9. Stefanis, L. (2005) Caspase-dependent and -independent neuronal death: two distinct pathways to neuronal injury, *Neuroscientist 11*, 50–62.

10. Cashman, N. R., Durham, H. D., Blusztajn, J. K., Oda, K., Tabira, T., Shaw, I. T., Dahrouge, S., and Antel, J. P. (1992) Neuroblastoma × spinal cord (NSC) hybrid cell lines resemble developing motor neurons, *Dev Dyn 194*, 209–221.

11. Lenaz, G., and Genova, M. L. (2010) Structure and organization of mitochondrial respiratory complexes: a new understanding of an old subject, *Antioxid Redox Signal 12*, 961–1008.

12. Condello, S., Curro, M., Ferlazzo, N., Caccamo, D., Satriano, J., and Ientile, R. (2010) Agmatine effects on mitochondrial membrane potential and NF-kappaB activation protect against rotenone-induced cell damage in human neuronal-like SH-SY5Y cells, *J Neurochem.*

13. Dedoni, S., Olianas, M. C., and Onali, P. (2010) Interferon-beta induces apoptosis in human SH-SY5Y neuroblastoma cells through activation of JAK-STAT signaling and down-regulation of PI3K/Akt pathway, *J Neurochem* *115*, 1421–1433.

14. Arnoult, D. (2008) Apoptosis-associated mitochondrial outer membrane permeabilization assays, *Methods* *44*, 229–234.

15. Pallotti, F., and Lenaz, G. (2007) Isolation and subfractionation of mitochondria from animal cells and tissue culture lines, *Methods Cell Biol* *80*, 3–44.

16. Pasinelli, P., Belford, M. E., Lennon, N., Bacskai, B. J., Hyman, B. T., Trotti, D., and Brown, R. H., Jr. (2004) Amyotrophic lateral sclerosis-associated SOD1 mutant proteins bind and aggregate with Bcl-2 in spinal cord mitochondria, *Neuron* *43*, 19–30.

17. Fieni, F., Parkar, A., Misgeld, T., Kerschensteiner, M., Lichtman, J. W., Pasinelli, P., and Trotti, D. (2010) Voltage-dependent inwardly rectifying potassium conductance in the outer membrane of neuronal mitochondria, *J Biol Chem* *285*, 27411–27417.

Chapter 3

Degenerating Processes Identified by Electron Microscopic Immunocytochemical Methods

Teresa A. Milner, Elizabeth M. Waters, Danielle C. Robinson, and Joseph P. Pierce

Abstract

The application of electron microscopic immunolabeling techniques to the identification and analysis of degenerating processes in neural tissue has greatly enhanced the ability of researchers to examine apoptosis and other degenerative disease mechanisms. This is particularly true for the early stages of such mechanisms. Traditionally, degenerating processes could only be identified at the ultrastructural level after significant cellular atrophy had occurred, when subcellular detail was obscured and synaptic relationships altered. Using immunocytochemical labeling procedures, degenerating neural and glial processes are first identified through the use of antibodies directed against a variety of degenerative markers, such as proapoptotic effectors (i.e., cytoplasmic cytochrome c), pathological components (i.e., beta amyloid deposits), or inflammatory agents (i.e., Iba1). Both the subcellular distribution of the marker within the process and the relationship of the labeled process to surrounding elements can then be carefully characterized. The information obtained can be further refined through the use of dual immunolabeling, which can provide additional data on the phenotype of the degenerating process and inputs to the process.

Key words: Immunocytochemistry, Ultrastructure, Immunogold, Pre-embedding immunoelectron microscopy, Antibodies, Quantitation

1. Introduction

The identification of degenerating processes in the brain at the electron microscopic (EM) level has traditionally relied on the use of distinct morphological criteria, such as the presence of electron-dense cytoplasm and swollen mitochondria (1, 2). For instance, we have identified apoptotic nuclei in the substantia nigra following injury to projections to the medial forebrain bundle (3), degenerating terminals in the hippocampus following lesions of afferents traveling through the fornix (4), and plaques in animal models of

Giovanni Manfredi and Hibiki Kawamata (eds.), *Neurodegeneration: Methods and Protocols*,
Methods in Molecular Biology, vol. 793, DOI 10.1007/978-1-61779-328-8_3, © Springer Science+Business Media, LLC 2011

Alzheimer's disease (5, 6). The concurrent development of EM immunocytochemical labeling methods and the identification of specific markers associated with both apoptosis and degenerative disease mechanisms have allowed high-resolution analysis of the morphology of degenerating processes at early time points when subcellular relationships are preserved. For example, EM immunocytochemistry has been used to label early proapoptotic effectors (e.g., cytochrome c released into the cytoplasm) (7), markers of pathology in models of disease (e.g., amyloid beta) (5, 6), or markers of inflammation (e.g., Iba1) (8, 9). Using ultrastructural analysis, one can then determine (1) the cell type of a labeled process (e.g., neuronal vs. glial cells), (2) the relationship of the labeled process to surrounding processes (e.g., making synaptic contacts or surrounded by glial processes), and (3) the subcellular localization of the marker (e.g., mitochondria vs. multivesicular bodies). In addition, dual-label EM methods can be used to determine the phenotype of the degenerating process and the nature of inputs to the process. Experimental approaches utilizing dual-labeling EM methods can, thus, provide valuable insight into the mechanisms underlying apoptosis and degenerative disease.

2. Materials (A Complete List of Materials with Catalogue Numbers Is Available Upon Request)

2.1. General Laboratory Materials

2.1.1. Chemicals

Contrad (dilute 10% with water for cleaning glassware; Polysciences, Warrington, PA).

Hydrochloric acid (HCl; make 70%) and sodium hydroxide (NaOH; 1 N). Store near pH meter.

Water purification filtration system.

2.1.2. Large Equipment

Fume hood equipped with a vacuum connection (flow rate ≥100 ft/min).

Rotating shaker table (VWR, West Chester, PA).

2.1.3. Small Equipment and Nondisposable Supplies

pH meter and stirrer/hot plate (Corning Life Sciences, Lowell, MA).

Pipette men with disposable tips (1–1,000 μL capacity), sharpie pens, stir bars, and timer.

2.1.4. Disposable Supplies	Colored tape, kimwipes, parafilm, squirt bottles, weigh paper, and boats (VWR).
	Plastic wrap (Costco Stretch-tite; for covering beakers).
	Scintillation vials (RPI Corps, Mt. Prospect, IL).
	Transfer pipettes (Biologixresearch, Lenexa, KS).
2.2. Materials and Solutions for Perfusion Fixation	Acrolein (Polysciences, Warrington, PA).
	Normal saline with heparin (Henry Schein, Melville, NY).
2.2.1. Chemicals	Paraformaldehyde, granular (Electron Microscopy Sciences (EMS), Fort Washington, PA).
	Sodium phosphate [dibasic and monobasic (VWR)].
2.2.2. Large Equipment	Balance (for weighing animals and organs).
	Enclosed pan for collecting perfusate (New Pig, Tipton, PA).
	Peristaltic Masterflex pump with silicon tubing (tubing, 14 gauge; the tubing should be connected to three-way stopcock; Fig. 1; Cole Parmer, Vernon Hills, IL).
2.2.3. Small Equipment and Nondisposable Supplies	Beakers (glass; 50 ml for saline; 500 ml for fixative) and plastic graduated cylinders (100 ml).
	18 gauge stainless steel tubing, 2″ long (custom cut 2″; Small Parts Inc., Miami Lakes, FL).
	Brain blocking mold (1 mm, stainless steel; Ted Pella, Redding, CA).
	Buchner funnel, Erlenmeyer 500-ml filtration flask, and spatula with pointed end (VWR).
	Connector for attaching small needle to output line (Small Parts Inc., Seattle, WA).

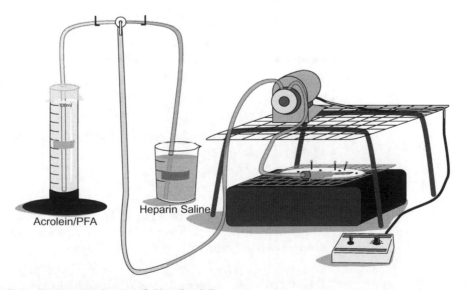

Fig. 1. Perfusion fixation setup for mice (Subheading 3.2).

Pins with large balls on ends and rack to place over pan (K-mart or Bed Bath and Beyond).

Pliers-type, vial decapper (Sigma-Aldrich, Milwaukee, WI).

Rongeurs and tissue forceps and scissors (FST, Foster City, CA). Silicon pad (remove plastic tray; EMS).

2.2.4. Disposable Supplies   Anesthesia (we use 150 mg/kg sodium pentobarbital), plastic bags (for appropriate disposal of carcasses; varies by institution), razor blades, syringes (1 cc for injecting animals; 5 cc for measuring acrolein), and Whatman #3 filter paper.

Microtubes (rinsed with heparin before collecting blood; Biologixresearch).

2.2.5. Solutions   0.2 M Phosphate Buffer Solution

1. Place 1 L of deionized water (dH$_2$O) into a beaker on a stir plate.

2. With the water gently stirring, add 21.8 g sodium phosphate dibasic followed by 6.4 g sodium phosphate monobasic to the beaker. Stir the solution until the phosphate buffer (PB) crystals are dissolved. The pH should be close to 7.4. DO NOT pH since this can affect the immunocytochemical reactions.

3. Make more than you need and dilute some of the 0.2 M phosphate buffer with dH$_2$O to make 0.1 M phosphate buffer solution for slicing. Store at 4°C.

0.1 M PB

Dilute 500 ml 0.2 M PB with 500 ml dH$_2$O. Alternatively, make PB as described above, except use 10.9 g sodium phosphate dibasic and 3.2 g sodium phosphate monobasic.

2% Paraformaldehyde

1. Add 250 ml dH$_2$O to a glass beaker containing a stir bar.

2. Heat the dH$_2$O to 55°C on a stir plate, and then turn the heat to a low setting.

3. Add 10 g paraformaldehyde (PFA); this does not dissolve immediately. Add 1–2.5 ml 1 M NaOH using a squirt bottle. Stir the solution until the PFA is dissolved.

4. Filter the solution using a vacuum filter with a Buchner filter and #3 filter paper.

5. Add 250 ml 0.2 M PB and adjust pH to 7.4 with HCl.

3.75% Acrolein in 2% PFA

1. Add 96.25 ml 2% PFA solution to a 100 ml graduated cylinder; you need about 250 ml/rat and 40 ml per mouse. Cover the cylinder with parafilm.

2. Fill a syringe (18- or 20-gauge, 1.5-in. long needle) with 3.75 ml acrolein. To measure the acrolein safely, hold the acrolein bottle inverted under the hood, puncture the rubber top

with the syringe needle, and remove 3.75 ml. Keep the hood sash pulled as low as possible.

3. Insert the needle through the parafilm on top of the cylinder containing the 2% PFA solution and add the acrolein slowly. Add an additional layer of parafilm and mix by inverting slowing three to four times.

2.3. Materials and Solutions for Sectioning the Brain

2.3.1. Chemicals

Agar and ethylene glycol (Sigma-Aldrich).

Sucrose (JT Baker, Phillipsburg, NJ).

2.3.2. Large Equipment

Vibrating microtome (Leica Microsystems, Deerfield, IL).

2.3.3. Small Equipment

Beakers, 1-L plastic (for storage solution).

Camel and sable hair brushes (Ted Pella Inc).

Fine forceps (for removing stray tissue and meninges; EMS).

Glass quadrant petri dishes (Corning Life Sciences).

Pipette aid (Drummond Scientific Co., Broomall, PA).

2.3.4. Disposable Supplies

Serological pipettes, 10 and 25 ml, and tissue culture dishes, 24-well (VWR).

Super glue (Loctite prism) and injector razor blades (EMS).

2.3.5. Solutions

Agar (2%)

1. Heat 100 ml water to 55°C on a stir plate.

2. Add 2 g agar to the water and stir until dissolved and bring to a gentle boil.

3. Pour about 15 ml of the agar solution into scintillation vials.

4. Store at 4°C (can keep up to 1 year).

5. To use stored agar, liquefy agar by loosening cap on vial and microwaving for 7–12 s, mix with a disposable plastic 1-ml pipette, and use.

Tissue Storage Solution (30% Sucrose and 30% Ethylene Glycol in 0.1 M PB)

1. Place a 500-ml plastic beaker on a stir plate.

2. Add 150 g sucrose and 150 ml of ethylene glycol to the beaker.

3. Bring the solution up to 500 ml with 0.1 M PB (see Subheading 2.2.5). Stir until the sucrose is dissolved. (This takes a while.)

4. Adjust the final pH of the solution to 7.4 using NaOH.

5. Store at −20°C (this is good for several years).

2.4. Materials and Solutions for Immunocytochemical Processing

6. Prior to use, stir solution at room temperature until close to 4°C.

Alcohol (ACS grade, 200 proof) and Trizma base (VWR).

Bovine serum albumin (BSA), diaminobenzidine (DAB), and sodium borohydride (Sigma-Aldrich).

2.4.1. Chemicals

Cold water fish gelatin, glutaraldehyde (25% EM grade) and sodium citrate, sodium hydroxide, 1 N (NaOH) (EMS).

Glycerin (glycerol), sucrose, and Triton X 100 (optional; 0.025% for EM incubations) (JT Baker).

2.4.2. Large Equipment

Plexiglass hood dedicated to DAB step containing a balance and DAB waste bucket.

2.4.3. Small Equipment

Beakers (Graduated Griffin, 1 L, 2 L), glass crystallizing dishes, glass petri dishes (9 cm), and pipette bulbs (VWR).

Beem capsules (for portioning DAB power), metal lids of staining dishes, and plastic mesh crucibles (EMS).

Camel and sable hair brushes (Ted Pella Inc).

Modeling clay (place on sides of trays to hold vials during incubations).

Punch tools for marking tissue (Small Parts Inc.).

Rubber mats (black; custom 1/32″ Neoprene; Garlock Rubber Technologies, Paragould, AR).

2.4.4. Disposable Supplies

Biotinylated secondary immunoglobulins (IgG; Jackson ImmunoResearch, West Grove, PA; Vector Laboratories, Burlingame, CA).

Gold (1 nm) conjugated IgG (EMS).

Microtubes (freestanding), microtube screw top lids with assorted colors, pasteur pipettes, 12-well tissue culture dishes, plastic beakers, 50 ml, 100 ml and plastic petri dishes, plastic petri dishes and wooden applicator sticks (VWR).

Silver IntenSE M kit (GE Healthcare, Piscataway, NJ).

Vectastain ABC kit (Vector).

2.4.5. Solutions for Immunocytochemical Processing, Part 1

Immunocytochemical processing is a multiday procedure. For all solutions, use deionized water that is 18.2 mΩ/cm at 25°C. This is especially important for the silver intensification procedure.

Sodium Borohydride Solution (Make Immediately Before Using)

1. Put 100 ml of 0.1 M PB (see Subheading 2.2b), with a stir bar, into a plastic beaker.

2. Add 1 g of sodium borohydride, and stir for 30 s. (Use immediately.)

0.1 M Tris–Saline Solution

1. To make 0.1 M Tris–saline (TS) solution, place about 975 ml dH$_2$O into a beaker on a stir plate.

2. With the water gently stirring, add 12.1 g Trizma base to the beaker.

3. pH the solution to 7.6 with HCl and bring the volume up to 1 L. Store the solution at 4°C.

0.5% BSA in 0.1 M TS (See Note 1)

Add 0.25 g of BSA to 50 ml of 0.1 M TS solution.

0.1% BSA in TS

Mix 10 ml 0.5% BSA with 40 ml 0.1 M TS.

Cryoprotectant Solution (for Optional Freeze–Thaw)

Mix 100 g sucrose, 40 ml of glycerin, 100 ml 0.2 M PB, and 260 ml dH$_2$O.

2.4.6. Solutions for Immunocytochemical Processing, Part 2

0.01 M Phosphate-Buffered Saline

1. Add 1 L of dH$_2$O to a plastic beaker. With the water gently stirring, sequentially add 1.09 g dibasic sodium phosphate, 0.32 g monobasic sodium phosphate, and 9 g NaCl.

2. pH the solution with HCl or NaOH to 7.4.

Washing Incubation Buffer

1. Measure 250 ml 0.01 M phosphate-buffered saline (PBS) into another beaker.

2. As the PBS is stirring, add 1.25 ml 40% gelatin stock and 2 g BSA. (Adjust the volume for the experiment.)

3. Readjust the pH to 7.4.

0.2 M Citrate Buffer

1. Measure 500 ml dH$_2$O to a plastic beaker. As the water is stirring, add 29.45 g sodium citrate.

2. Adjust the pH with citric acid (4.2 g of sodium citrate in 100 ml dH$_2$O) to 7.4.

2.5. Materials and Solutions for Embedding Tissue Sections for EM

2.5.1. Chemicals

Alcohol ACS grade, 200 proof.

2.5.2. Large Equipment

EMbed 812 embedding media kit osmium, tetroxide, 4% aqueous solution, and propylene oxide, ACS grade (EMS).

2.5.3. Small Equipment and Nondisposable Supplies

Oven, 60°C (not used for wet incubations).
Coors shallow multiwell white ceramic dishes (Sigma-Aldrich).

Large metal trays (Cole Parmer).

Rectangular steel weights (cold rolled steel type; Small Parts Inc.).

Rotary mixer (R2 model, Ted Pella).

2.5.4. Disposable Supplies   Aclar (Fluoropolymer film 33 C; Honeywell, Morristown, NJ).

Syringes, 30 cc and 60 cc (BD; Franklin Lakes, NJ).

2.5.5. Solutions   Osmium Stock Solution

1. Add 10 ml 0.2 M PB to a scintillation vial.

2. Wrap 10 ml snaptop 4% osmium vial with a paper towel, snap open under the hood, and add the osmium to the scintillation. Gently pipette up and down to mix. Osmium should be a pale yellow color.

3. Wrap the scintillation vial with foil and place in a container protected from light. Store the container at 4°C.

EMbed Resin

1. Remove the plunger of a 30-ml syringe and seal the tip with parafilm. Stand the syringe upright in a glass beaker to prevent it from tipping over. Sequentially pour the following into the syringe: 12 ml EM-bed 812 (to the 12-ml mark); 7.5 ml DDSA (to the 19.5-ml mark); 6.75 ml NMA (to the 26.25-ml mark). Wipe the lids of the bottles with a kimwipe before closing them. With a pipetman, add 600 μl DMP-30.

2. Carefully push the plunger into the syringe until it is secure, leaving a little air in the syringe so that the EMbed resin can mix.

3. Seal over the tip with another piece of parafilm and gently invert the syringe up and down ~ten times until the resin looks homogenous. Place syringe on a rotary mixer for 30 min.

Alcohols

30, 50, 70, and 95% ethanol for dehydration.

EMbed/Propylene Oxide (1:1) Mixture

1. 20 ml Embed and 20 ml propylene oxide in a conical tube.

2.6. Materials for Sectioning Tissue for EM   Acetone, ACS grade (EMS).

2.6.1. Chemicals

2.6.2. Large Equipment   Light microscope.

Ultramicrotome (Ultracut) and glass knife maker (Leica).

2.6.3. Small Equipment and Nondisposable Supplies   Antistatic gun (zerostat 3; Sigma Aldrich).

Beakers (10 ml, glass) and petri dishes (9 cm, glass) (VWR).

Diamond knife (2.7-mm edge), embedding capsule holder embedding molds, fine forceps, glass knife box, grids 400 mesh copper thin bar, plastic partition box, for storing blocks, silicone pads with grid marks, and super glue (Loctite prism) (EMS).

Eyelash brush (eyelash mounted on a 3-in. wooden applicator stick with nail polish; see Figure 7).

Light box and scissors.

2.6.4. Disposable Supplies   Beem capsules, sand paper, diamond knife cleaning rod, glass knife strips, and extra long razor blades (EMS).

Filter paper, 9 cm and 15 cm (Whatman #1) and syringes (3 cc).

2.7. Materials and Solutions for Counterstaining EM Grids

2.7.1. Chemicals   Lead nitrate, sodium hydroxide (1 N), and uranyl acetate (EMS).

Sodium hydroxide pellets (JT Baker).

Distilled water (for uranyl acetate stain; supermarket).

2.7.2. Small Equipment and Nondisposable Supplies   Dental wax, fine forceps, and grid boxes (EMS).

Kimax media storage bottles, 100 ml (Cole-Parmer).

Glass petri dish (95 mm) with foil-covered lid (VWR).

2.7.3. Disposable Supplies   Luer lock syringe filters, 13 mm, with 0.2 μm PTFE (VWR).

Square plastic petri dish; can be reused (EMS).

2.7.4. Solutions   Before counterstaining, prepare uranyl acetate and Reynold's lead citrate solutions. These can be stored for months at 4°C.

5% Uranyl Acetate Solution

1. Add 5 g uranyl acetate to 100 ml distilled H_2O (*do not* use Millipore deionized water or uranyl acetate precipitates) in a foil-covered 100-ml bottle.

2. Stir for 18–24 h until uranyl acetate is suspended; it does not dissolve completely.

Reynolds Lead Citrate (Important: Do Not Breathe Over the Bottle as This Causes Lead to Precipitate)

1. Add 2.66 g lead nitrate to a 100-ml bottle, add 60 ml dH_2O, close lid, and shake gently for 1 min.

2. Add 3.52 g sodium citrate to the bottle, close lid, and shake gently for 1 min. Shake every 2–3 min for 30 min.

3. Add 16 ml 1 N NaOH to the bottle, bring total volume up to 100 ml with dH_2O (~24 ml), and store at 4°C. Do not use the solution if a white precipitate is visible.

3. Methods

3.1. General Comments

Before initiating any pre-embedding EM immunocytochemical study, one must consider the question that one wishes to answer. The type of question determines the experimental design. The three most common questions are: (1) Which is the cellular (e.g., neurons or glia) or subcellular (e.g., mitochondria, multivesicular bodies) localization of substance X? (2) What is the relationship of cells containing substance X with cells containing substance Y? This would include: (a) Is substance X is colocalized with substance Y? and (b) Do cellular profiles (e.g., terminals) containing substance X contact cellular profiles containing substance Y (e.g., dendrites)? (3) Does the cellular or subcellular distribution of substance X, or the relationship of profiles containing substances X and Y, change after an experimental manipulation (e.g., ischemia, or other processes inducing degeneration, potentially in combination with neuroprotective manipulations)?

To answer questions like those posed above, we have used single- and dual-labeling EM immunocytochemical protocols. Using these protocols, antibodies can be identified using the avidin–biotin complex (ABC) peroxidase methods and/or immunogold methods (10, 11). The ABC method provides a sensitive approach to detecting antigens (12). The immunogold-silver method allows precise subcellular location of an antigen (10). For quantitative comparisons between groups, marked sections from each experimental condition are collected into a single vessel so that they are processed together through all steps. To prevent variability in immunolabeling due to day-to-day temperature variations, different reagents, etc., tissues from several groups are processed simultaneously (13). We and others have found that this is a reliable and reproducible method for quantitative comparisons (13–16).

3.2. Perfusion Fixation Methods

Sesack et al. (17) recently published a detailed perfusion fixation procedure in rats. Thus, the perfusion fixation procedure described here focuses on mice. For an illustration of the perfusion setup, see Fig. 1. All of the steps described below should be performed under a fume hood (minimum air flow 100 cycles/min) with the hood sash pulled down to 12–18 in.

1. *Make solutions*: *0.2 M PB, 0.1 M PB, 2% PFA, 3.75% acrolein.* Make enough 3.75% acrolein in 2% PFA to have about 40 ml/ mouse, as well as enough 0.2 M PB to make 0.1 M PB for sectioning (Subheading 3.3).

2. *Prepare the perfusion pump and tubing.* Flush all tubing in the perfusion pump with dH$_2$O, and then air. The tubing should be arranged so that there are two inputs separated by a stopcock, and one output (see Fig. 1). Place one input branch in the parafilm-covered graduated cylinder containing the acrolein/PFA

solution; use scissors to make a hole in the parafilm. Prime the tubing with the acrolein/PFA to avoid losing solution. Place the output branch into acrolein/PFA cylinder and run pump until there are no air bubbles. Then, turn stopcock to close off the acrolein/PFA branch and collect the remaining acrolein/PFA solution from the output tubing. Flush the other input branch with dH_2O and then load with saline/heparin until no air bubbles are visible. Finally, close the saline line and run about 2 ml of acrolein into the saline/heparin output tube; the saline/heparin is the first 5 ml of solution used in the perfusion.

3. *Organ and blood collection.* In addition to the brain, assessment of other organs can provide insight into the effect of the experimental manipulations. This must be accomplished without compromising the quality and rapidity of the perfusion and brain collection. A checklist can be used to observe the health of the coat, eyes, nose, and genitals of the animals. Organs from the thoracic and abdominal cavity can be collected after perfusion fixation for histology, immunocytochemistry, and in situ hybridization. In particular, organ weights can be a bioassay of the degree of stress and immune responses that have occurred (see Note 2).

   Blood collection must be performed rapidly to not compromise fixation of the brain. To minimize the time needed for blood collection, calculate in advance the minimum blood volume necessary for analysis. To collect blood before beginning the perfusion, open the thoracic cavity as usual and remove the pericardium. Hold the heart with forceps, and insert a heparinized 20-gauge needle and 1-ml syringe into the right ventricle at a 1–2-mm depth. Release the pressure on the forceps slightly and slowly draw the blood from the ventricle. Volumes between 0.2 and 0.5 ml can be drawn with minimal hemolysis.

4. *Perfusion procedure.* To begin the perfusion, deeply anesthetize the mouse (we use 150 mg/kg, i.p., sodium pentobarbital, about 0.1 cc per mouse). Check that the animal is deeply anesthetized by pinching the paw to test for a reflex. Place the mouse chest up onto the silicon pad and pin the paws. Lift the chest and cut the skin under the diaphragm without injuring the underlying organs. Cut through the diaphragm and through the skin and ribs on each side. Cut the rib cage off without cutting the thymus. Cut the right atrium and grab the heart with the hemostatic forceps. Insert the needle in the left ventricle without puncturing through. Adjust the pump to a speed of about 9 ml/min and then pull the hood door down as far as possible. Perfuse transcardially a total of 35 ml.

5. *Brain removal.* When the perfusion is complete, stop the perfusion pump and remove the needle from the heart. Turn the animal over and drain out any fixative in the carcass. Remove the eyes (this makes it easier to remove the brain later) and then

decapitate the carcass. Using rongeurs, remove the skull from the region overlying the brainstem. Continue to remove the skull in small pieces from the top and sides of the brain; do not insert the rongeurs so deep that you nick the brain. Place the brain in a scintillation vial containing about 5 ml of 1.87% acrolein/2% PFA (the perfusate diluted in half with 2% PFA) and place the vial on a shaker for 30 min at room temperature (the shaker can be outside of the hood). Under the hood, decant the fixative into the perfusion tray and rinse two times with 0.1 M PB. Store the brain in 0.1 M PB at 4°C until it is cut (no more than 2 days). Flush the perfusion tubing with dH_2O.

3.3. Sectioning the Brain

1. *Preparation for sectioning (can be done the day before perfusion).* Make color-coded 24-well culture plates for each animal; mark the lid and base of each tray with colored tape. Label the top of the tray with all pertinent information (date of the experiment, animal information, experimental manipulation, investigator's name, other relevant information). Fill each well of the tray with cold 0.1 M PB and keep refrigerated until needed. Prepare a color-coded experimental data sheet (sheet for 24-well dishes *provided upon request*) to accompany each tray to record cutting and experimental information. Finally, tape one long piece of tape (about 18 in.) per animal to a bench top. Label this tape with the pertinent information. The tape is used to seal the tray before it is put in the freezer.

2. *Set up the vibratome for cutting.* Check knife holder, buffer tray, and other moving parts for buffer residue, and rinse with dH_2O. Turn on the vibratome, lower the buffer tray holder all the way down, and retract the knife holder mechanism all the way back. Set the speed between 6 and 9 and the frequency between 7 and 9. Set the section thickness to 40 μm.

3. *Blocking the brain.* Pour the brain out of the scintillation vial into a weigh boat. Do not use forceps to remove the brain as this often damages it. Place the brain ventral side up into the brain mold for blocking. Cut the brain into two or three blocks (no more than 5-mm tall) with razor blades. We usually block the brain behind the hypothalamus (see Fig. 2) so that brain is in two pieces, which are glued side by side and cut simultaneously on the vibratome. However, if you are interested in studying the ventral tegmental area or dorsal raphe, block the brain appropriately to preserve these regions.

4. *Securing the brain for slicing.* Place the specimen disc on the table and orient it so that the notch is at the 7 o'clock position. Melt the agar solution in a microwave (if you overheat it, pipette the solution up and down to cool it off; hot agar affects antigenicity). Place the blocked brain on a paper towel, and blot the part of the brain that is glued until it is completely dry on the bottom. Place one or two drops of glue in the middle of

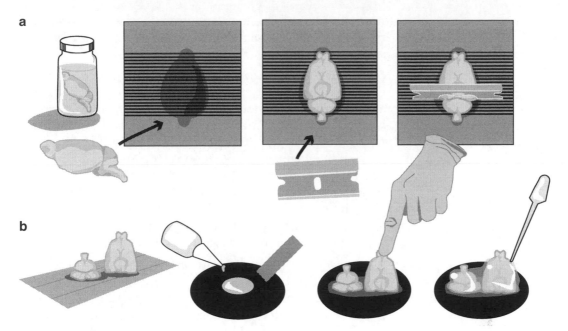

Fig. 2. Sectioning the brain on the vibratome. (**a**) Blocking the brain (Subheading 3.3, step 3). (**b**) Securing the brain for slicing (Subheading 3.3, step 4).

the stage and spread glue into a thin layer using the cardboard cover from a razor blade. Place the brain pieces close together on the glue spot, orienting them so that the cortex on the forebrain piece and ventral surface of the brainstem piece are placed away from you. Gently touch the tops of brain pieces to make sure that they are adhered completely to the specimen disc. This step is critical since the brain does not cut well if it is loose. Using a plastic pipette, surround the brain pieces with the warm agar solution.

5. *Preparing the vibratome for slicing.* Position the appropriate color-coded 24-well culture dish with chilled PB next to the vibratome and over a black background. Have the color-coded data sheet on hand. After the agar is solidified (about 1 min), place the stage with brain pieces into the buffer tray. Place the buffer tray on the tray holder. Lift the buffer tray up slightly on the holder before locking the buffer tray in place, so you do not hit the bottom of the vibratome range while sectioning. Gently pour cold PB into the buffer tray until it covers the brain. Eject a vibratome razor blade from the holder onto a paper towel. Squirt a small amount of 100% ethanol on the blade and gently dry it off with a kimwipe without touching the sharp edge. Put the blade in the knife holder and tighten into place. Place the knife holder in the vibratome, being careful not to hit the brain. Move the magnifying glass over the brain and adjust the position of the fiber-optic lights so that the brain is illuminated (for GFP mice, dim the lights). Move the razor blade toward

the brain with the up/down and forward/reverse knobs. Once the blade is about 2 mm above the brain, set the limits of the sectioning window.

6. *Sectioning.* Set the stroke on "continuous" and cut through the brain. Immediately after each section is cut, pick it up with a brush and place it systemically in a well of the tray. When you reach the end of the tray (e.g., well 24), mark the sheet and then start over at well 1. Repeat these steps until brain is cut as close to the stage as possible (about six passes for a full mouse brain). As you are cutting, make sure that the sections are cutting evenly and not tearing. Periodically, you need to hit the pause button and pull off any dura, blood vessels, or meninges that are sticking to the sides of the brain with a forceps. Adjust the size of the continuous sectioning window as needed.

7. *Storage of tissue sections.* Clean up vibratome and sectioning area. Once the sectioning is complete, note the well that contains the last section on the data sheet. Place the tray in a cold room or refrigerator until you are ready to add the storage solution. This should be done within a few hours of sectioning; do not leave sections overnight in PB as this severely diminishes antigenicity. To prepare the trays for storage at –20°C, pipette off PB completely a few wells at a time (without drying out the sections) and quickly add storage solution until the wells are about half full. Remove all PB or the sections freeze in storage and ruin the tissue. Once all the wells have been transferred to storage solution, seal the edges of the tray with the 18-in. color-coded, labeled tape that you placed earlier on the bench top. Let the sections sit in the storage solution for 15–30 min in the cold room before placing them flat in the –20°C freezer. Sections can be kept for at least 5 years in storage without any apparent loss in antigenicity or compromise to morphology. Sequential storage in 24-well plates allows for the selection of stereologically unbiased random systematic series of sections through a given region for analysis. For instance, selecting the tissue in one well provides a random systematic 1-in-24 series.

3.4. Immunocytochemical Processing, General Comments

For studies comparing different experimental manipulations, sections from all groups must be processed for immunocytochemistry simultaneously (13). For EM studies, a minimum of three sets of tissue should be used; each set should contain sections from the brains of animals from each experimental manipulation. The sections should be punched with different codes so that they can be pooled into single containers (see Fig. 3). Tissue from up to six different animals can be pooled into a single container.

The two-part dual-labeling EM protocol is described below: the first part involves preparing the tissue for incubation with the

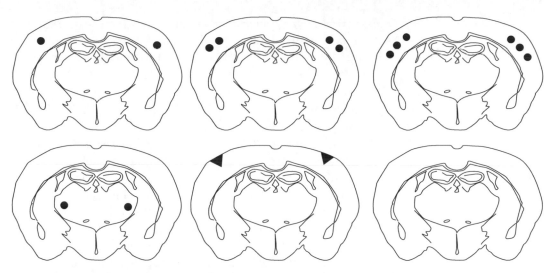

Fig. 3. Immunocytochemical processing procedures: Tissue selection. Examples of punch patterns used to distinguish groups within a crucible (Subheading 3.4).

primary antibody(s) and the second part involves secondary antibody(s) incubation and peroxidase and/or immunogold-sliver processing. Single-labeled studies can also be done for either peroxidase or immunogold-silver by processing the tissue through the appropriate sections of the dual-labeling protocol. For all steps, do not let the sections dry out when transferring them between solutions. Unless noted, all steps are carried out at room temperature with solutions at room temperature. All incubation steps and rinses are conducted on a shaker. Delicate pieces of glassware (e.g., crystallizing dishes, petri dishes, crucibles) should be cleaned in 10% Contrad solution.

3.5. Immunocytochemical Processing, Part 1

1. *Tissue selection*. First, select tissue from the brain region you wish to investigate. Remove the trays containing the stored tissue from the –20°C freezer. Gather divided petri dishes (one per tray), data sheets, "punch" dish (i.e., petri dish lined with black rubber or black electrical tape), punch tool, paint brushes, crucibles, and crystallizing dishes and make at least 1 L of PB. Sort the trays into the appropriate groups. Select tissue from one group at a time to avoid confusion. For each group of tissue, dry off the trays and carefully remove the tape sealing them. Save the long tape to reseal the tray later. Put PB into each quadrant of the divided petri. Place one petri dish in front of each tray. Fill crystallizing dishes with about 100 ml of PB (up to the bottom of the white label) and set aside. Place the tissue from the desired well into the petri dish and put the selected brain section into the adjacent division of the petri dish. Continue the selection process; each set has tissue selected

from one animal per experimental condition. Note removed sections on the data sheet. Place tissue sections from each animal into the dish lined with black rubber and mark them with a pattern (see examples in Fig. 3) using the hole-punch tool. Record your punch codes. Once all sections from one set are marked, pool them together in a color-coded crucible placed in a crystallizing dish containing PB.

2. *Removal of unbound aldehydes with sodium borohydride.* Once crucibles from all sets of animals are in the crystallizing dishes, unbound aldehydes are removed with sodium borohydride. For this, place the crystallizing dishes with crucibles on a shaker that is moving slowly. Mix the sodium borohydride solution, put it in a second crystallizing dish, transfer the crucibles into the sodium borohydride solution, and incubate on the shaker for 30 min. Gently squirt solution over the tissue during the incubation. At the conclusion of the incubation, rinse copiously with several changes of PB until there are no bubbles, and the sections sink in the crucibles. One technique is to lift the cups out of the dish and pour PB through them while gently swirling. For all immunocytochemistry steps, watch for lost sections.

3. *Optional freeze–thaw.* To enhance penetration of certain antibodies, sections can be quickly frozen and thawed (method adapted from Yoland Smith, see ref. 17). Place crucibles in a crystallizing dish containing cryoprotectant solution and incubate for 20 min at room temperature. Transfer each crucible to a 50-ml plastic beakers containing about 15 ml cryoprotectant solution, cover beakers with parafilm, and place in the –80°C freezer for 20 min. After the –80°C incubation, remove the beakers from the freezer and place them in crystallizing dishes with warm or room-temperature water. When the cryoprotectant begins to melt and the crucibles are free (less than 5 min), transfer the crucibles to a crystallizing dish with room-temperature cryoprotectant solution for a 10-min incubation. During this incubation, prepare 70, 50, and 30% cryoprotectant diluted in 0.1 M PB. Incubate crucibles for 10 min in each, and finally wash for an additional 10 min in 0.1 M PB.

4. *Rinse.* Transfer crucibles to crystallizing dish with TS. Rinse twice, 5 min each.

5. *Blocking step.* To block nonspecific staining, incubate the tissue in TS containing 0.5% serum. We use BSA to avoid cross-species reactivity with primary and secondary antibodies. However, normal goat or rabbit serums can be used instead. Add 0.5 g BSA to 100 ml TS in a plastic beaker and stir slowly. (You need about 100 ml per large crystallizing dish and 50 ml per small crystallizing dish.) Be careful not to overstir, as this cause the BSA to froth. Place the crucibles in the crystallizing dish with BSA solution and shake gently for 30 min.

6. *Rinse.* Transfer crucibles to crystallizing dish with TS. Rinse 2×, 5 min each.

7. *Make up primary antibody solutions during the rinses.* Primary antibody aliquots are diluted in 0.1% BSA in TS. For some antibodies, the addition of 0.025% Triton-X-100 to the 0.1% BSA in 0.1 M TS allows greater penetration of the antibody into the tissue. (This works best for nonmembrane-bound antigens.) To make the 0.025% Triton solution, make 0.25% Triton first and then dilute it. For factors to consider when working with antibodies and suggestions for antibodies to identify degenerating processes, see Note 3.
Mix up the total volume of primary antibody solution in one vial and divide if necessary. Using a pipetman, add the primary antibody(s) to the diluent and mix gently. Allow 2 ml of primary antibody solution per vial. This amount should be adjusted depending on the amount of tissue being placed in the vial and the availability of the antibody.

8. *Begin the primary antibody incubation.* Add tissue sections to the vials with a clean paintbrush (rinse bristles with hot water between different antibodies). Cap each vial and gently swirl the sections to be sure that none are stuck to the sides of the vial. Press the vials into a clay-lined staining dish lid at about a 10° angle. Place the tray with the vials on a shaker table and adjust the speed of the shaker so that the sections are slightly moving. (If the shaker is too fast, the sections will break apart or even dissolve.) After 20–24 h, move the tray with the vials to a shaker table in the cold room. Continue incubating the sections in the antibody diluent for 1–4 days (times vary by antibody).

3.6. Immunocyto-chemical Processing, Part 2

Peroxidase Labeling Procedure (Modified from Hsu et al. (11))

1. *Remove sections from primary antibody and rinse in TS.* Using a clean brush, transfer the tissue to appropriately labeled crucibles in TS in a crystallizing dish. Alternatively, swirl tissue in the scintillation vial and immediately dump the contents into a crucible over a waste beaker. Check for missing sections, and rinse tissue in TS 3×, 10–15 min each.

2. *Prepare secondary biotinylated antibody.* (The secondary antibodies should be aliquoted (25 μl) into plastic screw-top tubes and stored at −20°C or lower until use.) In a scintillation vial, dilute the secondary antibody to 1:400 with 0.1% BSA in TS (25 μl for each 10 ml). Mix the solution gently and aliquot into scintillation vials. Usually, 2 ml per vial is sufficient but should be adjusted depending upon how many tissue sections are there.

3. *Secondary biotinylated antibody incubation.* Using a brush, transfer the tissue to the scintillation vials containing the

secondary antibody. Place the scintillation vials on the tray in the shaker for 30 min.

4. *ABC solution preparation.* When the tissue is in the secondary antibody, prepare the ABC solution from an Elite Vectastain kit. Add two drops of solution A and two drops of solution B to each 10 ml of TS (no serum!) and mix *immediately* by vortexing or rapidly pipetting up and down. Let the solution sit undisturbed for 30 min prior to use.

5. *Rinse in TS.* At the conclusion of the secondary antibody incubation, transfer tissue to crucibles in TS in a crystallizing dish. Wash the tissue in TS 3×, 10–15 min each.

6. *ABC incubation.* Aliquot the ABC solution into scintillation vials, allowing about 2 ml/vial. Transfer the sections to vials and place the vials on a tray on the shaker. Incubate the sections in ABC solution for 30 min exactly. Time this carefully; over-incubation in the ABC solution increases the background peroxidase labeling.

7. *Defrost DAB aliquots during ABC incubation.* (We weigh out the DAB and place it into Beem capsules. The capsules are stored in a jar with desiccant in the –20°C freezer.) Defrost for 30 min; protect aliquot from light during thaw using a foil-covered beaker.

8. *Rinse in TS.* Once the ABC incubation is complete, transfer the sections to crucibles in TS in a crystallizing dish. Wash the tissue in TS 3×, 10 min each. Prior to the last wash, replace the crucibles with clean ones. Otherwise, you get ABC–DAB precipitate on your sections.

9. *Prepare the DAB solution under the hood during the last rinse.* DAB is carcinogenic and should be handled with gloves. All liquid and solid DAB waste should be collected and disposed of according to institutional guidelines. The first TS rinse after the DAB step should be considered DAB waste. To make the DAB solution, get 100 ml TS gently stirring in a disposable beaker. Carefully open a capsule containing DAB (22 mg) and release the DAB; then, drop the capsule into the solution. Use a stir bar dedicated to DAB (we use a yellow one). DAB is light sensitive, so cover the beaker with a larger foil-covered beaker. Just before using the DAB solution, add 10 µl of 30% H_2O_2 with a pipetman. Drop the pipette tip into the solution.

10. *DAB peroxidase labeling procedure.* Move the crystallizing dishes containing the crucibles under the DAB hood. Lay disposable plastic petri dishes (one per four crucibles) next to the crystallizing dishes. Pour the DAB solution into the petri dishes, place the crucibles into the DAB solution, and start the dedicated DAB-hood timer counting up. This can be done sequentially at timed intervals or four crucibles at a time. As

soon as the crucibles go into the DAB solution, start gently squirting the solution over tissue using a plastic disposable pipette. Continue squirting the tissue with the DAB solution until the conclusion of the incubation. The dilution of the primary antibody should be such that the optimum time for the tissue in the DAB solution is 6 min. However, DAB times can vary from different experiments due to room temperature, age of the ABC kit, or antibody concentrations. Thus, the reaction should be checked by wet mounting a section on a slide and looking at it under a microscope. The section can then be put back in the DAB solution and incubated longer if necessary.

11. *Rinse in TS.* Once the DAB reaction is complete, wash the tissue in TS 3×, 2 min each. Be sure to discard the first rinse as DAB waste.

12. *Rinse in PB.* Transfer the crucibles to PB and rinse 3×, 5 min each. If you are proceeding directly to the embedding step (3.7), be sure and rinse sections thoroughly in PB. Saline contamination ruins the osmium reaction.

Immunogold-Silver Labeling Procedure

1. *Rinse in 0.01 M PBS.* Wash the tissue in 0.01 M PBS for 5 min.

2. *Washing incubation buffer.* Place the crucibles containing the tissue sections in the washing incubation buffer solution on shaker for 10 min.

3. *Prepare gold-conjugated secondary.* While the tissue is incubating in the washing buffer, prepare gold-conjugated IgG solution. For this, add 20 µl IgG per ml incubation buffer (i.e., 1:50 dilution) in a scintillation vial. Two milliliter per vial is usually ideal.

4. *Gold-conjugated secondary incubation.* Using a brush, transfer the tissue into the scintillation vials containing the secondary antibody. Incubate the tissue for 2 h in the secondary antibody. During the incubation step, bring the silver intensification kit to room temperature on the bench top.

5. *Washing incubation buffer.* Transfer the tissue into crucibles in a dish containing the washing buffer. Wash the tissue in buffer 1× for 5 min.

6. *Rinse.* Wash tissue in 0.01 M PBS, 3× for 5 min.

7. *Glutaraldehyde incubation.* Under the fume hood, place the crucibles in a crystallizing dish containing 2% glutaraldehyde in 0.01 M PBS for 10 min. (To make the glutaraldehyde solution, add 1 ml 25% EM grade glutaraldehyde to 12.5 ml 0.01 M PBS.) Agitate the crucibles periodically; the tissue does not need to be on a shaker.

8. *Rinse in 0.01 M PBS.* Wash tissue in PBS for 1×, 2 min.

9. *Rinse in 0.1 M PB.* Wash tissue in 0.1 M PB for 2×, 2 min.

10. *Citrate buffer.* Transfer the tissue to a crystallizing dish containing citrate buffer and place the dish on a bench top next to the silver kit.

11. *Silver intensification procedure.* This step is tricky (see Fig 4). Do not use brushes to transfer tissue to and from the silver solution. Instead, use pointed sticks made by breaking applicator sticks in half (use your hands to break the sticks not razor blades as metal contaminates the silver solution, see Fig. 4a). Open a *sterile* 12-well plastic petri dish and fill the top and bottom row with 0.2 M citrate buffer. When you are ready to begin, place ten drops of reagent A followed by ten drops of reagent B (i.e., a 1:1 ratio) into one well in the center row. This solution can be used at least once. Watch for precipitate that appears as a silver film on top of the solution. Work on a dark surface so that you can watch for this precipitation and do not use a well that has precipitated.

Test one section from each antibody condition first. Using the stick, transfer test section to intense solution. With the tissue culture plate flat on the bench top, gently swirl the plate so

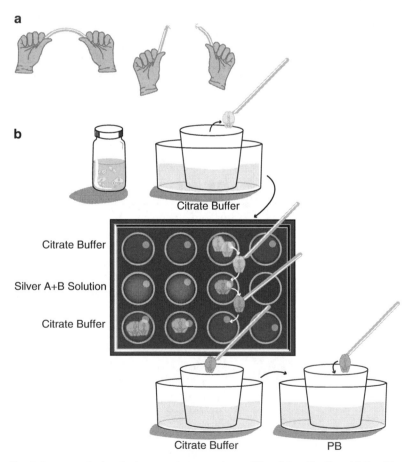

Fig. 4. Immunocytochemical processing procedures: Silver intensification. (**a**) Breaking wooden applicator sticks. (**b**) Silver intense procedure (Subheading 3.6, part 2 "immuno-gold-silver labeling procedure" steps 11–13).

that the section is constantly moved in the intense solution during the entire incubation step (usually, 4–7 min). To stop the intensification reaction, transfer the tissue citrate buffer, and note the length of the intensification. Wet mount the test section on a slide and exam it under the light microscope. Ideally, you should see small black dots (barely visible) over the cells that you expect to be labeled. If you intensify too long (over 8 min), nonspecific gold particles appear over the entire tissue section. If needed, the test section can be placed back in the silver solution to increase the incubation time. Once the right incubation time has been identified, repeat the procedure with the rest of the sections.

12. *Transfer tissue back to citrate buffer.* After all of the tissue sections have undergone silver intensification, use a wooden stick to transfer tissue to the crucibles in a crystallizing dish containing citrate buffer.

13. *Rinse in 0.1 M PB.* Transfer crucibles to 0.1 M PB for 3× 10-min washes. Saline contamination in the PB ruins the osmium reaction.

3.7. Embedding Tissue Sections for Electron Microscopy

All osmication and embedding procedures should be performed under a fume hood. Glassware contaminated with osmium should be deactivated with 70% ethanol prior to removing it from the hood. All liquid and solid osmium waste should be collected and disposed of according to institutional guidelines. Glassware used for EM embedding should be cleaned with Contrad.

1. *Separate tissue according to animal and/or group before beginning osmication.* Sections may crack or break during this step, making it impossible to separate groups. Label Coors dishes with tape to keep animals and/or groups separate. Place under the hood, add a small amount of PB into each well, and put two to four sections in each well. It is okay if the sections overlap, but it is important to unfold and flatten the sections in each well with a brush.

2. *Osmium incubation.* Slowly draw off the PB from each well with a glass pipette and replace with 2% osmium tetroxide in PB stock solution using another glass pipette. If necessary, use wooden applicators that have been broken in half to form a point to manipulate sections. Protect Coors dishes from light (we use the bottoms of plastic food storage containers that have been covered with foil) and incubate the sections for 1 h undisturbed.

3. *Embed preparation.* While the tissue is incubating in osmium, prepare the EMbed resin.

4. *PB rinse.* After the 1-h osmium incubation, uncover the Coors dishes, remove the osmium, and immediately add PB with a fresh pipette. Place the used osmium solution, as well as the

first PB rinse, in osmium waste. Rinse sections with PB, 3×
3 min each.

5. *Dehydration*. Dehydrate sections through 30, 50, 70, and 95%
   ethanol, 5 min each. Remove PB and add the alcohols with
   fresh glass pipettes, one row at a time, moving quickly so that
   the tissue does not dry out. Transfer the sections to scintillation vials preloaded with about 5 ml of 100% ethanol. (The
   100% ethanol should be from a bottle opened within 1 week,
   as opened bottles can absorb moisture from the air.) From this
   point on, replace solutions one vial at a time and keep the vials
   sealed. Any moisture in the solutions can ruin the hardening of
   the EMbed resin. If the room is above 40% humidity, use a
   dehumidifier.
   Incubate the sections in two changes of 100% ethanol, 10 min
   each. Then incubate the sections in two changes of propylene
   oxide, 10 min each. During the incubation steps, dilute the
   EMbed resin 1:1 with propylene oxide in a conical tube and
   invert to mix.

6. *EMbed/propylene oxide (1:1) mixture*. Working one vial at a
   time, remove all propylene oxide with a fresh glass pipette and
   add the 1:1 mixture to the vials. Try to keep the sections flat in
   the vials, as they are very fragile. Once the EMbed/propylene
   oxide mixture is added, gently swirl the vials so that the sections are suspended in the solution. Place vials on a slowly
   rotating mixer. Incubate the sections in the EMbed/propylene
   oxide mixture overnight at room temperature. The rotating
   mixer can be on the bench top.

7. *EMbed incubation*. The next day, prepare fresh EMbed
   (5–10 ml per vial). One vial at a time, replace the EMbed/
   propylene oxide with straight EMbed. Be sure and pull off all
   the EMbed/propylene oxide since propylene oxide can affect
   EMbed hardening. Gently rotate the vials by hand to suspend
   the sections in the EMbed. Replace the vials on the rotary
   mixer and allow the tissue to incubate in the EMbed for 2–4 h.
   (Do not go longer than 4 h as this makes the sections brittle
   and the EMbed does not cure properly.)

8. *Preparation for flat embedding*. While the sections are incubating in the EMbed, prepare for the flat embedding step. For
   this, clean the metal trays and with 100% ethanol. Forcefully
   wipe the Aclar sheets with 100% ethanol (this helps the Aclar
   peel off the Epon later) and let dry completely. Break wooden
   applicator sticks so that they form sharp points. Put the trays
   on blue pads under a hood. Set up a box of kimwipes, cotton
   applicators, and a plastic bag to collect waste near the trays.

9. *Flat embedding*. One vial at a time, tilt the vial and use the stick
   to remove the sections from the vial (see Fig. 5). Gently wipe
   off the EMbed from the sections on the lip of the vial as you

remove them. Place the sections about 1 cm apart on an Aclar sheet. Gently touch the tops of the sections with a cotton applicator or kimwipe to soak up excess EMbed. Once the sections from one vial are on the Aclar, slowly cover them with a smaller Aclar sheet. Use small pieces of Aclar as it is expensive. Once the sections are covered, gently roll your finger on each section several times to squeeze out EMbed and push the air bubbles off the sections. (Do not push too hard or the sections break). Try to squeeze out as much EMbed as possible as this makes the subsequent ultrathin sectioning much easier.

Place pieces of colored tapes identifying the tissue from a particular vial near, but not on top of the Aclar cover. Do not use marker as it is dissolved by any EMbed resin that oozes out of the cover. Place steel weights on top of the sections. Bake the sections in a 60°C oven for between 3 and 5 days.

10. *Storage of flat-embedded tissue.* Remove the tray from the oven and let cool a few minutes before removing the steel weights. To remove the weights, hold the top Aclar sheet in place and

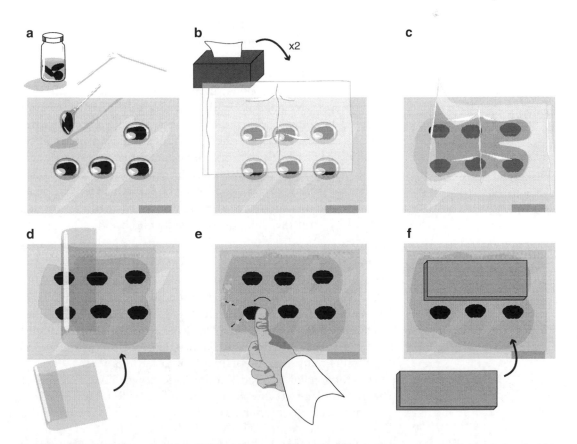

Fig. 5. Embedding tissue sections for electron microscopy: Flat embedding. (**a**–**f**) Sequence of steps for flat embedding (Subheading 3.5, step 9).

carefully lift the weights off. If the weights stick, gently rock them back and forth to dislodge them. Do not pull the weights off too quickly; this can pull the Aclar sheet and the sections off. Store the flat-embedded tissue in envelopes labeled with all the pertinent experimental details. The flat-embedded sections can be stored indefinitely.

3.8. Sectioning Tissue for EM

The ultratome should be on an anti-vibration table. The table should be in a room that has 30–40% humidity, little foot traffic, and away from direct ventilation. All items that contact the EM sections should be as clean as possible. The steps described below do not have to be performed on the same day. However, before preparing large quantities of tissue to be examined under the electron microscope, examine a test section to insure that the morphology and immunocytochemical labeling are satisfactory. Detailed instructions on use of the ultratome as well as videos can be found on the Leica Web site (http://www.leica-microsystems.com).

1. *Preparing EMbed chucks.* Make EMbed resin. We make two types of EMbed chucks. For the first, cut out preprinted labels (laser-printed text on white paper) and place them label side down in embedding molds. Fill the molds with EMbed and place in a 60°C oven until they are set (about 2 days). For the second, cut off the pointed ends of Beam capsules, close the lids, and place lid side down in an embedding capsule holder. Insert preprinted labels, label side out, into the top of the beam capsule (wrap the label around a syringe top to insert). Fill the capsules with EMbed, let them sit about a half hour to allow the bubbles to rise to the top, and bake at 60°C for 3–4 days. For both methods, do not take the capsules out until they are hard; they cannot be rebaked once they have cooled. Prior to gluing tissue on the blocks, remove the chucks from the molds and lightly sand the tops (do this under the hood).

2. *Mounting sections on EMbed chucks.* Select the region of the desired region of the brain section for EM analysis using a light microscope. Mark the section with a marker. Slowly peel one of the Aclar sheets off the "flat-embedded" sections. The sections may stick to the cover, bottom, or both sides of the Aclar. (Make note of which side they stick.) Lay the "flat-embedded" sections, sanded EMbed chucks, forceps (use a pair dedicated to gluing), and long razor blade on a light box (see Fig. 6). (Place the peeled EMbed/section side down.) Use the end of the razor blade to cut an irregular four-sided shape (about 2–3-mm wide) containing the desired tissue piece out of the flat-embedded tissue. (This way, if the tissue flips over, you can tell which side is up). Put a drop of glue on the EMbed chuck and thin it out with a piece of cardboard. Take the forceps and

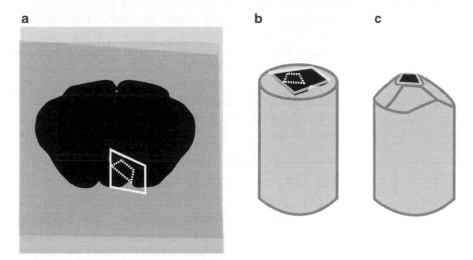

Fig. 6. Sectioning tissue for EM: Mounting sections on EMbed chunks (Subheading 3.6, steps 1–4). (**a**) Marking a trapezoid. (**b**) Gluing tissue to block. (**c**) Trimming tissue blocks on the ultratome

pick up the tissue section and place it on the glue (make sure that the EMbed side of the tissue, not Aclar, apposes the glue). If the Aclar falls off during the gluing process, do not touch the tissue surface with the forceps while placing it on the chuck. Use a light microscope to check that the glue completely surrounds the tissue. If not, add a little glue to the sides of the tissue. Do not worry if glue gets on top of the block; this comes off when the Aclar is peeled off later. Tissue blocks can be stored indefinitely.

3. *Cleaning grids and collection supplies.* Pour grids into a 10-ml beaker that contains about 2 ml of acetone. Swirl the grids around about ten times and then tilt the beaker so that the grids clump together. Carefully pick up the grids with forceps and lay them on two pieces of #1 filter paper in a glass petri dish. To prevent the grids from moving around due to static, tape a piece of filter paper on the top of the petri dish and moisten it with water. This is critical if the air is dry, e.g., in the winter. Next, take a rubber grid mat, squirt it with 100% ethanol and wipe it off with a paper towel (kimwipes leave lint), and place inside another clean petri dish. Pour about 10 ml of 100% ethanol in a scintillation bottle; this can be reused for several weeks. Prior to cutting and in between specimens, dip the eyelash brush and forceps in the 100% ethanol.

4. *Trimming tissue blocks on the ultratome.* Examine the block and glued section by placing it in the ultratome block holder under the microscope to check that the section is glued. If there are air bubbles under the tissue, add more glue and let dry. Draw (or photograph) the tissue piece in a logbook. Note distin-

guishing features, like blood vessels, cracks, or axon fields. Take the block and secure it in the ultratome block holder. Wear a disposable mask covering your nose and mouth for trimming process. Trim the block to a 1–1.5-mm long and 1-mm wide trapezoid shape (this shape helps orient the tissue on the electron microscope; Fig. 3.6). Do not touch the top of the block with your fingers while you are trimming; this can introduce contamination into the electron microscope later. If the Aclar piece covering the tissue falls off during trimming, gently push it off with the forceps before finagling your shape. Use a fresh razor blade to insure nice, clean edges on the tissue section. The top and bottom of the trapezoid should be parallel. If you are not going to section the block immediately, make the trapezoid slightly larger and do the final trim immediately before you section (otherwise, the sections may not form a ribbon during cutting).

5. *Aligning tissue block with the ultratome knife.* Immunoreaction products usually penetrate about 1–2 μm into a tissue section. Thus, ultrathin sections should be collected from the surface (i.e., tissue–plastic interface) of the vibratome section. The trapezoid should be aligned so that the block face hits the knife evenly during the sectioning procedure. Place the block in the specimen arm, and place a disposable glass knife in the knife holder. This knife can be reused several times. Illuminate the block with the top light only. Place the knife about 2 mm away from the face of the block. Using the shadow from the knife on the block face to guide you, align the left and right sides of the blade until they appear parallel to the block face. Do not touch the up and down control yet. Once the left/right alignment appears parallel, retract the knife 2 or 3 mm and move the block up and down with the handwheel. Using the shadow, get the top and bottom to look equidistant from the knife. Move the knife a little closer and repeat these steps until the shadow of the knife on the block face looks even at all points. If you saw a thick layer of EMbed on top of the tissue while you were trimming the block, cut 1 or 2 μm off the top of the block with the glass knife. This saves time later. If not, back up the knife holder about 2 cm from the block face. Remove the glass knife and replace it with the diamond knife.

6. *Cutting tissue sections.* Illuminate the block with both the top and bottom lights. Move the knife edge within about 0.5 mm from the block face. Fill the knife boat with dH_2O, adjust the water level so that it is just below the knife edge (the water should have a light silver reflection), and let sit for 5 min. Set the cutting window and then approach the block 1 μm at a time until you are as close as possible to the knife without taking a section (do not cut 1-μm sections with the diamond knife as this dulls it quickly). Decrease the approach interval to

Fig. 7. Sectioning tissue for EM: Collection supplies. Eyelash brush made for collecting thin sections (Subheading 3.6, step 7).

0.5 μm, then 0.4 μm, etc. until you cut the first section. At this point, turn on the automatic sectioning button (start with 70-nm thick) and allow the ribbon of sections to float in the boat (the sections should be of a silver or silver/gold color). Keep cutting until the vibratome ridges are no longer apparent (if the embedded vibratome sections are flat, this should be about 20 sections). Turn off the automatic sectioning button and place the block face below the knife edge. The tissue sections should stick together to create a ribbon. As the ultratome cuts, use the eyelash brush (see Fig. 7) to position the floating sections for collection.

7. *Collecting thin tissue sections.* Set up your workspace with one petri dish containing the acetone-cleaned grids, another dish with the rubber tissue mat, two pieces of 9-cm #1 filter paper, several more pieces of filter paper cut into one-eighth wedges, a pair of clean sharp forceps, and an eyelash brush. Pick up the most superficial sections first (the ones farthest from the knife-edge). Hold the forceps in your dominant hand and the eyelash brush in the other. Pick up a grid by the edge with the forceps, and hold it dull side up. Bring the grid to the knife boat and gently break the surface tension to submerge it in an area of the boat away from the sections you wish to collect. Once the grid is under the water, orient the grid about 15° from vertical, with the dull side facing up. Slowly move it under the floating sections using the eyelash brush to move the sections if necessary. Hold the sections in place with the brush and lift the grid up out of the water, maintaining the angle of the grid. As the top of the grid emerges from the water, adjust the angle of the grid to catch your sections. Remove the grid to the stacked filter paper, and blot gently on the bottom. Use

a wedge of the cut filter paper to push the grid onto the filter paper. Move the forceps holding the grid over the rubber mat. Without letting go of the forceps, insert the filter paper wedge between the tips of the forceps. Slowly push the wedge into the tips while releasing pressure on the forceps. This allows the grid to come off the forceps and land on the mat. Repeat these steps until all the sections are collected. Once the sectioning is complete, remove the knife, rinse with water, and dry with a canned duster. Take the specimen block and turn it upside down in a beam capsule (this way, you can cut more, if needed, at a later time). Wipe off and cover the ultratome.

3.9. Counterstaining EM Grids

Counterstain only every other grid or every third grid. This preserves at least one set of grids in case something goes wrong during the staining procedure. Stain only four to six grids at first to get a sense of the proper timing; if grids sit for longer than 7 min in the lead citrate, you will see precipitate on the tissue.

1. *Preparation for uranyl acetate step.* Take small amount (about 0.5 ml) of the uranyl acetate and lead citrate solutions out of the bottle with a 3-ml syringe. Cover the syringe containing the uranyl acetate with foil. Put a luer lock filter (13 mm, 0.2 μm, cellulose acetate membrane) on each syringe. Let the solutions warm up to room temperature (about 30 min).

2. *Uranyl acetate step.* Place a clean piece of dental wax into a foil-covered petri dish. Uranyl acetate is light sensitive, so keep the lid closed as much as possible during the incubation. Place small drops of uranyl acetate solution on dental wax, one drop for each grid to be stained. Float the grids, section side down, on drops. Wipe the forceps off with a paper towel in between grid transfers. Incubate the grids in the stain for 20 min, protected from light.

3. *Preparation for lead citrate step.* When there are 2 min left in the uranyl acetate incubation step, prepare for the lead citrate incubation step. Place a clean piece of dental wax in a petri dish and surround the wax with NaOH pellets (usually, 15–20 pellets). Place small drops of the lead citrate solution on the wax. Keep the lid closed as much as possible and try not to breathe over the lead citrate drops, as this causes the lead to precipitate. Arrange five 10-ml beakers filled with dH_2O between the uranyl acetate dish and the lead citrate dish. During the uranyl incubation, prepare for the drying step. Place a fresh set of five 10-ml beakers with dH_2O between the lead citrate and a stack of two pieces of Whatman (22 cm) #1 filter paper. Cut up eight wedges of 9-cm filter paper and set in a petri dish next to the stacked paper.

4. *First rinse and movement of grids from uranyl acetate to lead citrate.* Move grids to the lead citrate in the same order they

were placed in the uranyl acetate solution. Grab the first grid with the forceps. Rinse each grid by submerging it in the first beaker and moving it up and down about ten times while keeping the grid under the surface. Move through the series of five beakers in this way. At the end of the rinsing series, place the grid section side up inside the lead citrate drop. Close the lid of the petri dish between transfers. Incubate the grids in the lead citrate stain for 5–7 min.

5. *Second rinse and drying.* Remove the grids one by one and rinse them in the series of beakers as described in **step 3**. Minimize the transition time between the lead citrate and the first rinse to avoid lead precipitate on the grids. After rinsing, blot the bottom of each grid on filter paper and use a filter paper wedge to push the grid off of the forceps tips onto a filter paper-lined petri dish. Dry the grids for 30 min before putting into grid boxes. (If the room is dry, shoot the grid boxes with an antistatic gun prior to loading the grids. Otherwise, the grids will fly out of the box).

6. *Clean up.* Dispose of the drops and unused stains in the appropriate waste bottles. Rinse the beakers and wax well with dH_2O and set aside to dry.

3.10. Sampling Tissue Sections

1. *Controlling for the effects of antibody and label penetration*
   The immunolabeling protocols described preserve ultrastructural membranes through avoidance (or reduced use) of detergents, like Triton-X-100 (see Note 3). Because membranes are preserved, labeling is limited to a depth of 1–2 μm on either side of a section. Within that region, the extent of labeling decreases markedly deeper in the tissue. It is, therefore, critically important to control for the effects of penetration when sampling, both to ensure optimum labeling and to allow quantitative comparisons between different tissue samples. The best approach is to restrict sampling to the region that is immediately adjacent to the plastic/tissue interface (14). Every effort should be made to embed tissue sections as flat as possible, with relatively little EMbed plastic resin covering the tissue surface (see Subheading 3.5). However, tissue sections are never perfectly flat. As thin sections are cut through a block, tissue will likely begin to appear as islands surrounded by plastic (i.e., plastic/tissue interface). The interface is ridge-like or jagged in appearance due to the vibration of the Vibratome blade (see Fig. 8). Sampling should be limited to grid squares that are adjacent to this border and which contain no visibly damaged tissue. To select which grid squares to sample, first capture a low-magnification image of the entire thin section on the grid with an electron microscope and use this as a map. Then, increase magnification, adjust the contrast level so that

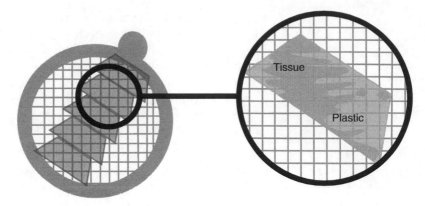

Fig. 8. EM analysis: Sampling. Detail of tissue/plastic interface (Subheading 3.8, step 1).

the plastic/tissue interface is apparent, and select a grid square that is next to the plastic/tissue interface. For systematic random sampling (18), identify the interface, randomly select a grid square, and then analyze a defined series of grid squares along the interface (e.g., every third grid square). Finally, increase magnification again, adjust contrast, and begin collecting micrographs.

2. *Mapping regional borders*

   Images of individual thin sections also are useful if one wants to limit analysis to specific tissue regions or subregions. Capturing a light microscopic image of the sectioned block face can help to identify borders between regions. Illuminate the block from below, preferably after sectioning when the face is smooth, and use this image to create a map of the regions included in the block. Scale the image and use as an overlay to define regional borders within a thin section. Thin sectioning itself can produce compression in the axis perpendicular to the block face. Compensate for this by using the "free transform" function found in graphic software packages. By using an overlay to define regional borders, ideal thin sections in which the plastic/tissue interface crosses the region of interest can then be identified.

3. *Sampling within grid squares and initial analysis*

   The manner in which one samples neuropil within selected grid squares depends on the goals of the experiment and the pattern of labeling. If the density of labeled processes is sparse, acquire images of all labeled processes in a selected grid square for analysis. However, if labeling density is high, then acquire images of a smaller, random field (e.g., a corner of the grid square) and analyze labeling in that field. Labeled process "profiles" are cross-sections of a labeled structure. Profiles can be categorized according to type (e.g., glial, or neuronal perikarya, dendrites or terminals (19)). Profiles also can be qualitatively described in terms of the pattern of labeling, including association with subcellular locations (e.g., mitochondria, endomembranes, or plasma membranes). Labeled neuronal somata and dendrites also can be classified by the type of synaptic input they receive and the presence or absence of immunoreactivity in contacting terminals. Labeled terminals can be examined for cross-sectional area, the types of synapses formed, and vesicle content. Such analysis can provide morphological evidence for the type of transmitter within the terminal (20). This type of analysis can yield information on the types of degenerating processes, their position within neural circuits, and the subcellular distribution of a specific degeneration marker.

3.11. Quantifying the Extent of Degeneration

Ultrastructural analysis can be used to quantify regional degeneration and/or to facilitate comparisons between experimental conditions or across time points. The most apparent measure is the number of labeled profiles per unit area (N_A). However, this value is dependent on both the actual number of degeneration processes and their size. Larger processes more frequently appear in random cuts through tissue than smaller ones. One solution to this problem is to apply stereological counting methods to determine N_A by using the physical disector (18). This requires the analysis of adjacent thin sections, but generates a number that is independent of size. Another approach is to determine the volume density (V_V), the fraction of the volume occupied by degenerating processes. This is equal to the fraction of neuropil area occupied by degenerating processes (A_A), the total area of labeled processes in a given field area. Of course, these types of analyses can be combined with the categorization schemes described in Subheading 3.10.8 to determine how various subgroups of degenerating processes are affected. It should be remembered that both of these measures, N_A and V_V, represent densities. If the overall volume of tissue changes, a change in density may appear because of the volume change. This could be problematic in later stages of degeneration. In this case, total tissue volume of the affected region should be analyzed.

3.12. Analysis of Dual-Labeled Material

Ultrastructural analytic approaches can be further refined through dual-labeling EM methods. For instance, immunoperoxidase labeling can be used to identify a marker of degenerating processes in parallel with immunogold labeling. The immunogold labeling can identify an antigen, which could identify the phenotype of the degenerating process, identify the phenotype of afferent inputs, or label an additional component associated with apoptosis or degenerative disease mechanisms. The selection of antigen labels influences the type of analysis that can be performed. Immunoperoxidase labeling is particularly sensitive, and can fill labeled processes, so that even small profiles of the process can be identified (particularly useful to measure V_v, for instance). However, it can obscure subcellular detail within the labeled process. Immunogold labeling is less sensitive, but more carefully identifies the subcellular location of the antigen, and provides a quantifiable particle. Silver-enhanced immunogold (SIG) particles can be used to determine the density of labeling along the plasmalemma or a subcellular membrane (particles/μm) (21) and/or the density of labeling within the cytoplasm (particles/μm²) (22–24). In addition, the fractional distribution of SIG particles between membranes and cytoplasm (22–26) or between specific subcellular organelles, such as mitochondria or at synapses also can be quantified (27). The relative proportions of dual-labeled versus single-labeled processes can also be determined. Methods utilizing dual-labeling EM methods provide valuable insight into the mechanisms underlying apoptosis and degenerative disease.

4. Notes

1. *Solutions with BSA.* Solutions containing BSA should be kept in the cold room and not used if they are more than 2 days old. These solutions grow bacteria quickly. Make only what you need as BSA is costly.

2. *Organ collection following perfusion fixation.* Chronic stress decreases thymus and spleen weight (28) and increases adrenal weight (29). Spleens are sensitive to immunosuppression (30) and infection (31), and can be further examined by histology (for review, see ref. 32). The liver is especially interesting because of its role in drug metabolism. Challenges to the metabolic function of the liver by drugs (33) or diet (34) can alter liver weight. Equally striking are histological changes, necrosis, and apoptosis present after liver damage (35). Gonads reflect changes in circulating hormone levels; high gonadal steroid levels correlate with increased uterine and seminal vesicle weight (36).

3. *Factors to consider while working with antibodies*

    (a) *Antibodies used for identifying degeneration.*
        Electron microscopy can be used to identify the subcellular structures associated with light-level immunolabeling. Characteristic morphological changes for apoptosis and autophagy, such as chromatin clumping, mitochondrial swelling, lysosome, endosome, or autophagosome alterations, autophagic vacuole accumulation, ubiquitin inclu-

sions, and nuclear or cellular membrane involution, can also be immunolabeled. A rise of neurodegeneration in specific cell types can be identified by a number of protein changes, such as translocation of cytoplasmic cytochrome c (7), ubiquitin inclusions (37), amyloid-beta accumulation (38), phosphorylation of Tau (38), and changes in inflammatory markers (9). Oxidative damage is associated with alterations in the levels of NAPH oxidase subunits (26). Organelles' composition changes while autophagic vacuoles are rare in the normal adult brain; calnexin-labeled autophagic vacuoles increase during neurodegeneration (39). In addition, compartmentalization of transgenes labeled with GFP can be examined with immunoelectron microscopy (8, 40, 41). A list of antibodies used to identify the degeneration-related processes by EM is presented in Table 1.

(b) *Determining optimal primary antibody parameters.* Incubation parameters vary with each antibody. For initial antibody dilutions, we typically start at 1:1,000 and incu-

Table 1
Primary antibodies for labeling degeneration-related processes

Antigen	Species	Catalog no.	Source	References
Amyloid-beta 42	Rabbit	AB507/8P	Chemicon	Takahashi et al. (38)
Calnexin (autophagic vacuoles)	Rabbit	SPA-860J	Stressgen	Nixon et al. (39)
Cytochrome c, cytoplasmic	Mouse			Alonso et al. (7)
GFP	Chicken	GFP-1020	Aves Lab	Bulloch et al. (8)
GFP	Rabbit	A11122	Invitrogen	Justice et al. (40)
Iba1 (microglia)	Rabbit	MCA497R	Wako	Bulloch et al. (8)
Tau phosphorylated at ser 202 and thr 205, clone AT8	Mouse	90343	Innogenetics	Takahashi et al. (38)
Tau phosphorylated at thr 231, clone AT180	Mouse	AT180	Endogen	Takahashi et al. (38)
NADPH oxidase P47 (ROS production)	Goat	sc-7660	Santa Cruz	Pierce et al. (26)
NADPH oxidase P22 (ROS production)	Goat	sc-11712	Santa Cruz	Pierce et al. (26)
Ubiquitin	Rabbit	Z0458	DAKO	Komatsu et al. (37)

bate the tissue for 1 day at room temperature and 1 day overnight. For antigens not located on the membrane (e.g., neuropeptides or nuclear proteins), we test the antibody in the presence and absence of 0.25% triton in the primary antibody diluent. Although *0.25% triton is not good for EM*, it is useful in determining if penetration enhancement will promote better labeling for EM studies. Thus, if better labeling is seen in the presence of 0.25% triton in the primary antibody diluent, future EM studies should use either the freeze–thaw method or 0.025% triton in the primary antibody diluent. If a primary antibody appears to yield the desired staining pattern, a dilution series should be performed with another set of test tissue to determine the optimal dilution of the antibody (see ref. 42 for details).

(c) *Antibody specificity tests.* If an antibody appears to yield the desired staining pattern, additional tests should be performed to determine specificity. These tests include (a) Western blots; (b) labeling in cells transfected with the antigen of interest; (c) lack of labeling in Western blots, cells, and/or tissue sections with primary antibody preadsorbed with the antigen; and (d) if available, absence of labeling in knock-out animals. For discussion of these issues, *see* refs. 43, 44.

(d) *Antibody storage.* Unless noted otherwise by manufacturer, primary antibodies store best in the –70°C freezer. Before freezing a "neat" solution of antibody, divide it into 25- or 50-µl aliquots into microtubes with screw-top lids. Label tubes with the name and species that the antibody was raised against, company, catalog number, aliquot date, and how much antibody is contained in the tube. We also store some antibodies diluted 1:10. For this, add 10 µl antibody to 90 µl of 1% BSA/TS. Gently mix the antibody with a vortex or by pipetting in and out, and aliquot as described above. Primary antibody diluents can also be saved and reused. For this, add 1–2 µl of 1% sodium azide solution to the diluent and store the vial at 4°C.

Acknowledgments

GRANT SUPPORT: NIH grants DA08259 (TAM), HL096571 (TAM; JPP), and DK07313 (EMW).

We thank Ms. June Chan and Mr. Eric Colago for technical advice and Ms. Louisa Thompson for help in preparing the manuscript.

References

1. Mugnaini, E. and Friedrich, V. L. (1981) in *Neuroanatomical Tract-Tracing Methods* (Heimer, L. and Robards, M. J., Eds.) pp 377–406, Plenum Press, New York.

2. Sloviter, R. S., Sollas, A. L., Dean, E., and Neubort, S. (1993) Adrenalectomy-induced granule cell degeneration in the rat hippocampal dentate gyrus: Characterization of an in vivo model of controlled neuronal death. *J. Comp. Neurol.* **330**, 324–336.

3. DeGiorgio, L. A., DeGiorgio, N., Milner, T. A., Conti, B., and Volpe, B. T. (2000) Neurotoxic APP C-terminal and b-amyloid domains colocalize in the nuclei of substantia nigra pars reticulata neurons undergoing delayed degeneration. *Brain Res.* **874**, 137–146.

4. Milner, T. A. and Veznedaroglu, E. (1993) Septal efferent axon terminals identified by anterograde degeneration show multiple sites for modulation of neuropeptide Y-containing neurons in the rat dentate gyrus. *Synapse* **14**, 101–112.

5. Takahashi, R. H., Milner, T. A., Li, F., Nam, E. E., Edgar, M. A., Yamaguchi, H., Beal, M. F., Xu, H., Greengard, P., and Gouras, G. K. (2002) Intraneuronal Alzheimer abeta42 accumulates in multivesicular bodies and is associated with synaptic pathology. *Am. J. Pathol.* **161**, 1869–1879.

6. Takahashi, R. H., Almeida, C. G., Kearney, P. F., Yu, F., Lin, M. T., Milner, T. A., and Gouras, G. K. (2004) Oligomerization of Alzheimer's beta-amyloid within processes and synapses of cultured neurons and brain. *J. Neurosci.* **24**, 3592–3599.

7. Alonso, D., Encinas, J. M., Uttenthal, L. O., Bosca, L., Serrano, J., Fernandez, A. P., Castro-Blanco, S., Santacana, M., Bentura, M. L., Richart, A., Fernandez-Vizarra, P., and Rodrigo, J. (2002) Coexistence of translocated cytochrome c and nitrated protein in neurons of the rat cerebral cortex after oxygen and glucose deprivation. *Neuroscience.* **111**, 47–56.

8. Bulloch, K., Miller, M. M., Gal-Toth, J., Milner, T. A., Gottfried-Blackmore, A., Waters, E. M., Kaunzner, U. W., Liu, K., Lindquist, R., Nussenzweig, M. C., Steinman, R. M., and McEwen, B. S. (2008) CD11c/EYFP transgene illuminates a discrete network of dendritic cells within the embryonic, neonatal, adult, and injured mouse brain. *J. Comp Neurol.* **508**, 687–710.

9. Ito, D., Imai, Y., Ohsawa, K., Nakajima, K., Fukuuchi, Y., and Kohsaka, S. (1998) Microglia-specific localisation of a novel calcium binding protein, Iba1. *Brain Res. Mol. Brain Res.* **57**, 1–9.

10. Chan, J., Aoki, C., and Pickel, V. M. (1990) Optimization of differential immunogold-silver and peroxidase labeling with maintenance of ultrastructure in brain sections before plastic embedding. *J. Neurosci. Methods* **33**, 113–127.

11. Hsu, S. M., Raine, L., and Fanger, H. (1981) Use of avidin-biotin-peroxidase complex (ABC) in immunoperoxidase techniques: a comparison between ABC and unlabeled antibody (PAP) procedures. *J. Histochem. Cytochem.* **29**, 557–580.

12. Pickel, V. M. (1981) in *Neuroanatomical Tract Tracing Methods* (Heimer, L. and Robards, M. J., Eds.) pp 483–509, Plenum Publishing, New York.

13. Pierce, J. P., Kurucz, O., and Milner, T. A. (1999) The morphometry of a peptidergic transmitter system before and after seizure. I. Dynorphin B-like immunoreactivity in the hippocampal mossy fiber system. *Hippocampus* **9**, 255–276.

14. Auchus, A. P. and Pickel, V. M. (1992) Quantitative light microscopic demonstration of increased pallidal and striatal met5-enkephalin-like immunoreactivity in rats following chronic treatment with haloperidol but not with clozapine: Implications for the pathogenesis of neuroleptic-induced movement disorders. *Exp. Neurol.* **117**, 17–27.

15. Gray, J. D., Punsoni, M., Tabori, N. E., Melton, J. T., Fanslow, V., Ward, M. J., Zupan, B., Menzer, D., Rice, J., Drake, C. T., Romeo, R. D., Brake, W. G., Torres-Reveron, A., and Milner, T. A. (2007) Methylphenidate administration to juvenile rats alters brain areas involved in cognition, motivated behaviors, appetite, and stress. *J. Neurosci.* **27**, 7196–7207.

16. Chang, P. C., Aicher, S. A., and Drake, C. T. (2000) Kappa opioid receptors in rat spinal cord vary across the estrous cycle. *Brain Res.* **861**, 168–172.

17. Sesack, S. R., Miner, L. H., and Omelchenko, N. (2006) in *Neuroanatomical Tract-Tracing 3: Molecules, Neurons, Systems* (Zaborszky, L., Wouterlood, F. G., and Lanciego, J. L., Eds.) pp 6–71, Springer, New York.

18. Mouton, P. R. (2002) *Principles and practices of unbiased stereology. An introduction for bi01scientists.* John Hopkins University Press, Baltimore.

19. Peters, A., Palay, S. L., and Webster, H. d. (1991) *The fine structure of the nervous system, 3rd ed.* Oxford University Press, New York.

20. Carlin, R. K., Grab, D. J., Cohen, R. S., and Siekevitz, P. (1980) Isolation and characterization of postsynaptic densities from various brain

regions: Enrichment of different types of postsynaptic densities. *J. Cell Biol.* **86**, 831–832.

21. Pierce, J. P., van Leyen, K., and McCarthy, J. B. (2000) Translocation machinery for synthesis of integral membrane and secretory proteins in dendritic spines. *Nat. Neurosci.* **3**, 311–313.

22. Znamensky, V., Akama, K. T., McEwen, B. S., and Milner, T. A. (2003) Estrogen levels regulate the subcellular distribution of phosphorylated Akt in hippocampal CA1 dendrites. *J. Neurosci.* **23**, 2340–2347.

23. Hara, Y. and Pickel, V. M. (2008) Preferential relocation of the N-methyl-D-aspartate receptor NR1 subunit in nucleus accumbens neurons that contain dopamine D1 receptors in rats showing an apomorphine-induced sensorimotor gating deficit. *Neuroscience.* **154**, 965–977.

24. Lane, D. A., Lessard, A. A., Chan, J., Colago, E. E., Zhou, Y., Schlussman, S. D., Kreek, M. J., and Pickel, V. M. (2008) Region-specific changes in the subcellular distribution of AMPA receptor GluR1 subunit in the rat ventral tegmental area after acute or chronic morphine administration. *J. Neurosci.* **28**, 9670–9681.

25. Torres-Reveron, A., Williams, T. J., Chapleau, J. D., Waters, E. M., McEwen, B. S., Drake, C. T., and Milner, T. A. (2009) Ovarian steroids alter mu opioid receptor trafficking in hippocampal parvalbumin GABAergic interneurons. *Exp. Neurol.* **219**, 319–327.

26. Pierce, J. P., Kievits, J., Graustein, B., Speth, R. C., Iadecola, C., and Milner, T. A. (2009) Sex differences in the subcellular distribution of angiotensin type 1 receptors and NADPH oxidase subunits in the dendrites of C1 neurons in the rat rostral ventrolateral medulla. *Neuroscience.* **163**, 329–338.

27. Milner, T. A., Ayoola, K., Drake, C. T., Herrick, S. P., Tabori, N. E., McEwen, B. S., Warrier, S., and Alves, S. E. (2005) Ultrastructural localization of estrogen receptor beta immunoreactivity in the rat hippocampal formation. *J. Comp Neurol.* **491**, 81–95.

28. Dominguez-Gerpe, L. and Rey-Mendez, M. (1997) Time-course of the murine lymphoid tissue involution during and following stressor exposure. *Life Sci.* **61**, 1019–1027.

29. Armario, A. (2006) The hypothalamic-pituitary-adrenal axis: what can it tell us about stressors? *CNS. Neurol. Disord. Drug Targets.* **5**, 485–501.

30. Beaulieu, J., Dupont, C., and Lemieux, P. (2007) Anti-inflammatory potential of a malleable matrix composed of fermented whey proteins and lactic acid bacteria in an atopic dermatitis model. *J. Inflamm. (Lond).* **4**, 6–16.

31. Lopez, M. C., Chen, G. J., Colombo, L. L., Huang, D. S., Darban, H. R., Watzl, B., and Watson, R. R. (1993) Spleen and thymus cell subsets modified by long-term morphine administration and murine AIDS–II. *Int. J. Immunopharmacol.* **15**, 909–918.

32. Cesta, M. F. (2006) Normal structure, function, and histology of mucosa-associated lymphoid tissue. *Toxicol. Pathol.* **34**, 599–608.

33. van Bezooijen, C. F. (1984) Influence of age-related changes in rodent liver morphology and physiology on drug metabolism – a review. *Mech. Ageing Dev.* **25**, 1–22.

34. Lyn-Cook, L. E. J., Lawton, M., Tong, M., Silbermann, E., Longato, L., Jiao, P., Mark, P., Wands, J. R., Xu, H., and de la Monte, S. M. (2009) Hepatic ceramide may mediate brain insulin resistance and neurodegeneration in type 2 diabetes and non-alcoholic steatohepatitis. *J. Alzheimers. Dis.* **16**, 715–729.

35. Savransky, V., Reinke, C., Jun, J., Bevans-Fonti, S., Nanayakkara, A., Li, J., Myers, A. C., Torbenson, M. S., and Polotsky, V. Y. (2009) Chronic intermittent hypoxia and acetaminophen induce synergistic liver injury in mice. *Exp. Physiol.* **94**, 228–239.

36. Romeo, R. D., Lee, S. J., and McEwen, B. S. (2004) Differential stress reactivity in intact and ovariectomized prepubertal and adult female rats. *Neuroendocrinology.* **80**, 387–393.

37. Komatsu, M., Waguri, S., Chiba, T., Murata, S., Iwata, J., Tanida, I., Ueno, T., Koike, M., Uchiyama, Y., Kominami, E., and Tanaka, K. (2006) Loss of autophagy in the central nervous system causes neurodegeneration in mice. *Nature.* **441**, 880–884.

38. Takahashi, R. H., Capetillo-Zarate, E., Lin, M. T., Milner, T. A., and Gouras, G. K. (2008) Co-occurrence of Alzheimer's disease beta-amyloid and tau pathologies at synapses. *Neurobiol. Aging.*

39. Nixon, R. A., Wegiel, J., Kumar, A., Yu, W. H., Peterhoff, C., Cataldo, A., and Cuervo, A. M. (2005) Extensive involvement of autophagy in Alzheimer disease: an immuno-electron microscopy study. *J. Neuropathol. Exp. Neurol.* **64**, 113–122.

40. Justice, N. J., Yuan, Z. F., Sawchenko, P. E., and Vale, W. (2008) Type 1 corticotropin-releasing factor receptor expression reported in BAC transgenic mice: implications for reconciling ligand-receptor mismatch in the central corticotropin-releasing factor system. *J. Comp Neurol.* **511**, 479–496.

41. Lazarenko, R. M., Milner, T. A., Depuy, S. D., Stornetta, R. L., West, G. H., Kievits, J. A., Bayliss, D. A., and Guyenet, P. G. (2009) Acid sensitivity and ultrastructure of the retrotrapezoid nucleus in Phox2b-EGFP transgenic mice. *J. Comp Neurol.* **517**, 69–86.

42. Harris, J. A., Chang, P. C., and Drake, C. T. (2004) Kappa opioid receptors in rat spinal cord: sex-linked distribution differences. *Neurosci.* **124**, 879–890.

43. Saper, C. B. and Sawchenko, P. E. (2003) Magic peptides, magic antibodies: guidelines for appropriate controls for immunohistochemistry. *J. Comp Neurol.* **465**, 161–163.

44. Lorincz, A. and Nusser, Z. (2008) Specificity of immunoreactions: the importance of testing specificity in each method. *J. Neurosci.* **28**, 9083–9086.

Part II

Neural Cell Culture Techniques

Chapter 4

Isolation and Culture of Neurons and Astrocytes from the Mouse Brain Cortex

Hyun Jeong Kim and Jordi Magrané

Abstract

Many experimental animal models of human neurodegenerative diseases have been developed to understand the events leading toward neuronal dysfunction and death. However, definitive comprehension of the molecular and cellular mechanisms in these animal models is problematic because of the complexity of the intact nervous tissue. Primary neuronal cultures prepared from rodent nervous tissues represent a powerful tool not only to study the individual contribution of different cell types (such as neurons or glia) to disease progression, but also to investigate the role of neuron–glia interactions during development and patho-genesis of disease. Here, we describe a method to isolate and culture neurons and astrocytes from the mouse cerebral cortex, and we also present a practical application for transfection and subsequent immunofluorescence.

Key words: Cortical neurons, Astrocytes, Coating, Immunofluorescence, Cellular markers

1. Introduction

Primary neuronal and glial cultures isolated from the rodent cerebral cortex are widely used to study brain function in a controlled in vitro environment. Primary neurons readily extend their neurites and establish synapses in culture 7–10 days after isolation. Thus, cellular and molecular events related to neurite development and maintenance, neuritic transport, and synaptic transmission can be studied (1–4). Unlike neurons, primary astrocytes are able to reenter the cell cycle; cultures are easy to prepare, reproducible, and can be obtained from any central nervous system region. Astrocytes play key roles in the survival and guidance of neurons during brain development, formation and preservation of the blood–brain barrier, and maintenance of neuronal homeostasis and plasticity (5–11).

Giovanni Manfredi and Hibiki Kawamata (eds.), *Neurodegeneration: Methods and Protocols*,
Methods in Molecular Biology, vol. 793, DOI 10.1007/978-1-61779-328-8_4, © Springer Science+Business Media, LLC 2011

In several human neurodegenerative diseases, such as Alzheimer's and Huntington's diseases, brain cortex is severely affected: synapses are lost, neuritic network is compromised (12–15), and neuroinflammatory events occur (16, 17). Therefore, cultures of isolated neurons or glia from cerebral cortex represent an accessible model to study neurodegeneration and a first step to test new therapeutic approaches.

2. Materials

2.1. Dissection Equipment and Reagents

1. Large standard scissors, large toothed forceps, and a small perforated spoon (Moria MC 17) for embryo removal and handling.

2. Small scissors (Moria MC 22), fine forceps (Dumont No. 55), a blunt spatula, scalpel, and blades (No. 15) for brain dissection.

3. A silicone base for dissection made by polymerizing Sylgard 184 silicone (Fisher Scientific, Pittsburgh, PA) in plastic tissue culture dishes. The addition of activated charcoal helps to increase the contrast with the brain tissue.

4. A stereoscopic dissecting microscope.

5. Dissection buffer (for cortical neurons): 0.6% glucose in 0.1 M PBS. Prepare fresh and place on ice.

6. Dissection buffer (for cortical astrocytes): Dulbecco's modified Eagle's medium (DMEM) supplemented with 10% heat-inactivated fetal bovine serum (FBS). Place on ice.

2.2. Cell Culture

1. Neuronal culture medium: Neurobasal I, 1% heat-inactivated horse serum (HS), 0.5 mM glutamine, 1% penicillin–streptomycin (P/S), 0.04% sodium bicarbonate, 33 mM glucose, 2% B27 supplement.

2. Glial culture medium: DMEM supplemented with 10% FBS and 1% P/S.

3. Trypsin solution: 0.25% trypsin–EDTA.

4. DNAse I solution: 1 mg/ml DNAse in cortical neuron dissection buffer.

5. 70- and 100-µm cell strainers.

6. Dounce glass tissue grinder pestle.

2.3. Coating of Tissue Culture Surfaces

1. Microscope cover slips (12 mm diameter, 0.13–0.17 mm thickness, Carolina Biological, Burlington, NC).

2. Coating solution: 3 µg/ml poly-D-ornithine (3 mg/ml stock in water) and 5 µg/ml laminin (1 mg/ml stock in PBS) in 0.1 M PBS (see Note 1).

3. For coverslip coating, prepare a sterile Petri dish with a piece of parafilm (sterile side up). Add 30 μl drops of coating solution, and put coverslips on top. Incubate them at 37°C for at least 2 h. Wash the coverslips twice in 0.1 M PBS and place them in a 24-well plate with the coated face up. Add PBS and leave them at room temperature inside the hood until used (see Note 2).

2.4. Transfection, Antibodies, and Immunofluorescence

1. For transfection of primary cortical neurons: Lipofectamine 2000 (LF2000, Invitrogen), OptiMEM I medium (Invitrogen).

2. Neuronal antigens: β-tubulin III protein (also known as Tuj 1; Sigma T3952). Glial antigen: GFAP (Dako M0761; Carpinteria, CA). Neuritic markers: tau (for axons; Chemicon MAB3420; Temecula, CA) and microtubule-associated protein 2 (MAP2, for dendrites; Chemicon AB5622).

3. Secondary antibodies: Anti-mouse and anti-rabbit IgG conjugated to either Cy2 or Cy3 (Jackson ImmunoResearch, West Grove, PA).

4. Fixative: 3% paraformaldehyde, 60 mM sucrose, 0.1 M PB. Washes: 20 mM glycine in 0.1 M PBS. Permeabilization: 0.1% Triton X-100 in PBS–glycine. Blocking and antibody buffers: 0.5% BSA in PBS–glycine.

3. Methods

There are a number of factors that may influence the quality and reproducibility of primary neuronal and glial cultures. A precise dissection of the tissue is critical to obtain homogeneous and reproducible cell populations among isolations. Moreover, to achieve healthy cultures, it is important to complete the dissection and isolation procedures quickly with no interruptions. Mechanical stress should be minimized by reducing unnecessary disruption of the brain tissue and excessive pipetting. Finally, it may be worth testing the effects of some of the reagents used on culture viability, especially when changing lot numbers or company sources.

3.1. Tissue Dissection

Brain dissection (for cortical neuron isolation)

1. Pour ice-cold dissection buffer into several Petri dishes: one large dish (150 mm) for the removed uterus, and several 100-mm dishes for embryo isolation and dissection. Place dishes on ice.

2. Euthanize the pregnant mouse (see Note 3) according to institutional guidelines (see Note 4).

3. Place the mouse on its back and thoroughly swab the abdomen with 70% ethanol to reduce the risk of contamination.

4. Cut the skin of the abdomen, grasp the peritoneum with forceps, and cut to expose the abdominal cavity (see Note 5).

5. Grasp the uterus below the oviduct and cut it free along the mesometrium. Place it into a Petri dish with dissection buffer.

6. Remove the fatty tissue from the uterine horns. Transfer to a new dish.

7. Cut along the muscle wall of the uterus to expose the amniotic sacs. Transfer to a new dish. The embryos are released using the tips of forceps and transferred to a new Petri dish.

8. Under a stereoscopic dissecting microscope (see Note 6), hold the embryo with forceps by gripping the neck. Insert the tip of a fine forceps between the eyes, and slide the tip underneath the skull backward along the midline toward the rear of head. Break open skin and skull by pulling the forceps up.

9. Separate the brain from the cerebellum to the olfactory bulbs with a blunt spatula. Move the brain to a Petri dish with a silicone base containing dissection buffer for further dissection (see Note 7).

10. Steady the brain with a scalpel and remove the cerebrum (see Note 8). A schematic representation of this and following steps is shown in Fig. 1.

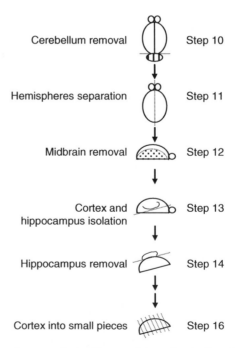

Fig. 1. Dissection of mouse brain. The numbers refer to the steps described in Subheading 3.1. "Brain dissection (for cortical neuron isolation)." Note that steps 10–15 are common for both neuronal and glial isolations.

11. Separate the cerebral hemispheres from each other by cutting along the midline with a scalpel.

12. Place the brain ventral side up. Place a spatula in the medial aspect of the ventral cortex and midbrain, and cut the cortices off. Discard the midbrain.

13. Dissect the hippocampus and the cortex: Place the cortex medial side up and place a spatula into the lateral ventricle pushing forward through the lateral aspect of the frontal cortex. Extend the cut through the cortex from rostral to caudal to isolate the posterior half of cortex and hippocampus. Discard the remainder.

14. Dissect the hippocampus from the cortex: Place a spatula in the lateral ventricle underneath the hippocampus. Make cuts to free the respective ends of the hippocampus. Roll out the hippocampus with the spatula and cut the hippocampus from the cortex near the dentate-entorhinal cortex junction.

15. Peel away the meninges from the surface and inner side of the cortex using fine forceps (see Note 9).

16. Place the cleaned cortices into a 35-mm Petri dish with ice-cold dissection buffer. Cut the tissue with a scalpel into small pieces (about 1 mm). Transfer the pieces into 15-ml tubes (see Note 10) and transport them into a biosafety hood in the tissue culture room.

Brain dissection (for astrocyte isolation)

1. Anesthetize 1- or 2-day-old neonatal mouse pups (P1–P2; see Note 11) according to institutional guidelines.

2. Once the pups have stopped moving, decapitate with scissors and put their heads into a 60-mm Petri dish with glial culture medium on ice (see Note 12).

3. Pin the head ventral side down on a 60-mm Petri dish containing glial culture medium by gripping the snout with fine forceps, and slide the small scissors between the scalp and the skull. Cut the skin from the base of the skull to the mid-eye area and fold back the skin flaps.

4. Cut the skull along the midline of the brain without damaging the brain tissue.

5. Separate and remove the raised skullcap from the surface of brain with forceps.

6. Scoop out the brain from the skull cavity from the olfactory bulbs to the cerebellum by using a spatula (see Note 7).

7. Follow the procedure described in the previous section "Brain dissection (for cortical neuron isolation)" from step 10 to 15 (see also Fig. 1).

8. Place the cleaned cortices into a 35-mm Petri dish with ice-cold glial culture medium. Transport them into a biosafety hood in the tissue culture room.

3.2. Cortical Neuron Isolation

1. (See Note 13) Once in the biosafety cabinet, let the cortex pieces settle down to the bottom of the 15-ml tubes for at least 2 min. Remove the extra dissection buffer using a pipette or aspirate with the vacuum line (see Note 14).

2. Add 1 ml prewarmed trypsin solution to each tube. Incubate in a water bath at 37°C for 15–20 min, carefully tipping every 5 min to agitate the tissue (see Note 15).

3. During the incubation time, start washing the coated plastic or glass surfaces with 0.1 M PBS (see Subheading 2.3, step 3).

4. Inactivate trypsin by adding 1/3 volume of prewarmed horse serum (other types of serum can be used). Carefully swirl the tissue and let it settle down.

5. Aspirate the supernatant, and add 1 ml prewarmed DNAse solution. Incubate for 5–10 min in a water bath at 37°C, carefully tipping after 5 min (see Note 16).

6. Dissociate cells by pipetting up and down (15–20 strokes) with a P1000 pipette (see Note 17). The pieces of cortices brake, and the solution gets cloudy. Wait for 2 min (chunks of tissue settle down), and then transfer all supernatant (containing dissociated neurons) to two new 15-ml tubes (see Note 18).

7. Centrifuge at 800–1,000×g for 5 min.

8. Resuspend each pellet gently (3–5 strokes) with 1 ml prewarmed neuronal culture medium.

9. Count the number of neurons (see Note 19) and plate them (see Note 20) at the desired density on coated dishes (see Note 21).

10. Change media every 2–3 days (see Note 22).

11. The result is an almost pure cortical neuron culture (see Note 23).

3.3. Astrocyte Isolation

1. Gently push all the cortex pieces through a 100-μm cell strainer by using a sterile glass tissue grinder pestle (see Note 24). A schematic representation of the steps described in this subheading is shown in Fig. 2.

2. Add 8-ml glial culture medium drop wise to the surface of the cell strainer, and collect the cell suspension in a 50-ml tube.

3. Pass the cell suspension through a 70-μm cell strainer, and collect the supernatant into a new 50-ml tube (see Note 25).

4. Centrifuge the cells at 800–1,000×g for 5 min and then aspirate the supernatant.

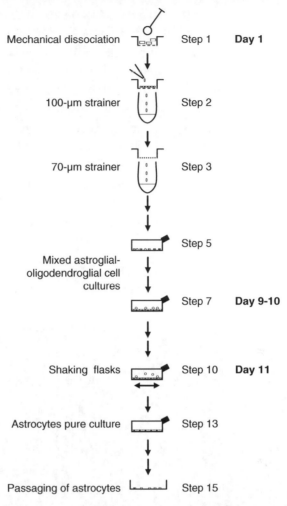

Fig. 2. Scheme for the isolation of astrocytes from the mouse cerebral cortex. The step numbers refer to the steps described in Subheading 3.3. Note that 11 days are needed to obtain a pure culture of astrocytes.

5. Resuspend the pellet in 1-ml glial culture medium, count the cells, and seed at a density of 3×10^5 cells/cm^2 into cell culture flasks with screw on caps.

6. Twenty-four hours after plating the cells, aspirate the culture medium to remove cell debris and feed the cultures with fresh glial culture medium (see Note 26).

7. After 7–9 days (see Note 27), aspirate medium, gently rinse the cultures three times with culture medium to remove floating cells, and add fresh glial culture medium.

8. Place the flasks back to the incubator and allow them to equilibrate for 2 h.

9. Tighten the flasks' caps, and carefully wrap each flask in parafilm.

10. Shake the flasks on a rocker platform at 250 rpm/min for 12 h at 37°C to separate oligodendrocytes from astrocytes.

11. Remove medium, rinse the cells with culture medium, and add 5 ml of glial culture medium.

12. Shake the flask by hand (see Note 28), remove the medium, and rinse the cultures five times with culture medium.

13. Trypsinize the cells with 1-ml trypsin solution; incubate at 37°C for 5 min. Add 9-ml glial culture medium to harvest the cell suspension.

14. Centrifuge the cells at 800–1,000×g for 5 min, resuspend the pellet with culture medium, and count the cells.

15. Plate them at the desired density for experiments. For subculture, plate 3×10^4 cells/cm² in culture medium (see Note 29).

3.4. Transfection and Immunofluorescence

1. Transfection of primary cortical neurons can be achieved using LF2000. For neurons growing on a coated coverslip placed on

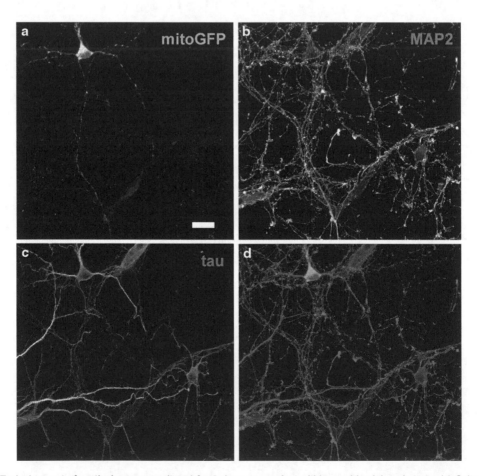

Fig. 3. Typical aspect of cortical neurons cultured for 9 days on a poly-ornithine and laminin substrate. (**a**) Culture was transfected with mitoGFP to label mitochondria (*in green*). (**b**) Dendrites and (**c**) axons were identified using MAP2 (*in red*) and tau (*in blue*) antibodies, respectively. (**d**) Combination of the three labelings. Scale bar, 20 μm.

a 24-well plate, mix 25 µl OptiMEM I medium with 1 µl LF2000 and incubate for 5 min; mix 25 µl OptiMEM I with 0.75–1 µg DNA. Combine both solutions and incubate at room temperature for 20 min. Slowly add the mixture to the neurons, and gently swirl the medium. Put the culture back to the incubator. Change medium after 4–6 h (see Note 30). An example of transfection of primary neurons with a mitochondrially targeted GFP (mitoGFP) cDNA is shown in Fig. 3.

2. For immunofluorescence, quickly wash the neurons on coverslips with 0.1 M PBS to completely remove medium components, and fix the cells for 15–30 min at room temperature. Wash three times with PBS–glycine (5 min each wash). Permeabilize, if required, for 7–10 min at room temperature. Wash again three times with PBS–glycine. Block for at least 10 min. Incubate with primary antibodies for either 45 min at 37°C or overnight at 4°C. Wash three times with PBS–glycine. Incubate with secondary antibodies for 30 min at 37°C. Wash again and mount (see Note 31). Examples of immunodetection of neuronal axons and dendrites and of astrocytes in culture are shown in Figs. 3 and 4, respectively.

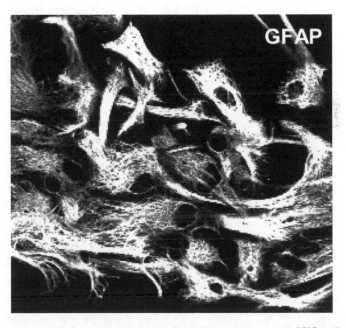

Fig. 4. Identification of astrocytes by immunofluorescence using an anti-GFAP antibody.

4. Notes

1. Other coating solutions can be used, such as 10 μg/ml poly-D-lysine in 0.1 M PBS or a mixture of poly-D-lysine and laminin. Test different coating solutions when setting up the neuronal isolation protocol. No coating is required for glial cultures.

2. Start coating the dishes before brain dissection. Wash them by the end of the cortical neuron isolation procedure (see Subheading 3.2).

3. The optimal embryonic age for primary cortical neuron isolation is E16-18. Younger embryos are more difficult to dissect. Older embryos or postnatal pups contain a higher proportion of contaminating glial cells.

4. Euthanasia and gross dissections should be performed in a room separated from the tissue culture room to maintain sterility of the latter. Once isolated, the embryos are transported to a horizontal laminar-flow tissue culture hood, and all subsequent steps of Subheading 3.1 are performed under a dissecting microscope.

5. All instruments are sterilized by dipping in 70% ethanol. Make sure that ethanol is completely evaporated from the tools before using them.

6. It is preferable to dissect the embryos in a horizontal laminar-flow tissue culture hood to avoid contamination problems. If a hood is not available, make sure that your working area is clean and not disturbed by airflow.

7. Brains should be completely covered in dissecting buffer to prevent tissue from drying out during all dissection procedures.

8. The use of a scalpel to dissect the tissue allows for a clean and fast cut. Using fine forceps may be more precise, but it is more time consuming and damages the tissue more easily.

9. Failure to completely remove the meninges results in fibroblast contamination.

10. It is important to distribute the brain pieces into several tubes (consider one tube every two embryos) to facilitate trypsinization of the tissue.

11. The use of 1- or 2-day-old mouse pups minimizes the number of neurons present in the cell suspension.

12. Deprivation of the blood supply in the brain causes degradation of the tissue. Therefore, pups should be sacrificed one at a time, and the time between collecting the brains and dissecting the cortices should be minimized.

13. All the following steps are carried out in a biosafety hood in the tissue culture room.

14. When using the vacuum line to aspirate liquids, decrease the vacuum flow by placing a micropipette tip to the tip of a glass pipette. This helps reducing the risk of accidental aspiration of the tissue.

15. Typically, after 7–8 min, the cortical pieces attach to each other (in a "beads on a string" appearance) due to DNA release caused by trypsinization. Insufficient trypsinization results in poor recovery of cortical neurons; too much digestion results in cortical neuron death.

16. It is also possible to add DNAse I directly to the tube containing the inactivated trypsin and serum. DNAse digestion is monitored by observing the break of the "beads on a string" appearance. Similarly to trypsin incubation, too little or too much DNAse I digestion affects the recovery and viability of the purified cortical neurons.

17. Pipetting should be performed slowly and avoiding air bubble formation. To prevent foaming, push the solution out against the side of the tube rather than into the liquid at the bottom.

18. The digestion step can be repeated up to three times if the chunks of tissue are still persistent.

19. Expect around 7–10 million cells per dissected embryo.

20. Always pipette cells up and down before plating them to avoid cell aggregate formation.

21. It is important to plate different cell densities to determine which one is more appropriate, when cortical neurons are isolated for the first time. Insufficient densities may result in poor neuronal survival; excessive densities may result in clustering of neurons.

22. Care must be taken when changing the media. It is recommended to remove or add new medium from the well walls without disturbing the neurons. Moreover, completely removing the medium may result in neuronal death due to prolonged drying; therefore, change only half of it. Also, minimize the time neurons are outside the incubator.

23. The neuronal culture medium conditions do not favor the growth of contaminating glia. However, if glial cells become a problem, use culture medium without serum and with reduced glutamine concentration. Alternatively, anti-mitotic agents (such as AraC) can be added to the culture.

24. This protocol is based on a method of McCarthy and de Vellis (18). The use of the mechanical sieving techniques destroys most of neurons, but allows the survival of small, undifferentiated astroglial precursor cells.

25. The filtering of the cell suspension through the cell strainer is necessary to remove chunks of tissue, including meningeal remnants, and blood vessels.

26. Culture medium is changed every 2–3 days until the final purification of astrocytes is performed.

27. Culture flasks contain an intact monolayer of astrocytes, which are flat fibroblast-like cells, and few oligodendrocytes scattered on the surface. To achieve sufficient stratification of astrocytes and oligodendrocytes, the initial culture period should be held between 7 and 9 days. Longer periods result in the clustering of astrocytes above the monolayer.

28. Detachment of oligodendrocytes should be confirmed by observation through a tissue culture microscope before rinsing the cells.

29. Because astrocytes proliferate faster than oligodendrocytes, subsequent trypsinization and passage steps remove contamination of the latter. However, in each passage, astrocytes change their biological and immunological characteristics. Therefore, care must be taken with the passage number at which astrocytes are used. Alternatively, dividing oligodendrocyte and microglia can be eliminated by treating the cultures with 5–10 mM cytosine arabinoside (Ara-C, Sigma-Aldrich) for 2–5 days. AraC should only be applied at high cell density, that is, when astrocytes enter a nonproliferative state induced by cell–cell interaction.

30. The efficiency of transfection is low (5–10%), but generally sufficient for microscopy-related studies. It is important to determine the optimal neuronal density that results in high efficiency of transfection with minimal toxic effects. Best results are achieved when cultures are less than a week old; older cultures can still be used, but efficiencies of transfection decrease. Transgene expression can be detected 24–36 h later.

31. This is a standard immunocytochemistry protocol; certain steps may need to be optimized for some specific antibodies (fixation, blocking, antibody incubation times). To prevent neurite damage, coverslips can be placed in bigger recipients and large wash volumes can be used; solutions should always be added to the well walls without disturbing the neurons. All steps are carried out at room temperature, except for antibody incubations. To minimize the amount of antibody used, place coverslips on a piece of parafilm, neurons facing up, and add a 40–50-µl drop on top.

Acknowledgments

This work is supported by grants from the Alzheimer's Association and the Muscular Dystrophy Association.

References

1. Dotti CG, Sullivan CA, Banker GA (1988) The establishment of polarity by hippocampal neurons in culture. J. Neurosci. **8**, 1454–68.

2. Fletcher TL, Banker G (1089) The establishment of polarity by hippocampal neurons: the relationship between the stage of a cell's development in situ and its subsequent development in culture. Dev. Biol. **136**, 446–455.

3. Craig AM, Graf ER, Linhoff MW (2006) How to build a central synapse: clues from cell culture. Trends Neurosci. **29**, 8–20.

4. Atluri PP, Ryan TA (2006) The kinetics of synaptic vesicle reacidification at hippocampal nerve terminals. J. Neurosci. **26**, 2313–2320.

5. Araque A, Carmignoto G, Haydon PG (2001) Dynamic signaling between astrocytes and neurons. Annu. Rev. Physiol. **63**, 795–813.

6. Doetsch F (2003) The glial identity of neural stem cells. Nat. Neurosci. **6**, 1127–1134.

7. Fields RD, Stevens-Graham B (2002) New insights into neuro-glia communication. Science. **298**, 556–562.

8. Janzer RC (1993) The blood-brain barrier: cellular basis. J. Inherit. Metab. Dis. **16**, 639–647.

9. Montgomery DL (1994) Astrocytes: form, functions, and roles in disease. Vet. Pathol. **31**, 145–167.

10. Parri R, Crunelli V (2003) An astrocyte bridge from synapse to blood flow. Nat. Neurosci. **6**, 5–6.

11. Verkhratsky A, Toescu EC (2006) Neuronal-glial networks as substrate for CNS integration. J. Cell. Mol. Med. **10**(4), 826–836.

12. Eriksson PS, Perfilieva E, Björk-Eriksson T, Alborn AM, Nordborg C, Peterson ST, Gage FH (1998) Neurogenesis in the adult human hippocampus. Nat. Med. **4**(11), 1313–1317.

13. Luebke JI, Weaver CM, Rocher AB, Rodriguez A, Crimins JL, Dickstein DL, Wearne SL, Hof PR (2010) Dendritic vulnerability in neurodegenerative disease: insights from analyses of cortical pyramidal neurons in transgenic mouse models. Brain Struct. Funct. **214**, 181–99.

14. Rozas JL, Gómez-Sánchez L, Tomás-Zapico C, Lucas JJ, Fernández-Chacón R (2010) Presynaptic dysfunction in Huntington's disease. Biochem. Soc. Trans. **38**, 488–92.

15. Knobloch M, Mansuy IM (2008) Dendritic spine loss and synaptic alterations in Alzheimer's disease. Mol Neurobiol. **37**, 73–82.

16. Dong Y, Benveniste EN (2001) Immune function of astrocytes. Glia. **36**, 180–190.

17. Sofroniew MW (2005) Reactive astrocytes in neuronal repair and protection. Neuroscientist. **11**, 400–407.

18. McCarthy KD, de Vellis J (1980) Preparation of separate astroglial and oligodendroglial cell cultures from rat cerebral tissue. J. Cell. Biol. **85**, 890–902.

Chapter 5

Isolation and Culture of Postnatal Spinal Motoneurons

Carol Milligan and David Gifondorwa

Abstract

Neuronal cultures, including motoneuron (MN) cultures, are established from embryonic animals. These approaches have provided novel insights into developmental and possibly disease mechanisms mediating cell survival or death. Motoneurons isolated from mouse models of disease, such as the SOD1^{G93A} mouse, demonstrate subtle abnormalities that may contribute to pathology. Nonetheless, in the animal model, pathological events become more prominent as the animal matures, but the ability to isolate individual cells to investigate these events is limited. Here, we describe a protocol derived and modified from previously published protocols to isolate motoneurons from mature animals. While the yield of cells is low, the ability to examine mature motoneurons provides a new platform to investigate pathological changes associated with motoneuron disease.

Key words: Neuronal cultures, Neurodegenerative diseases, Motoneuron disease, Trophic factor, Cell death/apoptosis, Mitochondria, Axonal transport

1. Introduction

The development of motoneurons (MNs), including their dependence on target-derived trophic support, has been characterized in the spinal cords of developing animals, most notably the chick embryo (1–3). The development and utilization of in vitro culture allow specific effects of trophic factors on neurite extension and cell survival to be determined (4–7). In vitro culture of MNs has provided insight into specific cell death pathways and stress responses (8–12). Furthermore, culture of embryonic MNs has been used to determine potential disease-specific processes related to the pathogenesis of motoneuron diseases, such as ALS. For example, caspases were shown to be involved in MN cell death during development both in vivo and in vitro (13), and subsequently they were suggested to also play a role in MN degeneration in the mouse model of ALS

Giovanni Manfredi and Hibiki Kawamata (eds.), *Neurodegeneration: Methods and Protocols*,
Methods in Molecular Biology, vol. 793, DOI 10.1007/978-1-61779-328-8_5, © Springer Science+Business Media, LLC 2011

(14, 15). Nonetheless, drawing conclusions on disease processes using embryonic cultures is riddled with experimental caveats, and the most obvious is that embryonic neurons generally do not exhibit the pathological events that occur much later in the adult animal.

Culturing MNs from postnatal, especially adult, animals was considered difficult, if not impossible, in part due to the length of the MN axon, cell–cell interactions, and thousands of synapses. Nonetheless, adult neurons and neurospheres can be isolated from adult animals and a detailed protocol was recently published (16). This approach involves a stepwise Optiprep gradient that allows separation of debris and myelin from neurons and glial cells. Our lab routinely isolates and cultures embryonic MNs (17). Using approaches from both of these protocols we have been able to culture MNs from older animals to investigate specific events observed in MNs in a mouse model of ALS. The results are encouraging and appear to provide a foundation for future studies to examine pathophysiological processes in mature MNs. The protocol described in this chapter has been used successfully to isolate viable postnatal day (P) 30 mouse MNs from the lumbar spinal cord.

2. Materials

2.1. Cleaned and Sterile 13-mm Diameter Glass Coverslips

Preparation of glass coverslips

1. Wash coverslips in 95% absolute ethanol and 5% acetic acid. Rinse several times in Milli-Q water.

2. Let coverslips dry on filter paper. It is necessary for coverslips to be completely dry before autoclaving; otherwise, they stick together and not come apart.

3. Transfer coverslips to a glass petri dish and autoclave.

4. Set coverslips in tissue culture plates. We have used grierner dishes with a small drop of vacuum grease to hold the coverslip in place (see Note 1 and Fig. 1). Alternately, coverslips can be placed into a 24-well tissue culture plate; however, this makes coating of them to limit where cells can attach more difficult (see Note 2).

5. Coverslips should be coated with poly-ornithine (3 μg/ml in sterile dH_2O) for several hours (at room temp) to overnight (at 4°C).

6. Poly-ornithine-coated coverslips can be stored at 4°C for several weeks.

7. Wash coverslips several times with $PBS + Ca^{+2}, + Mg^{+2}$.

8. Coat coverslips with laminin (3 μg/ml in $PBS + Ca^{+2}, + Mg^{+2}$) for 3–5 h at 37°C, 5% CO_2.

9. Before plating cells, wash coverslips several times with $PBS + Ca^{+2}, + Mg^{+2}$.

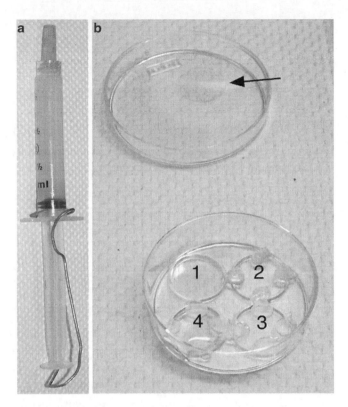

Fig. 1. (**a**) Image of syringe filled with vacuum grease and paper clip to prevent loss during autoclaving. (**b**) Shown is a Greiner dish. Well 1 is empty, 2 has vacuum grease applied, 3 has the coverslip, and 4 has the coverslip with 100 μl of liquid. *Arrow* indicates 13-mm glass coverslip.

2.2. 9" Siliconized Pasteur Pipettes

This procedure must be performed under a fume hood. Using a rubber pipette bulb, Silicoat (Sigma) is aspirated into the glass pipette several times until the majority of the interior of the pipette has been coated. Excess silicoat is expelled and the pipettes allowed to air dry on a paper towel overnight in the fume hood. Once completely dry, the pipettes are wrapped in clean paper towels and then foiled and autoclaved. Pipettes are stored at room temperature until use.

2.3. Dissections

Dissection instruments, including #1 dumont forceps, #5 dumont forceps, and micro-knife.

2.4. Tissue Culture Reagents

Specific reagents are listed below. While many of the reagents can be obtained from multiple sources, specific sources are provided because these have proven successful for MN culture.

1. B27 (Invitrogen).
2. 10% BSA (Tissue Culture grade; Sigma) in HABG (see Subheading 2.5).
3. Gentamycin (Invitrogen).
4. Glutamax (Invitrogen).

5. Hibernate A (Brain Bits).

6. Hibernate A w/o calcium (Brain Bits).

7. Horse Serum (Invitrogen).

8. Laminin (Sigma).

9. Neurobasal media (Invitrogen).

10. OptiPrep (Sigma or Accurate Chemical).

11. Papain (Worthington).

12. 1 mg/ml Poly-ornithine, made in 0.15 M boric acid (Sigma).

13. Trophic factors (CT-1, CNTF, GDNF available from R&D).

14. DNAse (Sigma D7291-2MG).

2.5. Tissue Culture Media

All reagents should be sterile filtered prior to use.

1. HABG is an artificial CSF media with supplements to promote neuronal survival: 98 ml HA, 2 ml B27, 10 µl gentamycin, 293 µl glutamax.

2. NB medium is formulated to promote neuronal growth and survival in culture: 48 ml NB, 1.0 ml B27, 146 µl glutamax, 5 µl gentamycin, 1 ml heat-inactivated horse serum. Trophic factors (1 ng/ml) are added immediately prior to plating cells.

3. Papain is a protease with endopeptidase, amidase, and esterase activities. It is used when a more gentle tissue dissociation is required: 12 mg papain, 6 ml HA without calcium, 17.6 µl glutamax.

2.6. Optiprep Gradient

For one 15-ml tube:

Layer	OptiPrep (µl)	HABG (µl)	Total (ml)
1 (Bottom layer)	173	827	1
2	127	876	1
3	99	901	1
4 (Top layer)	74	926	1

3. Methods

3.1. Dissection and Isolation of Spinal Cord

Dissections should be performed in a clean and draft-free area. Although dissections can be performed in a laminar-flow hood, we usually perform them on a clean lab bench that has been wiped down with 70% ethanol. All instruments and containers for buffers and tissue are sterile.

1. From an anesthetized mouse, isolate the lumbar spinal cord. Performing a dorsal lamenectomy prior to taking the spinal cord makes actual dissection easier.

2. Place lumbar spinal cord in cold PBS ($-Ca^{+2}/Mg^{+2}$). Dissect spinal cord from vertebrae and transfer to cold HABG. Keep tissue and media cold during dissections. Keep all reagents on ice and place petri dishes for dissections on an ice pack during dissection.

3. Using fine forceps, remove nerve roots and meninges.

4. Using a micro-knife, separate dorsal from ventral spinal cords and discard the dorsal portion.

5. Using the micro-knife, chop ventral spinal cord into 0.5-mm pieces of tissue.

6. Transfer the tissue pieces to a 50-ml tube containing ±10–15 ml HABG on ice. Place tissue from no more than three spinal cords/tube.

3.2. Dissociation to Single Cell Suspension

1. Allow pieces to settle and come to room temp. Remove HABG and replace with papain (at room temp).

2. Incubate tissue in papain at 30°C for 30 min in a shaking incubator or water bath. The tube should shake at a speed just high enough to keep the tissue pieces in suspension (e.g., ±170 rpm).

3. Allow tissue pieces to settle at room temp. Remove papain and replace with 1.9 ml HABG (at room temp) + 100 μl 10% BSA + 1 μl DNAse.

4. Using a fire-polished, silicon-coated glass pipette, triturate 8–10 times in approximately 45 s. Let intact tissue pieces settle. Remove suspension and transfer to a clean 15-ml tube.

5. Add 2 ml HABG to tissue pieces and repeat step 4.

6. Repeat with another 2 ml HABG.

3.3. Isolation of Neurons

1. To prepare the gradient, make each layer in an individual tube and then create the gradient by adding each layer on top of the previous layer, starting with layer 1 at the bottom. Be sure to add each layer very gently so not to disrupt the gradient (see Fig. 2a). A handheld glass pipette is useful.

2. Carefully apply cell suspension on top of the OptiPrep gradient (see Fig. 2b).

3. Centrifuge at $800 \times g$ (no brake) for 15 min at room temperature.

4. Remove debris layer and fractions 1 and 2 (see Fig. 2c). Collect fraction 3 and transfer to a 50-ml tube. Add +10 ml HABG to cells.

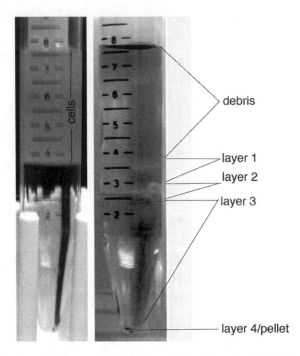

Fig. 2. (**a**) Optiprep density gradient with dissociated cells on *top*. (**b**) Individual layers of cells are visible after centrifugation. Layer 3 contains primarily neurons, including motoneurons. Layer 2 also contains neurons (16); however, we did not recover motoneurons from this layer.

5. Centrifuge cells at $228 \times g$ (no brake) for 10 min at room temp to remove OptiPrep.

6. Resuspend cell pellet in 5 ml HABG. Using a 9" glass pipette (not siliconized), add 4% BSA in HABG (1 ml) to the bottom of the tube to create a BSA cushion.

7. Centrifuge cells at 1,000 rpm (no brake) for 10 min. Additional cell debris is removed in the BSA cushion.

8. Resuspend cells in 1-ml NB media and determine cell yield using a hemocytometer.

9. Plate $1–4 \times 10^4$ cells in 100 μl/coverslip.

10. Allow cells to attach at 37°C, 5% CO_2, for at least 1 h. This insures that cells attach to the coverslip and are not washed into other areas of the dish.

11. Add NB + trophic factors to tissue culture plates (2 ml/35-mm dish) containing coverslips and return to 37°C/5% CO_2 incubator (see Note 3).

12. At this point, the investigator can move on to specific experiments.

**3.4. Final
Considerations**

We took the entire lumbar spinal cord for our initial culture and found that within 24 h, 57% of the cells were neurofilament (H + L) immunopositive and 18% of these cells were choline acetyl transferase (ChAT)-immunopositive. The cells had intact and healthy appearing nuclei (as determined by Hoechst 33342) and extensive neurite outgrowth. We next utilized only the ventral lumbar spinal cord. After 4 days in culture, approximately 70% of the cells are neurons and 93% of these cells are ChAT positive. Within 24 h, the cells extend numerous processes that are quite extensive by 4 days (see Fig. 3). The cells are healthy as determined by the Live/Dead

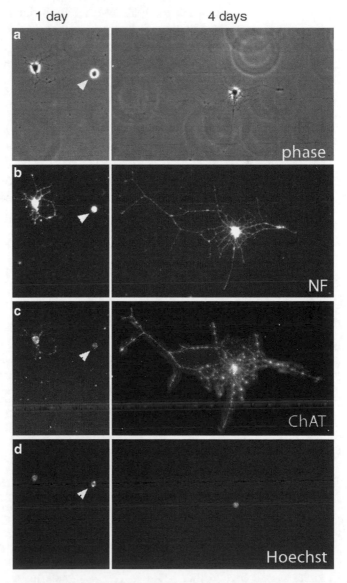

Fig. 3. Shown are P30 mouse MNs in culture for 1 or 4 days. Phase-contrast images (**a**), NF immunoreactivity (**b**), ChAT immunoreactivity (**c**), and Hoechst staining (**d**) are shown. The cells had intact and healthy appearing nuclei (as determined by Hoechst 33342) and extensive neurite outgrowth. *Arrowhead* indicates a dead cell.

kit (Molecular Probes), where calcein AM is taken up by healthy cells and hydrolyzed by esterases to a green-fluorescent product. No cell with this morphology had a red-fluorescent nucleus that would indicate disrupted membranes and incorporation of ethidium into DNA (not shown).

The yield for these cultures is very low. We estimate approximately 4×10^4 cells/ventral spinal cord. For this reason, characterization of the culture and subsequent experiments appear limited to assays that allow individual cell analysis. Nonetheless, this procedure has yielded relatively consistent results using postnatal day 30 mice. We were unsuccessful, however, in isolating motoneurons for P100 animals. We believe that the ability to examine mature motoneurons provides a new platform to investigate pathological changes associated with motoneuron disease. For these reasons, the benefits appear to outweigh the limitations.

4. Notes

1. Vacuum grease should be autoclaved before use in tissue culture; however, it can expand and leak out of the syringe during autoclaving. To prevent this, Leur Lok 1 or 3-ml syringes are filled with vacuum grease and a large paper clip is used to secure the plunger so that it is not pushed out in the autoclave (see Fig. 1).

2. To limit the ability of cells to plate only on the coverslips, coat coverslips with 100 µl only. The liquid does not reach to the edge of the coverslip creating a barrier for cells.

3. It is very difficult to view cells in culture. One approach that we use is to remove the coverslip from culture and place it (cell side down) onto a drop of PBS on a microscope slide. Once the coverslip is flipped onto a microscope slide, cells can be viewed with phase contrast. Note that this procedure is not sterile, so do not put the coverslip back in culture as it becomes contaminated.

References

1. Hamburger, V. (1977) *Neurosci Res Program Bull* **15 Suppl**, iii-37

2. Sendtner, M., Pei, G., Beck, M., Schweizer, U., and Wiese, S. (2000) *Cell Tissue Res* **301**, 71–84

3. Oppenheim, R. W. (1996) *Neuron* **17**, 195–197

4. Masuko, S., Kuromi, H., and Shimada, Y. (1979) *Proc Natl Acad Sci USA* **76**, 3537–3541

5. Henderson, C. E., Huchet, M., and Changeux, J. P. (1981) *Proc Natl Acad Sci USA* **78**, 2625–2629

6. Eagleson, K. L., and Bennett, M. R. (1983) *Neurosci Lett* **38**, 187–192

7. Bloch-Gallego, E., Huchet, M., el M'Hamdi, H., Xie, F. K., Tanaka, H., and Henderson, C. E. (1991) *Development* **111**, 221–232

8. Milligan, C. E., Oppenheim, R. W., and Schwartz, L. M. (1994) *J Neurobiol* **25**, 1005–1016

9. Milligan, C. E., Prevette, D., Yaginuma, H., Homma, S., Cardwell, C., Fritz, L. C., Tomaselli, K. J., Oppenheim, R. W., and Schwartz, L. M. (1995) *Neuron* **15**, 385–393

10. Barnes, N. Y., Li, L., Yoshikawa, K., Schwartz, L. M., Oppenheim, R. W., and Milligan, C. E. (1998) *J Neurosci* **18**, 5869–5880

11. Li, L., Oppenheim, R. W., and Milligan, C. E. (2001) *J Neurobiol* **46**, 249–264

12. Robinson, M. B., Tidwell, J. L., Gould, T., Taylor, A. R., Newbern, J. M., Graves, J., Tytell, M., and Milligan, C. E. (2005) *J Neurosci* **25**, 9735–9745

13. Li, L., Prevette, D., Oppenheim, R. W., and Milligan, C. E. (1998) *Mol Cell Neurosci* **12**, 157–167

14. Pasinelli, P., Borchelt, D. R., Houseweart, M. K., Cleveland, D. W., and Brown, R. H., Jr. (1998) *Proc Natl Acad Sci USA* **95**, 15763–15768

15. Li, M., Ona, V. O., Guegan, C., Chen, M., Jackson-Lewis, V., Andrews, L. J., Olszewski, A. J., Stieg, P. E., Lee, J. P., Przedborski, S., and Friedlander, R. M. (2000) *Science* **288**, 335–339

16. Brewer, G. J., and Torricelli, J. R. (2007) *Nat Protoc* **2**, 1490–1498

17. Taylor, A. R., Robinson, M. B., and Milligan, C. E. (2007) *Nat Protoc* **2**, 1499–1507

Chapter 6

Converting Human Pluripotent Stem Cells to Neural Tissue and Neurons to Model Neurodegeneration

Stuart M. Chambers, Yvonne Mica, Lorenz Studer, and Mark J. Tomishima

Abstract

Human embryonic stem cells (hESCs) and the related induced pluripotent stem cells (hiPSCs) have attracted considerable attention since they can provide an unlimited source of many different tissue types. One challenge of using pluripotent cells is directing their broad differentiation potential into one specific tissue or cell fate. The cell fate choices of extraembryonic, endoderm, mesoderm, and ectoderm (including neural) lineages represent the earliest decisions. We found that pluripotent cells efficiently neuralize by blocking the signaling pathways required for alternative cell fate decisions. In this chapter, we detail methods to direct hESCs or hiPSCs into early neural cells and subsequently postmitotic neurons.

Key words: Human pluripotent stem cells, Neurodegeneration, Neural, Disease modeling, Dual SMAD inhibition

1. Introduction

Human pluripotent stem cells (embryonic and induced pluripotent; hPSCs) offer a renewable source of clinically relevant cells for treating diseases. Alternatively, there is a growing appreciation for using pluripotent cells to model disease. Instead of using hiPSCs directly for cell replacement therapy, patient-specific cells can be reprogrammed into pluripotent cells before being directed into neural tissue with the goal being to study pathogenic events underlying the disease process. Nearly unlimited quantities of patient-specific neural tissue can be grown, creating an ideal scenario for drug, gene, or shRNA screens designed to slow or stop the neurodegenerative mechanisms underlying disease. Disease modeling using hPSCs also provides a distinct advantage over vertebrate models,

Giovanni Manfredi and Hibiki Kawamata (eds.), *Neurodegeneration: Methods and Protocols*,
Methods in Molecular Biology, vol. 793, DOI 10.1007/978-1-61779-328-8_6, © Springer Science+Business Media, LLC 2011

where additional knowledge regarding disease etiology may be needed to engineer mutations leading to a complete pathology.

A central challenge of using pluripotent stem cells is directing the differentiation of such cells to the desired cell type of interest. There are numerous developmental decisions occurring as pluripotent stem cells differentiate into postmitotic neurons. The pluripotent state needs to transition through a primitive ectodermal stage before adopting a neurectodermal fate. We previously published a method for high efficiency neurectoderm conversion called "dual SMAD inhibition," since both the TGF beta/Activin/Nodal and BMP branches of SMAD signaling need to be inhibited to achieve efficient neural induction (1).

After conversion to neurectodermal tissue, additional differentiation cues are required for efficient conversion into neurons. This differentiation is achieved by passaging the confluent neurectoderm. Neural rosettes are an intermediate neural stem cell population typically observed after passage (for review, see ref. 2). Such rosettes are polarized epithelial structures that proliferate rapidly and are capable of being patterned into neural tissue specific to the different regions of the nervous system (3–7). Postmitotic neurons emerge from rosettes, although they are also capable of being born directly from neurectoderm. In this chapter, we include a discussion of the methods needed to generate postmitotic neurons from in vitro-derived neurectoderm.

While directing neural fate specification is a first step for human neurodegenerative studies, it is important to keep in mind the caveats for the use of in vitro disease models. Many neurodegenerative diseases do not strike until late in life, and not all cell types found in vivo are represented in a tissue culture dish (e.g., immune cells, stromal cells). It is currently unclear how tractable this system will be for understanding such diseases since the neural tissue/neurons made using these techniques are likely to be developmentally equivalent to embryonic neural tissue. So far, two studies have shown that iPSCs can model neural diseases but both reports focused on the early onset disorders spinal muscular atrophy and familial dysautonomia (8, 9). Another challenge for iPSC-based disease modeling is the contribution of environmental factors that are typically not present and difficult to model in vitro. Further work is needed to better understand how effective the iPSC approach will be at recapitulating the broad range of neurodegenerative diseases.

2. Materials

2.1. Equipment

1. Cell culture disposables: Cell culture dishes, multiwell plates, centrifuge tubes, pipettes, pipette tips, sterile filter units.
2. Nylon mesh cell strainers, 45-μm pore size.

3. Glass Pasteur pipettes, sterilized by autoclaving and dry heat monitored with indicator tape.

4. Bench-top laminar flow hood with an HEPA filter.

5. Inverted microscope.

6. Picking hood.

7. Dissecting microscope.

8. CO_2 incubator with CO_2, humidity, and temperature control.

9. Cell culture centrifuge.

10. Glass hemocytometer.

2.2. hPSC Culture

1. Human embryonic stem cells (hESCs) or induced pluripotent stem cells (hiPSCs) (see Note 1).

2. MEF CF-1 mitomycin-C treated mouse embryonic fibroblasts (MEFs; GlobalStem, Inc.).

3. MEF medium: 900 mL of Dulbecco's modified Eagle medium (DMEM), 100 mL fetal bovine serum. Filter sterilize.

4. Recombinant FGF2 (R&D) dissolved in 1× DPBS containing 0.1% BSA to 100 μg/mL. See Note 2 regarding the preparation of growth factors.

5. hESC media: DMEM/F12 medium (Invitrogen) with 20% Knockout serum replacement (KSR; Invitrogen), 1 mM L-glutamine (Invitrogen), 0.1 mM MEM nonessential amino acids (MEM NEAA; Invitrogen), and 55 μM 2-mercaptoethanol (Invitrogen). The medium is sterile filtered in a hood and FGF2 is added after filtration to a final concentration of 6 ng/mL.

6. Sterile 1× DPBS (Gibco/Invitrogen).

7. Gelatin-coated dishes made by adding enough gelatin to coat the bottom of the dish (0.1% gelatin in PBS). Allow gelatin to incubate for at least 15 min prior to plating cells.

8. Neutral protease/Dispase (Worthington Biochemical).

9. MEF conditioned media (CM) is harvested from MEF-coated flasks. MEFs are plated at a density of 50,000 cells/cm² in a T225 flask in MEF media. The next day, the cells are washed once with PBS before adding 100 mL of hESC media. Incubate media with MEFs for 24 h before removal. The medium is now known as "conditioned media" and can be directly used or stored at 4°C for less than 2 weeks. Additional hESC media can be conditioned daily for up to 10 days on the same flask of feeders. Just before using, FGF2 is added to CM to a final concentration of 10 ng/mL, hereafter called complete CM (cCM).

2.3. Neural Differentiation

1. Accutase (Innovative Cell Technologies).

2. Nylon mesh cell strainers, 45-μm pore size (BD Falcon).

3. Matrigel Basement Membrane Matrix (BD Bioscience: we only use lots that contain over 10 mg/mL protein). Thaw the frozen vial of Matrigel on ice overnight in a 4°C refrigerator. Prepare 1 mL aliquots in a 50-mL centrifuge tube using chilled pipettes and freeze at −20°C. Matrigel must be thawed slowly to prevent gelatinization. Chilled pipettes and 50-mL centrifuge tubes should be used when making aliquots of the Matrigel.

4. KSR media: Knockout DMEM (Invitrogen) with 15% KSR (Invitrogen), 2 mM L-glutamine, 0.1 mM MEM NEAA solution, and 55 μM 2-mercaptoethanol (Invitrogen).

5. N2 media: DMEM/ F12 powder, 1:1 (Gibco/Invitrogen) is resuspended in 550 mL of distilled water. Add: 1.55 g of glucose (Sigma), 2.00 g of sodium bicarbonate (Sigma), putrescine (1 mL aliquot of 1.61 g dissolved in 100 mL of distilled water; Sigma, cat. no. P5780), progesterone (20 μL aliquot of 0.032 g dissolved in 100 mL 100% ethanol; Sigma), sodium selenite (60 μL aliquot of 0.5 mM solution in distilled water; Bioshop Canada), and 100 mg of transferrin (Celliance/Millipore) are added. 25 mg of powdered insulin (Sigma) is added to 10 mL of 5 mM NaOH and is shaken until completely dissolved. The solubilized insulin is added to the media, and double-distilled water (with a resistance of 18.2 MΩ) is added to a final volume of 1,000 mL before sterile filtration.

6. SB431542 (Tocris Bioscience) dissolved in 100% ethanol to 10 mM (1,000× stock).

7. Y-27632 dihydrochloride (Tocris Bioscience) dissolved in filtered water to 10 mM (1,000× stock).

8. Recombinant Mouse Noggin/Fc Chimera (see Note 3; R&D) dissolved in 1× DPBS containing 0.1% BSA to 100 μg/mL (500× stock if using at 200 ng/mL).

2.4. Neural Cell Culture and Neuron Differentiation

1. Recombinant Human BDNF (R&D Systems) dissolved in 1× DPBS containing 0.1% BSA to 10 μg/mL (500× stock if using at 20 ng/mL).

2. Ascorbic acid (Sigma). Add 1.76 g/100 mL sterile water and filter sterilize.

3. Recombinant Mouse Sonic Hedgehog (C25II) N-Terminus (R&D Systems; see Note 4) dissolved in 1× DPBS containing 0.1% BSA to 100 μg/mL (2,000× stock if using at 50 ng/mL).

4. Retinoic acid (Sigma) dissolved in DMSO to 10 mM. Do not filter sterilize: DMSO will degrade membrane. Primary storage should be at −80°C while working stocks can be kept at −20°C. Protect from light, and only freeze–thaw three times before discarding.

5. Recombinant mouse Fgf-8b (FGF8; R&D Systems) dissolved in 1× DPBS containing 0.1% BSA to 100 μg/mL (500× stock if using at 200 ng/mL).

6. Recombinant human GDNF (PeproTech) dissolved in 1× DPBS containing 0.1% BSA to 10 μg/mL (500× stock if using at 20 ng/mL).

7. Recombinant human TGF-β3 (R&D Systems) dissolved in 4 mM HCl containing 0.1% BSA to 20 μg/mL (20,000× stock if using at 1 ng/mL).

8. Dibutyryl cAMP sodium salt (cAMP; Sigma) dissolved in sterile water to 100 mM (200× stock if using at 500 μM).

9. Recombinant Delta-like 4 (R&D Systems) dissolved in 1× DPBS containing 0.1% BSA to 200 μg/mL (400× stock if using at 500 ng/mL).

3. Methods

Two transitions are necessary to move pluripotent cells to neurons: the first is neural induction, where pluripotent cells are directed to neurectoderm; the second is the maturation of neurectoderm to postmitotic neurons. Neural induction is complete by day 10 of this procedure (although see Note 5). At this stage, primitive neural progenitors are packed into a tight sheet of cells that only poorly differentiate into neurons. Therefore, the neural tissue typically needs to be passaged before it will differentiate further into postmitotic neurons.

In this chapter, we cover two basic methods for day 10 passage, yet there are many variations on this basic protocol that can be made depending on the circumstances of the experiment. Neural rosettes, a later neural stem cell stage appear after passage and can be further patterned into different neuronal subtypes (see Note 6) (1, 2). In general, the patterning and growth factors that one provides can cause expansion of rosettes/neural stem cells while withdrawing these factors causes differentiation. It often helps to keep survival factors around during the differentiation phase. The composition of such factors depends on the target neuronal subtype.

3.1. Grooming and Preparation of hPSCs for Differentiation

1. Groom pluripotent cells (hESCs or hiPSCs) using the picking hood, objective microscope and 200-μL pipette with sterile filter tips. Remove any pluripotent colonies that have the appearance of differentiated cells, irregular borders, or transparent centers.

2. Aspirate hESC media and add minimal Accutase to coat the dish and let sit at 37°C until all colonies are rendered to single

cells (approximately 30 min). Amount used: 1 mL for a 6-well dish, 2 mL for a 6-cm dish, or 5 mL for a 10-cm dish.

3. Avoiding bubbles, triturate the cells in the dish using a pipette with additional hESC media until there is a single cell suspension and filter using a 45-μm nylon cell strainer to remove clumps.

4. Wash and centrifuge cells ($200 \times g$ for 5 min) twice in hESC media to remove all traces of Accutase.

5. While washing, prepare Matrigel-coated dishes. Add 19 mL of DMEM:F12 to the frozen Matrigel aliquot (see equipment setup) and pipette until thawed. Work quickly and do not let the Matrigel warm up or it will polymerize. A 45-μm nylon cell strainer can be used to remove any insoluble clumps. Coat culture dishes with the diluted Matrigel solution and incubate for 1 h in the hood.

6. After washing hPSCs, resuspend the cells in hESC media containing 10 μM Y-27632 and plate on a gelatin-coated dish of the same size as step 2 (e.g., 10 cm gelatin-coated dish for a 10-cm dish of Accutase-treated cells).

7. Incubate dish at 37°C for 30 min in a cell incubator.

8. After 30 min, collect the nonadherent cells and wash the dish with hESC media containing 10 μM Y-27632 and centrifuge cells.

9. Resuspend the cells in cCM + 10 μM Y-27632.

10. Determine the cell concentration using a hemocytometer and add cCM + 10 μM Y-27632 to the appropriate cell concentration.

11. Aspirate Matrigel solution and rinse culture dishes once with DMEM:F12 prior to plating cells.

12. Plate pluripotent cells on Matrigel-coated dishes at a density appropriate for the neural cell type desired (usually between 20,000 and 50,000 cells/cm² for CNS neurectoderm, but see Note 7).

13. Twenty-four hours after plating, aspirate media and add fresh cCM + 10 μM Y-27632.

14. Forty-eight hours after plating, aspirate the media and add fresh cCM – from this point on, Y-27632 is no longer necessary. Cells can be maintained for additional days in cCM until the ideal differentiation density is obtained (~90% confluence for CNS cells, 60% confluence for neural crest and CNS; see Fig. 1 for densities).

3.2. Neural Induction of hPSCs

1. *Neural differentiation.* To initiate differentiation, add KSR containing 10 μM SB4\31542 and 200 ng/mL Noggin (day 0; Fig. 2).

2. On day 1 of differentiation, aspirate the KSR and add fresh KSR containing 10 μM SB431542 and 200 ng/mL Noggin.

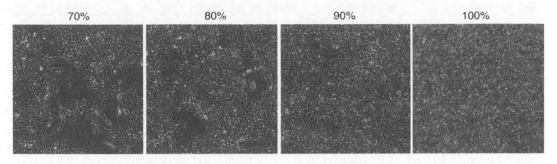

Fig. 1. Range of starting densities on day 0 – The neural differentiation of hPSCs can be started at a range of densities (50–100% confluent). At lower densities (~50–70% confluence), more cells are observed expressing neural crest markers, and at maximal density primarily CNS derivatives are observed. Below 50% confluence, cells fail to efficiently neuralize.

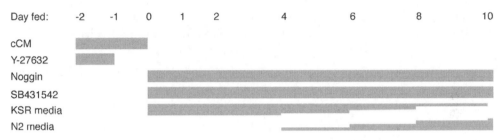

Fig. 2. Differentiation Scheme – hPSCs are plated in the presence of cCM and Y-27632, to promote single cell survival. Y-27632 is withdrawn after 1 day and the cells are cultured in cCM and can be grown for additional days until the ideal confluence is observed. Differentiation is initiated by addition of Noggin and SB431542 in KSR media. Cells are fed on the days indicated and N2 media is introduced in 25% increments starting on day 4, reaching 100% N2 on day 10.

3. On day 2 of differentiation, aspirate the KSR and add fresh KSR containing 10 µM SB431542 and 200 ng/mL Noggin.

4. On day 4 of differentiation, aspirate the KSR and add KSR/N2 (3:1) media containing 10 µM SB431542 and 200 ng/mL Noggin (final concentration to the combined KSR/N2 mixture).

5. On day 6 of differentiation, aspirate the KSR/N2 and add KSR/N2 (1:1) media containing 10 µM SB431542 and 200 ng/mL Noggin.

6. On day 8 of differentiation, aspirate the KSR/N2 and add KSR/N2 (1:3) media containing 10 µM SB431542 and 200 ng/mL Noggin.

7. On day 10 of differentiation, cells can be passaged en bloc or as single cells onto Matrigel-coated dishes (although see Note 5).

3.3. Passage of Neural Cells En Bloc

1. Mechanically dissociate thickened neurectoderm using a 200-µL pipette into small pieces.

2. Plate the blocks of tissue onto Matrigel-coated dishes in N2 containing the appropriate growth factors (see Subheading 3.5; see Note 6).

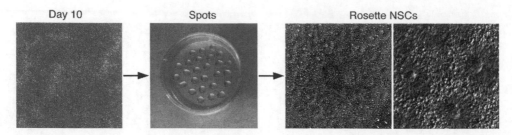

Fig. 3. Cell passage – upon day 10 the tissue will appear dramatically thicker with densely packed nuclei and phase-bright ridges. This early neurectoderm can be manually passaged using onto Matrigel-treated dishes en bloc or using Accutase as single cells in 10–20 μL drops (shown). With appropriate conditions, rosette neural stem cells will form within a few days.

3.4. Passage of Neural Cells as Single Cells

1. Aspirate differentiation media and add minimal Accutase to coat the dish and let sit at 37°C until all cells are rendered to single cells (approximately 30 min).

2. Avoiding making bubbles, triturate the cells in the dish using a pipette with additional N2 media until the cells are in single cell suspension and filter using a 45-μm cell strainer.

3. Wash and centrifuge cells ($200 \times g$ for 5 min) twice in N2 media.

4. Resuspend the cells in N2 media.

5. Determine the cell concentration using a hemocytometer.

6. Resuspend the cells in N2 media to a cell concentration of 5×10^6/mL.

7. Prepare Matrigel-coated dish by carefully aspirating all liquid from the dish, taking care not to touch the surface. Let dish dry for 15 min before plating drops (see Note 8).

8. Spot plate with 20 μL drops of the cell suspension (1×10^5 cells; Fig. 3) and let stand in the hood for 20 min before slowly adding N2 containing the appropriate growth factors (see Subheading 3.5; see Note 6). Move dish carefully to the incubator.

3.5. Differentiation of Neural Cells into Neurons

1. Minimal supportive media for differentiating neural cells into neurons is N2 containing 20 ng/mL of BDNF and 200 μM ascorbic acid. Differentiations are usually carried out for at least 1 (and sometimes 2) weeks with media changes every 2–3 days.

2. To enrich for motor neurons, 20 ng/mL BDNF, 200 μM ascorbic acid, 50 ng/mL SHH (C25II), and 1 μM retinoic acid are added to the N2 base medium.

3. To enrich for dopamine neurons, 20 ng/mL BDNF, 200 μM ascorbic acid, 50 ng/mL SHH (C25II), and 100 ng/mL FGF8 are added to the base N2 media for 1 week, followed by

20 ng/mL BDNF, 200 μM ascorbic acid, 20 ng/mL GDNF, 1 ng/mL TGF-β3, and 500 μM cAMP for all subsequent weeks.

4. To enrich for rosettes, use high-density replating, 50 ng/mL SHH (C25II) and 500 ng/mL Dll4 (7).

4. Notes

1. Most pluripotent stem cell lines respond appropriately to many different neural induction protocols; however, some pluripotent stem cell lines do not work well with particular protocols. Some lines that resist neural differentiation using a serum-free EB-based protocol (10) differentiate well with the dual SMAD inhibition method described here (Gist Croft, personal communication). Conversely, other lines work well with serum-free EB protocols but not with the dual SMAD inhibition protocol (Agnete Kirkeby, personal communication). For each new line, this must be empirically determined. Moreover, it has been our general experience that hiPSCs do not differentiate as well as hESCs.

2. Unless otherwise noted, our growth factors are made up in sterile 1× DPBS containing 0.1% BSA (Sigma, cat. no. A4503). We usually purchase bulk orders that arrive lyophilized. After resuspension to the appropriate concentration, we aliquot the factor and store at −80°C. Recombinant proteins are not filter sterilized since protein can stick to the filters, reducing the effective concentration. Working stocks are kept at 4°C to avoid freeze–thaws. We prefer to use thawed factors within 2 weeks.

3. The original protocol (1) uses 500 ng/mL Noggin. Noggin is expensive, so we have used a few variations to reduce the cost. (1) *Lower Noggin concentrations*. Titration experiments have shown that as little as 125 ng/mL Noggin work well (S.M.C. and L.S., unpublished data). We recommend 200 ng/mL here as a more affordable but still excess concentration of Noggin. (2) *Small-molecule BMP inhibitors*. We have successfully used two small-molecule inhibitors: LDN-193189 (100 nM Stemgent, cat. no. 04-0019) and Dorsomorphin (600 nM; Tocris, cat. no. 3093). These inhibitors have off target effects (such as inhibiting VEGFR and AMPK) and exhibit some level of toxicity (11). While these inhibitors work in the Dual-SMAD inhibition protocol, more work is needed to address potential differences compared to Noggin exposure. For example, the use of small-molecule inhibitors instead of Noggin could impact the results of downstream differentiation experiments.

Therefore, it might be helpful to try these inhibitors side-by-side with Noggin for the initial studies, at least until it becomes clear that the chemical inhibitors can be used successfully to support derivation of the cell type of interest. The identification of newer, more specific BMPRI inhibitors have been described and are likely to be commercially available soon (11).

4. SHH can be purchased that is either mouse or human (although they are 92% identical at the amino acid level). Furthermore, it can be purchased with engineered isoleucines on the N-terminus: these modifications make the protein more hydrophobic and likely act as a membrane tether and intercellular transport mechanism. The engineered modifications phenocopy the cholesterol and palmitate modifications that occur in mammalian cells and are necessary since bacterial expression of the protein does not provide these mammalian modifications. Most of our previous work uses conventional SHH (such as R&D Systems cat. # 461-SH), but we have found that it is more economical to use tenfold less of the isoleucine-modified version.

5. The end user should assess which day is best to end the differentiation. Different neural derivatives (e.g., central vs. peripheral nervous system) may require a longer or shorter differentiation to achieve optimum yield. Cells can be maintained in N2 containing 10 µM SB431542 and 200 ng/mL Noggin until the desired length of time is reached. Media should be changed every 2 days, although the high cell densities at this point in the differentiation may require daily changes.

6. Neural tissue receives signals during or shortly after specification that help define the positional identity of the cells that ultimately develop – a process called neural patterning. We (and others) have previously published that neural rosettes are composed of a novel type of neural stem cell capable of being patterned (reviewed in ref. (2)). More recent studies have demonstrated that neural tissue can be patterned even earlier than previously thought (1, 12). In Fasano et al. (12), evidence is provided that patterning can occur as early as 48 h after the start of neural induction. Therefore, the patterning window for a target cell population must be determined for each cell type in question, and could start much earlier than previous work had suggested. We generally believe that patterning should begin even before rosette generation for maximum cell yield.

7. Initial plating density determines the relative amounts of central vs. peripheral (neural crest) neural cells produced, with lower densities making more neural crest (see Fig. 1 for examples of plating densities). Plating efficiencies vary between pluripotent cell lines. To determine the ideal plating conditions for generating neural crest or homogenous neurectoderm, cell confluence should be examined prior to the addition of KSR.

A confluence of 50–70% will give rise to a mixture of neurectoderm and neural crest by day 10 after adding KSR, whereas a confluence of >90% creates near homogenous neurectoderm. Densities below 50% do not result in an efficient differentiation toward neural tissue and many of the cells will die.

8. Do not over dry the Matrigel-coated dish. We believe that over dried Matrigel is less effective at cell adhesion than correctly dried Matrigel. On the other hand, nice droplets will not form if too much moisture remains on the surface of the dish. The surface tension of the drop will not be maintained and the drop will rapidly spread out over the surface of the dish, lowering the cell density. If this technique becomes problematic, an alternative approach is to simply plate in a smaller well, such as a 96-well dish. This could make downstream applications more difficult, so it is important to choose the correct plate format for your experiment.

Acknowledgments

The authors would like to thank NYSTEM and The Starr Foundation for support of our research. SMC is a Starr Foundation, Tri-Institute Stem Cell Initiative Fellow.

References

1. Chambers SM, Fasano CA, Papapetrou EP et al (2009) Highly efficient neural conversion of human ES and iPS cells by dual inhibition of SMAD signaling. *Nat Biotechnol* **27**(3): 275–80.

2. Elkabetz Y, Studer L. (2008) Human ESC-derived neural rosettes and neural stem cell progression. *Cold Spring Harb Symp Quant Biol* **73**:377–87.

3. Perrier AL, Tabar V, Barberi T et al (2004) Derivation of midbrain dopamine neurons from human embryonic stem cells. *Proc Natl Acad Sci USA* **101**(34):12543–8.

4. Yan Y, Yang D, Zarnowska ED et al (2005) Directed differentiation of dopaminergic neuronal subtypes from human embryonic stem cells. *Stem Cells* **23**(6):781–90.

5. Li XJ, Du ZW, Zarnowska ED et al (2005) Specification of motoneurons from human embryonic stem cells. *Nat Biotechnol* **23**(2):215–21.

6. Lee G, Papapetrou EP, Kim H et al (2009) Modelling pathogenesis and treatment of familial dysautonomia using patient-specific iPSCs. *Nature* **461**(7262):402–6.

7. Elkabetz Y, Panagiotakos G, Al Shamy G et al (2008) Human ES cell-derived neural rosettes reveal a functionally distinct early neural stem cell stage. *Genes Dev* 2008 **22**(2):152–65.

8. Ebert AD, Yu J, Rose FF Jr et al (2009) Induced pluripotent stem cells from a spinal muscular atrophy patient. *Nature* **457**(7227):277–80.

9. Lee H, Shamy GA, Elkabetz Y et al (2007) Directed differentiation and transplantation of human embryonic stem cell-derived motoneurons. *Stem Cells* **25**(8):1931–9.

10. Zhang SC, Wernig M, Duncan ID et al (2001) In vitro differentiation of transplantable neural precursors from human embryonic stem cells. *Nat Biotechnol* **19**(12):1129–33.

11. Hao J, Ho JN, Lewis JA et al (2010) In vivo structure-activity relationship study of dorsomorphin analogues identifies selective VEGF and BMP inhibitors. *ACS Chem Biol* **5**(2): 245–53.

12. Fasano CA, Chambers SM, Lee G et al (2010). Efficient derivation of functional floor plate tissue from human embryonic stem cells. *Cell Stem Cell* **6**(4):336–47.

Chapter 7

Neural Differentiation of Induced Pluripotent Stem Cells

Mark Denham and Mirella Dottori

Abstract

The great potential of induced pluripotent cells (iPS) cells is that it allows the possibility of deriving pluripotent stem cells from any human patient. Generation of patient-derived stem cells serves as a great source for developing cell replacement therapies and also for creating human cellular model systems of specific diseases or disorders. This is only of benefit if there are well-established differentiation assay systems to generate the cell types of interest. This chapter describes robust and well-characterized protocols for differentiating iPS cells to neural progenitors, neurons, glia and neural crest cells. These established assays can be applied to iPS cell lines derived from patients with neurodegenerative disorders to study cellular mechanisms associated with neurodegeneration as well as investigating the regenerative potential of patient derived stem cells.

Key words: Neural differentiation, Neural progenitor, iPS, Neurons, Glia, Neural crest, PA6, Noggin, Laminin

1. Introduction

One of the most significant advantages of iPS technology is that it is now possible to derive pluripotent stem cells from any patient suffering from any disease condition. This allows opportunities and novel approaches for establishing cellular models of all neurodegenerative conditions, particularly those that lack appropriate animal models. The creation of human cellular models of neurodegeneration from iPS cells requires a well-characterized assay system of neural differentiation. To date the methodology of neural differentiation has been well established in human embryonic stem (ES) cells and commonly used approaches involve the treatment of ES cells with BMP inhibitors, coculture with stromal cell lines, or the

Giovanni Manfredi and Hibiki Kawamata (eds.), *Neurodegeneration: Methods and Protocols,*
Methods in Molecular Biology, vol. 793, DOI 10.1007/978-1-61779-328-8_7, © Springer Science+Business Media, LLC 2011

formation of embryoid bodies (1–10). It has been shown that these methods can also be applied to iPS cells (4, 6, 7).

Here, we describe three different methods of neural induction of iPS cells and human ES cells based on using Noggin treatment, PA6 cocultures or direct neural induction on laminin substrate (see Fig. 1). All three methods generate populations of early neural progenitors, as shown by expression of neural stem cell markers, including Pax6, Sox1, and/or Sox2 (1, 2, 5, 8–11). Neural progenitors

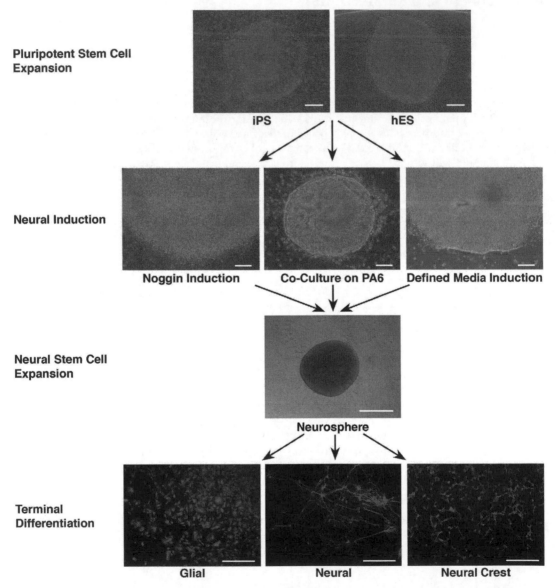

Pluripotent Stem Cell Expansion

iPS   hES

Neural Induction

Noggin Induction   Co-Culture on PA6   Defined Media Induction

Neural Stem Cell Expansion

Neurosphere

Terminal Differentiation

Glial   Neural   Neural Crest

Fig. 1. Neural differentiation of iPS and hESC cells. Pluripotent iPS cells undergo neural induction via noggin treatment, PA6 coculture system, or directly on laminin substrate in defined neural induction media. Following neural induction, neural progenitors are mechanically harvested and expanded as neurospheres. From a neurosphere state, progenitors can be plated on substrate conditions and media to bias their differentiation to mature neurons, glia or neural crest cells, as shown by expression of S100β, β-III tubulin and p75, respectively. Images taken by Mark Denham and Jessie Leung.

can then be maintained as neurospheres and further differentiated into mature neurons, glia and neural crest cells (2, 4, 5, 9–12). A protocol for each stage and lineage of neural differentiation is described below. Having a system that defines different stages of differentiation is very important as it allows room to introduce modifications so that the method can be adapted to generate specific lineages. Indeed, supplementation of additional growth factors during the early neural induction phase is often used to bias differentiation toward specific neuronal subtypes, such as motor neurons and dopaminergic neurons (4, 6, 7, 11, 13, 14). These differentiation assays serve as useful model systems for analysing the regenerative potential of diseased iPS cell lines as well as for investigating cellular mechanisms of neurodegeneration.

2. Materials

2.1. Supporting Cells

1. Mitotically inactivated mouse embryonic fibroblasts (MEFs) are derived from 129/SV E12.5 eviscerated fetuses (see Subheading 3.1).
2. PA6 cells (Riken Bioresource Centre Cell Bank #RCB1127: MC3T3-G2/PA6).

2.2. Growth Factors

1. EGF: Human recombinant epidermal growth factor (see Note 1).
2. bFGF: Human recombinant basic fibroblast growth factor (see Note 1).
3. Noggin (see Note 1).
4. PDGF-AA: Human recombinant platelet-derived growth factor (see Note 1).

2.3. Plasticware

1. T75 flask (75 cm^2).
2. Organ tissue culture dishes (BD), surface area 2.89 cm^2.
3. Flat bottom ultra low attachment 96-well plates.

2.4. Reagents

1. B-27 Supplement (Invitrogen) (see Note 2).
2. Dispase.
3. DMEM: Dulbecco's modified Eagle's medium (Invitrogen).
4. DMEM/F12 (Invitrogen).
5. EDTA: ethylenediaminetetraacetic acid.
6. FCS: fetal calf serum (Hyclone Laboratories) (see Note 3).
7. Gelatin Type A from porcine skin.
8. GMEM: Glasgow Minimum Essential Medium (Invitrogen).
9. L-Glut: L-Glutamine.
10. Fibronectin, human.

11. ITS-A: insulin-transferrin-selenium solution (Invitrogen). ITS-A supplement consists of 1 g/l insulin, 0.67 mg/l sodium selenite, 0.55 g/l transferrin, and 11 g/l sodium pyruvate.

12. KSR: knockout serum replacement (Invitrogen).

13. Laminin, mouse.

14. 2β-mercaptoethanol.

15. MEM: minimum essential medium alpha (Invitrogen) consists of: L-glutamine, ribonucleosides, and deoxyribonucleosides.

16. Mitomycin C.

17. N-2 Supplement (Invitrogen): 1 mM human transferrin, 86.1 μM insulin recombinant full chain, 2 μM progesterone, 10.01 mM putrescine, 3.01 μM selenite.

18. NBM: Neurobasal medium (Invitrogen) (see Note 2).

19. NEAA: Nonessential amino acids solution (Invitrogen): 750 mg/l glycine, 890 mg/l L-alanine, 1,320 mg/l L-asparagine, 1,330 mg/l L-aspartic acid, 1,470 mg/l L-glutamic acid, 1,150 mg/l L-proline, and 1,050 mg/l L-serine.

20. Phosphate-buffered saline solution without calcium and magnesium (PBS−): 2.67 mM potassium chloride, 1.47 mM potassium phosphate monobasic, 137.93 mM sodium chloride, and 8.06 mM sodium phosphate dibasic.

21. Phosphate-buffered saline solution with calcium and magnesium (PBS+): 0.901 mM calcium chloride, 0.493 mM magnesium chloride, 2.67 mM potassium chloride, 1.47 mM potassium phosphate monobasic, 137.93 mM sodium chloride, and 8.06 mM sodium phosphate dibasic.

22. Pen/Strep: penicillin–streptomycin solution with 5,000 U/ml penicillin and 5,000 μg/ml streptomycin.

23. Poly-D-lysine.

24. Trypsin.

25. Y27632.

2.5. Specific Equipment

1. Stereomicroscope (Leica MZ6, Leica Microsystems, Wetzlar, Germany).

2. Microscope stage (Leica Microsystems, Wetzlar, Germany).

2.6. Preparation of Culture Media

1. *Mouse Embryonic Feeder Culture Medium (F-DMEM)*
   1× DMEM supplemented with 10% heat-inactivated FCS, 2 mM L-Glut, and 0.5% Pen/Strep. Mix well and sterilize using a 0.22-μm filter.

2. *iPS Cell Culture Medium*
   1× DMEM supplemented with 20% heat-inactivated FCS, 0.1 mM nonessential amino acids solution, 2 mM L-Glut, 0.5%

Pen/Strep, 0.1 mM 2β-mercaptoethanol, and 1% ITS-A. Mix well and sterilize using a 0.22-μm filter.

3. *PA6 Culture Medium*
   1× MEM alpha supplemented with 10% FCS and 0.5% Pen/Strep.

4. *Coculture Medium*
   1× GMEM supplemented with 8% KSR, 1% NEAA, 2 mM L-Glut, 1 mM pyruvate, 0.1 mM β-mecaptoethanol, and 0.5% Pen/Strep.

5. *Defined Neural Induction Medium (N2B27)*
   Combine equal amounts of NBM with DMEM/F12 and supplement with: 0.3% glucose, 2 mM L-Glut, 1% B-27 supplement, 1% ITS-A, 1% N-2 supplement, and 0.5% Pen/Strep.

6. *Neurosphere Medium*
   1× NBM supplemented with 2% B-27 supplement, 1% ITS-A, 1% N2 supplement, 2 mM L-Glut, and 0.5% Pen/Strep. Mix well and sterilize using a 0.22-μm filter.

2.7. Preparation of Solutions

1. *Mitocycin C Solution*
   Dissolve 2 mg Mitomycin C (Sigma-Aldrich, St. Louis, MO) in 4 ml of distilled H$_2$O and filter sterilize using 0.22-μm filter. Keep solution protected from light and store at 4°C. When ready to use, dilute to 10 μg/ml in F-DMEM (see Note 4).

2. *Trypsin-EDTA Solution*
   A stock solution of 0.4% EDTA and 2.5% trypsin is made in PBS–, followed by filter sterilization. Aliquots can be stored at –20°C. Once an aliquot is thawed, it can remain active at 4°C for up to a week. Working solutions (0.5% trypsin–EDTA) are diluted from stock aliquot with PBS–.

3. *Dispase Solution*
   A 10 mg/ml solution of dispase is made in prewarmed iPS culture media. The solution is incubated at room temperature for 5 min, gently inverted, followed by 15 min incubation at room temperature. The solution is then sterilized using a 0.22-μm filter and can be used or stored at 4°C for up to 2 days.

2.8. Preparation of Adhesive Substrates

1. *Gelatinized Plates*
   Stocks of 1% gelatin can be made in distilled H$_2$O and stored at –20°C. For gelatinizing plates, stock solutions are diluted to 0.1% using distilled H$_2$O and filter sterilized. Plates are coated with 0.1% gelatin and kept at room temperature for at least 1 h. Gelatin solution is aspirated and plates are ready to use for plating cells.

2. *Laminin-Coated Plates*
   Plates are covered with poly-D-lysine solution (10 μg/ml in PBS+), and kept at room temperature for at least 30 min. This solution is aspirated and the plates are washed three times with PBS⁺.

Plates are then coated with laminin solution (5 µg/ml in PBS+) and incubated overnight at 4°C. The next day, the laminin solution is removed, followed by three washes with PBS+. Culture media can be added for plating cells/neurospheres.

3. *Fibronectin-Coated Plates*

Plates are covered with poly-D-lysine solution (10 µg/ml in PBS+), and kept at room temperature for at least 30 min. This solution is aspirated and the plates are washed three times with PBS+. Plates are then coated with fibronectin solution (10 µg/ml in PBS+) and incubated overnight at 4°C. The next day, the fibronectin solution is removed, followed by three washes with PBS+. Culture media can be added for plating cells/neurospheres.

3. Methods

All tissue culture procedures described below are performed using aseptic techniques in Class II biological safety cabinets. All of the cell cultures are maintained at 37°C with 5% CO_2 in a humidified incubator and media changed every 2–3 days.

3.1. Preparation Mouse Embryonic Fibroblast Feeder Layer

MEFs are derived from E12.5 to E14 embryos. Usually, about three embryos are used to obtain a confluent T75 flask of fibroblasts.

3.1.1. Derivation of Mouse Embryonic Fibroblasts

1. Embryos are extracted from placental tissue.
2. Head and visceral tissue (which includes fetal liver) are removed.
3. Transfer remaining fetal tissue to a clean dish and wash three times with PBS – by transferring tissue from dish to dish.
4. Mince the tissue using either scissors or scalpel blade.
5. Add 2 ml of 0.5% trypsin and continue mincing.
6. Add an additional 5 ml of trypsin and incubate for 20 min at 37°C.
7. Pipette the embryos in the trypsin until few chunks remain and then incubate for further 10 min at 37°C. The aim is to get a single cell suspension.
8. Add 10 ml F-DMEM to neutralize the trypsin and transfer contents to a 50-ml tube.
9. Mix well and transfer all to T75 flask (see Note 5).
10. The next day, the media is changed to remove cell debris and dead cells.
11. When the flask is 70–80% confluent, the cells are frozen down and labeled as passage 0. At this stage, the cells grow very vigorously and so can be expanded 1:5 (freeze at 1:5). They cannot

be used past passage 4 as their growth rate begins to slow down significantly and may develop morphological changes.

To use these cells as feeder layers in iPS cultures, it is necessary to passage them a few times prior to use to minimize contamination with residual nonfibroblast cells. Thus, to use these cells as feeder layers, one needs to:

1. Thaw passage 0.
2. Culture for 2 days until confluent, then trypsinize and expand (becomes p1).
3. Culture p1 cells for 2 days, then trypsinize and expand again (p2).
4. Culture p2 cells for 2 days, then trypsinize and freeze at passage 3 (p3).
5. Cells at p3 are used for feeder layer. The cells are mitotically inactivated either by Mitomycin C treatment or irradiation.

3.1.2. Mitomycin C Treatment of MEF

1. Aspirate existing media from flask containing MEF cultures.
2. Add mitomycin C (10 μg/ml) made in warm F-DMEM to each flask.
3. Incubate at 37°C for 2.5 h.
4. Aspirate media containing mitomycin C.
5. Wash flask once with F-DMEM and twice with PBS–.
6. Add 0.5% trypsin–EDTA solution (see Note 6) and incubate at 37°C for 3 min.

3.2. Culture of iPS Cells

iPS cell lines are cultured as colonies on mitotically inactivated MEF feeder layer, in a gelatin-coated organ tissue culture dish (2.89 cm^2) (see Subheading 3.1). The MEF feeder layer density is 6.0×10^4 cells/cm^2 (1.7×10^5 cells/organ culture dish). iPS cells are passaged weekly by mechanical slicing of the colonies using either glass pulled pipettes or a 27 gauge needle (see Fig. 2).

3.2.1. Passage of iPS Cells

iPS cell lines are routinely passaged once a week. The method used for iPS passage is by mechanical dissociation of colonies, combined with dispase treatment, to dissect the colony into "pieces." Dissection of colony is performed using a stereomicroscope with a 37°C warm microscope stage. Each colony "piece" is then transferred onto a fresh feeder layer (see Fig. 2). Detailed description of this method is described:

1. Prepare 2 × 6-cm Petri dishes with PBS+ for two washes.
2. Make dispase solution in iPS media and filter sterilize.
3. Change the media of iPS culture to PBS+.
4. Working with a stereomicroscope, cut iPS colonies into pieces using fine glass pulled pipettes or 27 gauge needles. Avoid cutting

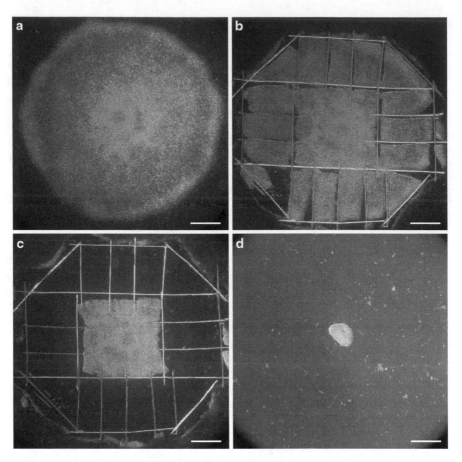

Fig. 2. Propagation of iPS colonies using mechanical dissection and dispase. (**a**) iPS colony cultured for 7 days after transfer, visualized under a stereomicroscope. (**b**) After day 7, undifferentiated regions of colonies are cut into small pieces, avoiding areas showing differentiation which often occurs in the central regions of the colony. Colonies are then treated with dispase until the edges of the pieces begin to lift and the pieces are then mechanically transferred to a PBS wash dish. (**c**) iPS colony after mechanical removal of undifferentiated pieces. (**d**) Colony pieces are then transferred to an organ culture dish with freshly plated MEF. Scale bars = 400 μm. Images taken by Mark Denham and Jessie Leung.

differentiated portions of partially differentiated colonies (see Note 7).

5. Aspirate the PBS+.

6. Add dispase solution and leave for 3 min at 37°C on heating stage, or until edges of the cut colony pieces start to detach.

7. Using a p29 pipetman, nudge the corner of the colony piece until it completely detaches, then collect and transfer to the PBS+ wash dish.

8. Once all the colony pieces have been collected and transferred to the wash dish, transfer them to the second PBS+ wash dish.

9. Finally, transfer the pieces to an organ culture dish containing a fresh feeder layer of mitotically inactivated MEF and iPS media. Usually, about eight pieces are placed per organ culture dish, evenly spaced and neatly arranged in a circle.

3.3. Neural Induction of iPS

3.3.1. Neural Induction by Noggin Treatment

For neural induction, 500 ng/ml recombinant noggin is added to iPS media at the time of iPS colony transfer onto a fresh feeder layer of MEFs (day 0) (see Note 8). Cells are cultured for 14 days without passage and noggin-media is replaced every other day. Colonies are then ready to be dissected into fragments in PBS + for neurosphere formation (see Subheading 3.4).

3.3.2. Neural Induction Using PA6 Coculture

Seed 6×10^4 PA6 cells onto gelatinised organ culture plates in PA6 media the day prior to neural induction (see Note 9). On the day of neural induction, PA6 cells are washed in PBS + and then 1 ml of coculture media supplemented with noggin (500 ng/ml) is added to organ culture plates (see Note 8). Approximately 8–10 freshly dissected iPS colony pieces (see Subheading 3.2) are plated per organ culture plate of PA6 cells and media changed every 2–3 days. After 6 days, noggin is removed from the media. On day 12–14 of differentiation, cells will have formed a tight rosette colony (see Fig. 1). Colonies are then ready to be dissected into fragments in PBS + for neurosphere formation (see Subheading 3.4).

3.3.3. Neural Induction in Defined Medium

Pieces of iPS colonies (see Subheading 3.2 are plated onto laminin-coated organ culture plates in N2B27 medium supplemented with noggin (500 ng/ml) (see Note 8). Following 6 days after differentiation, media is changed to N2B27 medium supplemented with 20 ng/ml bFGF (no noggin supplementation) for a further 6 days, until a monolayer of neural epithelial cells are present. Colonies are then ready to be dissected into fragments in PBS + for neurosphere formation (see Subheading 3.4).

3.4. Maintenance of iPS-Derived Neural Progenitors as Neurosphere Cultures

After neural induction colonies are dissected into pieces as performed for normal iPS passage but without dispase treatment (see Subheading 3.2 omitting steps 2, 6 and 9) and transferred to individual wells in a nonadherent 96-well plate to allow neurosphere formation (see Note 10). Neurospheres are cultured in suspension in neurospheres medium supplemented with 20 ng/ml EGF and 20 ng/ml bFGF. Contaminating undifferentiated iPS cells that may be transferred to the neurosphere culture conditions either do not survive or form cystic-like structures. The latter are easily identified and removed from culture dish. A more detailed protocol of transfer to neurosphere cultures follows:

1. Rehydrate the surface of the 96-well plate with neurosphere medium without any growth factors for at least 15 min in the incubator.

2. Aspirate and add neurosphere medium supplemented with 20 ng/ml bFGF and 20 ng/ml EGF (100–200 µl/well) and leave in the incubator until ready to plate the pieces.

3. Prepare a Petri dish with PBS+.

4. Wash the "pieces" of iPS-derived neural progenitors with PBS+.

5. Using a p20 pipette, pick up pieces and transfer them to the PBS dish.

6. Transfer one piece per well.

7. By 3–5 days in culture, the pieces will form a smooth sphere, indicating neurosphere formation (see Fig. 1). Those that do not form neurospheres will either disintegrate or form cysts.

8. Maintain neurosphere cultures at 37°C with 5% CO_2 in a humidified incubator.

Change media every 2–3 days. Neurospheres may be subcultured by dissection by using glass needles in neurosphere medium. Dissected neurosphere fragments are returned to fresh 96-well plates and cultured as described above.

3.5. Differentiation to Committed Lineages

3.5.1. Differentiation to Neurons

For neuronal differentiation, whole neurospheres are plated onto laminin substrates in neurosphere medium without bFGF and EGF supplements (see Note 11).

3.5.2. Differentiation to Glia

For differentiation biased toward glial lineages, whole neurospheres are plated onto fibronectin substrates in neurosphere medium with 20 ng/ml bFGF, 20 ng/ml EGF, and 20 ng/ml PDGF-AA supplements for 1 week, followed by neurospheres medium without supplements (see Note 12).

3.5.3. Differentiation to Neural Crest

Differentiation to neural crest is carried out by plating day 4 old neurospheres onto a feeder layer (6×10^4 cells/cm^2) of mitomycin-treated MEFs in neurosphere medium supplemented with 20 ng/ml bFGF and 20 ng/ml EGF for 24 h. After 24 h, the medium is replaced with fresh medium supplemented with 20 ng/ml bFGF, 20 ng/ml EGF and the small molecule, Y27632 at a final concentration of 25 μM. Neural crest cells migrating away from the neurospheres colony can be observed 24–48 h after Y27632 treatment (see Note 13).

4. Notes

1. Growth factors (bFGF, EGF, Noggin, and PDGF-AA) are reconstituted according to manufacturer's recommendations to stock concentrations of 50–100 μg/ml. Long-term storage of factors is at –80°C. Stocks can be kept short term (up to 3 months) at –20°C, or for 1–2 weeks at 4°C.

2. The components of neurobasal A and B-27 supplements are proprietary.

3. FCS for iPS cultures needs to be batch tested for support of iPS maintenance over at least four passages.

4. Mitomycin C is cytotoxic, mutagenic, and carcinogenic; appropriate safety measures must be adhered to for use and disposal of this agent.

5. In primary cultures of mouse embryo fibroblast preparations, blood cells, as well as large intact fragments of tissue, may be present along with fibroblasts. Generally, these contaminating cell types disappear following subculture as the fibroblasts overgrow them.

6. The volume of trypsin/EDTA solution used for a T75 (75 cm^2)–T175 (175 cm^2) flask is 2–5 ml, respectively.

7. By 7 days after passage, iPS colonies show some regions of cell differentiation, particularly within the centre of the colony. These regions are morphologically distinguished by cystic-like structures extending up from the colony.

8. The small-molecule Smad inhibitor SB431542 has been shown in other protocols to increase neural induction along with noggin (4).

9. Neural induction with PA6 cells is most effective with low passage PA6 cells (<25 passages).

10. After neural induction, the colonies are easily detached by mechanical dissociation from the feeder layer, and thus additional enzymatic treatment is not required.

11. On laminin substrates, neuronal differentiation with axonal outgrowths can be observed usually within 5 days of culturing. Expression of early neuronal markers, such as beta-tubulin III, can be observed at this time.

12. On fibronectin substrates, migrating glia expressing S100β and glial fibrillary acidic protein (GFAP) can be observed after 2 weeks of culture.

13. Migrating neural crest cells can be identified by expression of neural crest markers, including p75, HNK1, and Sox10.

References

1. Chiba, S., Lee, Y. M., Zhou, W., and Freed, C. R. (2008) Noggin enhances dopamine neuron production from human embryonic stem cells and improves behavioral outcome after transplantation into Parkinsonian rats. *Stem Cells.* **26**, 2810–2820.

2. Pera, M. F., Andrade, J., Houssami, S., Reubinoff, B., Trounson, A., Stanley, E. G., Ward-van Oostwaard, D., and Mummery, C. (2004) Regulation of human embryonic stem cell differentiation by BMP-2 and its antagonist noggin. *J Cell Sci.* **117**, 1269–1280.

3. Pomp, O., Brokhman, I., Ben-Dor, I., Reubinoff, B., and Goldstein, R. S. (2005) Generation of peripheral sensory and sympathetic neurons and neural crest cells from human embryonic stem cells. *Stem Cells.* **23**, 923–930.

4. Chambers, S. M., Fasano, C. A., Papapetrou, E. P., Tomishima, M., Sadelain, M., and Studer, L.

(2009) Highly efficient neural conversion of human ES and iPS cells by dual inhibition of SMAD signaling. *Nature Biotechnol.* **27**, 275–280.

5. Gerrard, L., Rodgers, L., and Cui, W. (2005) Differentiation of human embryonic stem cells to neural lineages in adherent culture by blocking bone morphogenetic protein signaling. *Stem Cells.* **23**, 1234–1241.

6. Karumbayaram, S., Novitch, B. G., Patterson, M., Umbach, J. A., Richter, L., Lindgren, A., Conway, A. E., Clark, A. T., Goldman, S. A., Plath, K., Wiedau-Pazos, M., Kornblum, H. I., and Lowry, W. E. (2009) Directed differentiation of human-induced pluripotent stem cells generates active motor neurons. *Stem Cells.* **27**, 806–811.

7. Cooper, O., Hargus, G., Deleidi, M., Blak, A., Osborn, T., Marlow, E., Lee, K., Levy, A., Perez-Torres, E., Yow, A., and Isacson, O. Differentiation of human ES and Parkinson's disease iPS cells into ventral midbrain dopaminergic neurons requires a high activity form of SHH, FGF8a and specific regionalization by retinoic acid. *Mol Cell Neurosci.* **45**(3), 258–266.

8. Vazin, T., Chen, J., Lee, C. T., Amable, R., and Freed, W. J. (2008) Assessment of stromal-derived inducing activity in the generation of dopaminergic neurons from human embryonic stem cells. *Stem Cells.* **26**, 1517–1525.

9. Davidson, K. C., Jamshidi, P., Daly, R., Hearn, M. T., Pera, M. F., and Dottori, M. (2007) Wnt3a regulates survival, expansion, and mainte-nance of neural progenitors derived from human embryonic stem cells. *Mol Cell Neurosci.* **36**, 408–415.

10. Pomp, O., Brokhman, I., Ziegler, L., Almog, M., Korngreen, A., Tavian, M., and Goldstein, R. S. (2008) PA6-induced human embryonic stem cell-derived neurospheres: a new source of human peripheral sensory neurons and neural crest cells. *Brain Res.* **1230**, 50–60.

11. Denham, M., Thompson, L. H., Leung, J., Pebay, A., Bjorklund, A., and Dottori, M. (2010) Gli1 is an inducing factor in generating floor plate progenitor cells from human embryonic stem cells. *Stem Cells.* **28**(10), 1805–1815.

12. Hotta, R., Pepdjonovic, L., Anderson, R. B., Zhang, D., Bergner, A. J., Leung, J., Pebay, A., Young, H. M., Newgreen, D. F., and Dottori, M. (2009) Small-molecule induction of neural crest-like cells derived from human neural progenitors. *Stem Cells.* **27**, 2896–2905.

13. Li, X. J., Du, Z. W., Zarnowska, E. D., Pankratz, M., Hansen, L. O., Pearce, R. A., and Zhang, S. C. (2005) Specification of motoneurons from human embryonic stem cells. *Nat. Biotechnol.* **23**, 215–221.

14. Yan, Y., Yang, D., Zarnowska, E. D., Du, Z., Werbel, B., Valliere, C., Pearce, R. A., Thomson, J. A., and Zhang, S. C. (2005) Directed differentiation of dopaminergic neuronal subtypes from human embryonic stem cells. *Stem Cells.* **23**, 781–790.

Part III

Models of Neurodegeneration

Chapter 8

Developing Yeast Models of Human Neurodegenerative Disorders

Alejandro Ocampo and Antoni Barrientos

Abstract

Neurodegenerative diseases represent one of the most devastating types of diseases in older populations in our time. Significant efforts have been made over the last 20 years to understand the molecular, biochemical, and physiological alterations underlying these diseases. However, in most cases, little is known about their pathological mechanisms due to their high complexity and involvement of a multiplicity of cellular pathways. To gain insight into this group of disorders and to devise potential therapeutic approaches, cellular and animal models of neurodegenerative proteinopathies have been created. Among them, the yeast *Saccharomyces cerevisiae* has been one of the most popular model organisms due to the degree of conservation of many biological pathways from yeast to human as well as its ease of use. Here, we describe how to create yeast models of neurodegenerative proteinopathies by ectopic expression of human proteins and how to perform a basic characterization of these models by analyzing cellular toxicity and protein aggregation.

Key words: Neurodegeneration, Proteinophaties, Yeast model, Protein aggregation

1. Introduction

The understanding and treatment of age-associated neurodegenerative diseases is probably one of the biggest challenges of modern medicine in developed countries. Neurodegenerative proteinopathies are a group of neurodegenerative diseases in which a single protein or a set of proteins misfold and aggregate, thus resulting in a progressive and selective loss of anatomically or physiologically related neuronal systems. Prototypical diseases that belong to this group are autosomal dominant diseases, including Alzheimer's disease (AD), Huntington's disease (HD), and other polyglutamine (polyQ) diseases, synucleinopathies, such as Parkinson's disease (PD), amyotrophic lateral sclerosis (ALS), and prion diseases.

Giovanni Manfredi and Hibiki Kawamata (eds.), *Neurodegeneration: Methods and Protocols,*
Methods in Molecular Biology, vol. 793, DOI 10.1007/978-1-61779-328-8_8, © Springer Science+Business Media, LLC 2011

For many years, the study of these devastating diseases relied on the pathological analysis of postmortem patients' tissues. In the last decades, researchers have identified mutations in genes that encode proteins that misfold and aggregate, ultimately leading to neuronal death, and thus establishing a link to these neurodegenerative diseases. These discoveries have encouraged the creation of cellular and animal models to gain more insight into the pathological mechanisms behind these devastating diseases and to design therapeutic strategies aiming for future treatments.

The unicellular yeast *Saccharomyces cerevisiae* has been extensively exploited in different areas of biotechnology and biomedicine. *S. cerevisiae* has been used as a valuable organism for studying the principles of microbiology, characterizing biochemical pathways, and understanding the biology of more complex eukaryotic organisms (1). In the last decade, this yeast has become one of the cellular models most extensively used to understand basic mechanisms underlying neurodegenerative proteinopathies. Although yeast is clearly less complex than neurons, it possesses several features that make it a good model system to learn about human neurodegenerative diseases. Cellular activities conserved from yeast to mammals include DNA replication, recombination, and repair; RNA transcription and translation; and enzymatic activities of general metabolism (1). More specifically, the degree of conservation of protein folding, quality control and degradation, mitochondrial function and biogenesis, and intracellular trafficking and secretory pathways make yeast an extremely relevant model to study neurodegeneration (2, 3). In addition, yeast has certain properties that make it particularly suitable for biological studies, including rapid growth, the ability to be maintained in a haploid or diploid state, ease of mutant isolation, a well-defined genetic system, and a highly versatile DNA transformation system. Finally, a whole array of biological and bioinformatics tools is today available for researchers to work with yeast in different scientific areas, making it a perfect candidate as a model organism.

When modeling a human neurodegenerative disease in yeast, two main approaches are generally followed, depending on whether the protein or proteins potentially responsible for the disease have a yeast homologue. In the case that a yeast homologue exists, the yeast gene/s can be disrupted, mutated, replaced by the human gene or overexpressed to evaluate the phenotype of the loss or gain of function. Instead, if the human gene has no clear or potential yeast homologue and the disease results at least in part from a "gain of function" of the protein involved, the strategy used to create a yeast model involves the overexpression of the wild type or mutant gene/s. In this line, yeast models of the most common neurodegenerative diseases, such as AD (4, 5), HD (6–8), PD (9–11), and ALS (12) caused by gain of function of a particular protein have been generated. These models are based on the overexpression of one or several proteins and recapitulate the crucial events preceding cell death that manifest during the course of the human disorder, including protein misfolding and aggregation. Because we have gained significant experience in modeling this kind of common human neurodegenerative diseases, in this chapter we focus on describing the creation of yeast models of gain-of-function neurodegenerative diseases based on yeast ectopic overexpression of human proteins.

2. Materials

2.1. Cloning

1. Primers for PCR amplification (see Note 1).
2. High fidelity polymerase and buffers.
3. DNA Gel Extraction Kit.
4. Required DNA restriction enzymes and the corresponding digestion buffers.
5. T4 DNA ligase and buffer.
6. Chemically competent *Escherichia coli* cells.

All enzymes must be kept at –20°C.

2.2. Yeast Transformation

Yeast Strains and Yeast Growth Media

1. Yeast strain W303-1A (MAT-A *ade2-1 his3-1,15 leu2-3,112 trp1-1 ura3-1*).
2. The compositions of the yeast growth media routinely used (13) are described below.

   (a) Minimum Medium (WOD): 0.67% (w/v) yeast nitrogen base without amino acids, 2% glucose.

   (b) Complete media containing glucose (YPD: 2% (w/v) glucose, 1% (w/v) yeast extract, 2% (w/v) peptone), or galactose (YPGal: 2% (w/v) galactose, 1% (w/v) yeast extract, 2% (w/v) peptone), or raffinose(YPRAF: 2% (w/v) raffinose, 1% (w/v) yeast extract, 2% (w/v) peptone), or ethanol plus glycerol (YPEG: 2% (w/v) ethanol, 3% (w/v) glycerol, 1% (w/v) yeast extract, 2% (w/v) peptone) as the carbon sources.

   Agar (20 g/l) is added when required to obtain solid media. All culture media are prepared in distilled water. Media are autoclaved for 20 min at 20 lbs/sq. inch prior to use. The media must be taken out from the autoclave promptly to avoid excessive burning of the sugars.

Yeast Transformation Using Lithium Acetate

1. Lithium acetate solution (TEL): 10 mM Tris–HCl, pH 7.5, 1 mM ethylenediaminetetraacetic acid (EDTA), 100 mM lithium acetate. Sterilize and store at room temperature.

2. Polyethylene glycol solution (PEG): 40% (w/v) PEG 4,000 in TEL. Sterilize by filtering (do not autoclave). Because PEG easily degenerates, it is advised to prepare the solution freshly for each experiment or to store it at room temperature for a short period of time.

3. DNA carrier solution: 2 mg/ml deoxyribonucleic acid sodium salt type III from Salmon testes in sterile TE solution (10 mM Tris–HCl, pH 8.0, 1 mM EDTA). Disperse the DNA into solution by passing it up and down repeatedly in a 10-ml pipette and mix vigorously in a magnetic stirrer until it dissolves completely. Aliquot and store at –20°C. Denature the DNA prior using it by boiling an aliquot at 95°C for 5 min and cool it down in ice.

2.3. Serial Dilution Growth Test

Yeast Growth Media
Complete liquid and solid media containing glucose, raffinose, galactose, or ethanol plus glycerol as previously described in Subheading 2.2 with the addition of the appropriate inducer to activate expression of human protein (see Note 2).

2.4. Preparation of Samples for Protein Aggregation Studies

1. Complete liquid and solid media containing glucose, raffinose, galactose, or ethanol plus glycerol as previously described in Subheading 2.2 with the addition of the indicated inducer to activate expression of human protein (see Note 2).

2. 1 M potassium phosphate (KPO_4).

3. 1.2 M sorbitol.

4. 1.2 M sorbitol, 20 mM KPO_4 pH 7.5.

5. Cell wall digestion buffer: 1.2 M sorbitol, 20 mM K_3PO_4, pH 7.4, 0.6 mg/ml zymolyase-20T (see Note 3).

6. Lysis buffer: 40 mM Hepes, pH 7.5, 50 mM KCl, 1% (v/v) Triton X-100, 2 mM DTT, 5 mM EDTA, 1 mM PMSF.

2.5. Western Blot Analysis of Protein Aggregation Samples

1. Separating polyacrylamide gel: 12% (v/v) polyacrylamide, 375 mM Tris–Cl pH 8.8, 0.1% (w/v) SDS, 0.033% (v/v) ammonium persulfate, 0.08% (v/v) TEMED.

2. Stacking polyacrylamide gel: 4.5% (v/v) polyacrylamide, 125 mM Tris–Cl pH 6.8, 0.1% (w/v) SDS, 0.05% (v/v) ammonium persulfate, 0.15% (v/v) TEMED.

3. Sample buffer (4×): 200 mM Tris–Cl pH 6.8, 4% (w/v) SDS, 40% (v/v) glycerol, 4% (v/v) β-mercaptoethanol, 0.08% (w/v) bromophenol blue.

4. Running buffer (10×): 250 mM Tris, 1.92 M glycine, 1% (w/v) SDS, pH 8.3.

5. Transfer buffer (1×): 200 mM glycine, 25 mM Tris, 20% (v/v) methanol.

6. Rinse buffer (1×): 10 mM Tris pH 8, 1 mM EDTA, 150 mM NaCl, 0.1% (v/v) Triton X-100.

7. Ponceau red: 0.5% (w/v) Ponceau red; 1% (v/v) acetic acid.

8. Nonfat dried milk.

9. Primary antibody (see Note 4).

10. Secondary antibody conjugated to horseradish peroxidase (HRP).

11. Enhanced chemiluminescent substrate for detection of HRP.

12. X-ray film.

2.6. Fluorescence Microscopy Analysis

1. Complete liquid containing glucose, raffinose, galactose, or ethanol plus glycerol as previously described in Subheading 2.2 with the addition of the appropriate inducer to activate expression of human proteins (see Note 2).

2. Phosphate saline buffer (PBS).

3. Microscope glass slides and covers.

4. Immersion oil for fluorescence microscopy.

3. Methods

The creation of yeast models of neurodegenerative proteinopathies by ectopic expression of human proteins requires two basic steps. The first step involves the design and creation of the model. Concerning the design, important points should be considered, such as the gene or genes to be expressed, regulation of protein expression, the yeast vector to be used and potential protein tags for easy detection of the expressed protein. Due to the wide range of possibilities available, we describe details concerning model design in the notes section (see Note 1 and Fig. 1). Regarding the construction of the model, this involves basic molecular biology techniques, including cloning and transformation as described in this section. Once the model has been generated, the second step involves its validation by performing a basic characterization of the phenotypes obtained by expression of the human protein and the evaluation of the relevance of the model to the human disease. Once the model has been created and validated, it can be used to study in more depth the pathological mechanism and biological pathways involved in the human disease at the cellular level.

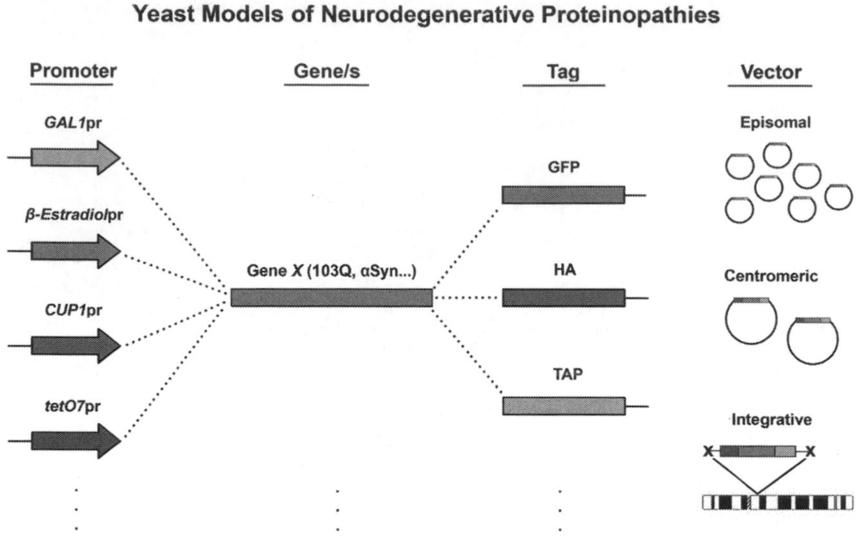

Yeast Models of Neurodegenerative Proteinopathies

Fig. 1. Basic design of yeast models of human neurodegenerative proteinopathies. See explanation in the text.

3.1. Creation of Constructs Expressing Proteins Responsible for Human Neurodegenerative Disorders

1. Amplify by PCR the gene of interest responsible or candidate for the target human disease from human genomic cDNA using the appropriate primers.

2. Amplify by PCR the inducible promoter of choice using primers according to the selected strategy for protein expression (see Note 1).

3. Digest the PCR products and a suitable yeast vector with the appropriate restriction enzymes (see Note 1).

4. Purify the digested DNA from a 0.8% agarose gel using a DNA Gel Extraction Kit.

5. Ligate the purified DNA fragments at 16°C overnight with T4 DNA ligase, transform the ligation into *E. coli*, and plate the bacteria on selective medium.

6. Test the transformant clones for the presence of the correct construct by PCR and/or appropriate restriction digestion of isolated plasmid minipreps.

7. Grow the positive clones overnight and isolate the constructs by standard plasmid DNA maxipreps.

8. In the case of integrative yeast vectors, linearize the construct by digestion with an enzyme that cuts only once and within the selectable prototrophic marker.

9. Transform the linearized integrative construct into the yeast knockout strain by the lithium acetate method as described below (see Subheading 3.2).

**3.2. Yeast
Transformation**

1. Grow 10 ml preculture of wild-type cells overnight in complete YPD medium at 30°C with constant shaking at $200 \times g$.

2. Transfer an aliquot of the preculture to a fresh 10-ml flask of YPD to obtain a final OD^{600} of ~0.1–0.2. Grow the cells until the culture reaches confluency at OD^{600} of ~0.4–0.5 (approximately 2–3 h) (see Note 5).

3. Transfer 1.5 ml of the culture to a microcentrifuge tube and pellet the cells at $1,500 \times g$ for 5 min at room temperature.

4. Wash the cells in 1 ml of sterile TEL and resuspend them in 0.1 ml of sterile TEL.

5. Denature the salmon sperm carrier DNA (2 mg/ml) by boiling an aliquot at 95°C for 5 min and cool it in ice.

6. To the cell suspension, add 5 µl of salmon sperm carrier DNA premixed with 1–4 µg of transforming DNA in a volume of less than 15 µl. Include a negative control (sample without transforming DNA) to check for contaminations and the possible reversion of the marker locus and a positive control (sample transformed with an empty circular plasmid of known concentration) to evaluate the transformation efficiency.

7. Incubate for 30 min at 30°C without shaking.

8. Add 0.7 ml of sterile 40% (w/v) PEG prepared in TEL and mix it by pipetting (see Note 6).

9. Incubate for 30–60 min at room temperature without shaking.

10. Heat-shock the cells for 10–15 min at 42°C and place them on ice for 2 min.

11. Pellet the cells by centrifugation at $10,000 \times g$ for 1 min at room temperature and wash them in 0.5 ml of sterile dH_2O.

12. Resuspend the cells in 0.1 ml of sterile dH_2O and plate them on selective WOD solid medium lacking the appropriate amino acid or nucleobase.

13. Incubate the plates at 30°C. After 3–4 days single colonies start to appear. Transfer a few transformant clones on selective WOD solid media to obtain small patches of cells. Allow them to grow at 30°C.

**3.3. Serial Dilution
Growth Test**

1. Grow overnight the yeast strain of interest (the wild-type control and wild type transformed with the construct for the expression of the human protein responsible for the disease) in YPEG liquid medium at 30°C with constant shaking at $200 \times g$. The cultures should reach late logarithmic phase of growth.

2. Estimate cellular concentration by chamber counting or by spectrophotometric measurement of absorbance at 600 nm (see Note 7).

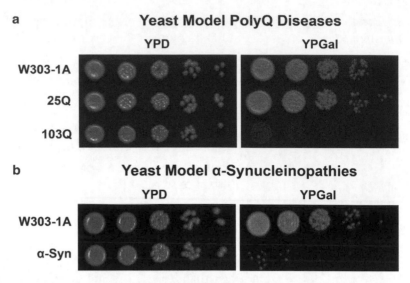

a   Yeast Model PolyQ Diseases

Fig. 2. Yeast models of polyQ diseases and α-synucleinopathies. Growth properties of cells expressing 103Q domains or α-synuclein. Tenfold serial dilutions of W303 carrying an integrative plasmid expressing mutant 103Q (**a**) or α-Syn (**b**) fused to GFP under the control of the *GAL1* promoter were spotted on media containing glucose (YPD) or galactose (YPGal) to induce protein expression. Pictures were taken after 48 h of incubation at 30°C.

3. Prepare for each strain 3–5 serial 10× dilutions in sterile dH$_2$O. The samples concentration should be between 10^7 and 10^3 cells/ml.

4. Deposit a 5–10 μl drop of each sample on solid medium in an ordered grid to obtain a series of spots of known cellular concentration (see Note 8).

5. Repeat the step on different solid media: YPD, YPEG, or YPRAF used as loading controls, YPGAL or YPEG, YPRAF plus galactose, β-estradiol, or appropriate inducer to activate expression of human protein (see Note 2).

6. Incubate the plates at 30°C for 2–6 days depending on the strains doubling time in the different media.

On YPD, YPEG, or YPRAF media in the absence of induction all the strains should grow equally. On YPGAL, the media containing galactose or alternative inducer strains expressing the human protein have reduced or abolished growth compared to wild type in cases, where the protein being expressed is toxic for the cell. An example is shown in Fig. 2.

3.4. Preparation of Samples for Protein Aggregation Studies

1. Grow overnight the yeast strains of interest (the wild-type control and wild type transformed with the construct for the expression of the human protein responsible for the disease) in YPEG liquid medium at 30°C with constant shaking at 200×*g*. The cultures should reach late logarithmic phase of growth.

2. Estimate cellular concentration by chamber counting or by spectrophotometric measurement of absorbance at 600 nm.

3. Inoculate cells into 50 ml of fresh medium containing galactose or the appropriate inducer to activate protein expression (see Note 2). Allow protein expression for the desired amount of time by growing the cells at 30°C with constant shaking at $200 \times g$.

4. Pellet the cells by centrifugation at $1,500 \times g$ for 10 min and wash them with 1.2 M sorbitol using 15 ml per gram of cells.

5. Resuspend the washed cells in cell wall digestion buffer using 3 ml for every gram of cells and incubate for 30–60 min at 30°C with gentle shaking.

6. Following 30 min of incubation, the cells should be checked for conversion to spheroplasts. This can be done by adding 50 µl of cells to 2 ml of water and a control sample of 50 µl of cells to 2 ml of 1.2 M sorbitol. The water sample should have 10–20% reduction in absorbance at 600 nm as compared to the control if most of the cells have been converted to spheroplasts (see Note 9).

7. After ~80% of the cells have been converted to spheroplasts, cells are diluted by addition of 1.2 M sorbitol, 20 mM KPO_4 pH 7.5, and pelleted by centrifugation at $5,500 \times g$ for 10 min.

8. Wash two additional times with 1.2 M sorbitol, 20 mM KPO_4 pH 7.5.

9. Resuspend the spheroplasts in lysis buffer using 20 ml per gram of cells (see Note 10) and transfer to 15-ml tubes.

10. Incubate on ice for 1 h.

11. Transfer the top part of the supernatants carefully to new tubes (T, total).

12. Measure protein concentration in supernatant and adjust volumes so the concentration is similar in all samples.

13. Centrifuge supernatant at $2,000 \times g$ for 10 min.

14. Separate supernatant (S, supernatant) and pellet (P, pellet). Keep supernatant and wash pellet an additional time with lysis buffer and resuspend in appropriate volume of water.

3.5. Western Blot Analysis of Protein Aggregation Samples

1. Determine protein concentration in the different samples and prepare samples to a similar final protein concentration by diluting them in water and adding the appropriate volume of 4× Laemmli sample buffer.

2. Prepare a 12% SDS-PAGE separating gel, overlay it with isopropanol to obtain a flat gel surface after polymerization, which takes approximately 30–45 min. After polymerization of the running gel, discard the isopropanol, wash out its traces with

abundant distilled water, carefully dry the gel surface with a piece of filter paper and overlay a 4.5% SDS-PAGE stacking gel. Insert the appropriate comb and wait approximately 30 min for the stacking gel to polymerize (see Note 11).

3. Mount the gel on the electrophoresis apparatus and add the 1× running buffer to both anode and cathode chambers. Check that the upper chamber is not leaking and for the absence of air bubbles between the gel and the lower chamber. Remove the comb of the gel and rinse the wells twice with running buffer. Load the samples. If you have empty wells, fill them with 25 μl of 1× sample buffer.

4. Connect the chamber to a power supply and run the gel at 100–150 V for approximately 2 h or until the loading dye is just about running off the bottom (see Note 12).

5. Disassemble the electrophoresis glass plate sandwich. **Do not remove the stacking gel**. This is a crucial step since insoluble protein aggregates are detected in the stacking gel.

6. Blot the proteins onto a nitro-cellulose membrane using a standard Western blotting technique using either a wet or a semidry blotting apparatus and 1× transfer buffer.

7. After finishing the transfer, stain with Ponceau red for a few minutes, wash off excess stain and mark migration of molecular weight markers.

8. Block the nitro-cellulose membrane with 5% (w/v) milk prepared in 1× rinse buffer for 30 min under rocking movement.

9. Incubate the membrane with primary antibody against the appropriate protein or protein tag used in the construction of the strains prepared in 5% (w/v) milk in 1× rinse buffer from 1 h to overnight depending in protein concentration, antibody specificity and strength (see Note 4).

10. Save the antibody solution for future reuse with additional membranes and wash the membrane twice for 15 min each with 1× rinse buffer.

11. Incubate membrane with secondary antibody (mouse or rabbit) conjugated to HRP prepared in 5% (w/v) milk in 1× rinse buffer for 1 h.

12. Perform detection by using a chemiluminescent substrate for HRP following instructions from manufacturer.

13. Place nitro-cellulose membrane on glass plate, cover it with plastic wrap and expose to X-ray film for the appropriate time. An example result is shown in Fig. 3.

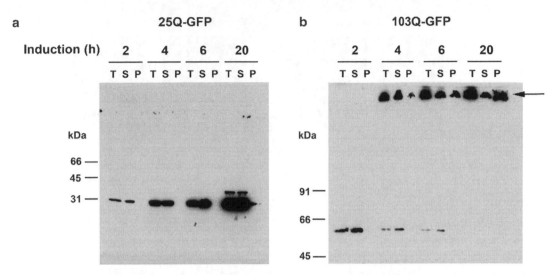

Fig. 3. Detection of polyQ–GFP fusion proteins. Wild-type W303-1A cells carrying integrative plasmids expressing 25Q-GFP (a) or 103Q-GFP (b) fusion proteins under the control of a *GAL1* promoter were grown overnight in media containing ethanol and glycerol (YPEG). Next day, cells were transferred to fresh media containing galactose (YPGal) to induce protein expression for the indicated amount of time. After different times of induction, cultures were processed as described above (Subheading 3.4). PolyQ–GFP fusion proteins were detected by Western blot analysis using equivalent amounts of the total (T), supernatant (S), and pellet (P) fractions. The fusion proteins were detected using an anti-GFP monoclonal antibody.

3.6. Fluorescence Microscopy Analysis of Yeast Cells Expressing Proteins Responsible for Human Neurodegenerative Disorders Fused to Fluorescence Tags

1. Grow overnight the yeast strain of interest (the wild-type control and wild type transformed with the construct for the expression of the human protein responsible for the disease) in YPEG liquid medium at 30°C with constant shaking at $200 \times g$. The cultures should reach late logarithmic phase of growth.

2. Estimate cellular concentration by chamber counting or by spectrophotometric measurement of absorbance at 600 nm.

3. Inoculate cells into 10 ml of fresh medium containing galactose or the appropriate inducer to activate protein expression (see Note 2). Allow protein expression for the desired amount of time growing the cells at 30°C with constant shaking at $200 \times g$.

4. Centrifuge 1 ml of sample and wash cells once with phosphate-buffered saline (PBS).

5. Resupend cells in 5–10 µl of PBS and mount 2 µl of sample on a glass slide. Place glass cover and add one drop of immersion oil for fluorescence microscopy.

6. Visualize cells under fluorescence microscope to detect protein expression and aggregation. An example result is shown in Fig. 4.

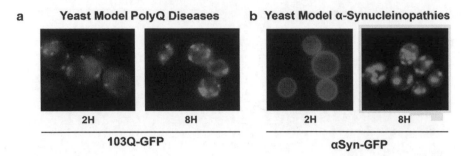

Fig. 4. Yeast models of polyQ diseases and α-synucleinopathies. Visualization of cells expressing 103Q domains or α-synuclein. W303 cells carrying an integrative plasmid expressing mutant 103Q (**a**) or α-Syn (**b**) fused to GFP under the control of the *GAL1* promoter were grown overnight in media containing ethanol and glycerol (YPEG). Next day, cells were transferred to fresh media containing galactose (YPGal) to induce protein expression for 2 or 8 h. Cells were mounted and visualized by a fluorescence microscopy.

4. Notes

1. The design of yeast models of human neurodegenerative proteinopathies should encompass four crucial aspects: A. Human gene or genes to be expressed in yeast; B. Regulation of protein expression by different promoters; C. Fusion protein with protein tags; and D. Yeast vector used.

   (a) At present, many genes responsible for different neurodegenerative diseases have been identified, but some remain yet to be discovered. In cases of diseases of known cause, the gene or genes responsible for the disease are expressed in yeast cells in order to generate the yeast model. This is the case of yeast models of polyQ disease, including Huntington's disease, where exon1 of huntingtin with different polyQ length is expressed or in another example, yeast models of Parkinson's disease where wild-type or mutant alpha-synuclein are expressed. In the hypothetical case of other proteinophaties where the gene responsible for the disease is unknown, different gene candidates can be tested to identify potential toxic phenotypes in the yeast cells.

   (b) Since many of the proteins responsible for human neurodegenerative proteinophaties are toxic when expressed in yeast cells, a tight regulation of expression is required when creating yeast models. If a toxic protein is constitutively expressed, the cells would readily die or could accumulate suppressor mutations that would alter the phenotypic characterization of the model. Regulatable expression can be achieved by using different promoters, where expression can be activated by addition of diverse inducers. In this way, cells can be maintained or grown in the absence of protein expression, and expression can be activated in the desired moment and for the desired amount of time. Among the different options, the *GAL1* inducible promoter is one of the most frequently used due to its tight regulation by addition of galactose and strong level of expression. In the last years, β-estradiol inducible promoters have also been used. In this case, protein expression levels can be regulated accurately by addition of different amounts of the human hormone β-estradiol. Finally, copper inducible promoters, such as *CUP1* and tetracycline inducible promoters, are among other options. A more detailed description of these different systems can be found in a recent report, where we reviewed the properties of some of these promoters and evaluated their suitability for the creation of yeast models of neurodegenerative proteinopathies (14). Although inducible promoters are available today, researchers have also created yeast models where protein expression is placed under the control of a constantly activated constitutive promoter, which is absolutely not recommended for reasons explained above.

(c) To allow for an easy detection or visualization of the human protein expressed in yeast, proteins can be fused to different tags, including green fluorescence protein (GFP), human influenza hemagglutinin (HA) or tandem affinity purification (TAP). GFPs and many of the different color variants have been extensively used because they allow fusion proteins to be easily detected by fluorescence microscopy. This becomes extremely important in the case of human proteins with a tendency to misfold and aggregate because the aggregation process can easily be monitored. HA or TAP tags are also useful in situations, where antibodies against the proteins of interest are not commercially available so anti-HA or anti-TAP antibody can be used. It is, in general, convenient to create strains with two versions of the protein of interest, one native untagged and one tagged version to analyze the effect that the presence of the protein tag may have in the folding or function of the protein under study.

(d) Finally, an appropriate yeast vector must be chosen for yeast expression of the human proteins. Three main groups are generally used in yeast: episomal, centromeric, and integrative vectors. The main difference among them relies on the copy number. In episomal plasmids, ranges from 200–300 copies per cell like in the case of the YEp vector series used in our laboratory, centromeric plasmids will be present in 1–3 copies per cell and integrative plasmids, where only one copy is expected to integrate into the genome as in the case of the YIp series. Because of the way the plasmids are maintained in the cell, cells containing self-replicative plasmids like episomal and centromeric must be kept under selection to prevent plasmid loss that in these cases will be very frequent due to the high toxicity of the proteins being expressed. Therefore, we and others have systematically used integrative vectors for the creation of yeast models of neurodegenerative diseases since the expression construct, once integrated in the genome will be maintained even in the absence of selection.

2. As explained in Note 1, different promoters can be used to regulate protein expression in yeast. Consequently, depending on the promoter used for the creation of the model, different inducers will be used to activate protein expression. In the case of *GAL1* promoter, expression will be activated by growing the cells in media containing galactose; when a β-estradiol inducible system is used, addition of β-estradiol will turn on expression. Copper will activate expression of *CUP1* promoter while doxycycline will induce expression from tetracycline-inducible promoters. Different amounts of the different inducers can be added to induce different levels of expression, although some promoters will not be as regulatable as others in terms of expression levels.

3. Digestion buffers containing zymolyase-20T must be prepared immediately before using to maintain the maximum activity of the enzyme.

4. The primary antibody used will depend on the protein to be detected. Commercial antibodies are now available for the detection of many human proteins, including for example huntingtin and alpha-synuclein. In cases where antibodies against the protein of interest are not commercially available, they can be raised in mouse or rabbit using the services of specialized companies. As an alternative, antibodies against protein tags, such as GFP, HA, or TAP, can also be used for the detection of fusion proteins (see Note 1).

5. Reinoculation of cells into fresh media is crucial because only cells in the mid logarithmic phase of growth will become significantly competent for transformation.

6. Because the PEG solution is very viscous, it is essential to mix the cells well after adding the solution.

7. The conversion absorbance to cellular concentration should be empirically determined in the laboratory and is expected to be approximately of one OD^{600} equal to 2×10^7 cells/ml.

8. Sample concentration and drop volume are chosen to obtain a series of spots containing between 10^5 and 10 cells/spot. Drops of volume smaller than 5 μl increase the pipetting error while drops of volume higher than 10 μl cannot be easily absorbed on the solid media.

9. The digestion of the yeast cell wall with zymolyase is a crucial step. Overdigestion with zymolyase could cause not only digestion of the yeast cell wall, but also plasma membrane. On the other hand, underdigestion will decrease the yield of isolated proteins.

10. In cases where protein concentration obtained is very low, a lower volume per gram of cells of lysis buffer can be used. Alternatively, a classical protein precipitation step using TCA followed by resuspension in smaller final volume can be performed to increase protein concentration.

11. We use 10 cm × 10 cm glass plates with 1.5-mm spacers. A 4 cm stacking gel with 1.5-cm deep comb allows us to easily load up to 10–20 µl of sample.

12. The gel of the sizes proposed in note 11 usually takes 1.5–2 h to run at constant voltage of 100–150 V. Do not use higher voltages to avoid overheating the gel and altering the migration pattern.

References

1. Botstein, D. (1991) Why yeast *Hosp. Pract. (Off. Ed.)* **26**, 157–61.

2. Miller-Fleming, L., Giorgini, F., and Outeiro, T. F. (2008) Yeast as a model for studying human neurodegenerative disorders *Biotechnol. J.* **3**, 325–38.

3. Khurana, V., and Lindquist, S. Modelling neurodegeneration in Saccharomyces cerevisiae: why cook with baker's yeast? *Nat. Rev. Neurosci.* **11**, 436–49.

4. Middendorp, O., Luthi, U., Hausch, F., and Barbcris, A. (2004) Searching for the most effective screening system to identify cell-active inhibitors of beta-secretase *Biol. Chem.* **385**, 481–5.

5. Caine, J., Sankovich, S., Antony, H., Waddington, L., Macreadie, P., Varghese, J., and Macreadie, I. (2007) Alzheimer's Abeta fused to green fluorescent protein induces growth stress and a heat shock response *FEMS Yeast Res.* **7**, 1230–6.

6. Krobitsch, S., and Lindquist, S. (2000) Aggregation of huntingtin in yeast varies with the length of the polyglutamine expansion and the expression of chaperone proteins *Proc. Natl. Acad. Sci. USA* **97**, 1589–94.

7. Meriin, A. B., Zhang, X., He, X., Newnam, G. P., Chernoff, Y. O., and Sherman, M. Y. (2002) Huntington toxicity in yeast model depends on polyglutamine aggregation mediated by a prion-like protein Rnq1 *J. Cell Biol.* **157**, 997–1004. Epub 10 June 2002.

8. Giorgini, F., Guidetti, P., Nguyen, Q., Bennett, S. C., and Muchowski, P. J. (2005) A genomic screen in yeast implicates kynurenine 3-monooxygenase as a therapeutic target for Huntington disease *Nat. Genet.* **37**, 526–31.

9. Outeiro, T. F., and Lindquist, S. (2003) Yeast cells provide insight into alpha-synuclein biology and pathobiology *Science* **302**, 1772–5.

10. Chen, Q., Thorpe, J., and Keller, J. N. (2005) Alpha-synuclein alters proteasome function, protein synthesis, and stationary phase viability *J. Biol. Chem.* **280**, 30009–17. Epub 7 June 2005.

11. Dixon, C., Mathias, N., Zweig, R. M., Davis, D. A., and Gross, D. S. (2005) Alpha-synuclein targets the plasma membrane via the secretory pathway and induces toxicity in yeast *Genetics* **170**, 47–59.

12. Johnson, B. S., McCaffery, J. M., Lindquist, S., and Gitler, A. D. (2008) A yeast TDP-43 proteinopathy model: exploring the molecular determinants of TDP-43 aggregation and cellular toxicity *Proc. Natl. Acad. Sci. USA* **105**, 6439–44.

13. Myers, A. M., Pape, L. K., and Tzagoloff, A. (1985) Mitochondrial protein synthesis is required for maintenance of intact mitochondrial genomes in Saccharomyces cerevisiae *EMBO J.* **4**, 2087–92.

14. Ocampo, A., and Barrientos, A. (2008) From the bakery to the brain business: developing inducible yeast models of human neurodegenerative disorders *Biotechniques* **45**, vii–xiv.

Chapter 9

Modeling Dopamine Neuron Degeneration in *Caenorhabditis elegans*

Michelle L. Tucci, Adam J. Harrington, Guy A. Caldwell, and Kim A. Caldwell

Abstract

Ongoing investigations into causes and cures for human movement disorders are important toward the elucidation of diseases, such as Parkinson's disease (PD). The use of animal model systems can provide links to susceptibility factors as well as therapeutic interventions. In this regard, the nematode roundworm, *Caenorhabditis elegans*, is ideal for age-dependent neurodegenerative disease studies. It is genetically tractable, has a short life span, and a well-defined nervous system. Fluorescent markers, like GFP, are readily visualized in *C. elegans* as it is a transparent organism; thus the nervous system, and factors that alter the viability of neurons, can be directly examined in vivo. Through expression of the human disease protein, alpha-synuclein, in the worm dopamine neurons, neurodegeneration is observed in an age-dependent manner. Furthermore, application of a dopamine neurotoxin, 6-hydroxy-dopamine, provides another independent model of PD. Described herein are techniques for *C. elegans* transformation to evaluate candidate neuroprotective gene targets, integration of the extrachromosomal arrays, genetic crosses, and methods for dopamine neuron analysis that are applicable to both types of neurotoxicity. These techniques can be exploited to assess both chemical and genetic modifiers of toxicity, providing additional avenues to advance PD-related discoveries.

Key words: *Caenorhabditis elegans*, Parkinson's disease, Alpha-synuclein, Dopamine, Neurodegeneration, Neurotoxicity

1. Introduction

The sense of urgency to define causes and cures for neurodegenerative diseases cannot be overstated in terms of the devastation these disorders have on individuals, families, and our society. Among the most devastating diseases that result from degeneration and loss of neurons within the brain include both Alzheimer's disease and Parkinson's disease (PD). While extensive studies on the genetics and cellular pathology involved with these diseases

Giovanni Manfredi and Hibiki Kawamata (eds.), *Neurodegeneration: Methods and Protocols*,
Methods in Molecular Biology, vol. 793, DOI 10.1007/978-1-61779-328-8_9, © Springer Science+Business Media, LLC 2011

have been conducted, effective therapies to prevent associated progressive neurodegeneration remain elusive (1, 2). This highlights the need to investigate means to more rapidly accelerate the discovery process. In this context, establishment of methods to evaluate factors impacting neurodegeneration in an experimentally reliable and inexpensive model organism that can be easily genetically altered to identify potential therapeutic targets for treatment of these diseases represents an attractive strategy.

The use of the nematode *Caenorhabditis elegans* as a model system has allowed the study of various pathologies associated with diseases in a simple, multicellular organism (3). Notably, the transparent anatomy of this animal allows for dynamic visualization of cells, including neurons, through expression of fluorescent proteins (i.e., GFP) in living animals. Here, we describe two methods used to study a key pathological hallmark of PD, dopaminergic (DA) neuron degeneration, in *C. elegans*. DA neurodegeneration can be achieved by genetic overexpression of the PD-associated gene, alpha-synuclein (α-syn), or treatment of worms with the DA neuron-specific neurotoxin, 6-hydroxy-dopamine (6-OHDA) (4, 5). Additional methods described here include microinjection of candidate gene targets, integration of extrachromosomal transgenic arrays, and genetic crosses to evaluate potential genetic modifiers of neurotoxicity. Strategies designed to optimize effective scoring of DA neuron survival in the nematode model are also discussed. These assays can be used to identify both genetic and chemical modifiers of DA neurotoxicity in vivo (6).

2. Materials

Materials below are separated based on the sequential steps of the methods described.

Common materials used throughout experimentation

35-, 60-, and 100-mm nonvented Petri dishes.

2.1. Microinjection

1. Calibrated pipettes with aspirator (needle loader).
2. Borosilicate glass capillaries (injection needles).
3. Standard mouth pipetter tubing (included with borosilicate glass capillaries).
4. Young adult/gravid worms N2 (wild type (WT)) worms.
5. Series 700 Halocarbon oil.
6. Narishige PP-830 needle puller.
7. Microinjection system with microscope manipulator, DIC, or Hoffman modulation contrast optics is required.
8. Compressed nitrogen gas tank.

9. 2% agarose injection pads (agarose and 22×50-mm coverslips) (see Note 1).

   (a) Prepare a 2.0% agarose solution with water and microwave until molten.

   (b) Using a Pasteur pipette, place a small drop of agarose on a 22×30-mm coverslip.

   (c) Immediately place a second coverslip 90° to the first coverslip and gently press down.

   (d) Wait ~30 s for the agarose to solidify and then remove the second coverslip.

   (e) Repeat this process to make multiple injection pads.

   (f) Allow the coverslips to dry overnight on the benchtop and then store in the original coverslip box.

2.2. Integration

UV cross-linker.

2.3. Genetic Crosses

1. *C. elegans* N2 males.

2. Transgenic hermaphrodites containing genes of interest.

2.4. Neuro-degeneration Analysis

1. Worm strains used in neurodegeneration experiments (see Note 2):

BY200 *vtIs1*(P_{dat-1}::GFP, pRF4(*rol-6*(*su1006*)))

UA44 *baIn11*(P_{dat-1}::α-synuclein, P_{dat-1}::GFP)

2. *Escherichia coli* OP50 strain (*Caenorhabditis* Genetics Center).

3. Nematode growth medium (NGM) for plates.

975 mL	ddH_2O
2.5 g	Peptone
3.0 g	NaCl
17.0 g	Bacto agar

Autoclave media to sterilize and allow to cool, and then add the following sterile solutions aseptically:

1.0 mL	Cholesterol (dissolve 0.125 g in 25 mL 100% ethanol)
1.0 mL	1 M $CaCl_2$
1.0 mL	1 M $MgSO_4$
25.0 mL	1 M KPO_4, pH 6.0
0.567 mL	Streptomycin sulfate (0.36 g/mL)
1.0 mL	Nystatin (11.2 mg dissolved in 1.0 mL 100% ethanol; prepare this immediately before adding to media)

After NGM plates have been poured and solidified, *E. coli* OP50 bacteria are seeded onto the plates. OP50 bacteria

Table 1
NGM plate volumes

Plate size	35 mm	60 mm	100 mm
NGM media volume	4 mL	10 mL	25 mL
Amount of OP50 bacteria seeded onto each plate	50 μL	100 μL	750 μL

should be inoculated into LB liquid media the evening before they are needed. See Table 1 for appropriate media volumes and amount of LB broth + OP50 bacteria per NGM plate used for seeding. Once seeded with LB broth + OP50 bacteria, incubate overnight at 37°C (12–18 h). Store inverted at 4°C until use (see Note 3).

4. 5-fluoro-2′-deoxyuridine (FUDR) worm plates:
Prepare NGM media as above. After autoclaving, however, do not add streptomycin sulfate. Add 0.04 mg/mL FUDR directly to the media. FUDR is light sensitive. Concentrated OP50 (see below) is used to seed FUDR plates.

5. LB broth:

10.0 g	Tryptone
5.0 g	Yeast extract
10.0 g	NaCl

Add up to 1,000 mL with ddH$_2$O and adjust pH to 7.0. Aliquot into 75-mL working volumes and autoclave to sterilize; store at room temperature.

6. Production of concentrated OP50 using 2XTY

16.0 g	Tryptone
10.0 g	Yeast extract
5.0 g	NaCl

Add up to 1,000 mL with ddH$_2$O and adjust pH to 7.0. Autoclave to sterilize and allow the solution to cool to room temperature. Upon reaching room temperature, add the following to the 2XTY medium:

1.0 mL	Cultured LB + OP50 bacteria
3.0 mL	1 M MgSO$_4$ (sterile)
0.567 mL	Streptomycin sulfate (0.36 g/mL)

Incubate at 37°C with shaking for 12–18 h before harvesting the cells. Aseptically aliquot the cultures into four, sterile, 250-mL centrifuge tubes. Centrifuge at $3,800 \times g$ for 10 min. Pour off supernatant and add 25.0 mL M9 buffer (see item 5 in Subheading 2.5). Resuspend pellet by gentle vortex and store at 4°C until time of use.

7. Levamisole.

8. Glass microscope slides.

9. Glass coverslips (22×30 mm).

10. 2% agarose.

2.5. Chemical Manipulation

1. Sterilized 10.0-mL glass conical tubes.

2. Sterilized glass Pasteur pipettes.

3. 6-OHDA.

4. Ascorbic acid.

5. M9 buffer:

22 mM	KH_2PO_4
90 mM	Na_2HPO_4
85.6 mM	NaCl

Add 3.0 g KH_2PO_4, 12.8 g Na_2HPO_4, and 5.0 g NaCl to 900 mL of ddH_2O, bring volume to 1 L, and stir for 1 h at room temperature. Aliquot into 75-mL working volumes and autoclave to sterilize. Cool to room temperature and add 75 µL of separately autoclaved 1 M $MgSO_4$. Store at room temperature.

6. 20% bleach/NaOH solution: Add 1.0 mL 5 N NaOH and 2.0 mL bleach (sodium hypochlorite; preferably from a fresh bottle) to 7 mL of ddH_2O. This solution is light sensitive. Prepare fresh before each use.

3. Methods

3.1. PD-Related Genes (Genetic Manipulation)

Several genetic factors have been linked to PD (7). Most notably, α-syn was identified to have a detrimental effect on PD patients through either multiplication of the α-syn gene locus or through genetic mutations (8, 9). Although *C. elegans* lacks conservation of this gene, multiple worm PD models have been established through overexpression of the human α-syn gene in worm DA neurons or expressed pan-neuronally (10–12). Furthermore, these models have been used in large-scale genomic screens to identify genetic factors that attenuate α-syn-induced toxicity. These genetic factors have worm orthologs and can be further studied to identify mechanisms underlying neuroprotection (6, 13, 14). The methods below describe the experimental process of injecting *C. elegans* with the gene of interest, integration of extrachromosomal DNA, genetic crossing into a GFP-illuminated DA neuron strain expressing human α-syn (P_{dat-1}::α-syn + P_{dat-1}::GFP), and analysis of the DA neurons. *C. elegans* transformation can also be performed by ballistic methods (15).

The methodology for cloning the gene of interest (either worm or human cDNA) into the appropriate expression vector is assumed in this protocol. Although traditional cloning is an acceptable approach, recombinational cloning using Gateway technology is available for a variety of vectors that enable neuronal subtype-specific expression in the worm community. The *C. elegans* ORFome developed by Marc Vidal has facilitated the cloning of approximately 19,000 *C. elegans* ORFs into Gateway vectors (ORFeome v3.1) (16). Gateway technology uses the lambda recombination system to move the gene of interest from one vector to the next. Fragments that contain recombination sites use recombinase machinery to make the cloning process simple (17). In this study, the expression vector utilized consists of the dopamine transporter (*dat-1*) promoter driving expression in the eight worm DA neurons.

3.1.1. Microinjection

The procedures described in this section can be viewed in video format from the *Journal of Visualized Experiments* (JoVE) (18).

1. To generate transgenic animals, two plasmids are injected simultaneously. One is an expression vector that contains the gene of interest and the second serves as an injection marker plasmid (see Note 4).

2. Prior to microinjection, it is advisable to purify the DNA. In this regard, DNA isolated from the Qiagen mini-prep DNA isolation kit can be used (see Note 5).

3. Quantitate both plasmids and mix together in a 1:1 ratio of 50 ng/µl each for a final concentration of 100 ng/µl (see Note 6).

4. To prepare needle loaders, take the calibrated pipettes and heat the middle over an open flame.

5. Rapidly pull the ends apart to approximately twice the total length, removing from the flame as you pull them apart.

6. Snap the two halves apart once the capillary tube cools.

7. Store upright and out of the way to prevent personal harm and/or damage to capillary tubes.

8. To prepare injection needles, place the borosilicate glass capillaries into a needle puller. For the Narishige PP-830 needle puller, a heat setting of 24.8 is recommended (see Note 7).

9. Prior to loading DNA into loading needles, centrifuge injection DNA at $13,100 \times g$ 10–30 min to pellet any particulate matter that may clog the injection needle.

10. Using a standard mouth pipetter tubing, insert the needle loader and draw up approximately 1–2 µl of injection mixture.

11. The needle loader is placed at the back end of the injection needle and inserted all the way until the tip comes in contact with the internal needle tip.

12. Gently blow with the mouth pipetter to release the DNA solution into the tip of the injection needle tip. Remove the loader carefully (see Note 8).

13. To prepare equipment, open the main valve of the compressed nitrogen tank that is connected to the microinjection arm and holder. The regulator valve must be closed to allow the gas to enter the line. The pressure should be about 35 psi. The foot pedal is used to release the pressure each time it is stepped on.

14. Place the injection needle in the microinjection arm. Position the needle in the center of the viewing field at 4× under bright-field illumination.

15. Place an injection pad on the stage and focus the edge of the pad to the center of the viewing field.

16. Increase the magnification to 40×, keeping both the needle and edge of agarose injection pad in the center and in focus.

17. Slowly lower the injection needle until it is in view and gently move the needle down to touch the agarose injection pad.

18. To break the needle, gently press the foot pedal down (this causes a slight movement in the needle) and slowly apply the needle to the edge of the agarose injection pad. Repeat this until a small amount of injection mix visibly expels. For a proper break, a small, but visible, amount of injection mix expels each time the pedal is pressed (see Note 9).

19. Do not change the X or Y positions on the microinjection arm of the scope, but raise the needle arm up (Z position) to provide room to remove the agarose injection pad. Slide the stage out from underneath the needle.

20. Prepare the agarose injection pad by placing a small amount of halocarbon oil on the surface (see Note 10).

21. With a small amount of halocarbon oil on the worm pick (acts as glue), transfer a worm onto the agarose injection pad and gently stroke it to adhere the worm to the pad. The worm should not move.

22. Quickly place the agarose injection pad back onto the stage and center the worm in the field of view, with the gonad distal tips toward you.

23. Lower the needle into the same plane of focus as the worm. Use the fine focus to position the injection needle to the same plane as the gonad.

24. If using the microinjection microscope system, switch the scope filters to Hoffman Illumination (high contrast). As an alternative, DIC may be used to increase contrast.

25. Using the 40× objective, focus on the syncytial center of the gonad (it should appear "grainy"). This is below the "honeycomb" pattern of the germ nuclei.

26. Use the fine focus to position the tip of the injection needle to the same plane as the gonad.

27. Insert the needle slowly and apply a short pulse of gas pressure. A proper injection looks like a small wave of liquid spreading across the distal gonad (see Note 11).

28. If the other gonad arm is readily accessible to the needle and less than 30–45 s have passed, repeat steps 27 and 28. Injection of the worm (one or both gonads) should not take longer than 1 min (see Note 12).

29. After one or both gonad arms are injected, quickly remove the worm by placing a small drop of M9 buffer on top of the worm using a mouth pipetter. The liquid "drops" underneath the worm and release the worm from the injection pad.

30. Using the mouth pipetter, slowly draw up the worm into the loading needle and dispense onto a seeded agar plate.

31. Place the worm near the lawn of food to allow them to recover.

32. Repeat this injection process until you have injected roughly 30–50 worms successfully, with each injected worm segregated out onto individual 35-mm plates.

33. Allow the P0 to reproduce for 2–3 days and then search for F1 progeny expressing the injection marker.

34. After 2–3 days, segregate out single F1 worms expressing the injection marker to individual small plates and wait for an additional 3–4 days to allow for the F2 generation to grow (see Note 13).

35. Filter through the plates and keep the ones that have a high penetrance of the transgenic marker. Worms expressing the transgenic array are considered a "stable" line (see Note 14).

36. Obtain at least three stable lines to control for the variability in the gene copy number among transgenic animals.

37. Maintain each stable line separately and analyze the lines for the phenotype of interest (see Subheading 3.1.4). In this example, the phenotype is neurodegeneration (see Note 15).

3.1.2. Integration

Due to the fact that stable transgenic worm lines consist of extra-chromosomal arrays, random segregation of the transgenic DNA occurs during cell division processes, resulting in variable copy number within the progeny. Integration of the transgenic constructs within the chromosomes avoids this issue and enables establishment of isogenic populations for analysis. Various methods of DNA integration have been used, such as gamma and UV irradiation and integration by injection (19). Irradiation causes chromosomal breaks, whereby the injected DNA is ligated during repair. Mutations can arise with exposure to irradiation, thus outcrossing to wild-type animals is necessary (see Subheading 3.1.3). The procedure below describes UV irradiation. It is advisable to analyze the three stable lines prior to integration to obtain preliminary phenotypes of interest (in this example, enhanced neurodegeneration or neuroprotection) (see Subheading 3.1.4). Once this is completed, select the line that is most representative of all the stable lines and begin integration.

1. Grow and select for approximately 200 L4-stage worms (see Note 16). These worms should all express the genetic marker that indicates expression of the injected constructs (such as fluorescence).

2. Wash the worms with M9 buffer. Using a sterilized glass Pasteur pipette, rinse the worms off the NGM plate with ~1.0 mL M9 buffer into a sterile 10.0-mL glass conical tube (see Note 17).

3. Centrifuge in a clinical centrifuge at $145 \times g$ for 1 min.

4. Remove supernatant and suspend worm pellet in 1.0 mL of M9 buffer. Repeat centrifugation and wash (steps 2–4) until supernatant is clear of bacteria (approximately three times).

5. Pipette the worms onto unseeded NGM plates and allow M9 buffer to absorb into the plate (see Note 18).

6. Place the plate inside the cross-linker with the Petri dish lid off (see Note 19).

7. Close the cross-linker door, press the energy button, and input energy level to 400–450 J/m^2 (see Note 20).

8. Press start and when the irradiation is finished, remove the Petri plate from cross-linker. Add 100 μL of concentrated OP50, cover with lid, and allow to incubate at 20°C for at least 4 h, but do not exceed 12 h (see Notes 21 and 22).

9. Transfer two healthy irradiated worms to 60-mm plates. The dead worms are visible on the plate (see Note 23).

10. Repeat step 7 until 50 plates (100 worms) are transferred. Label the plates 1–50.

11. Grow until starved +2 days to allow dauer formation. At 20°C, this takes approximately 10 days.

12. If a fluorescent marker was used, check under a fluorescent stereomicroscope and take a small chunk from each plate that contains a high concentration of worms carrying the transgenic

marker. This should be performed using a sterilized inoculating loop. Transfer the chunk to a 60-mm-sized plate (repeat for the remaining 49 plates). Approximately 100–500 transgenic larvae should be transferred onto each plate. Label the plates 1–50 to correspond to the plates of origin.

13. At room temperature, grow until almost starved (2–4 days).

14. From each plate, pick eight young larvae transgenic animals, one worm per 35-mm plate. Keep track of which worms came from the original 50 plates (label individual plates A–H, along with the original plate number). This allows one to determine independent lines (see Notes 24 and 25).

15. After progeny have developed (~4 days), check for 100% transmission of the transgenic marker.

16. If successful, clone three animals to single plates and confirm successful integration by examining the next generation for 100% transmission of the transgenic marker.

17. Once an integrated line is obtained, outcrossing must be performed to remove any extraneous mutations due to the irradiation (see Note 26).

3.1.3. Genetic Crosses

The procedures described in this section can be viewed in video format from JoVE (20).

Males arise in a WT population at a very low frequency via chromosomal nondisjunction (~0.2%). To generate *C. elegans* males, L4 hermaphrodites can be placed for an extended "heat shock" in 34°C incubator for 2–4 h, and then back to room temperature to allow for propagation. Once males are obtained, a simple genetic cross should be set up to maintain a stock, whereby each plate contains a single hermaphrodite with three to five males. Following a successful cross, ~50% of the F1 population is male.

The process given below describes the cross of the transgenic lines (stable or integrated) that express the gene of interest for neuronal analyses and the transgenic line expressing α-syn in the DA neurons.

1. Place three to four L4-staged transgenic (integrated or stable) hermaphrodites expressing your gene of interest on a 35-mm Petri dish with a small bacterial lawn (see Note 27). Set up individual plates for each independent line expressing the gene of interest.

2. Place 10–15 males expressing P_{dat-1}::GFP + P_{dat-1}::α-syn on the mating plates.

3. Propagate worms for 2 days and then remove the males to prevent the P0 males from interacting with the F1 generation.

4. Monitor the F1 generation and look for male progeny. If males are present, the cross was successful.

5. On five 35-mm Petri dishes, clone out individual L4 hermaphrodites from each cross that expressed both the P_{dat-1}::GFP marker and the injection marker used with the gene of interest (see Note 28).

6. Allow to self-fertilize and produce F2 progeny. This takes 2–3 days.

7. Similar to step 5, clone out the F2 generation (~5–10 animals) to their own individual plates.

8. Monitor the F3 generation to search for plates with 100% worms that express both GFP and the transgenic marker for the gene of interest.

3.1.4. Dopamine Neurodegeneration Analysis

The procedures described in this section can be viewed in video format from JoVE (20).

1. Synchronize worms by transferring ~50 gravid adult hermaphrodites to 100-mm seeded NGM plates, and place worms at 20°C for 5 h for egg laying (see Note 29).

2. Remove gravid adult worms from the plates, leaving only the eggs. Grow worms at 20°C for 4 days.

3. Four days posthatching, transfer ~100 transgenic young adult worms to 60-mm seeded FUDR plates (see Note 30).

4. On days of DA neuronal analysis, prepare molten 2.0% agarose solution (see Note 31).

5. Prepare agarose pads for analysis by placing a piece of tape on two microscope slides to act as a spacer. A drop of molten agarose is deposited on a third microscope slide that is placed between the two "spacer" slides. A fourth microscope slide is placed on to the top of and perpendicular to the agarose slide, pressing down slightly, to create a pad. See Fig. 1 (see Note 32). When the agarose has solidified (for a few seconds), the perpendicular slide should be carefully removed so that the exposed agarose pad can be used for mounting worms.

6. Transfer 35–40 worms to an 8.0 µL drop of 3 mM levamisole (an anesthetic) on a 22 × 30-mm cover glass. Invert cover glass and place it onto the 2% agarose pad prepared in step 5. Repeat this process for each worm strain that is analyzed.

7. Score the six anterior dopaminergic neurons (4 CEP and 2 ADE) as normal (i.e., wild-type) or degenerative for 30 worms/strain (Fig. 2). Degenerating neurons may exhibit several morphological defects, including neurite retraction and complete cell loss (see Fig. 3a–c) (see Notes 33 and 34).

8. Repeat each experiment in triplicate (30 worms/screen × 3 screens = 90 total worms analyzed/strain).

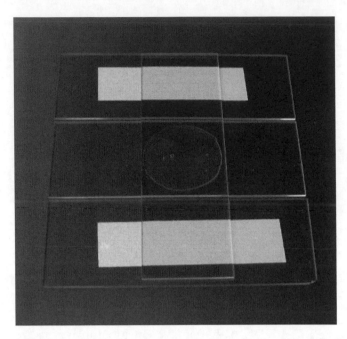

Fig. 1. Image of the microscope slide arrangement used when creating agarose pads. Removal of the top slide yields a pad that worms can be mounted on for microscopy analysis.

Screen:　　　　　　　　　　　　　　　　　　**Date:**

Transgenic line:

worm no.	WT neurons	Degenerating neurons		
		ADE	CEP ventral	dorsal
1	◯	◯◯	◯◯	◯◯
2	◯	◯◯	◯◯	◯◯
3	◯	◯◯	◯◯	◯◯
4	◯	◯◯	◯◯	◯◯
5	◯	◯◯	◯◯	◯◯

Fig. 2. Example of a scoring sheet that can be used in the analysis of the anterior DA neurons in *C. elegans*. Each neuron within an animal exhibiting degenerative changes can be individually noted. This allows for quantitative data on both the total number of neurons degenerating in a population (each worm has six anterior DA neurons × number of worms analyzed), as well as the number of worms within a population that exhibit any degenerating neurons.

3.2. 6-OHDA-Induced Toxicity (Chemical Manipulation)

Alternative methods to induce DA neurotoxicity may also be used to observe neuronal cell death. The DA analogue, 6-OHDA, is readily taken up into dopaminergic neurons through the dopamine transporter, DAT-1, where it forms free radicals and causes oxidative stress, resulting in cell death (4, 5). For this degeneration assay,

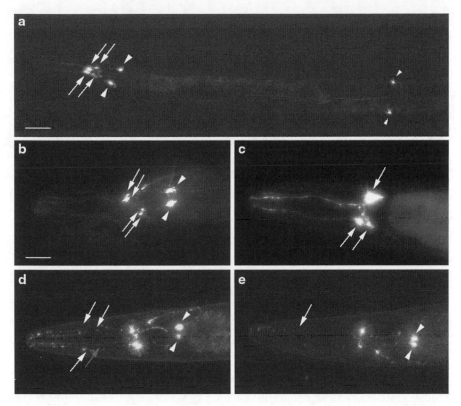

Fig. 3. *C. elegans* DA neurons are shown using GFP driven from the DA transporter promoter (P$_{dat-1}$::GFP). (**a**) The 6 anterior DA neurons include two pairs of cephalic (CEP) neurons (*arrows*) and one pair of anterior deirid neurons (ADEs; *large arrowheads*); the cell bodies and processes are highlighted. Two posterior deirid neurons are also present in each hermaphrodite (PDEs; *small arrowheads*). (**b**). The anterior region of *C. elegans* is magnified, displaying the six anterior-most DA neurons. The cell bodies of the four CEP neurons and the two ADE neurons are labeled with *arrows* and *arrowheads*, respectively. (**c**) A 7-day-old worm coexpressing GFP and α-syn in DA neurons; most worms within this population are missing anterior DA neurons when they are adults. In this example, three of the four CEP neurons (*arrows*) remain while there are no ADE neurons. (**d, e**) Following exposure to 6-OHDA, CEP and ADE neurons progressively degenerate as shown in these examples. (**d**) In this example, three of four CEP cell processes are present, but degenerating (*white arrows*). The fourth CEP neuronal process is not yet degenerating. The two ADE neurons in this animal are still intact (*arrowheads*). (**e**) This worm exhibits further degeneration, whereby only one CEP cell process remains; the two ADE neurons in this animal are still intact (*arrowheads*). Scale bar in (**a**) = 100 μM. Scale bar for (**b–e**) = 50 μM.

late L3–L4 worms expressing GFP under the *dat-1* promoter [BY200, *vtIs1*(P$_{dat-1}$::GFP), pRF4(*rol-6*(*su1006*))] are treated with various concentrations of 6-OHDA. Younger worms should not be treated due to increased lethality while older worms have slightly higher resistance to the toxin than L4-stage worms.

1. Worms should by synchronized either through bleaching or egg laying (see Note 35).

2. To bleach, perform the following steps:

   (a) Wash gravid adult hermaphrodites off a 60-mm NGM plate with 1.0 mL ddH$_2$O and transfer them into a 10.0-mL glass conical tube.

(b) Centrifuge tube at $145 \times g$ for 1 min and decant supernatant, leaving worm pellet at bottom.

(c) Add an additional 1.0 mL ddH$_2$O to the worm pellet, centrifuge, and decant supernatant. Repeat twice.

(d) Add 1.0 mL 20% bleach solution to each tube of worms and incubate at room temperature for 5–7 min with gentle agitation.

(e) After bleaching, centrifuge tube and decant supernatant.

(f) Wash worms with 1.0 mL M9 buffer, centrifuge, and decant supernatant. Repeat wash for three additional times.

(g) After last centrifugation, leave ~100 µL M9 buffer in each tube, resuspend pellet (worm eggs), and transfer eggs to 100-mm seeded NGM plate using a Pasteur pipette. Store at 20°C.

3. Allow worms to grow at 20°C for 65–70 h, until they reach the late L3–L4 larval stage.

4. Wash worms off the 100-mm plates with 3.0 mL ddH$_2$O into 10.0-mL glass conical tubes, centrifuge, and decant supernatant.

5. Wash worms with 1.0 mL ddH$_2$O, centrifuge, and decant supernatant. Repeat wash for three additional times (see Notes 36 and 37).

6. Decant ddH$_2$O, add 1.0 mL 6-OHDA solution (see Table 2) to each tube, and incubate for 1 h at 20°C with gentle agitation every 10 min (see Note 38).

7. After the 1-h incubation, centrifuge worms and decant supernatant.

8. Wash worms with 1.0 mL M9, centrifuge, and decant supernatant. Repeat wash for three additional times.

9. After the last centrifugation, leave ~100 µL M9 buffer in each tube, resuspend worm pellet, and transfer worms to 60-mm seeded NGM plates. Place the worm plates at 20°C until time of analysis (see Table 2).

Table 2
6-OHDA concentrations and times of analysis

6-OHDA concentration	10 mM 6-OHDA, 2 mM ascorbic acid	30 mM 6-OHDA, 6 mM ascorbic acid	50 mM 6-OHDA, 10 mM ascorbic acid
Analysis times	2, 6, 24, 48, and 72 h	2, 6, 12, 24, and 48 h	1, 2, 3, and 6 h

Scoring of the anterior dopaminergic neurons is conducted using the same procedure as mentioned above. Degenerating neurons may exhibit several morphological defects, including neurite blebbing, cell body rounding, and cell loss/death (see Fig. 3a, b, d, e). Degeneration from 6-OHDA is most often seen in the CEP neurons while the ADE neurons are less sensitive to this type of toxicity (see Note 39).

4. Notes

1. The injection pads have a long shelf life once made. However, if too dry, worms will desiccate.

2. For general worm maintenance, see ref. 21.

3. To grow LB + OP50 for seeding NGM plates, inoculate 75 mL of LB broth with *E. coli* OP50 and 50 µL of streptomycin sulfate (0.36 g/mL). Incubate in 37°C incubator for 12–18 h. Store at 4°C.

4. There are a variety of injection marker plasmids that provide a readily visible phenotype for successful transgenic animal production. Examples include fluorescent tissue markers, such as pharyngeal muscle (P_{myo-2}::mCherry) and body wall muscle (P_{unc-54}::GFP). Other phenotypic markers include the plasmid pRF4 which encodes a dominant collagen mutation (*rol-6*(*su1006*)) that elicits a roller phenotype. Care should be taken, however, to avoid choice of a marker that may obfuscate scoring of dopaminergic neurons.

5. A good DNA prep should have a 260/280 absorbance ratio range between 1.7 and 2.0.

6. Injection marker plasmids that express RFP (i.e., mCherry, dsRed) may need to be injected at a lower injection due to lethality with high expression. The injection plasmid P_{myo-2}::mCherry is one example, where the injection concentration should be 1–5 ng/µl.

7. Multiple injection needles can be made and stored in a small box. These needles are very fragile and are easily broken if not stored properly; one method is to lay them on top of molding clay.

8. Several injection needles should be loaded per transgenic construct to save time. After several worm injections, the needle can break or become clogged, thus a new needle is required. Since bubbles can prevent proper plasmid flow, allow time for any bubbles to float out of the needle tip. This is best done when the needle is held with the tip facing down.

9. If the needle opening is too large, this may cause damage to the worm. Conversely, too small of a break (and opening) prevents proper flow of DNA.

10. Injection pads that are too dry can desiccate the worm quickly. To prevent this, breathe on the pad one time. If the pad is too moist (immediate use after making the injection pads), the worms will have difficulty adhering.

11. Injecting too much can cause the liquid to flow into the proximal bend of the gonad arm and shut down oocyte production.

12. If the worm is on the injection pad too long, the worm will desiccate and die.

13. The DNA is not integrated in the chromosome of the worm and it exists as an extrachromosomal array, which segregates randomly; this can be inherited (also randomly).

14. Usually, 10% of the F1 progeny yield a stable transgenic line.

15. Since gene copy can be variable across transgenic lines, when selecting a line for chromosomal integration (see Subheading 3.1.2), it is important to choose a transgenic worm line that is representative of the other (≥ 2) stable lines and has >50% transmission frequency.

16. Upon UV irradiation, half of the L4 population may die.

17. *C. elegans* have a propensity for sticking to plastic tubes, thus glass tubes are recommended.

18. Bacteria can act as a shield against the UV irradiation.

19. The plastic of the Petri dish can act as a shield against the UV irradiation.

20. Repetitive use of the UV bulbs causes the intensity to decrease over time, and therefore this energy level may need to be incrementally increased by ~50–100 J/m^2, if chromosome integration is no longer observed (and bulbs should be replaced accordingly).

21. The machine self-monitors until the dose is complete; this takes approximately 20 s. However, as described in Note 20, the time to completion is dependent on bulb usage.

22. After irradiation, sufficient time should be allowed for worms to recover because transferring worms too early (before ~4–6 h) causes extra stress on the animals and results in lethality.

23. Healthy worms can be identified by the worms moving in a wild-type manner. Other worms are sluggish and typically die over time.

24. At least three independent lines are necessary to account for variation in gene copy number.

25. It is critical at this stage that only a single worm is transferred/plate and that there are no small larvae or eggs attached to the single worm.

26. By crossing with an N2 WT animal, recombination through the genetic cross helps reduce the number of these mutations. For genetic cross procedures, (see Subheading 3.1.3).

27. A small bacterial lawn forces closer contact between the hermaphrodite and males.

28. The cloned hermaphrodite animals need to be at the L4 stage to ensure that they have not mated with the P0 males.

29. If stable transgenic lines are used in crosses instead of chromosomally integrated lines, the transmittance rate of the transgene within these stable lines will dictate how many adults are needed. For example, low transmittance lines may require more adults to lay eggs in order to obtain ~100 worms for analysis.

30. FUDR blocks DNA synthesis and does not allow the eggs that have been laid to develop, thus eliminating the need to transfer animals throughout their life span prior to analysis.

31. The day you choose to analyze is dependent on your hypothesis for the overexpressed gene. For example, if the coexpressed gene might enhance neurodegeneration, then analysis should begin on day 6 of development. However, if the overexpressed gene might be neuroprotective, then analysis should be performed on days 7 and 10 (see Fig. 4).

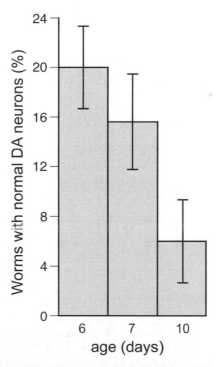

Fig. 4. DA neurons within *C. elegans* expressing both P$_{dat-1}$::GFP + P$_{dat-1}$::α-syn exhibit age-dependent degeneration. When these worms are 6 days old, approximately 20% of the population displays a full complement of normal DA neurons while only 5% of the population has normal DA neurons in 10-day-old animals. This information can be used to examine the consequences of expressing potential neurodegenerative or neuroprotective proteins (in 6 and 10-day-old animals, respectively).

32. Do not make the pad too thick as this could cause a microscope objective to place too much pressure on the pad during analysis, causing the worms to burst and die.

33. Do not score dead worms or those that have exploded, as the neurons degenerate quickly and skew the results.

34. It might be useful to score ventral and dorsal CEPs separately because the dorsal CEP neurons synapse onto the ADE DA neurons while the ventral CEP neurons do not. In P*dat-1*::α-syn worms, dorsal CEPs degenerate significantly more often than ventral CEPs. ADEs degenerate significantly more than either dorsal or ventral CEP neurons (Hamamichi and Caldwell, personal communication) (see Fig. 5).

35. Synchronizing by egg laying increases the percentage of transgenic animals; this is important when working with stable (and not chromosomally integrated lines) because bleaching a total population of transgenic and nontransgenic animals results in nontransgenic offspring that unnecessarily overcrowd the plate.

36. Make sure to wash off most of the bacteria from the worms. The presence of bacteria increases the oxidation of 6-OHDA and decreases the potency of the chemical.

37. Do not use M9 buffer to wash bacteria off; it oxidizes 6-OHDA quickly.

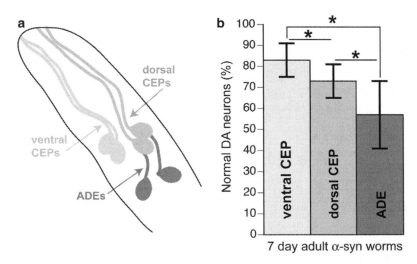

Fig. 5. Anterior DA neurons exhibit differential sensitivity to α-syn-induced degeneration. (a) There are six DA neurons in the anterior region of a *C. elegans* hermaphrodite; these neurons can be subclassified in pairs as two dorsal CEP neurons, two ventral CEPs, and two ADE neurons. The dorsal CEPs are postsynaptic to the ADE neurons while the ventral CEPs are not postsynaptic to the ADEs. (b) In worms expressing both P_{dat-1}::GFP + P_{dat-1}::α-syn, dorsal CEPs degenerate significantly more often than ventral CEPs. ADEs degenerate significantly more than either dorsal or ventral CEP neurons in this strain ($P < 0.05$; Fisher Exact Test).

38. Fresh 6-OHDA solution should be made immediately prior to adding the solution to the worms. If the solution turns pink immediately, the chemical will not cause DA neurodegeneration, as it is already oxidized. In general, this usually occurs after the hour incubation. When making the solution, mix water and ascorbic acid before adding the 6-OHDA because the ascorbic acid helps stabilize 6-OHDA in solution.

39. It is important to use the same 6-OHDA chemical supplier and lot number, as changes to either of these factors may elicit differences in the amount of degeneration observed. Each batch or lot number of 6-OHDA should be tested for efficiency to induce neurotoxic effects before a large-scale experiment is performed.

Acknowledgments

We would like to thank all members of the Caldwell laboratory, especially Songsong Cao, Shusei Hamamichi, and Laura Berkowitz, for their contributions to the development of the methods described herein. Research on movement disorders in the Caldwell lab is supported by grants from the Howard Hughes Medical Institute (GAC), National Science Foundation (KAC), and QRxPharma, Ltd. (GAC and KAC).

References

1. Dauer W and Przedborski S (2003) Parkinson's disease: mechanisms and models. *Neuron* **39**:889–909.

2. Fahn S (2003) Description of Parkinson's disease as a clinical syndrome. *Ann N Y Acad Sci* **991**:1–14.

3. Caldwell GA and Caldwell KA (2008) Traversing a wormhole to combat Parkinson's disease. *Dis Model Mech* **1**:32–6.

4. Cao S, Gelwix CC, Caldwell KA, and Caldwell GA (2005) Torsin-mediated neuroprotection from cellular stresses to dopaminergic neurons of *Caenorhabditis elegans*. *J Neurosci* **25**:3801–12.

5. Nass R, Hall DH, Miller DM III, and Blakely RD (2002) Neurotoxin-induced degeneration of dopamine neurons in *Caenorhabditis elegans*. *Proc Natl Acad Sci USA* **99**:3264–9.

6. Hamamichi S, Rivas RN, Knight AL, Cao S, Caldwell KA, Caldwell GA (2008) Hypothesis-based RNAi screening identifies neuroprotective genes in a Parkinson's disease model. *Proc Natl Acad Sci USA* **105**:728–33.

7. Harrington AJ, Hamamichi S, Caldwell GA, and Caldwell KA (2010) *C. elegans* as a model organism to investigate molecular pathways involved with Parkinson's disease. *Dev Dyn* **239**:1282–95.

8. Singleton AB, Farrer M, Johnson J, Singleton A, Hague S, *et al* (2003) Alpha-synuclein locus triplication causes Parkinson's disease. *Science* **302**:841.

9. Polymeropoulos MH, Lavedan C, Leroy E, Ide SE, Dehejia A, *et al* (1997) Mutation in the alpha-synuclein gene identified in families with Parkinson's disease. *Science* **276**:2045–7

10. Cao P, Yuan Y, Pehek EA, Moise AR, Huang Y, *et al* (2010) Alpha-synuclein disrupted dopamine homeostasis leads to dopaminergic neuron degeneration in *Caenorhabditis elegans*. *PLoS ONE* **5**:e9312.

11. Cooper AA, Gitler AD, Cashikar A, Haynes CM, Hill KJ, *et al* (2006) Alpha-synuclein blocks ER-Golgi traffic and rab1 rescues neuron loss in Parkinson's models. *Science* **313**:324–8.

12. Kuwahara T, Koyama A, Gengyo-Ando K, Masuda M, Kowa H, *et al* (2006) Familial Parkinson mutant alpha-synuclein causes dopamine neuron dysfunction in transgenic *Caenorhabditis elegans. J Biol Chem* **281**:334–40.

13. van Ham TJ, Thijssen KL, Breitling R, Hofstra RM, Plasterk RH, and Nollen EA (2008) *C. elegans* model identifies modifiers of alpha-synuclein inclusion formation during aging. *PLoS Genet* **4**:e1000027.

14. Kuwahara T, Koyama A, Koyama S, Yoshina S, Ren CH, *et al* (2008) A systematic RNAi screen reveals involvement of endocytic pathway in neuronal dysfunction in alpha-synuclein transgenic *C. elegans. Hum Mol Genet* **17**:2997–3009.

15. Wilm T, Demel P, Koop HU, Schnabel H, Schnabel R (1999) Ballistic transformation of *Caenorhabditis elegans. Gene* **229**: 31–5.

16. Lamesch P, Milstein S, Hao T, Rosenberg J, Li N, *et al* (2004) *C. elegans* ORFeome Version 3.1: Increasing the coverage of ORFeome resources with improved gene predictions. *Genome Res* **14**:2064–9.

17. Invitrogen Gateway Cloning (http://www.invitrogen.com/site/us/en/home/Products-and-Services/Applications/Cloning/Gateway-Cloning.html)

18. Berkowitz LA, Knight AL, Caldwell GA, and Caldwell KA (2008) Generation of stable transgenic *C. elegans* using microinjection. *JoVE* **18**: pii: 833. doi: 10.3791/833

19. Mello CC, Kramer, JM, Stinchcomb D, Ambros V (1991) Efficient gene transfer in *C. elegans*: extrachromosomal maintenance and integration of transforming sequences. *EMBO J.* **10**: 3959–70.

20. Berkowitz LA, Hamamichi S, Knight AL, Harrington AJ, Caldwell GA, and Caldwell KA (2008) Application of a C. *elegans* dopamine neuron degeneration assay for the validation of potential Parkinson's disease genes. *JoVE* **17**:pii: 835. doi: 10.3791/835.

21. Caldwell GA, Williams SN, and Caldwell KA (2006) Integrated Genomics: A discovery-based laboratory course. John Wiley & Sons, Ltd, Chichester, England:1–225.

Chapter 10

Role of the Proteasome in Fly Models of Neurodegeneration

Chun-Hung Yeh, Marlon Jansen, and Thomas Schmidt-Glenewinkel

Abstract

Most neurodegenerative disorders are associated with aggregates of ubiquitinated proteins, such as Lewy bodies in Parkinson's disease and neurofibrillary tangles in Alzheimer's disease. Although the etiology of the sporadic forms of these disorders remains elusive, these observations support our idea that proteasome impairment is an important risk factor in neurodegeneration. Proteasome dysfunction is, thus, expected to be a pivotal link between environmental and genetic factors that are implicated in triggering neurodegeneration. Here, we discuss the rationale for the use of *Drosophila* as a model system for the study of neurodegeneration. As an example of a specific application of this model system, we provide experimental methodology for the assessment of proteasome function by a nondenaturing gel assay, by Western blotting, as well as measurement of ATP levels which are critical for proteasome function. In addition, we discuss immunocytochemical approaches for the study of both the larval and adult *Drosophila* nervous system.

Key words: Neurodegenerative diseases, *Drosophila*, Proteasome, Immunocytochemistry, In-gel assay, Western blotting, ATP measurement

1. Introduction

The human nervous system can be afflicted by a variety of neurodegenerative diseases, many of which are sporadic with little or no insight into the fundamental cause of the disease. Most prominent in this group, we find Alzheimer's disease (AD) (1, 2), Parkinson's disease (3), trinucleotide repeat diseases (4) containing either coding trinucleotide repeats as in polyglutamine diseases – e.g., Huntington's disease – or noncoding trinucleotide repeats, e.g., CGG repeats in Fragile X-syndrome, and amyotrophic lateral sclerosis (5), but more than 30 additional neurodegenerative diseases have been described in humans.

Potential progress in understanding the causative factors of these diseases has been made through an extensive search for genetic

Giovanni Manfredi and Hibiki Kawamata (eds.), *Neurodegeneration: Methods and Protocols*,
Methods in Molecular Biology, vol. 793, DOI 10.1007/978-1-61779-328-8_10, © Springer Science+Business Media, LLC 2011

mutations as a cause of these diseases. In case of Alzheimer's disease, this search has allowed the identification of four genes – amyloid precursor protein (APP), two presenilin genes (PSEN1 and PSEN2), and apolipoprotein E (APO E) – in which certain mutations can give rise to Alzheimer's disease. A recent meta-analysis has further revealed at least 20 potential foci in the human genome which might have a significant effect on AD risk (6). However, no mutation has been identified so far in the tau protein which is also considered to be a causative factor in Alzheimer's disease. It is also important to note that these mutations have a role in the pathogenesis of Alzheimer's disease in only about 5% of the cases.

In Parkinson's disease, mutations in the PINK and PARKIN genes are known to cause the disease presumably because of the required interaction of the two proteins for mitophagy (7, 8). Mutations in the genes DJ-1 and leucine-rich repeat kinase 2 (LRRK2) have also been demonstrated to play a role in the development of Parkinson's disease, but the underlying mechanism in their pathogenicity of the disease is less clear.

In case of amyotrophic lateral sclerosis (ALS), a certain association of the disease has been found with mutations in the gene encoding Cu/Zn superoxide dismutase (SOD1) (5). Again the frequency is small – only 5–10% of all cases are inherited and only 10% of the inherited cases are caused by mutations in SOD1.

It is obvious from this brief summary description of neurodegenerative diseases that experimental models, which can be easily manipulated genetically and have a short life cycle, would offer a tremendous advantage for further investigation of these diseases. Such models have been established in yeast, the nematode *Caenorhabditis elegans* (9), and the fruit fly *Drosophila melanogaster* (10–13). In particular, *Drosophila* offers many advantages for the study of neurodegenerative diseases.

1. Forward-genetic screens for the detection of new genes involved in neurodegeneration.

2. Disruption of gene expression using an RNAi approach: A library of 22,247 transgenic flies carrying inducible UAS-RNAi against >80% of protein-coding genes of the *Drosophila* genome is available in the Vienna RNAi stock center (http://www.vdrc.at/rnai-library/).

3. Transposition of any DNA construct into the genome of *Drosophila* by P-element transformation (14).

4. Similarity of the *Drosophila* genome to the human genome in many important biological pathways.

5. Availability of many neurodegenerative disease models – e.g., Alzheimers disease, Parkinson's disease, ALS, and polyglutamine diseases – which allows the detailed examination of the causes and mechanism of human neurodegenerative diseases (10).

Many of these neurodegenerative disorders are associated with formation of protein aggregates, resulting ultimately in proteinaceous inclusions, such as Lewy bodies in Parkinson's disease and neurofibrillary tangles in Alzheimer's disease. While the composition of these inclusion bodies shows some variation with the disorder, a general feature is that these aggregates contain ubiquitinated proteins. Thus, although selective sets of neurons are affected in different neurodegenerative disorders, they are associated with an inability to degrade ubiquitinated proteins. In general, high levels of ubiquitinated proteins do not accumulate in healthy cells, as they are rapidly degraded by the ubiquitin–proteasome pathway. The inability to break down ubiquitinated proteins is likely due to proteasome impairment known to be associated with aging, mitochondrial dysfunction, and oxidative stress. For these reasons, we discuss experimental approaches in fly models of neurodegeneration, which allow assessing the state of the proteasome by in-gel assays (15), Western blotting, and ATP measurements and the general state of the nervous system by immunocytochemistry and microscopy.

2. Materials

2.1. Drosophila Stocks

1. Wild-type Oregon R flies are reared on standard corn-meal agar medium. Flies are passed to fresh vials every 4–6 days and maintained in humidified temperature-controlled environmental chambers at 25°C and 60% relative humidity throughout the experiments.

2. Oregon R and many other stocks can be obtained from the *Drosophila* Stock Centers in Bloomington, IN (http://www.flystocks.bio.indiana.edu/). The Web site of the stock center has also available excellent information about the composition and preparation of fly food as well as information about general fly husbandry.

2.2. Stock Preparation for Lysis Buffer

1. 0.5 M EGTA, pH 8.0: 9.5 g EGTA dissolved in water. Titrate pH to 8.0 and bring to final volume of 50 ml. Store at 4°C.

2. 100 mM Na_3VO_4 stock: 0.184 g Na_3VO_4 dissolved in 10 ml water. Store at –20°C.

3. 0.5 M NaF: 0.21 g NaF dissolved in 10 ml water. Store at –20°C.

4. 100 mM β-glycerophosphate: 0.22 g β-glycerophosphate dissolved in 10 ml water. Store at –20°C.

5. 100 mM $Na_4P_2O_7$: 0.45 g $Na_4P_2O_7$ dissolved in 10 ml water. Store at –20°C.

6. 100 mM PMSF: 0.174 g PMSF dissolved in 10 ml 2-propanol. Store at –20°C.

7. 5 M NaCl:14.61 g NaCl dissolved in 50 ml water. Store at 4°C.

8. 1 M Tris–HCl, pH 7.5: 60.57 g Tris–OH dissolved in water. Titrate to pH 7.5 and bring to final volume of 500 ml. Store at 4°C.

2.3. Fly Lysis Buffer for Western Blotting

1. 2× lysis buffer base: 40 mM Tris–HCl, pH 7.5, 274 mM NaCl, 2 mM EGTA, 20% glycerol. Mix 87.65 ml water with 25 ml glycerol, 0.5 ml 0.5 M EGTA, 6.85 ml 5 M NaCl, and 5 ml 1.0 M Tris–HCl, pH 7.5. Store at 4°C.

2. Lysis buffer: 20 mM Tris–HCl, pH 7.5, 137 mM sodium chloride, 1 mM EGTA, 10% glycerol, 1 mM sodium orthovanadate (Na_3VO_4), 1 mM phenylmethylsulfonylfluoride (PMSF), 1 mM β-glycerophosphate, 2.5 mM sodium pyrophosphate ($Na_4P_2O_7$), 50 mM sodium fluoride (NaF),1% Nonidet P-40 (NP40)/Igepal, and protease inhibitor cocktail. Mix 1.575 ml water with 2.5 ml lysis buffer base (2×), 100 μl protease inhibitor cocktail (Sigma), 50 μl 100 mM Na_3VO_4, 50 μl 100 mM PMSF, 50 μl 100 mM β-glycerophosphate, 125 μl $Na_4P_2O_7$, 500 μl NaF, and 50 μl Igepal.

2.4. SDS-Polyacrylamide Gel Electrophoresis

1. Resolving gel buffer stock: 3.0 M Tris–HCl, pH 8.8. Titrate to pH 8.8 and bring to final volume of 250 ml. Store at 4°C.

2. Stacking gel buffer stock: 0.5 M Tris–HCl, pH 6.8. Titrate to pH 6.8 and bring to final volume of 250 ml. Store at 4°C.

3. 10× reservoir buffer: 0.25 M Tris–OH, 1.92 M glycine. Store at 4°C.

4. Running buffer: Add 200 ml of 10× reservoir buffer and 20 ml 10% SDS to 1,780 ml water. Store at 4°C.

5. Transfer buffer (25 mM Tris–OH, 192 mM glycine, and 15% methanol): Add 200 ml of 10× reservoir buffer stock, 300 ml methanol, and 1,500 ml water. Store at 4°C.

6. Washing buffer: 1 packet of phosphate-buffered saline (PBS) and 2 ml Tween 20 dissolved in 2,000 ml water. Store at 4°C.

7. 2× Laemmli buffer: Mix 50 μl water with 200 μl 0.5 M Tris–Cl, pH 6.8, 400 μl 10% SDS, 200 μl glycerol, 100 μl 0.5% bromophenol blue solution, and 50 μl 2-mercaptoethanol. Store at 4°C.

8. 0.5 M Tris–HCl, pH 6.8:15 g Tris–OH dissolved in water. Titrate to pH 6.8 and bring to final volume of 250 ml. Store at 4°C.

9. 10% SDS: 50 g SDS dissolved in 500 ml water. Store at room temperature.

2.5. Fly Lysis Buffer for Nondenaturing Gel Electrophoresis (In-Gel Assay)

1. Harvesting buffer: 50 mM Tris–HCl, pH 7.4, 5 mM $MgCl_2$, 10% glycerol, 5 mM ATP, and 1 mM DTT. 3,880 μl of buffer A (see below), add 100 μl of 200 mM ATP and 20 μl of 200 mM DTT. Store at 4°C.

2. Buffer A: 50 mM Tris–HCl, pH 7.4, 5 mM $MgCl_2$, and 10% glycerol. 6.057 g Tris–OH and 1.01 g $MgCl_2$ are dissolved in water and titrated to pH 7.4. 100 ml glycerol is added to solution prior to adjusting volume to 1 l. Store at 4°C.

3. 200 mM ATP stock solution: 0.3306 g ATP dissolved in 3 ml buffer A. ATP solution is prepared fresh. Store at 4°C (see Note 2).

4. 200 mM DTT stock solution: 0.0924 g DTT dissolved in 3 ml water. DTT solution is prepared fresh. Store at 4°C (see Note 2).

2.6. Nondenaturing Gel for In-Gel Assay

1. Running buffer: 0.18 M boric acid, 0.18 M Tris–OH, 5 mM $MgCl_2$, 1 mM ATP, and 1 mM DTT. To 400 ml buffer B add 2 ml of 200 mM ATP stock solution, and 2 ml of 200 mM DTT stock solution. Store at 4°C.

2. Buffer B: Dissolve 11.1 g boric acid, 21.8 g Tris–OH, 1.01 g $MgCl_2$ in 1 liter of water. Store at 4°C.

3. Gel preparation buffer: The same recipe as buffer B.

4. Proteasome activity detection buffer: Buffer A with 1 mM ATP, 1 mM DTT, and 400 μM proteasome substrate. 14,250 μl of Buffer A, add 75 μl of 200 mM ATP, 75 μl of 200 mM DTT, and 600 μl of 10 mM of the fluorogenic substrate Suc-LLVY-AMC (Enzo Life Sciences). Substrate is added before proteasome activity detection. Store at room temperature (see Note 5).

5. Transfer buffer: 25 mM Tris–OH, 192 mM glycine, and 15% methanol. Add 200 ml of 10× reservoir buffer stock, 300 ml methanol, and 1,500 ml water. Store at 4°C.

6. 10× reservoir buffer: 0.25 M Tris–OH, 1.92 M glycine. 30.3 g Tris–OH and 144 g glycine in 1 l water. Store at 4°C.

7. 1.5-mm thick cassette, to form a three-step gradient gel (from bottom up) 5, 4, and 3% containing rhinohide polyacrylamide strengthener, is used to assess proteasome immunoreactivity (16).

8. 5% gel: Mix 8.888 ml buffer B with 2,100 μl 30% acrylamide, 400 μl rhinohide solution, 600 μl 1.5% ammonium persulfate, and 9 μl N,N,N',N' tetramethylethylene diamine (TEMED).

9. 4% gel: Mix 9.388 ml buffer B with 1,680 μl 30% acrylamide, 320 μl rhinohide solution, 600 μl 1.5% ammonium persulfate, and 9 μl TEMED.

10. 3% gel: Mix 9.888 ml buffer B with 1,260 μl 30% acrylamide, 240 μl rhinohide solution, 600 μl 1.5% ammonium persulfate, and 9 μl TEMED.

2.7. Western Blot for Determination of Proteasome Level

1. Transfer buffer: 25 mM Tris base, 192 mM glycine, and 15% methanol. Add 200 ml of 10× reservoir buffer stock, 300 ml methanol, and 1,500 ml water. Store at 4°C.

2. 10× reservoir buffer: 0.25 M Tris, 1.92 M glycine. 30.3 g Tris–OH and 144 g glycine in 1 l water. Store at 4°C.

3. Washing buffer: 1 packet of PBS and 2 ml Tween 20 dissolved in 2,000 ml water. Store at 4°C.

2.8. SDS-PAGE for Actin

1. 12.5× loading buffer: Mix 5 ml glycerol, 5 ml 2-mercaptoethanol, and few drops of a solution of 0.5% bromophenol blue. Store at room temperature.

2. 0.1 M Tris–EDTA, pH 7.5: 12.1 g Tris–OH and 8 g EDTA dissolved in 20 ml water. Store at 4°C.

3. Resolving gel buffer stock: 3.0 M Tris–HCl, pH 8.8. 90.75 g Tris–OH dissolved in water. Titrate to pH 8.8 and bring to final volume of 250 ml. Store at 4°C.

4. Stacking gel buffer stock: 0.5 M Tris–HCl, pH 6.8. 15 g Tris–OH dissolved in water. Titrate to pH 6.8 and bring to final volume of 250 ml. Store at 4°C.

5. 10× reservoir buffer: 0.25 M Tris, 1.92 M glycine: 30.3 g Tris–OH and 144 g glycine in 1 l water. Store at 4°C.

6. Running buffer: Add 200 ml of 10× reservoir buffer stock and 20 ml 10% SDS to 1,780 ml water. Store at 4°C.

7. Transfer buffer: 25 mM Tris–OH, 192 mM glycine, and 15% methanol. Add 200 ml of 10× reservoir buffer stock, 300 ml methanol, and 1,500 ml water. Store at 4°C.

8. Washing buffer: 1 packet of PBS and 2 ml Tween 20 dissolved in 2,000 ml water. Store at 4°C.

2.9. Dissection of Larval and Adult Brain

The process of dissecting both larval and adult brain requires practice and does not work for the untrained person in the first attempt. There are several good movies available on the Internet which demonstrate in great detail the dissection process. The Web sites are as follows: http://www.jove.com/index/details.stp?ID=1936, http://www.youtube.com/watch?v=j4rVa7JCzdg, http://www.youtube.com/watch?v=aGcnJeqEVEk.

1. Microscope coverslips (22×40×0.15 mm) and microscope slides.

2. 10× PBS, pH 7.4.

3. Paraformaldehyde stock solution: 16% stock solution (Electron Microscopy Service). Prepare a 4% (v/v) working solution with 1× PBS for each experiment.

4. Wash solution: 0.3% (v/v) Triton X-100 in PBS.

5. Primary antibody: Anti-Tyrosine Hydroxylase (ImmunoStar).

6. Secondary antibody: Cy3-conjugted AffiniPure Goat Anti-Mouse IgG (Jackson ImmunoResearch).

7. Mounting medium: Vectashield (Vector Laboratories).

2.10. ATP Determination

1. White Cliniplate 96-well (Thermo Scientific).

2. BCA Protein Assay Reagent A (Thermo Scientific).

3. Pierce BCA Protein Assay Reagent B (Thermo Scientific).

4. Immulon B Flat Bottom 1×12 Strip Assemblies 96 wells (Thermo Electron Corporation).

5. ATP Determination Kit (Invitrogen).

6. Buffer A: 160 ml H_2O + 1,600 μl of 1 M Tris–Cl, pH 8.0.

3. Methods

3.1. Preparation of Samples for Western Blotting

Western blotting allows the visualization and semiquantification of the three forms of the proteasome: 26S (2): 20S with two 19S subcomplexes, 26S (1): 20S with one 19S subcomplex, and the 20S core. The ratio of these three forms of the proteasome changes as a function of proteolytic activity in the cell and as a function of age.

1. Thirty male flies are prepared and harvested with 150 μl lysis buffer.

2. Flies are placed in microcentrifuge tubes and homogenized on ice with a Teflon pestle (60 up and down strokes).

3. The homogenate is then sonicated (4×5 s in 5-s intervals, power settings 2) using a Heat Systems Ultrasonic, Model W185 sonicator equipped with a microtip. Following the sonication, the sample is centrifuged ($19,000 \times g$, 15 min at 4°C), and the clear supernatant is collected into a new microcentrifuge tube for protein concentration measurement (see Note 3).

4. Protein concentration is determined in duplicate using the BCA Protein Assay using a bovine serum albumin (BSA) standard curve.

5. Prior to boiling at 95° for 5 min, clear supernatant containing 90 μg of protein is mixed with an equal volume of 2× Laemmli buffer.

6. Following the boiling, samples are cooled down on ice. Samples are ready for SDS-polyacrylamide gel electrophoresis (PAGE) assay.

3.2. Western Blotting

1. Prepare a 1.0-mm thick 12% gel by mixing 7.96 ml water with 8.2 ml 30% acrylamide, 2.6 ml 3.0 M Tris–HCl, pH 8.8, 210 μl 10% SDS, 1.03 ml ammonium persulfate, and 20 μl TEMED.

2. Pour the gel, leaving space for a stacking gel, and overlay with water. The gel should polymerize in 30 min.

3. Pour off water, and remove the remaining water by using filter paper.

4. Prepare a stacking gel by mixing 3.66 ml water with 1.03 ml 30% acrylamide, 2.07 ml 3.0 M Tris–HCl, pH 8.8, 830 μl 10% SDS, 1.03 ml ammonium persulfate, and 6 μl TEMED.

5. Pour the stacking gel and insert a comb. The stacking gel should polymerize in 30 min.

6. Once the stacking gel polymerizes, carefully remove the comb and use running buffer to wash the wells.

7. Add 400 ml running buffer to the chamber of the gel unit.

8. Load 90 µg of protein per sample for each well. Load 10 µl prestained molecular weight markers to the rightmost well.

9. Assemble the gel unit and connect to a power supply. The gel can run at 180 V for 1 h at room temperature.

10. While the gel is running, soak the PVDF membrane in methanol for 20 s. Rinse the PVDF membrane with water and then soak it in the transfer buffer.

11. After samples are separated by the 12% gel, disconnect the gel unit from the power supply and then disassemble the gel unit.

12. The stacking gel is removed and discarded from the separating gel. Put a wet, thick blotting paper on three wet sponge pads, and then the separating gel is laid on the wet, thick blotting paper.

13. Put the PVDF membrane on the gel and de-bubble it by rolling a 10-ml plastic pipette over it. Cover the membrane with wet, thick blotting paper. Another two wet sponge pads are laid on the filter paper and the transfer cassette closed.

14. The cassette is placed in the transfer chamber. Cover the transfer chamber with a lid. Transfer can be accomplished at 110 mA for 2 h in the cold room.

15. Once the transfer is completed, the cassette is removed from the transfer chamber. The PVDF membrane is taken out and incubated with 10 ml blocking buffer at 37°C for 30 min on a rocking platform.

16. The blocking buffer is discarded prior to addition of the antibody. We use an antibody against the core $\beta 5$ subunit, which allows detection of all three forms of the proteasome. The anti-$\beta 5$ antibody is used in a 1:4,000 dilution (16) and the incubation carried out overnight in the cold room (see Note 6).

17. After incubation with anti-$\beta 5$ antibody, the antibody is then removed and the membrane is washed three times for 10 min each with washing buffer.

18. The secondary antibody (1:10,000 dilution, Anti-Rabbit) in blocking buffer is added to the membrane for 30 min at 37°C.

19. The secondary antibody is then discarded and the membrane is washed three times for 10 min with washing buffer (see Note 7).

20. Once the final wash is removed from the PVDF membrane, the black and white ECL reagents are mixed in a 1:1 ratio and

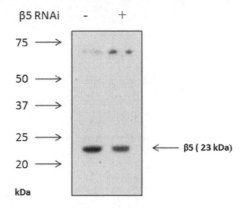

Fig. 1. 1–2 days male flies were prepared and analyzed. 90 μg of proteins were loaded per lane. β5 proteins were detected by immunoblotting. One group of transgenic flies contained an inducible RNAi construct against the *Drosophila* proteasome subunit β5 mRNA, resulting in a reduction of expression of the β5 subunit.

then added to the membrane. The membrane is rotated on a rocking platform for 5 min.

21. The membrane is removed from ECL reagents, wrapped in plastic, and exposed to film, for example (see Fig. 1).

3.3. Preparation of Samples for In-Gel Assay

The in-gel assay allows the detection of the two active forms [26S (2cap) and 26S (1cap)] of the 26S proteasome using as fluorogenic substrate Suc-LLVY-AMC.

1. Thirty male flies are prepared and harvested with 150 μl harvesting buffer.

2. Flies are placed in microcentrifuge tubes and homogenized on ice with a Teflon pestle (60 up and down strokes).

3. Following the centrifugation ($19,000 \times g$, 15 min at 4°C), cleared supernatant is collected to new microcentrifuge tubes for protein concentration measurement (see Note 3).

4. Protein concentration is determined in duplicates using the Bradford assay using a BSA standard curve.

5. 0.5% bromophenol blue is added into sample before loading.

3.4. In-Gel Assay (Nondenaturing Gel)

1. Pour the 5% gel for 2-cm height of gel cassette and overlay with water. The gel should polymerize in 10 min.

2. Pour off water, and remove residual water by using filter paper.

3. Pour the 4% gel for 3-cm height of gel cassette and overlay with water. The gel should polymerize in 10 min. Pour off water, and remove residue water by using filter paper.

4. Pour the 3% gel to cover the remaining space of the cassette and insert a comb. The gel should polymerize in 10 min.

5. Once the stacking gel polymerizes, carefully remove the comb and use running buffer to wash the wells.

6. Add 400 ml running buffer to the chambers of the gel unit.

7. Load 90 µg of protein per sample for each well (see Note 4).

8. Assemble the gel unit and connect to a power supply. The gel is run at 150 V for 3 h in the cold room.

9. After samples are separated by three-step gradient gel, disconnect the gel unit from the power supply and then disassemble the gel unit.

10. Gel is dislodged and then incubated with Proteasome Activity Detection Buffer in the chamber of UV transilluminator (Foto/Phoresis, Fotodyne) at room temperature.

11. Proteasome bands are visualized by exposure to UV light (360 nm) via a UV transilluminator and photographed with a digital camera, equipped with a green filter. (We use a Nikon Cool Pix 8700 camera with a 3-4219 fluorescent green filter.) Pictures are taken in short intervals as the fluorescence develops.

12. Proteins present in the nondenatured gel can be identified by transferring the gel to the PVDF membrane as follows.

13. After proteasome activity detection, the gel is soaked and washed with transfer buffer.

14. The protocols of protein transfer to the membrane and film development are described as in Subheading 3.2.

15. The primary antibody (anti-β5 antibody, 1:4,000 dilution) and the secondary antibody (anti-rabbit, 1:10,000 dilution) are used to assess proteasome level. An example of the results is shown in Fig. 2.

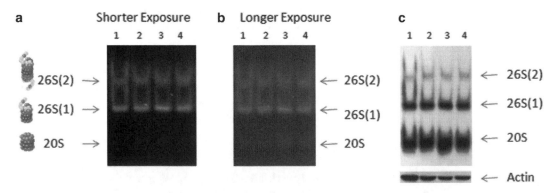

Fig. 2. (a) 1–2-day-old male flies were prepared and homogenized. Cleared fly lysates were subjected to nondenaturing gel electrophoresis. 90 µg of proteins were loaded per lane. The chymotrypsin-like activity was assessed with Suc-LLVY-AMC by the in-gel assay using shorter (a) and longer (b) exposure. (c) 26S and 20S proteasomes in fly extracts were detected by immunoblotting with anti-dß5 antibody. Total protein pattern was established by actin immunoblotting of denatured gels. 20S: core particle of proteasome. 26S(1): 20S with one 19S regulatory particle. 26S(2): 20S with two 19S regulatory particles.

3.5. SDS-PAGE for Actin

Actin levels in the cell are considered to be invariable. This allows the use of actin as an internal standard for comparison of different protein samples. A reference protein pattern is established by actin immunoblotting.

1. Cleared supernatant, containing 90 μg of protein from fly homogenates is mixed with an equal volume of buffer containing 2% SDS and 0.1 M Tris–EDTA, pH 7.5.

2. Prior to heating samples at 95°C, 12.5× loading buffer is added for a final concentration of 1×. After heating, samples are cooled down on ice and loaded on 12% SDS-PAGE.

3. The SDS-PAGE is ran at 150 V for 1 h at room temperature and then transferred into PVDF membrane at 110 mA for 2 h in the cold room.

4. The protocols of protein transfer to the membrane and film development are described as in Subheading 3.2.

5. The primary antibody (anti-Actin antibody, 1:5,000 dilution) and the secondary antibody (anti-rabbit, 1:10,000 dilution) are used to assess actin protein level. An example of the results is shown in Fig. 2.

3.6. Dissection of Third Instar Brain Lobes

Drosophila is a well-established model system for studying neurodegenerative diseases. Its nervous system has a well-characterized small network of dopaminergic neurons which can be visualized in their entirety using whole-mount preparations. The dopaminergic network has been used for the study of Parkinson's disease (3). The method described here allows the study of dopaminergic neurons at two developmental stages: third instar larvae and adult flies.

1. Third instar larvae are collected from fly bottle with a small #2 paintbrush and transferred to a 12-well spot plate containing 1× PBS (see Note 8).

2. Using Dumont #5 forceps, each larva is pulled apart at its middle (lengthwise).

3. The anterior segment is then gently turned inside out by passing the mouth hooks through the hollow insides of the dissected larva. This step exposes the central nervous system (CNS) to the solutions used in subsequent steps (see Note 9).

4. The larval CNS is kept attached to the chitin for the duration of the procedure and removed during mounting.

5. Samples are then washed 1× in PBS for 5 min to remove any food particles before fixation (see Note 10).

6. Dissected larvae are fixed in 4% paraformaldehyde in 1× PBS for 20 min at room temperature in a 1.5-ml microcentrifuge tube on a rocker.

7. Samples are then subjected 3×15-min washes using PBS Triton X-100 before primary antibody incubation.

8. Incubate with the primary antibody anti-TH overnight at 4°C on a shaker. Wash samples $3 \times 15'$ with PBST after overnight incubation. Incubate with secondary antibody Cy3-anti-mouse for 2 h at room temperature on a shaker covered with aluminum foil to protect from light (see Notes 11 and 12).

9. After secondary antibody incubation, samples are washed 3×15 min with PBST.

10. After the final wash, remove all the PBST from the 1.5-ml microcentrifuge with a pipette and replace with 100 µl mounting medium.

11. Transfer several of the chitin/CNS specimens to a microscope slide with a micro pipette, making sure to cut the tip of the pipette beforehand.

12. Under the dissecting microscope, use a Dumont #5 forceps to gently separate the chitin from the CNS. Mount one or more dissected brain lobes on a slide, cover with a coverslip, and seal with clear nail polish (see Notes 13 and 14).

13. The slides can be viewed with an inverted fluorescent confocal microscope. Excitation at 543 nm generates the red fluorescence of Cy3 and allow for visualization of dopaminergic neurons in the third instar larval brain.

3.7. Dissection of Adult Brain

1. Fill the wells of dissection dishes with 200–400 µl of cold 1× PBS. Prepare multiple dishes and keep them on ice (see Notes 15–17).

2. Anesthetize flies with CO_2 and place into cold PBS.

3. Place dissection dish under a dissecting microscope (30–50× magnification). Orient flies ventral side up, hold flies with forceps at the thorax, and remove head by pulling gently on the proboscis.

4. Retain the head in cold PBS while disposing of the body.

5. Grip the proboscis firmly with one Dumont #5 forceps while using the other to carefully tear away the right and left eyes. The brain can then be separated from the proboscis quite easily.

6. Carefully remove the brain and place into a dissection dish containing cold PBS.

7. Dissect as many brains as you can for about 30 min before stopping to fix the samples in 4% paraformaldehyde for 20 min (see Note 18).

8. After fixation, the brains are washed three times for 15 min each with PBS plus 0.3% Triton X-100 (PBST).

9. After each wash, remove carefully the washing buffer with a micropipette.

10. The brains are then incubated overnight with the anti-TH antibody.

11. For primary antibody incubation, place brains into a well on dissection dish filled with 400 µl of PBST containing the diluted antibody.

12. The primary antibody anti-TH is diluted 1:1,000 in PBST.

13. This dissection dish is then covered with aluminum foil and kept at 4°C without rocking.

14. The following day, the brains are washed 3× for 15 min each at room temperature with PBST.

15. After each wash remove carefully the washing buffer with a micropipette.

16. Alternatively, the brains can be transferred into a 1.5-ml microcentrifuge tube and then washed with 1 ml of PBST.

17. The dissection dish or the 1.5-ml microcentrifuge tube is placed on a rocker for gentle agitation.

18. Incubate with the secondary antibody for 2 h on the rocker. Cover with aluminum foil to protect the secondary antibody from light.

19. The Cy3-conjugated anti-mouse antibody is diluted 1:100 in PBST.

20. After secondary antibody incubation, wash the samples 3× for 15 min each with PBST. Keep covered with aluminum foil to minimize exposure to light.

21. Samples are now ready for mounting.

22. One drop of mounting medium is placed on a microscope slide. Under a dissecting microscope, place a brain in the mounting medium anterior side up.

23. A coverslip is placed at a 45° angle to the microscope slide and slowly lowered over the brain. One can use the forceps to hold the coverslip as it is being lowered. To conserve time and reagents, multiple brains can be mounted on a single microscope slide.

24. The excess mounting medium is blotted away using Kim wipes and clear nail polish is used to seal the coverslip to the slide.

25. Keep the newly prepared slides protected from light until the nail polish is dried. Slides can then be stored at 4°C for up to 1 month.

26. The slides can be viewed with an inverted fluorescent microscope or under confocal microscope. Excitation at 543 nm generates the red fluorescence typical of Cy3 and allow for visualization of dopaminergic neurons in the adult brain (see Fig. 3).

3.8. ATP Assay

The activity of the 26S proteasome is ATP-dependent for its initial assembly from the 20S and 19S subcomplexes. Proteolytic degradation of proteins by the proteasome requires also large amounts of ATP. The availability of ATP might constitute a rate-limiting

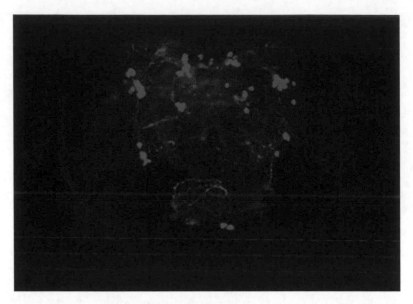

Fig. 3. Whole mount immunostain of the *Drosophila* brain showing dopaminergic neurons of 7–10-day-old fly. Primary antibody anti-tyrosine hydroxylase 1:1,000. Secondary antibody Cy3 anti-mouse 1:100. (×100-fold magnification).

step for the operation of the proteasome in light of the fact that cellular ATP levels decline significantly in the nervous system with advanced aging in parallel with the decline of proteasome activity.

1. Approximately 30 flies are homogenized on ice in 4% trichloroacetic acid (TCA) at a ratio of 1 fly per 5 μl of 4% TCA (see Notes 19–21).

2. Place homogenates on ice immediately. Do not centrifuge.

3. Into appropriately labeled tubes, add 4 μl 1 M Tris–HCL, pH 10.3, and 20 μl of the homogenate (see Note 22).

4. Cut the tip of the pipette before transferring the homogenate.

5. Vortex and place on ice for 10 min. Remaining samples are also kept on ice.

6. Add 216 μl of 10 mM Tris–HCL, pH 8.0, to the sample for an additional 1–10 dilution.

7. Into freshly labeled tubes, add 100 μl of the diluted sample and 100 μl of 10 mM Tris–HCl, pH 8.0, for a final 1–2 dilution.

8. The samples are now ready for the BCA protein assay. The final dilution factor for the protein assay is 24× (see Note 23).

9. Prepare a set of protein standards in 10 mM Tris–HCl, pH 8.0. Load standards onto a 96-well plate by pipetting 25 μl of each standard in duplicate into labeled wells on the plate.

10. Add 200 μl of BCA working reagent to each well and incubate at 37°C for 30 min.

11. To the samples prepared for the protein assay, add 400 μl of the BCA working reagent.

12. Incubate this mixture for 25 min at 37°C.

13. After incubation, vortex the microcentrifuge tubes and spin for 5 min at $4,500 \times g$.

14. Add 225 μl of each solution in duplicate to the 96-well plate with the protein standards.

15. Read plate at 562 nm using a 96-well plate reader.

16. ATP measurements are conducted with an ATP Determination Kit from Molecular Probes.

17. Spin down the remaining aliquot of the homogenate at 4°C and $19,000 \times g$.

18. Remove the supernatant and place into labeled microcentrifuge tubes.

19. Take 15 μl of this sample and add 15 μl of TCA and place into another set of labeled microcentrifuge tubes.

20. Into a third set of labeled microcentrifuge tubes, place 796 μl buffer A and 4 μl of the TCA diluted sample from step 19.

21. The sample is now ready for ATP measurement.

22. Prepare a set of ATP standards (0–5 mM) using 10 mM Tris–HCL, pH 8.0, as a diluent.

23. 10 μl of ATP standards and diluted samples are pipetted in duplicate onto a white cliniplate.

24. 100 μl of luciferase reagent, prepared according to manufacturer's instructions, is added to each well and the plate is read in a luminometer (see Notes 24 and 25).

25. Values obtained are in pmol units and are normalized to pmol/ug of protein (see Note 26).

4. Notes

1. Unless stated otherwise, all solutions should be prepared in water having at least a resistivity of 18 MΩ-cm.

2. For in-gel assay, because proteasome activity is ATP-dependent, both ATP and DTT stock should be prepared freshly.

3. For protein concentration measurement, the fly supernantant should be clear. If it is not, repeat the spin of the fly lysate for a longer time.

4. For the in-gel assay, to prevent occurrence of proteasome disassembly, we use freshly prepared fly samples.

5. Prior to preparation of proteasome activity detection buffer, buffer A should be warmed up to room temperature.

6. We have found that the anti-dβ5 antibody has a stronger signal if it is prepared in the super blocking buffer instead of 5% non-fat milk.

7. For reasons of economy, the primary and secondary antibodies can be saved for subsequent experiments by storing them at −20°C. These antibodies have been used for three to four times. Longer exposure of film to membrane may be required if the efficiency of antibodies declines.

8. Moisten a #2 paintbrush with 1× PBS for easy transfer of larvae from bottle to spot plate.

9. When turning the larvae inside out, keep the forceps closed to avoid tearing the sample.

10. Use multiple 1× PBS-filled wells on the spot plate. Place 8–12 larvae per well.

11. During the overnight incubation with the primary antibody, cover the shaker with aluminum foil to keep the microcentrifuge tubes in place.

12. Keep samples covered with aluminum foil for the secondary antibody incubation and for all subsequent steps.

13. When mounting the samples, angle the fiber-optic light downward and behind the slide for improved contrast.

14. After dissecting the CNS from the chitin, remove the debris from the slide before adding the coverslip.

15. Dissecting the adult fly brain may seem difficult at first. Initially, it may take more than 10 min to dissect one brain.

16. During this initial phase, many of the dissected samples might be damaged.

17. After several weeks of practice, it is possible to dissect more than five brains within 10 min. Over time, you also notice that there is less damage to the dissected sample.

18. Make sure that the #5 forceps are sharp and not bent in any way. If possible, maintain a second pair as a backup. A good pair of forceps decreases the difficulty of the dissection.

19. Groups of 30 adult male or female flies are anesthetized with carbon dioxide and placed into 1.5-ml microcentrifuge tubes.

20. Tubes are kept on ice before, during, and after homogenization.

21. Place the microcentrifuge tube into a small beaker filled with a mixture of ice and water. Hold beaker with one hand over the homogenizer and use the other hand to stabilize the microcentrifuge tube. Each sample is homogenized for 1 min at $2,000 \times g$.

22. Microcentrifuge tubes can be labeled before by hand to save time.

23. All solutions and standards needed for the BCA and ATP measurements should be prepared in advance.

24. When preparing the luciferase reagent, do not vortex as this may denature the luciferase enzyme. Instead, invert the master mix several times.

25. Protect the luciferase reagent from light at all times.

26. To save time, it is possible to set up and run the ATP reaction during the 30-min incubation required for the BCA assay.

Acknowledgments

This research was supported by NIH [NIGMS 1SC3GM086323 and CTSC GRANT #UL1-RR024996] to T.S-G] and RR03037 to Hunter College and PSC-CUNY 61853–00 39 to T.S.-G and the Graduate Center of the City University of New York.

References

1. Mattson, M.P. (2004) Pathways towards and away from Alzheimer's disease. Nature, **430**, 631–639.

2. Selkoe, D.J. (2001) Alzheimer's disease: genes, proteins, and therapy. Physiological reviews, **81**, 741–766.

3. Pienaar, I.S., Gotz, J. and Feany, M.B. (2010) Parkinson's disease: Insights from non-traditional model organisms. Progress in neurobiology, **92**, 558–571.

4. Bauer, P.O. and Nukina, N. (2009) The pathogenic mechanisms of polyglutamine diseases and current therapeutic strategies. Journal of neurochemistry, **110**, 1737–1765.

5. Perry, J.J., Shin, D.S. and Tainer, J.A. (2010) Amyotrophic lateral sclerosis. Advances in experimental medicine and biology, **685**, 9–20.

6. Bertram, L. and Tanzi, R.E. (2008) Thirty years of Alzheimer's disease genetics: the implications of systematic meta-analyses. Nat Rev Neurosci, **9**, 768–778.

7. Vives-Bauza, C., de Vries, R.L., Tocilescu, M. and Przedborski, S. (2010) PINK1/Parkin direct mitochondria to autophagy. Autophagy, **6**, 315–316.

8. Ziviani, E., Tao, R.N. and Whitworth, A.J. (2010) Drosophila parkin requires PINK1 for mitochondrial translocation and ubiquitinates mitofusin. Proc Natl Acad Sci USA, **107**, 5018–5023.

9. Partridge, L. (2009) Some highlights of research on aging with invertebrates, 2009. Aging Cell, **8**, 509–513.

10. Bilen, J. and Bonini, N.M. (2005) Drosophila as a model for human neurodegenerative disease. Annual review of genetics, **39**, 153–171.

11. Lessing, D. and Bonini, N.M. (2009) Maintaining the brain: insight into human neurodegeneration from drosophila melanogaster mutants. Nature reviews, **10**, 359–370.

12. van Ham, T.J., Breitling, R., Swertz, M.A. and Nollen, E.A. (2009) Neurodegenerative diseases: Lessons from genome-wide screens in small model organisms. EMBO molecular medicine, **1**, 360–370.

13. Lu, B. (2009) Recent advances in using drosophila to model neurodegenerative diseases. Apoptosis, **14**, 1008–1020.

14. Bachmann, A. and Knust, E. (2008) The use of P-element transposons to generate transgenic flies. Methods Mol Biol, **420**, 61–77.

15. Elsasser, S., Schmidt, M. and Finley, D. (2005) Characterization of the proteasome using native gel electrophoresis. Methods in enzymology, **398**, 353–363.

16. Vernace, V.A., Arnaud, L., Schmidt-Glenewinkel, T. and Figueiredo-Pereira, M.E. (2007) Aging perturbs 26S proteasome assembly in drosophila melanogaster. FASEB J, **21**, 2672–2682.

Chapter 11

Modeling Neurodegenerative Diseases in Zebrafish Embryos

Angela S. Laird and Wim Robberecht

Abstract

Although the zebrafish (*Danio rerio*) has been used extensively for many years in neurodevelopmental studies, use of this teleost to study neurological diseases has evolved only recently. Being a vertebrate, this animal offers advantages for the study of human disease over other small animals, such as the fly or worm. Genetic, as well as nongenetic, disorders can be modeled in both the adult organism and the embryo. Genetic manipulation of the embryo to generate stable and transiently expressing transgenic fish, and to knockdown genes to study loss of their function, can be easily achieved. Because of large offspring numbers screening studies can also be readily performed. Here, we describe some of the protocols useful for modeling neurodegenerative disease in zebrafish embryos, with particular emphasis on models to study motor neuron phenotypes.

Key words: Zebrafish, Motor neuron, Amyotrophic lateral sclerosis, Neurodegeneration

1. Introduction

Use of zebrafish to study human diseases has recently become more common due to the advantages that this model offers over other in vivo experimental models. As a vertebrate, *Danio rerio* offers advantages over other small animals, such as *Drosophila* and *Caenorhabditis elegans*. Logistically, zebrafish are small (4 mm long at 7 days post fertilization, dpf, and up to 40 mm as adults), allowing large numbers to be kept in a small space. A female zebrafish can spawn up to 250 eggs per week, allowing large sample sizes to be obtained in a short time frame, a perfect credential for screening studies.

There are extensive similarities between the zebrafish and human genomes. The zebrafish genome is, however, about half the size of the human genome, and therefore simpler to work with.

Giovanni Manfredi and Hibiki Kawamata (eds.), *Neurodegeneration: Methods and Protocols*,
Methods in Molecular Biology, vol. 793, DOI 10.1007/978-1-61779-328-8_11, © Springer Science+Business Media, LLC 2011

Zebrafish are particularly useful for modeling disease-related genes because they develop ex-utero allowing injection of genetic material directly into the recently fertilized embryo. The effect of knockdown of zebrafish orthologues can be examined through injection of antisense oligonucleotides (morpholinos) that prevent mRNA translation by binding to the start codon or 5′UTR region of the gene target or by preventing correct pre-mRNA splicing (1). Morpholino injection produces effective protein knockdown for up to 5 dpf, but this knockdown is usually diminished by 2–3 dpf. Transient overexpression of one or more genes can be achieved by injecting the coding mRNA into the fertilized egg. Expression of this mRNA can be detected up to 4 dpf. Furthermore, molecular tools have been developed to generate fish that stably overexpress wild-type and mutant human and fish genes. Nongenetic diseases can also be modeled by adding compounds to the water that the fish live in, an exposure that can be easily quantified.

Zebrafish are particularly suited to the study of neurological diseases. Their nervous system is well-developed and contains a telencephalon, diencephalon, mesencephalon, pons and rhombencephalon, cerebellum, spinal cord containing ascending and descending tracts, cranial nerves, plus motor and sensory spinal nerves. The zebrafish spinal motor system is particularly well-characterized, making the fish an excellent tool to study motor neuron disorders, such as spinal muscular atrophy (SMA) and amyotrophic lateral sclerosis (ALS).

Many of the techniques applied to characterizing phenotypes that develop in traditional in vivo models of neurodegeneration can also be adapted for use in zebrafish. Traditional protocols can be used to study the morphological hallmarks of neurodegeneration, such as immunofluorescence on fixed sectioned samples. However, because the zebrafish larvae remain transparent until around 30 hpf, immunofluorescence techniques can also be used on whole mount specimens and stable transgenic lines expressing proteins with fluorophore tags allow in vivo anatomical observation. Acridine orange staining of living zebrafish embryos has also been used to examine and quantify neuronal cell death (2, 3). Traditional western blot techniques can be used to examine the level of an endogenously expressed, knocked down, or overexpressed protein within the zebrafish. Identification of any post-translational modifications relevant to disease signatures, such as phosphorylation, can also be made. Filter retention assays can be used to identify protein species that result in protein aggregates in vivo (2, 4). Sophisticated electrophysiological evaluations, both of the central and peripheral nervous system, can be performed (5) and tools have been developed to quantitatively assess even subtle behavioral phenotypes in embryos, larvae, and adult fish (6, 7).

Using these techniques to study neurodegeneration, zebrafish models of ALS, SMA, Parkinson's disease, frontotemporal lobe degeneration (FTLD), Huntington's disease, and hereditary spastic paraplegia have been developed in adult fish and in embryos as early as 30 hpf (8–13).

This chapter describes many of the protocols necessary for the production and characterization of zebrafish models of neurodegenerative diseases, with particular reference to zebrafish embryos whose small size and large numbers often prove to be attractive. Because many of the characterization protocols are adaptations of traditional techniques used in higher models, we have aimed to highlight and expand on any steps specific to zebrafish models.

2. Materials

2.1. In Vitro Transcription of mRNA with Disease-Causing Mutations

1. cDNA template of your gene of interest downstream of the T7, T3, or SP6 polymerase promoter, store 1 µg/µl stock plasmid at –20°C. For the mutagenesis reaction, prepare an 80 ng/µl dilution of the stock plasmid.

2. Mutagensis primers (see Note 1): Primers can be stored as 1 µg/µl stock solution in ddH$_2$O as well as 100 ng/µl working solution at –20°C.

3. QuikChange Lightning Site-Directed Mutagenesis Kit; store at –20°C.

4. Chemi-compotent cells (e.g., TOP10 Chemi-compotent *Escherichia coli* cells, Invitrogen); store at –80°C.

5. LB medium: 1% bacto tryptone, 0.5% yeast extract, 1% NaCl made in H$_2$O (pH 7.5); autoclave to sterilize and store at 4°C.

6. LB agar plates: Prepare LB medium with 1.5% agar (Bacto agar), autoclave. When medium is below 65°C, e.g., when the bottle is cool enough to be handled, add appropriate antibiotic, e.g., 1 ml of ampicillin (100 mg/ml) or kanamycin (50 mg/ml).

7. RNase-free restriction enzyme and appropriate buffer, e.g., Pvull and HindIII; store at –20°C.

8. Nuclease-free water.

9. 0.5 M ethylenediaminetetraacetic acid (EDTA): Make up in nuclease-free water and store nuclease-free at room temperature.

10. 3 M sodium acetate: Make up in nuclease-free water and store nuclease-free at room temperature.

11. 100% ethanol (store nuclease-free at room temperature).

12. mMESSAGE mMACHINE™ Transcription Kit (T7, T3, or SP6, Ambion); store at –20°C.

13. MEGAclear™ RNA purification kit (Ambion); store at 4°C.

2.2. Zebrafish Embryo Injection

1. Zebrafish embryos (AB strain): Embryos are collected from setups of 4 males with 7–8 females (housed on a 14-h light: 10-h dark cycle, 28.5°C) by adding the males to the tank 1–2 h before commencement of the dark cycle. The fish should spawn on commencement of the light cycle in the morning, at which time the embryos can be collected and the male and female adult zebrafish separated.

2. mRNA of your gene of interest (see Subheading 2.1) or morpholino for knockdown experiments: Morpholinos can be obtained from Gene Tools, LLC. Prepare a 2 mM stock solution through addition of nuclease-free H_2O, heat at 65°C for 5 min (or until fully reconstituted), aliquot in 1–5 μl volumes, and store at –80°C. On the day of injection, remove the aliquot from the –80°C freezer and heat to 65°C for 1 min (see Note 2).

3. 0.5% w/v phenol red, prepared as in nuclease-free PBS.

4. FemtoJet programmable microinjector, including foot pedal, Grip head 0 (for capillaries with an outer diameter of 1.0–1.1 mm), and positioning aid (e.g., Eppendorf).

5. Stereo microscope for visualizing the embryos during injection (e.g., Zeiss Stemi 2000-C, with zoom of 0.63-5X).

6. Microinjection pipettes: 1 mm glass capillary tubes pulled with a micropipette puller.

7. E3 medium: 5 mM NaCl, 0.17 mM KCl, 0.33 mM $CaCl_2$, 0.33 mM $MgSO_4$, and 0.1% methylene blue to prevent fungal growth.

8. Agar injection channel dishes: Boil 25 ml of agar solution (1% ultrapure agarose, w/v in E3 medium), pour into a 10-cm petri dish, and leave to set. Boil another 25 ml of solution, pour a small amount on top of already set agar, place plastic channel mold (6-well zebrafish injection molds, I-An Manufacturing) on top of this layer of agar, and pour the remaining solution around the edges of the mold. Allow dish to set at 4°C. Store at 4°C until use and then use for 3 weeks or until damaged.

9. 1–10 μl pipette and gel-loading tips (0.5–10 μl capacity).

10. Micrometer for droplet calibration (1 mm scale containing 0.01 mm divisions).

11. Mineral oil.

12. Glass pipette for embryo transfer.

13. Rust-free Dumont #5 watchmaker forceps.

2.3. Treating Embryos with Small Compounds

1. Dimethyl sulfoxide (DMSO).

2.4. Protein Expression Quantification

1. Deyolking buffer: 25 ml ddH_2O, 0.3 mM phenylmethylsulfonyl fluoride (PMSF, solubilized in isopropanol), 1 mM EDTA; store at 4°C.

2. Lysis buffer: T-PER Tissue Protein Extraction Reagent (Thermo Scientific) with one tablet of Complete Protease Inhibitor Cocktail tablet (Roche) per 25 ml. Store frozen in 500 μl aliquots. Keep defrosted aliquots at 4°C for 1 week. Add PMSF (solubilized in isopropanol) for a final concentration of 1 mM immediately before use.

3. Glass pipette for deyolking: Heat the small end of a glass Pasteur pipette with the flame of a Bunsen burner to melt the glass to a smaller diameter.

4. Protein concentration assay kit.

5. Hand homogenization pestle for 1.5 ml Eppendorf tubes.

6. 5× SDS sample buffer: 250 mM Tris–HCl, pH 6.8, 10% beta mercaptoethanol, 10% sodium dodecyl sulfate (SDS), 50% glycerol, 0.02% bromophenol blue.

2.5. Filter Retention Assay for Aggregates

1. BioRad Bio-Dot Microfiltration Apparatus.

2. Membrane (cellulose acetate or nitrocellulose) with a pore size of 0.2 μm unless your protein is >70 kDa when 0.45 μm might be preferential.

3. Assay buffer: 1% SDS in PBS.

2.6. Immunofluorescent Staining of Whole Mount Embryos

1. 10× PBS: 1.37 M NaCl, 27 mM KCl, 100 mM Na_2HPO_4, 18 mM KH_2PO_4, pH 7.4, dilution to 1× for use.

2. 4% (w/v) paraformaldehyde in PBS: Heat PBS to 65°C to dissolve paraformaldehyde powder, allow to cool, and then store 1 ml aliquots at –20°C (see Note 3).

3. 1% Triton X-100 in PBS.

4. Acetone.

5. Blocking solution: 1% bovine serum albumin (BSA) and 1% DMSO in PBS. Store at 4°C for 3 weeks.

6. Mouse anti-synaptic vesicle 2 (SV2, Developmental Studies Hybridoma Bank, University of Iowa, Iowa City, IO, USA).

7. Donkey anti-mouse Alexa Fluor 555 or other anti-mouse fluorescent antibody.

8. Glass slides.

9. Fluorescence microscope and camera.

10. Analysis software.

2.7. Immuno fluorescence on Sectioned Zebrafish Embryos

1. 20% sucrose: Prepared in PBS and stored at 4°C for 2 weeks.

2. Tissue-Tek O.C.T. Compound (Sakura Finetek) and liquid nitrogen for snap freezing.

3. Cryostat capable of cutting sections at 10 μm.

4. Gelatinized slides.

5. PBS with 0.1% Triton-X100 (PBST).

6. 50% methanol solution.

7. Hydrophobic PAP pen (e.g., Vector Laboratories).

8. Blocking solution: Prepare PBST with 5% normal horse serum and 5% BSA.

9. Primary antibody against your protein of interest and suitable fluorescent antibody.

10. Mounting medium with DAPI; store at 4°C in opaque box to prevent exposure to light.

2.8. Acridine Orange Staining

1. 1 μg/ml acridine orange in E3 buffer.

2. Tricaine (3-amino benzoic acid ester, also called ethyl *m*-aminobenzoate).

3. Ethanol stored at −20°C.

4. 96-well plate and Fluorometry Plate Reader.

3. Methods

3.1. In Vitro mRNA Transcription

To enable modeling of genes with disease-causing mutations, site-directed mutagenesis must first be performed using a polymerase chain reaction (PCR) with primers designed to insert the mutation (again, see Note 1).

1. Prepare the reaction samples as follows:

   1 μl (80 ng) of DNA template containing your gene of interest

   1.25 μl (125 ng) of forward primer #1

   1.25 μl (125 ng) of reverse primer #2

   5 μl of 10× reaction buffer

   1 μl of dNTP mix

   1.5 μl of QuikSolution reagent

   39 μl of ddH_2O

2. Add 1 μl of QuikChange Lightning Enzyme.

3. Run the PCR in thermal cycler with the following program:

Step 1	95°C	2 min
Step 2	95°C	20 s
Step 3	60°C	10 s
Step 4	68°C	30 s/kb of template length
Repeat Step 2 18 times		
Step 5	68°C	5 min
Step 6	4°C	Until stop

4. Following the PCR reaction, add 2 μl of the supplied DpnI enzyme to each reaction vial, mix well, and incubate at 37°C for 5 min to digest any nonmutated template DNA.

5. Thaw one vial of chemically competent *E. coli* cells per mutated DNA species. It is important to thaw the cells slowly through constant contact of the vial with ice.

6. Transform 2 μl of the mutagenesis product in 100 μl of chemi-competent *E. coli* cells and grow on LB agar plates containing suitable antibiotic (e.g., ampicillin or kanamycin depending on the antibiotic-resistance gene contained within the expression vector).

7. Collect plates from the incubator the next morning, avoiding delay. Use a sterile pipette tip to transfer an isolated bacterial colony from the plate into a flask of LB broth. Place the flask on a shaker at 37°C for vigorous shaking overnight.

8. Isolate plasmid DNA from the bacterial broth using a filter column, e.g., Invitrogen MidiPrep or Qiagen MaxiPrep kit. Sequence the isolated DNA using a suitable sequencing primer, identified from the vector map of the vector containing your gene of interest, e.g., M13F, T7, T3. On confirmation of the correct clone, continue to step 9.

9. Prior to in vitro mRNA transcription, the DNA template needs to be linearized. Prepare the following in a 1.5 ml Eppendorf tube (see Note 4):

   5 μg template DNA

   7 μl restriction enzyme

   10 μl reaction buffer

   Make volume up to 100 μl with RNase-free H_2O.

   Distribute reaction volume over four Eppendorf tubes (i.e., 25 μl per tube).

   Incubate at 37°C for 2 h.

10. The linearized DNA is then pelleted by centrifugation in a microfuge ($33,000 \times g$) for 15 min at 4°C. Remove the supernatant, centrifuge again without any solution added, and remove further supernatant. Resuspend the pellet in 6 μl of RNase-free H_2O. To purify the linearized DNA, add the following to each vial:

    1.25 μl of 0.5 M EDTA

    2.5 μl 3 M sodium acetate

    50 μl 100% ethanol

    Vortex and incubate at −20°C for 1 h.

11. To perform in vitro mRNA transcription on the linearized, purified DNA, we routinely use Ambion mMESSAGE mMACHINE transcription kits (suitable for T7, T3, or SP6 promoters). Firstly, defrost the components of the appropriate mMESSAGE mMACHINE kit by placing the RNA polymerase enzyme mix and 2× NTP/CAP on ice and leaving the 10× reaction buffer at room temperature.

12. Assemble the reaction:

   1 µg linear template DNA in 6 µl nuclease-free water (all 6 µl from above step)

   10 µl 2× NTP/CAP

   2 µl 10× reaction buffer

   2 µl enzyme mix

13. Vortex and do short spin in centrifuge. Incubate at 37°C for 4 h.

14. Following the incubation, digest the DNA by adding 1 µl DNAse 1, mix well with vortex, and incubate at 37°C for 15 min.

15. Combine the contents of all four Eppendorf tube samples (a total volume of 80 µl), vortex well, and purify the mRNA through the use of an Ambion MEGA Clear RNA purification kit using filter columns to obtain superior mRNA yield and purification. Store the reagents and reaction tube on ice during the protocol. Store the purified mRNA at −80°C (see Note 5).

3.2. Zebrafish Embryo Injection

1. A typical injection setup is shown in Fig. 1a.

2. Prepare injection samples using either the mRNA of your gene of interest for overexpression experiments or morpholino for knockdown (see Notes 6 and 7).

   An example sample recipe for morpholino injection is as follows:

Morpholino (200 µM)	0.75 µl of 800 µM dilution
Phenol red	0.5 µl
H_2O	1.75 µl

   Spin down the samples in a microfuge, $16,800 \times g$ for 30 s.

3. Switch on the microinjector ensuring that the tubing is detached. Set the injection pressure to 10–20 psi and the balance pressure to 0.2–0.4 psi.

4. Take a pulled needle and break the end off using a pair of fine forceps (Dumont #5) (see Note 8).

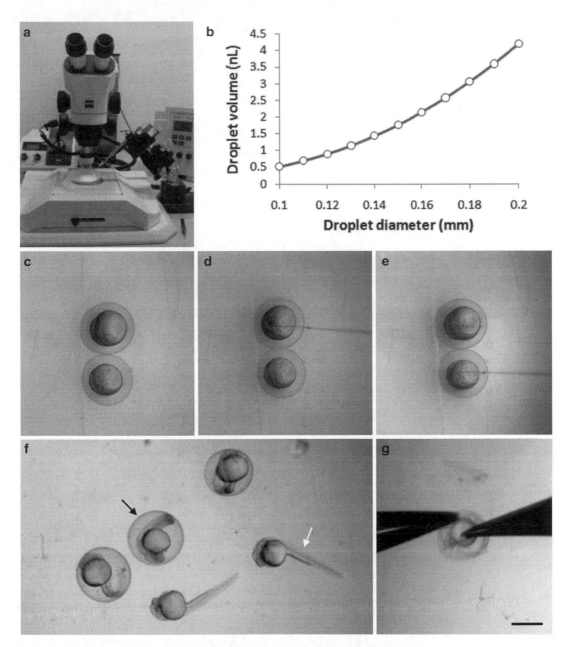

Fig. 1. Zebrafish embryo injection techniques. (a) Injection setup apparatus, including Femtojet injection system, manipulator, and fiber optic lamp; (b) measurement of the diameter of the injection droplet allows calculation of the injected volume through use of the equation for the volume of a sphere, the relationship is shown in graph form for easy reference; (c–e) injecting one-cell-stage embryos (c) in the yolk, (d) and cell (e); (f) 30 hpf embryos, both within their chorion (*black arrow*) and external to the chorion, following dechorionation procedure (*white arrow*); (g) technique of dechorionation using two forceps and a pinching movement. Scale bar represents 500 μm.

5. Backfill mRNA/morpholino/phenol red solution into the injection needle using a gel-loading tip (see Note 9).

6. Insert the needle into the micromanipulator and tighten securely, but not enough to break the needle.

7. Add a drop of mineral oil to the scale bar on the micronometer and focus the microscope onto the scale bar using the 5× magnification.

8. Press "Clean" on the microinjector, a large droplet is made, deposit this droplet to the side of the scale bar so as not to obscure the view of the scale bar.

9. Use the micromanipulator to position the end of the needle in the middle of the scale bar; press "INJECT" or use a foot pedal to trigger an injection. Adjust the "Pressure" value on the microinjector to produce the desired injection volume depending on the application (see Note 10). A graph showing the effect of droplet diameter on the droplet volume is shown in Fig. 1b.

10. Arrange the fertilized embryos to be injected into a line on the injection agar dish. Shift the magnification to 2× and focus on the embryos (see Fig. 1c).

11. Using the micromanipulator, insert the needle into the yolk of the embryo and press the foot pedal to trigger injection (see Fig. 1d). This injection technique into the yolk relies on cytoplasmic streaming to transport the injection bolus into the cell. This technique works for mRNA injections up to eight-cell stage, at which point a membrane forms between the yolk and the cell, preventing the bolus from reaching the cell. When injecting morpholino, it is best to inject one-cell-stage embryos because DNA expression tends to result in a more mosaic expression pattern than mRNA. Injections into the cell itself (for injection of DNA, particularly for germ-line transmission) can be made by inserting the needle through the yolk into the cell, as shown in Fig. 1e.

12. Remove the needle from the egg using the manipulator in a reverse motion. Shift the dish manually with your hand to position the next embryo to be injected.

13. After injecting the required number of embryos, remove any embryos that do not contain a phenol red marker, where the phenol red has turned yellow or where the mark seems to be of incorrect size. Also remove any embryos that had developed too far prior to injection.

14. Transfer the injected embryos to a 10 ml petri dish by washing the injection dish with E3 solution. Incubate the embryos at 27–28.5°C.

15. After the desired period of development, e.g., 24–30 hpf (see Fig. 1f), the sac surrounding the developing embryo, the chorion, can be manually removed using forceps in each hand and a pinching technique (see Fig. 1g).

3.3. Treating Embryos with Small Compounds

Exposing zebrafish embryos to chemicals to induce toxicity or for treatment is made simple by the fact that zebrafish can absorb chemicals applied to their fish water (often even through the chorion layer), rather than requiring injections or oral dosing. Compounds can often be dissolved in DMSO and then further diluted in the E3 solution that the embryos are incubated in. Fortunately, embryos can tolerate DMSO levels of up to 0.01% in the fish water without detectable toxicity (14), but higher concentrations should be avoided. Compounds can be added to the water immediately after fertilization or at later time points to reduce toxicity. If larger, less transmissible compounds that may be deterred by the chorion layer are to be used, the chorion can be punctured from 5 hpf or removed completely from 10 hpf to increase penetration (see Note 11).

3.4. Processing of Embryos for Protein Expression Quantification

1. Dechorionate embryos at the required time point post fertilization, e.g., 24 hpf.

2. Transfer one group of embryos (use 15–75 embryos, depending on the expected level of protein expression) to a petri dish containing 10 ml of deyolking buffer. Suck up and expel the embryos in this solution using a glass deyolking pipette, and repeat twice.

3. Use another thin glass pipette to transfer the deyolked embryos to Eppendorf tube, and remove excess supernatant.

4. Add 1 μl of lysis buffer per embryo and pipette the solution up and down to disperse the embryos completely. Leave embryos on ice to lyse for 10 min.

5. Using a hand homogenizer, grind the embryos against the wall of the Eppendorf tube to homogenize.

6. Centrifuge the sample in a microfuge at $10,750 \times g$ for 3 min, and then transfer the supernatant to a fresh Eppendorf tube.

7. Quantify the protein concentration in each sample by running a protein concentration assay to ensure equal loading of samples. Calculate the required volume of sample to produce desired amount of protein, e.g., 20–60 μg, depending on the prevalence of your protein of interest in the lysate. Add 5× SDS sample buffer, boil at 95°C for 5 min, and run on SDS-PAGE using traditional western blotting protocols.

3.5. Filter Retention Assay for Aggregates

1. Prepare zebrafish lysates in the same way as for western blot analysis (see above), including a series of sample dilutions with concentrations 10–50 μg.

2. Dilute samples to 50 μl with PBS, add a further 50 μl of 2% SDS/100 mM DTT, and boil samples at 95°C for 5 min.

3. Assemble the filtering apparatus covering any unused wells with tape or parafilm (see Fig. 2a).

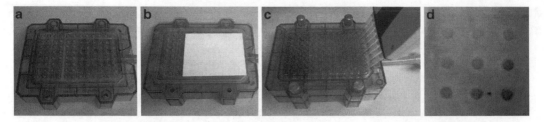

Fig. 2. Filter assay for aggregate retention. (**a**) Cover all unused wells of the filtration apparatus with tape or parafilm; (**b**) apply a prewetted membrane above the plastic liner and assemble the lid of the filtration apparatus; (**c**) apply buffer to each well using a multichannel pipette, filter, and then apply each sample in a similar manner; (**d**) an example membrane processed through traditional HRP detection methods is shown.

4. Cut membrane to required size, and wet the membrane in the assay buffer for 10 min. Let the buffer solution drain onto filter paper, and assemble the membrane into the filtering apparatus (see Fig. 2b).

5. Apply 200 μl assay buffer to all sample wells to be used (Fig. 2c). Set to dual air/vacuum setting and gently filter the buffer. As soon as the assay buffer drains from all the wells, adjust the flow valve to air; otherwise, halo marks develop on your membrane around the wells.

6. Vortex samples well and pipette samples into appropriate wells, marking down the order of loading. Remove air bubbles with a small pipette tip. Set the apparatus to dual air/vacuum setting to filter the sample and return to air setting once the sample has drained.

7. Apply 200 μl of assay buffer to each well and filter; repeat again.

8. Dismantle apparatus and remove membrane, marking the upper right corner with a notch to help orientate samples. Wash membrane in assay buffer for 10 min.

9. Continue as if doing a western blot by blocking in 10% milk for 1 h and then incubating in primary and secondary antibodies. An example filter assay result is shown in Fig. 2d.

3.6. Immuno-fluorescent Staining of Embryos for Motor Axon Outgrowth Measurement

1. Following dechorionation of the embryos, transfer the embryos to an Eppendorf tube, remove all excess fluid, and add 4% paraformaldehyde. Incubate overnight (or >6 h) at 4°C.

2. Remove the paraformaldehyde and add 150 μl of PBST solution until you are ready to stain the embryos; store at 4°C (see Note 12).

3. Replace the PBST solution on the embryos with 150 μl acetone and incubate at −20°C for 1 h (see Note 13).

4. Remove the acetone from each vial and add 150 μl BSA blocking solution and incubate at room temperature for 1 h.

5. Replace the blocking solution with primary antibody (e.g., 1/200 anti-SV2 in PBS containing 1% DMSO) to stain primary motor neurons. Incubate at room temperature for 5 h.

6. Wash the embryos 3×20 min with BSA blocking solution incubations (you can store overnight at 4°C at this point).

7. Incubate in the secondary antibody (e.g., 1/500 AlexaCy55 anti-mouse in PBS containing 1% DMSO) for 2 h at room temperature (see Note 14).

8. Give the embryos five PBST washes to reduce nonspecific background staining.

9. Store embryos protected from light at 4°C until you are ready to image or analyze them.

10. Axonal length can be measured using a quantification program through placement of an anchor mark at the site of axon exit from the spinal cord and another at the end point of the axon (see Fig. 3). The length of the axon can then be calculated by

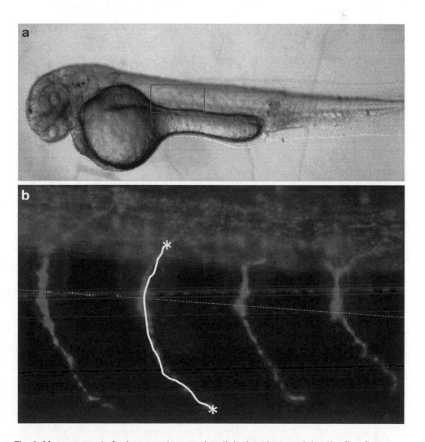

Fig. 3. Measurement of primary motor axon length is done by examining the first five axons after the yolk sac (box, **a**) in embryos stained with antisynaptic vesicle 2 (**b**, enlargement of *box* in **a**). Length measurement is made by tracing from the envisioned exit point (*asterisk*) to the end point of the axon (*asterisk*), as shown by a solid white line. A *dotted white line* indicates the choice point zone; axons branching proximal to this line are considered affected by aberrant branching.

a program via spatial mathematics. The length of multiple axons per embryo (e.g., five axons) can be measured and an average for each embryo calculated (see Notes 15 and 16).

11. In our studies, an embryo is scored as being affected by aberrant branching if more than two of its axons display branching at a site proximal to the choice point zone (see Fig. 3).

3.7. Sectioning Embryos for Staining

1. Transfer embryos fixed with 4% paraformaldehyde to a 20% sucrose solution and incubate at 4°C until they settle to the bottom of the Eppendorf tube (usually 24 h).

2. Apply a line of Tissue Tek O.C.T. medium to a foil tray and transfer embryos to this solution, arrange in a longitudinal manner, and submerge in liquid nitrogen to snap freeze.

3. Apply the frozen specimen to the cryostat chuck and freeze in place.

4. Cut sections at 10 μm and mount on glass slides; allow to dry.

5. Incubate slides in a 50% methanol solution for 3 min.

6. Draw around the edges of the slides with a hydrophobic PAP pen, working quickly to prevent the slide drying out.

7. Incubate slides in blocking solution for 30 min.

8. Drain the blocking solution off the slides, apply antibody solution (e.g., 1/250 primary antibody solution in PBST), and incubate at 4°C overnight.

9. Wash slides in PBST for 10 min, three times.

10. Incubate slides in secondary antibody solution (e.g., 1:500 Alexa Fluor fluorescent antibody raised against the species that the primary was made in) for 1 h at room temperature.

11. Give the slides five good washes in PBST, standing up in slide rack.

12. Mount a coverslip on the slides using mounting medium containing DAPI and store at 4°C until ready to examine (see Fig. 4 for an example of cross-sectional anatomy of the spinal cord of a 30 hpf embryo).

3.8. Detection of Cell Death in Zebrafish Embryos Using Acridine Orange Staining

Neuronal cell death can be visualized in live zebrafish embryos by incubation in an acridine orange solution which stains nucleic acids in dying cells.

1. Dechoriate embryos at appropriate time point (e.g., 24–30 hpf) and transfer to a dish containing a prewarmed solution of acridine orange. Incubate at 28°C for 30 min.

2. Transfer embryos to a clean petri dish containing E3 solution, incubate for 10 min, and repeat transfer to new E3 solution four times to wash thoroughly. A mesh strainer can be helpful for this.

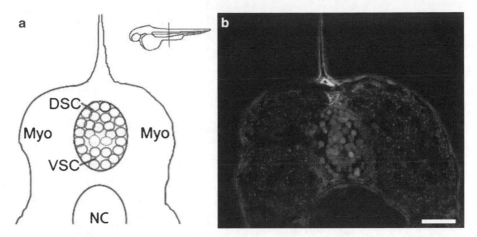

Fig. 4. (**a**) A diagrammatic representation of the neuroanatomy of the zebrafish embryo sectioned through the spinal cord mid body, as shown in the inset diagram. *DSC* Dorsal spinal cord, *VSC* Ventral spinal cord, *Myo* Myotomes, *NC* Notochord. (**b**) Immunofluorescence image of the neuroanatomy of a zebrafish embryo sectioned as shown in (**a**) showing DAPI staining of nuclei shown; the scale bar is 50 μm.

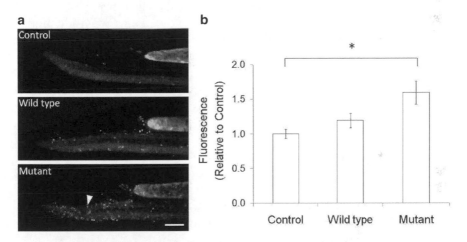

Fig. 5. (**a**) Acridine orange staining of apoptopic cells (*white arrowhead*) in control (noninjected), wild-type mRNA injected and mutant mRNA injected embryos. Scale bar is 10 μm. (**b**) Quantification of the amount of fluorescence of individual embryos within a 96-well plate, expressed relative to the mean fluorescence of the control group. The asterisks represents that the embryos expressing mutant protein had a higher fluorescence level than controls ($p < 0.05$).

3. Anesthetize embryos with tricaine anesthetic.

4. Observe and image embryos under a fluorescent microscope (see Fig. 5a for sample images).

Parng et al. have described an adapted assay protocol to allow quantification of acridine orange staining (15), as follows.

1. Distribute embryos into individual wells of a 96-well plate; drug or compound treatment may be done in these plates also.

2. Stain the embryos with acridine orange at 24 hpf, as described above, however keeping the embryos separated in the 96-well plate.

3. Following the E3 solution washes, drain all solution from each well using a pipette. Add 50 µl of 100% ice-cold ethanol to each well immediately following removal of the E3 solution. Incubate at –20°C for 30 min.

4. Add 50 µl of distilled H_2O to each well and incubate at 25°C for 10 min.

5. Measure fluorescence at 490 nm (see Fig. 5b for sample quantification results).

3.9. Conclusions

The protocols described within this chapter are just a sample of what can be done using zebrafish to study neurodegeneration. Other techniques, such as production of stable transgenic lines, gamma irradiation, chemical mutagenesis, TILLING, zinc finger nucleases, and surgery can also be used to induce neurological models. Further, many more techniques than just those described here can be used to characterize phenotypes that develop in these models, including in situ hybridization, behavioral testing, electrophysiology, and pharmacological challenges.

In addition to the many advantages of modeling diseases in zebrafish described here, zebrafish research also holds inevitable caveats. The counter of the benefits of its simplicity is that the nervous system of zebrafish is more primitive than the human system and in embryos this simplification is even greater. The zebrafish genome is smaller than the human genome, but many human genes have multiple zebrafish orthologues due to duplication of the genome during evolution. Although research using zebrafish embryos is fast and can be performed in relatively large replicates, modeling progressive, age-dependent diseases in zebrafish embryos also has obvious limitations. Therefore, zebrafish models of neurodegeneration are best used in parallel with other complementary models, where the relevance of findings to humans can be confirmed.

4. Notes

1. Primers should be between 25 and 45 bases in length with the desired mutation in the middle. It is best that the primer melting temperature (Tm) is ≥78°C and the GC content ≤40%. The primer should terminate in a C or G base.

2. It is best to prepare morpholino under RNase-free conditions in case you want to coinject the morpholino with mRNA in the future.

3. Use precautions when using paraformaldehyde because it is a suspected carcinogen.

4. Ensure that the restriction enzyme that you choose for the linearization cuts downstream of any stabilization signal (e.g., polyA).

5. Make 5 μl aliquots of the mRNA to prevent frequent freeze/thawing because it can result in mRNA degradation and loss of activity.

6. It is best to prepare injection samples in a designated RNase-free area (frequently cleaned with 0.5% SDS and 3% H_2O_2) using gloves and lab coat, filtered pipette tips, and RNase-free reagents that are only used within this area.

7. When starting morpholino experiments, try injecting a range of concentrations, including 100, 200, and 400 μM to test for toxicity (e.g., embryo death and abnormal gross morphology).

8. Remember that too fine a needle bends rather than entering the embryo, but too thick a needle damages the embryo and leaks injection fluid.

9. Only pipette half of your sample volume, e.g., 1.5 μl, so that if your needle breaks or is cut at the wrong size you still have some sample to make a second attempt.

10. We inject 2–3 nl when injecting mRNA or morpholinos into the egg yolk and 1 nl when injecting transgenesis constructs (DNA/mRNA combination) into the cell.

11. When it is necessary to dechorionate embryos early in development, for example to allow penetration of a compound, you can line the bottom of the dish containing the embryos with parafilm to decrease damage/rupture of the embryos' yolk sac.

12. Do not delay staining too long as it decreases the quality of the embryos.

13. To pipette the acetone easily without droplets falling from the pipette tip as you transfer from the bottle to the vial, pipette up and down in the acetone solution first to saturate the air within the tip with acetone vapors.

14. Cover the vials in foil wrapping to prevent photobleaching during the secondary antibody incubation and for storage.

15. The length of primary motor axons decreases as you move caudally along the spinal cord, so inclusion of all axons in a calculation of average axonal length increases variation.

16. When measuring under the fluorescence microscope, the embryos dry out with time and this causes their anatomical features, e.g., motor axons, to shrink. To prevent this, you should add drops of PBS to the slide over time.

Acknowledgments

This work was supported by grants from the K.U.Leuven. WR is supported through the E von Behring Chair for Neuromuscular and Neurodegenerative Disorders and by the IUAP program P6/43 of the Belgian Federal Science Policy Office.

References

1. Bill, B. R., Petzold, A. M., Clark, K. J., Schimmenti, L. A., and Ekker, S. C. (2009) A primer for morpholino use in zebrafish. *Zebrafish*. **6**, 69–77.

2. Schiffer, N. W., Broadley, S. A., Hirschberger, T., Tavan, P., Kretzschmar, H. A., Giese, A., Haass, C., Hartl, F. U., and Schmid, B. (2007) Identification of anti-prion compounds as efficient inhibitors of polyglutamine protein aggregation in a zebrafish model. *J Biol Chem*. **282**, 9195–9203.

3. Paquet, D., Bhat, R., Sydow, A., Mandelkow, E. M., Berg, S., Hellberg, S., Falting, J., Distel, M., Koster, R. W., Schmid, B., and Haass, C. (2009) A zebrafish model of tauopathy allows in vivo imaging of neuronal cell death and drug evaluation. *J Clin Invest*. **119**, 1382–1395.

4. van Bebber, F., Paquet, D., Hruscha, A., Schmid, B., and Haass, C. Methylene blue fails to inhibit Tau and polyglutamine protein-dependent toxicity in zebrafish. *Neurobiol Dis*.

5. Saint-Amant, L., and Drapeau, P. (2003) Whole-cell patch-clamp recordings from identified spinal neurons in the zebrafish embryo. *Methods Cell Sci*. **25**, 59–64.

6. Creton, R. (2009) Automated analysis of behavior in zebrafish larvae. *Behav Brain Res*. **203**, 127–136.

7. Blaser, R., and Gerlai, R. (2006) Behavioral phenotyping in zebrafish: comparison of three behavioral quantification methods. *Behav Res Methods*. **38**, 456–469.

8. McWhorter, M. L., Monani, U. R., Burghes, A. H., and Beattie, C. E. (2003) Knockdown of the survival motor neuron (Smn) protein in zebrafish causes defects in motor axon outgrowth and pathfinding. *J Cell Biol*. **162**, 919–931.

9. Lemmens, R., Van Hoecke, A., Hersmus, N., Geelen, V., D'Hollander, I., Thijs, V., Van Den Bosch, L., Carmeliet, P., and Robberecht, W. (2007) Overexpression of mutant superoxide dismutase 1 causes a motor axonopathy in the zebrafish. *Hum Mol Genet*. **16**, 2359–2365.

10. Kabashi, E., Lin, L., Tradewell, M. L., Dion, P. A., Bercier, V., Bourgouin, P., Rochefort, D., Bel Hadj, S., Durham, H. D., Vande Velde, C., Rouleau, G. A., and Drapeau, P. Gain and loss of function of ALS-related mutations of TARDBP (TDP-43) cause motor deficits in vivo. *Hum Mol Genet*. **19**, 671–683.

11. Ramesh, T., Lyon, A. N., Pineda, R. H., Wang, C., Janssen, P. M., Canan, B. D., Burghes, A. H., and Beattie, C. E. A genetic model of amyotrophic lateral sclerosis in zebrafish displays phenotypic hallmarks of motoneuron disease. *Dis Model Mech*.

12. Flinn, L., Mortiboys, H., Volkmann, K., Koster, R. W., Ingham, P. W., and Bandmann, O. (2009) Complex I deficiency and dopaminergic neuronal cell loss in parkin-deficient zebrafish (Danio rerio). *Brain*. **132**, 1613–1623.

13. Bai, Q., Garver, J. A., Hukriede, N. A., and Burton, E. A. (2007) Generation of a transgenic zebrafish model of Tauopathy using a novel promoter element derived from the zebrafish eno2 gene. *Nucleic Acids Res*. **35**, 6501–6516.

14. Hallare, A., Nagel, K., Kohler, H. R., and Triebskorn, R. (2006) Comparative embryotoxicity and proteotoxicity of three carrier solvents to zebrafish (Danio rerio) embryos. *Ecotoxicol Environ Saf*. **63**, 378–388.

15. Parng, C., Ton, C., Lin, Y. X., Roy, N. M., and McGrath, P. (2006) A zebrafish assay for identifying neuroprotectants in vivo. *Neurotoxicol Teratol*. **28**, 509–516.

Chapter 12

Practical Considerations of Genetic Rodent Models for Neurodegenerative Diseases

Kindiya Geghman and Chenjian Li

Abstract

In recent years, the rapid advances in human genetics have identified many genes that are responsible for inherited diseases. In the effort to understand pathogenic pathways and develop therapeutics, researchers have developed genetic animal models as critical tools for the purpose of neurodegenerative disease study. In this overview chapter, we do not intend to cover all the transgenic models for particular diseases, their phenotypical analyses, and their specific usage in this research field. Instead, we outline methods of genetic manipulation and their usage in disease modeling, explain the underlying principles of these methods, and compare the advantages and pitfalls to these different transgenic methods.

Key words: Transgenic, Animal models, Neurodegenerative disease, Alzheimer's disease, Parkinson's disease, Amyotrophic lateral sclerosis, Huntington's disease, Bacterial artificial chromosome

1. Genetic Models of Inherited Human Diseases

The demand for developing genetic animal models has been closely linked to the revolution in human genetics. Many neurodegenerative diseases are either purely genetic or have genetic components. For example, Huntington's disease (HD) and spinocerebellar ataxia type 1 (SCA1) are 100% genetic, whereas Alzheimer's disease (AD), Parkinson's disease (PD) and amyotrophic lateral sclerosis (ALS) are genetically inherited in 10% of patients and sporadic in 90%. For purely genetic diseases, the need to establish and study genetic animal models is self-evident. However, it is also important to study the genetic forms of diseases that are mostly sporadic and partially genetic; this is because the genetic and sporadic forms of these diseases are often similar pathogenically.

Giovanni Manfredi and Hibiki Kawamata (eds.), *Neurodegeneration: Methods and Protocols*,
Methods in Molecular Biology, vol. 793, DOI 10.1007/978-1-61779-328-8_12, © Springer Science+Business Media, LLC 2011

For these reasons, the study of purely genetic or genetic forms of neurodegenerative diseases has been the focus of many researchers. Genetic animal models are clearly among the most important tools for research.

2. Three Types of Disease-Causing Mutations and Methods to Model Them

In human genetic diseases, there are two common modes of transmission: recessive inheritance or dominant inheritance. The recessively inherited diseases require both alleles to be mutated, and are usually caused by a loss of function of genes and the proteins they encode. The dominantly inherited diseases need only one chromosome to carry the mutant allele, and are commonly caused by a gain of new function or hyperactivity of the original function of the mutant protein.

A third type of dominant inheritance can arise due to haploinsufficiency. Here, a genetic mutation disrupts a gene in one allele, and the resultant reduction in protein levels is sufficient to produce a disease phenotype.

In neurodegenerative diseases, both dominant and recessive disease inheritances are common. For example, in PD, Park 1- and Park 8-type PD are dominantly inherited due to mutations in α-synuclein and LRRK2 genes; whereas Park 2, Park5, Park 6, and Park 7 are recessively inherited due to the loss of Parkin, UCHL-1, DJ-1, and PINK 1, respectively.

These different types of human hereditary diseases require different strategies to establish genetic rodent models (see Fig. 1).

Strategies for genetic modification in mice

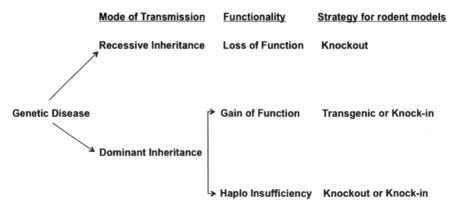

Fig. 1. Diagram illustrating the procedural flowchart that helps deciding which type of genetic modification is most appropriate for each kind of inherited disease.

For genetic loss of function mutations, the gene targeting (i.e., knockout) approach is the most common method of choice. Gene targeting allows the gene of interest to be deleted from the genome of experimental animals, resulting in a complete loss of the gene product.

For dominant genetic diseases, the transgenic approach is the most common method of choice. In this case, a mutated gene is introduced as a transgene into the genome of the experimental animals. Alternatively, a knock-in approach can be used. In this case, the endogenous rodent gene is engineered to carry a human disease-causing mutation. The comparison of these two methods is discussed later in this chapter.

These different methods of genetic manipulation allow us to fully or partially recapitulate a variety of disease phenotypes in living mammals.

3. Practical Considerations of Gene Targeting

There are two approaches to gene targeting: the conventional knockout and the conditional knockout (see Fig. 2). Briefly, the conventional knockout simply deletes a gene usually at its ATG initiation exon or exons encoding critical functional domains. Since this manipulation is done to the genome, all cells in the knockout animal have a complete loss of the gene and gene product from embryogenesis.

Currently, an alternative approach is becoming more common: the conditional knockout method. This method requires that two LoxP sites are inserted into the gene of interest usually surrounding the ATG initiation exon. Animals carrying this genetic manipulation (also called "floxed allele") have the gene of interest normally expressed. When the animal is injected with Cre viruses or crossed to a mouse expressing Cre recombinase, the floxed chromosomal

Conditional knockout

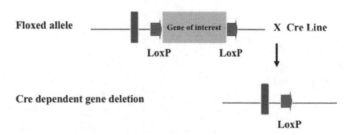

Fig. 2. Schematic representation of the steps involved in obtaining conditional excision (i.e., conditional knockout) of a gene of interest.

region is deleted by Cre-mediated recombination. Thus, the spatial and temporal patterns of Cre recombinase determine the deletion of the gene of interest.

There are advantages and disadvantages of both conventional and conditional knockout: the conventional knockout method is straightforward and less time consuming. In addition, there has been a huge effort by individual investigators, NIH, Wellcome Trust, and pharmaceutical/biotech companies to generate knockout mice. Because of this, there are a large number of ES cells and knockout mice readily available for investigators to use. For references, please see:

http://www.nih.gov/science/models/

http://www.jaxmice.jax.org/index.html

http://www.emmanet.org/

http://www.genetargeting.com/

The conditional knockout method requires more careful design and more effort in generating the conditional alleles. However, once the conditional allele is generated, there will be a much more versatile usage because one can choose Cre lines with particular spatial and temporal patterns to delete the gene of interest. This is extremely useful for at least two types of experiments: (a) to circumvent the situation in which a complete gene deletion is lethal and (b) to investigate the consequence of gene deletion in a tissue- and cell type-specific manner.

It should be mentioned that once the conditional allele is obtained, it can always be easily converted to a conventional full knockout by using the germ-line Cre lines. For references of germ-line Cre mice, please see: http://www.cre.jax.org/strainlist.html.

If an investigator is starting to analyze a new gene, a practical consideration and planning can be in the following sequence: (a) Check if conventional knockout allele exists. If it does, studying the knockout mice will allow to quickly gain some knowledge on the gene function. (b) If there is no readily available knockout allele, the investigator is better off going directly to a conditional knockout model because a slightly bigger effort will yield a more versatile tool.

4. Complete Loss of the Proteins vs. Mutant Proteins

Often, knockout mice that model loss of function mutations do not precisely recapitulate the cardinal phenotypes of their intended human conditions. Among the possible explanations, the main hypothesis is that mice and human have diverged enough in evolution, therefore the loss of genes and proteins impacts humans and mice very differently.

Interestingly, in some cases, there might be another factor that plays a role. In the study of Parkin, the disease gene for Park 2-type PD, investigators found that a complete knockout of Parkin in mice does not recapitulate the cardinal PD phenotypes. However, mice carrying mutated Parkin protein developed some PD-like phenotypes (1). These results were a surprise in the PD community, as human Park2 is considered to result from a loss of function of Parkin. Why the phenotype of a complete knockout is less severe in mice? One plausible explanation for this is that in a complete knockout mouse the absence of Parkin is compensated by other molecules and pathways, whereas in mice that carry mutant Parkin the nonfunctional mutant Parkin proteins still exist, occupying the spatial positions and protein complexes so that compensation does not occur. This scenario raises an interesting point: it is possible that in disease modeling, a straight knockout mouse may be less efficient in generating desired phenotypes than a mouse with mutant proteins.

5. Practical Considerations of Transgenic Approaches

As the knockout approach is used to model loss-of-function mutations, transgenic or knock-in methods are used to model gain-of-function mutations.

The knock-in method has the same principles as the knockout by homologous recombination, except that the gene of interest carries disease-causing mutations rather than deletions, resulting in a mouse protein that contains human mutations.

Differently from knock-in, the generation of transgenic mice involves construction of DNA vectors carrying gene of interest. The DNA vector is subsequently microinjected into fertilized mouse eggs. In about 5–15% of injected eggs, DNA vectors integrate into the mouse genome as a transgene, and therefore transgenic mouse founder lines are obtained.

There are some important differences between transgenic and gene knock-in animals. In transgenics, the integration sites of the transgenes are random, whereas in knock-in integration is targeted to the homologous chromosomal sites. The copy number of the transgene is random, ranging from a single copy to 1 or 200. A direct implication of the transgene random integration sites and variable copy numbers is that the expression of transgenes may vary widely among the founder lines, both in spatial and temporal patterns, as well as in abundance. The random insertion of transgene into mouse genome may disrupt other genes in some cases. Therefore, in practical experiments, studies with transgenic mice typically require at least two transgenic lines to ensure that the phenotypes are authentically caused by the transgenes.

The methods of generating transgenic mice have developed in increasing sophistication and become more powerful and versatile. A method worth special mentioning is the bacterial artificial chromosome (BAC)-mediated transgenesis (2). The traditional transgenic method uses a plasmid vector with a small capacity of less than 20 kb of DNA insert. It has to use an artificial promoter, such as prion, CMV, Thy1, etc. The small capacity leaves no space for genes' regulatory elements, such as enhancers, suppressors, and insulators, all of which are important for gene regulation. Because of their small size, when the transgenes integrate into the mouse genome, they are influenced strongly by positional effects caused by the adjacent chromosomal areas. These positional effects often cause spatial and temporal misexpression of the transgenes. In this aspect, BAC-mediated transgenesis has several advantages over the traditional method. First, the conventional method only allows for 10–15 kb DNA inserts while BAC vectors accommodate up to 400 kb DNA as transgenes, making them large enough for most of the mammalian genes. Second, the large BAC capacity allows the transgene to be insulated from positional effects of insertion sites on the chromosome and to be regulated by its native promoters, enhancers, and suppressors, intron–exon structures, and 5′ and 3′ untranslated regions (UTRs). Third, the BAC method allows for all splicing variants to be transcribed in a spatial and temporal manner that can be decided by the host cell. In recent years, BAC transgenesis has been used successfully to generate highly useful models in neurodegenerative disease research (3).

The inducible tet-on/tet-off system is another powerful tool. It allows inducible expression or suppression of transgenes. An elegant example of using this method was a study which demonstrated the possibility of reversing pathology in HD mice by suppressing the expression of mutant huntingtin (4).

6. Practical Considerations on Using Mouse vs. Human Genes

In transgenic or knock-in approaches, the choice of transgenes deserves some discussion. There are two options: the disease gene could either be a human gene with human disease mutations or the mouse homologous gene with human mutations. Both empirically and conceptually, the former works better in recapitulating the phenotypes of human disease. Empirically, for example, in modeling Huntington's disease, the model carrying mouse huntingtin with 111 CAG repeats developed much milder phenotypes than another model with human huntingtin of similar CAG repeats (5,6). In α-synuclein mouse models for Parkinson's disease, human α-synuclein gene in transgenic mouse was able to recapitulate some disease pathology while mouse α-synuclein gene failed (7).

Conceptually, the most straightforward explanation is that evolution has accumulated enough divergence that a mutation in human protein may or may not cause the same pathogenic conformational change in a mouse homolog. The most amazing example is that in fact the human A53T mutation in α-synuclein caused Park 1 type of Parkinson's disease, whereas in mice the 53T is a wild-type amino acid sequence. Hence, it seems that human genes are more appropriate for modeling human diseases in transgenic mice.

7. Increasingly Sophisticated Genetic Approaches

There is no doubt that transgenic mice are major tools in modeling neurodegenerative disease. Currently, in addition to mimicking the genotypes of human diseases, researchers are generating and using transgenic mice in sophisticated genetic manipulations. For example, mice with specific mutations in huntingtin were generated to test the role of caspase-mediated proteolysis in HD pathogenesis (8), and mice expressing mutant HD in particular brain regions and cell types were used to investigate important questions of cell-autonomous vs. noncell-autonomous degeneration (9). This approach was also used in the study of ALS and AD, which yielded important insight of the pathogenic processes. As a result of many researchers' efforts, there are many transgenic mouse models for each neurodegenerative disease, and collectively they have become powerful tools for research in their fields.

8. The New Field of Rat Genetics

Despite the overall success of using transgenic mice to model neurodegenerative diseases, a common problem is that mice are much more resilient and many times did not exhibit the cardinal phenotype of neuronal cell death. For example, in PD and AD research, mouse models were unable to recapitulate this most important aspect of pathology. It is worth mentioning that to combat this obstacle, transgenic rat approach may be a worthy alternative to bring new opportunities.

Although closely related in evolution, rats are different from mice and are closer to human in many physiological aspects. In many pharmacological studies, rats have much more similar pharmacokinetics to human. It is well-known that cardiovascular diseases are modeled in rats and are unable to be done in mice. In Huntington's disease modeling, rat models recapitulated striatal neuronal death with 57 CAG repeat length similar to human mutation (10) while mouse models needed much stronger insults of about

100 CAG repeats to elicit mutant phenotypes. In chemical models for PD, rats are much more sensitive to rotenone than mice (11). In α-synuclein study, viral expression of α-synuclein in rat substantia nigra resulted in loss of dopaminergic neurons, which was not observed in mouse models (12, 13). These data indicated that rat may be more sensitive to some pathological insults than mice.

In addition to their sensitized pathological aspects, rats are also more suitable for many studies for additional reasons. First, they are more suitable for behavioral studies because rats can perform much more complex tasks, including ones that involve more sophisticated learning and memory. Second, due to their larger size, many more physical manipulations including surgeries and neuroimaging are possible.

In recent years, both transgenic and knockout methods have been developed for utilizing rat as an experimental organism for genetic studies (14).

9. Conclusions

Genetic mouse and rat models have provided researchers with unprecedented powerful tools for mechanistic dissection of pathogenesis as well as development of therapeutics. The methodology, techniques, and usage of transgenics are continuously improving. These models give more reasons for optimism that a greater understanding of neurodegeneration eventually helps us fight these human diseases.

References

1. Lu, X. H., Fleming, S. M., Meurers, B., Ackerson, L. C., Mortazavi, F., Lo, V., Hernandez, D., Sulzer, D., Jackson, G. R., Maidment, N. T., Chesselet, M. F. & Yang, X. W. (2009) *J Neurosci* **29**, 1962–76.

2. Gong, S., Yang, X. W., Li, C. & Heintz, N. (2002) *Genome Research* **12**, 1992–8.

3. Li, Y., Liu, W., Oo, T. F., Wang, L., Tang, Y., Jackson-Lewis, V., Zhou, C., Geghman, K., Bogdanov, M., Przedborski, S., Beal, M. F., Burke, R. E. & Li, C. (2009) *Nat Neurosci* **12**, 826–8.

4. Yamamoto, A., Lucas, J. J. & Hen, R. (2000) *Cell* **101**, 57–66.

5. Wheeler, V. C., Auerbach, W., White, J. K., Srinidhi, J., Auerbach, A., Ryan, A., Duyao, M. P., Vrbanac, V., Weaver, M., Gusella, J. F., Joyner, A. I. & MacDonald, M. E. (1999) *Hum Mol Genet* **8**, 115–22.

6. Hodgson, J. G., Agopyan, N., Gutekunst, C. A., Leavitt, B. R., LePiane, F., Singaraja, R., Smith, D. J., Bissada, N., McCutcheon, K., Nasir, J., Jamot, L., Li, X. J., Stevens, M. E., Rosemond, E., Roder, J. C., Phillips, A. G., Rubin, E. M., Hersch, S. M. & Hayden, M. R. (1999) *Neuron* **23**, 181–92.

7. Chandra, S., Gallardo, G., Fernandez-Chacon, R., Schluter, O. M. & Sudhof, T. C. (2005) *Cell* **123**, 383–96.

8. Graham, R. K., Deng, Y., Slow, E. J., Haigh, B., Bissada, N., Lu, G., Pearson, J., Shehadeh, J., Bertram, L., Murphy, Z., Warby, S. C., Doty, C. N., Roy, S., Wellington, C. L., Leavitt, B. R., Raymond, L. A., Nicholson, D. W. & Hayden, M. R. (2006) *Cell* **125**, 1179–91.

9. Gu, X., Li, C., Wei, W., Lo, V., Gong, S., Li, S. H., Iwasato, T., Itohara, S., Li, X. J., Mody, I., Heintz, N. & Yang, X. W. (2005) *Neuron* **46**, 433–444.

10. von Horsten, S., Schmitt, I., Nguyen, H. P., Holzmann, C., Schmidt, T., Walther, T., Bader, M., Pabst, R., Kobbe, P., Krotova, J., Stiller, D., Kask, A., Vaarmann, A., Rathke-Hartlieb, S., Schulz, J. B., Grasshoff, U., Bauer, I., Vieira-Saecker, A. M., Paul, M., Jones, L., Lindenberg, K. S., Landwehrmeyer, B., Bauer, A., Li, X. J. & Riess, O. (2003) *Human Molecular Genetics* **12,** 617–24.

11. Alam, M., Mayerhofer, A. & Schmidt, W. J. (2004) *Behavioural Brain Research* **151,** 117–24.

12. Yamada, M., Iwatsubo, T., Mizuno, Y. & Mochizuki, H. (2004) *Journal of Neurochemistry* **91,** 451–61.

13. Lo Bianco, C., Schneider, B. L., Bauer, M., Sajadi, A., Brice, A., Iwatsubo, T. & Aebischer, P. (2004) *Proceedings of the National Academy of Sciences of the United States of America* **101,** 17510–5.

14. Geurts, A. M., Cost, G. J., Freyvert, Y., Zeitler, B., Miller, J. C., Choi, V. M., Jenkins, S. S., Wood, A., Cui, X., Meng, X., Vincent, A., Lam, S., Michalkiewicz, M., Schilling, R., Foeckler, J., Kalloway, S., Weiler, H., Menoret, S., Anegon, I., Davis, G. D., Zhang, L., Rebar, E. J., Gregory, P. D., Urnov, F. D., Jacob, H. J. & Buelow, R. (2009) *Science* **325,** 433.

Chapter 13

Modeling Focal Cerebral Ischemia In Vivo

Katherine Jackman, Alexander Kunz, and Costantino Iadecola

Abstract

Ischemic stroke is among the leading causes of mortality and long-term disability in the western world. Despite enormous research activities in the last decades, current therapeutic options for acute stroke patients are still very limited. Reliable and realistic in vivo animal models represent sine qua non for successful translation from bench to bedside. To date, several animal models of focal and global cerebral ischemia have been developed to mimic the clinical situation in humans as accurately as possible. This chapter focuses on models of focal cerebral ischemia, in particular on the most commonly used model: the intraluminal filament model of middle cerebral artery occlusion. The main objective is to provide a detailed instruction manual for researchers interested in learning this technique.

Key words: Brain ischemia, Focal ischemia, Transient ischemia, Reperfusion, Middle cerebral artery occlusion, String model, Mice, Laser Doppler flowmetry

1. Introduction

Cerebral ischemia has been studied experimentally in animals for more than 60 years. While the effects of cardiac arrest on the brain are replicated in animals using models of global cerebral ischemia, animal models of focal cerebral ischemia have been designed to replicate the occlusion of the middle cerebral artery (MCA), the artery most often occluded in human stroke (1). Over the years, a number of different models of focal ischemia have been developed, each one trying to better mimic the disease. The most popular models have been summarized in Table 1. At present, the most frequently applied technique for inducing focal cerebral ischemia is the intraluminal filament model of MCA occlusion (MCAO) in rats and mice.

Giovanni Manfredi and Hibiki Kawamata (eds.), *Neurodegeneration: Methods and Protocols*, Methods in Molecular Biology, vol. 793, DOI 10.1007/978-1-61779-328-8_13, © Springer Science+Business Media, LLC 2011

Table 1
Popular models of focal cerebral ischemia

Model	Basic procedure	Benefits	Disadvantages
Intraluminal filament	Filament advanced via the internal carotid artery to occlude MCA	Less invasive; Reproducible infarcts; Produce transient and permanent focal ischemia	Poststroke hyperthermia with longer ischemic periods; Subarachnoid hemorrhage (less likely with flow monitoring by laser Doppler)
Endothelin-1 (ET-1)	Direct application of vasoconstrictor ET-1 to the MCA	Less invasive	Magnitude and duration of ischemia is highly variable
Direct surgical	Ligation, cauterization or clipping of the MCA	Control of the site of occlusion, i.e., proximal vs. distal MCA	Invasive; No reperfusion with cauterization of MCA; Can induce brain trauma
Autologous clot embolism	Injection of autologous blood clots into ICA	More closely mimics human embolic stroke; Allows study of thrombolytic agents	Highly variable infarct volumes; Uncontrollable reperfusion
Nonclot embolism	Injection of artificial "clots" (e.g., microspheres) into ICA or CCA	No poststroke hyperthermia; Slow lesion development (increased therapeutic window)	Multiple, scattered infarcts; No reperfusion
Photothrombosis	Injection of photoactive dye (e.g., rose bengal), irradiation of cerebral cortex with laser, microvascular coagulation in irradiated area	Produce infarcts of defined location and size; Less invasive	Does not produce ischemia in a clinically relevant arterial territory, i.e., MCA; Infarct size usually small; Fails to produce typical penumbra

Intraluminal filament MCAO was first described in rats using silicon-coated filaments by Koizumi et al. (2) and then later using heat-blunted (3) and poly-L-lysine coated filaments (4). An important modification of intraluminal filament MCAO was its adaptation to mice in 1997 (5), as this allowed the study of transgenics and subsequent advances in the understanding of the molecular pathogenesis of ischemic stroke. In comparison to alternative models of focal ischemia, intraluminal filament MCAO has a number of advantages. It is less invasive, requiring only a scalp and midline neck incision, results in relatively low mortality and reproducible cerebral infarcts and can be used to induce both transient and permanent focal cerebral ischemia. Accidental vessel rupture and subsequent SAH is a common problem associated with intraluminal filament MCAO, however in our experience, and of others (6, 7), this is largely overcome with the use of laser Doppler flowmetry for guiding filament insertion.

In this chapter, we provide detailed instructions on how to perform intraluminal filament MCAO in mice using heat-blunted nylon sutures. In addition, methods for assessing outcome, including functional impairment and ischemic lesion size, are described.

2. Materials

1. Mice.

   Male C57Bl6 mice, 6 weeks, weight ~20–22 g (Jackson Laboratories, USA).

2. Heat-blunted filament.

   (a) 6–0 Nylon suture (Look Sutures 915B).

   (b) Geiger cautery tool (Model-100, Geiger Medical Technologies, USA).

   (c) Colored marker.

3. Surgical equipment.

General.

   (a) Dissecting microscope.

   (b) Heated surgical surface (heated water bath, rubber tubing, pump).

   (c) Rectal temperature probe and temperature controller (Digi-Sense Temperature Controller R/S, Cole-Parmer, USA).

   (d) Veterinary warming pad [such as Snuggle Safe (Lenric C21, UK)].

Femoral artery cannulation.

    (a) 6–0 Silk suture and 5–0 nylon suture with reverse cutting.

    (b) PE-10 polyethylene tubing.

    (c) Surgical tools:

- Standard scissors.
- Dumont #7 forceps (curved; two pairs).
- Vannas spring scissors (3-mm blade).

MCA occlusion.

    (a) Laser Doppler flowmeter (Periflux System 5010, Perimed, Sweden).

    (b) Fiberoptic probes (MT B500-0 L240, Perimed).

    (c) Perisoft for Windows Version 2.10 (Perimed).

    (d) Instant krazy glue and accelerator.

    (e) 4–0 and 6–0 silk suture and 5–0 nylon suture with reverse cutting.

    (f) PE-50 polyethylene tubing.

    (g) Surgical tools:

- Vannas spring scissors (3-mm blade).
- Standard scissors.
- Dumont #7 forceps (curved; two pairs).
- Crile Hemostat.

4. Devices for monitoring physiological parameters.

    (a) Blood gas analyser [such as Chrion 238 (Siemens, USA)].

    (b) Blood glucose meter [such as OneTouch UltraMini (LifeScan Inc., USA)].

    (c) Grass-Telefactor PT300 pressure transducer (Grass Instruments Co., USA).

    (d) Chart 5 Pro (AD Instruments).

5. Functional assessment.

    (a) Stainless steel wire (~1 mm diameter).

    (b) Two vertical stands.

    (c) Two pieces of cardboard (angled at ~30°).

6. Infarct quantification.

Sectioning.

    (a) Embedding Matrix [Shandon M-1 (Thermo Electron Corporation, USA)].

    (b) Cryostat [such as Leica CM1850 (Leica Microsystems, Germany)].

Staining reagents.

(a) 0.5% Cresyl violet solution (cresyl violet acetate).

(b) Distilled water (dH_2O).

(c) 0.1 M Phosphate buffer (PB, composed of sodium phosphate monobasic and dibasic).

(d) 4% Paraformaldehyde (PFA; diluted with 0.1 M PB, pH 7.4).

(e) Ethanol (100, 95, 75, and 50%).

(f) Xylenes.

(g) Dpx mounting media.

Imaging.

(a) Nikon camera and CCD (MTI CCD-72S).

(b) Light box [such as Model R95 (Imaging Research Inc., GE Healthcare, UK)].

Quantification.

MCID Elite 7.0 image analysis software (Imaging Research Inc.).

3. Methods

1. *Femoral artery cannulation*

Anesthesia is induced using 4% isoflurane in oxygen and maintained with 1.5–2% isoflurane in an oxygen/room air mix. Mice are kept at $37 \pm 0.5°C$ during the surgical procedure using a heated surgical surface.

Surgical Procedure:

(a) An ~2 cm incision is made in the femoral region of the thigh. The wound is sutured open using 5–0 nylon suture.

(b) The femoral artery is exposed and separated from the femoral vein. Using 6–0 silk suture, the artery is ligated loosely proximal and distal. Pressure is applied to the proximal ligature.

(c) A small incision is made in the femoral artery, pressure is released from the proximal ligature and a section of PE-10 tubing inserted into the femoral artery. This is tied in position using the proximal ligature.

(d) Approximately 50–60 µL of arterial blood is collected for the measurement of blood glucose and blood gases both before and after induction of ischemia (see Note 1, 2). For the measurement of blood pressure, the catheter is connected to PT300 pressure transducer and blood pressure recorded using Chart 5 Pro software (see Note 3).

2. *Filament preparation*

6–0 Nylon suture is cut into segments approximately 3 cm in length.

The tip of the filament is melted uniformly using a high temperature cautery. Temperature is adjusted to ensure even melting without overheating/burning.

Total length of the tip head is ~0.8–1 mm and diameter of tip head ~150–180 μm (see Note 4) (8, 9). Using a colored marker, a line is made 1 cm from the tip of the filament. This is used as a guide for filament insertion.

3. *Induction of cerebral ischemia*

Anesthesia is maintained during the procedure of MCAO using 1.5–2% isoflurane in an oxygen/room air mix. Mice are kept at $37.0 \pm 0.5°C$ during the surgical procedure using a heated surgical surface and until the animal regains consciousness using a veterinary warming pad (see Note 5).

Using instant krazy glue and accelerator, the laser Doppler fiberoptic probe is attached to the parietal bone 2 mm posterior and 5 mm lateral to bregma. The fiberoptic probe is then connected to laser Doppler flowmeter and cortical perfusion recorded (see Note 6).

Surgical Procedure (see Note 7):

Under a dissecting microscope:

(a) Using standard scissors, make a midline neck incision (~2 cm) and visualize the right common carotid artery (CCA).

(b) Clear the external carotid artery (ECA) and place two 6–0 silk suture loops at two positions around the ECA – one directly behind the CCA bifurcation and the other ~5 mm behind the CCA bifurcation (Fig. 1). Make sure that the distal suture ligates the ECA, while the proximal suture forms a loose loop around the proximal ECA.

(c) Clear the CCA of connective tissue and feed a ~8 cm length of 4–0 silk suture underneath the CCA. Be careful not to damage the vagus nerve, which runs parallel to the CCA. When in position, apply pressure to the 4–0 suture bridge using a hemostat until the blood flow through the CCA is interrupted.

(d) Using spring scissors, make a small incision in the ECA between the two 6–0 sutures and feed a heat-blunted 6–0 filament, via the ECA, into the CCA. Once the filament is

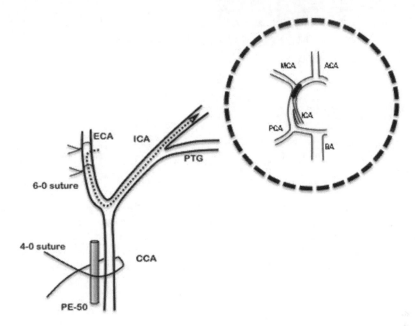

Fig. 1. Intraluminal filament MCAO. Diagram depicting the technique of intraluminal filament MCAO. The heat-blunted filament is inserted between the two ECA 6–0 suture knots (one proximal and one distal to the CCA bifurcation), and advanced along the ICA (*dotted line*) until it reaches the origin of the MCA (*see insert*). Once in position, the *CCA* is ligated with 4–0 suture and PE-50 tubing. *ACA* anterior cerebral artery, *BA* basilar artery, *CCA* common carotid artery, *ECA* external carotid artery, *ICA* internal carotid artery, *MCA* middle cerebral artery, *PCA* posterior communicating artery, PTG pterygopalatine artery.

within the lumen of the CCA, tighten the proximal ECA suture. Afterward, release pressure on the CCA 4–0 suture bridge and form a loose knot. Make sure that the blood flow suspension in the CCA is maintained for a maximum of 30 s.

(e) Advance the filament into the internal carotid artery (ICA) by gently bending the ECA in a posterior direction and "flicking" the filament around the bifurcation. As the filament approaches the pterygopalatine (PTG) artery, gently guide the filament tip slightly toward the left, to ensure that it is advanced into the ICA. If you are in the correct position, minimal resistance should be felt. Using the 1 cm mark on the filament as a guide, continue to advance the filament along the ICA until a sharp reduction in cerebral blood flow is observed. The filament is now occluding the MCA, ligate the CCA using the 4–0 suture knot and a ~2 cm section of PE-50 tubing (Fig. 1). The tubing allows easier removal of the knot upon reperfusion. A >85% reduction in CBF must be observed over the 30 min ischemic period (Fig. 2; see Note 8). During MCA occlusion, volatile anesthesia should be maintained at the smallest

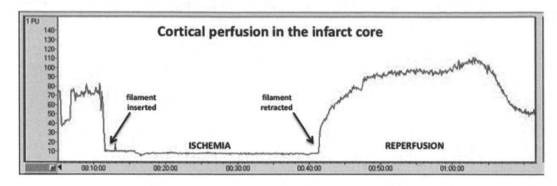

Fig. 2. Neocortical perfusion during intraluminal filament MCAO. Representative laser Doppler flow trace obtained using Perisoft for Windows. A transient reduction in blood flow is detected when pressure is applied to the CCA using the 4–0 suture bridge. When the heat-blunted filament is inserted, it occludes the MCA, and a sharp reduction in cortical blood flow (85%) is detected. This is sustained for the ischemic period (30 min). Upon retraction of the filament, blood flow recovers to >80% within 10 min. Reactive hyperemia is observed, followed by delayed oligemia (postischemic hypoperfusion).

dosage possible while closely monitoring the level of anesthesia.

(f) Following 30 min ischemia, remove the 4–0 suture knot from the CCA, and retract the filament. A >80% recovery of CBF must be observed within 10 min of induction of reperfusion (Fig. 2). After 30 min reperfusion, remove the laser Doppler flow probe, close the head and neck wounds using 5–0 nylon suture with reverse cutting and return the animal to a temperature-controlled recovery cage (veterinary warming pad) until it regains complete consciousness.

4. *Evaluation of Functional Impairment*

One to several days (3 days in our laboratory) after induction of cerebral ischemia-reperfusion, functional impairment is assessed using the following tests:

Modified Bederson score

Mice are assigned a score from 0 to 4 according to the following criteria (10):

0 = normal motor function;

1 = flexion of the torso and forelimb contralateral to MCA occlusion when mouse suspended by the tail;

2 = circling to the side contralateral to MCA occlusion, when mouse held by tail on a flat surface;

3 = leaning to the contralateral side at rest;

4 = no spontaneous motor function.

Hanging-wire test

Forepaw strength is assessed using the hanging wire test (11, 12). Using their forepaws, mice are made to grasp the middle of a wire that is ~60 cm long and suspended ~36 cm above a padded surface. Latency to fall is recorded in seconds, with a maximum time of 60 s. Mice that reach the support poles are automatically assigned the maximum time of 60 s. Three trials are performed at 5 min intervals and scores from the three trials are averaged. A shorter time suspended from the wire corresponds to a greater degree of brain injury.

Corner test

Sensory motor integration is assessed using the corner test (13, 14). The mouse being tested is placed between two boards angled at 30° with a small opening along the joint between the boards to encourage entry into the corner. As the mouse enters the narrowed corner, both sides of the vibrissae are stimulated causing the mouse to rear onto its hindlimbs and turn to the opposite end. The direction that the mouse turns is recorded. A total of ten trials are performed and the total percentage turns in the right (ipsilateral) direction is calculated. Normal mice not subjected to cerebral ischemia have on average a 50% chance of turning to the right (ipsilateral) side. However, mice subjected to cerebral ischemia have a >50% chance of turning to the right.

5. *Quantification of cerebral infarction and edema*

   Sectioning:

   Frozen brains are submerged in embedding matrix to prevent breakage of the tissue during cutting. Commencing at ~2.8 mm anterior to bregma, 30 µm coronal sections are collected at 600 µm intervals using a cryostat.

   Staining:

   Sections are stained using cresyl violet for visualization of cerebral infarction (see Note 9).

   Staining protocol:

   1. 4% PFA (10 min).

   2. 0.1 M PB (5 min).

   3. dH$_2$O (10 min).

   4. 0.5% cresyl violet (10–15 min, depending on desired intensity of stain).

   5. dH$_2$O (1–2 min).

   6. Ethanol (50, 75, 95, and 100%; 10 s each).

   7. Xylene (2 min)×2.

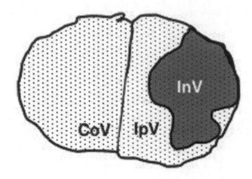

Fig. 3 Diagram of depicting regions for infarct quantification. InV infarct volume, IpV total ipsilateral hemisphere volume, CoV total contralateral hemisphere volume.

Coverslip with Dpx mounting media. Allow to dry in fume hood for a minimum of 24 h.

Quantification:

Images of coronal sections are captured using a CCD camera mounted above a light box.

Using MCID image analysis software, the following areas are quantified:

- Contralateral cortex and basal ganglia.
- Ipsilateral cortex and basal ganglia.
- Infarcted cortex and basal ganglia.

Total and regional volumes are calculated by multiplying individual areas by the intersection distance. Infarct volume is corrected for edema by subtracting the difference between the total ipsilateral and contralateral hemisphere volumes from the measured infarct volume (Fig. 3) (15, 16), according to the formula:

$$cInV = InV - (IpV - CoV)$$

cInV, corrected infarct volume; InV, infarct volume; IpV, total ipsilateral hemisphere volume; CoV, total contralateral hemisphere volume.

4. Notes

1. *Blood glucose*

   Due to the influence of blood glucose on brain injury following cerebral ischemia – hyperglycemia exacerbates ischemic damage (31) – it should be monitored both before and after cerebral ischemia. This can be performed in experimental groups in a subset of animals using a blood glucose meter. Furthermore, blood glucose levels can be controlled, via presurgery fasting, administration of glucose or by ensuring postsurgery access to food.

2. *Blood gases*

   In our laboratory, we do not artificially ventilate mice during the procedure of intraluminal filament MCAO, as the surgical procedure is relatively short (1 h and 15 min) and respiratory failure is very rarely observed. Despite this, we confirm that blood gases remain within normal limits (pH 7.35–7.45; PaO_2 120–140 mmHg; $PaCO_2$ 35–45 mmHg) both before and after induction of cerebral ischemia when studying new experimental groups. As the removal of even small volumes of blood (e.g., 50–60 μL) from the mouse can influence postsurgery survival, this is performed in an independent group of nonsurvival mice.

3. *Blood pressure*

   As variations in mean arterial pressure can influence stroke outcome, via changes in cerebral perfusion, we monitor mean arterial pressure via a femoral artery line connected to a pressure transducer. This is performed during the entire MCAO surgery, in the same group of nonsurvival mice used for blood gases and blood glucose measurements, and allows us to ensure consistency between experimental groups by excluding mice with abnormal arterial pressure traces.

4. *Occluding filament*

   In our laboratory, we use heat-blunted 6–0 nylon filaments with 150–180 μm tips to induce MCAO in mice. These produce both reproducible reductions in MCA territory cortical perfusion (>85%) and corresponding cerebral infarctions in mice aged between approximately 6–8 weeks. Mice from different colonies, genetic backgrounds or of different age/weight may require filaments with differing tip diameters. For these reasons, it is important that pilot studies be performed when studying a specific group of mice for the first time.

   Despite previous reports of the superiority of silicon-coated filaments (17), we observe an extremely high success rate, low variability in infarct volume and low rate of SAH with heat-blunted nylon filaments. This is largely due to the use of transcranial laser Doppler flowmetry, as it allows us to exclude mice in which cerebral blood flow does not meet our criteria (>85% reduction during ischemia; >80% increase upon reperfusion). Furthermore, the 1 cm marker on the occluding filament indicates the approximate distance that the filament must be inserted in order to occlude the MCA, and therefore overinsertion and subsequent vessel rupture are less likely.

5. *Core temperature*

Due to the confounding influence of temperature on the outcome of experimental stroke studies – protection with hypothermia and worsening with hyperthermia (28, 29) – temperature must always be controlled within physiological limits both during surgery and until the animal regains consciousness. The measurement of brain temperature can be performed directly by inserting a thermistor into the brain, requiring an invasive and technically difficult surgical procedure. Alternatively, it can be measured indirectly via placement of a needle thermistor in the temporal muscle. As body temperature has been shown to correlate with brain temperature in the rat (30), and can be monitored less invasively, we routinely measure and control for core temperature during MCAO using a rectal thermistor and heated surgical surface.

For controlling core temperature in mice during surgery, we use a heated surgical surface that we construct using polystyrene foam and rubber tubing connected to a pump and hot water bath. The tubing is laid inside the foam, just beneath the surface and the surgical area covered with plastic. This is an extremely effective way of regulating temperature in mice during MCAO and can be made both easily and inexpensively.

6. *Cerebral blood flow*

For the measurement of core blood flow in C57BL/6 mice, we position the laser Doppler probe 2 mm posterior and 5 mm lateral to bregma. At this position, we detect a >85% reduction in cortical cerebral blood flow upon occlusion of the MCA (18–24), resulting in a reproducible cerebral infarct. However, these coordinates are not fixed for all strains of mice and may require adjustment in order to accurately detect flow in the infarct core (e.g., detect a >85% reduction in flow upon occlusion of the MCA).

In contrast, the location of the ischemic penumbra can be estimated by measuring the distance between the midline and the border of the infarct in cresyl violet stained coronal sections. In C57BL/6 mice, we position the penumbral flow probe 0 mm posterior and 2 mm lateral to bregma, representing the region that is protected from infarction by neuroprotective interventions, such as lipopolysaccharide (LPS) preconditioning (22). A reduction in penumbral blood flow of 50–70% should be detected upon occlusion of the MCA (13, 22).

7. *Intraluminal filament MCAO in mice*

   Total duration of the procedure of intraluminal filament MCAO in mice, from initial induction of anesthesia to placement in the recovery chamber, should take approximately 1.5 h. With experience, surgery time will decrease and a corresponding increase in reproducibility of infarction and decreased mortality will be observed. Generally, it takes a minimum of 2 months daily practice to become competent in the technique of intraluminal filament MCAO in mice.

8. *Duration of ischemia and reperfusion*

   In our laboratory, depending on the surgeon, male C57BL/6 mice aged 6–8 weeks require between 20 and 35 min occlusion of the MCA in order to achieve a total, edema-corrected infarct volume of approximately 50 mm^3 (18–25). A standard infarct volume of 50 mm^3 is commonly chosen by laboratories as both increases and decreases in infarction with different treatment strategies can be detected with this level of baseline brain injury. Interestingly, using mice of the same background, different laboratories have reported ischemic periods of upto 60 min and even 2 h in order to achieve infarct volumes of similar size (26, 27). It is therefore apparent that the susceptibility of mice to ischemic injury is highly variable even within strains, although differences in experimental conditions are also likely to play a role. Again, these observations highlight the importance of performing pilot studies.

9. *Infarct quantification*

   Cresyl violet and triphenyl tetrazolium chloride are the two most commonly employed histological staining techniques for the quantification of brain injury following experimental cerebral ischemia. While the two methods have been reported to result in similar values of infarct volume (32), we find cresyl violet to have a number of benefits over TTC. Firstly, TTC is a marker of metabolic dysfunction while cresyl violet provides histological evidence of tissue damage. Secondly, cresyl violet allows clear visualization of different anatomical regions, such as the cortex and basal ganglia and provides clear demarcation of the boundaries between viable and necrotic tissue, improving the efficiency of infarct volume analysis. And thirdly, in contrast to TTC, cresyl violet staining is performed in frozen tissue allowing long-term storage of samples.

References

1. De Freitas, G. R., Christoph Dde, H., and Bogousslavsky, J. (2008) Chapter 22 Topographic classification of ischemic stroke. *Handb Clin Neurol.* **93**, 425–452.

2. Koizumi, J., Yoshida, Y., Nakazawa, T., and Oneda, G. (1986) A new experimental model of cerebral embolism in rats in which recirculation can be introduced in the ischemia area. *Jpn J of Stroke.* **8**, 1–8.

3. Longa, E. Z., Weinstein, P. R., Carlson, S., and Cummins, R. (1989) Reversible middle cerebral artery occlusion without craniectomy in rats. *Stroke.* **20**, 84–91.

4. Belayev, L., Alonso, O. F., Busto, R., Zhao, W., and Ginsberg, M. D. (1996) Middle cerebral artery occlusion in the rat by intraluminal suture. Neurological and pathological evaluation of an improved model. *Stroke.* **27**, 1616–1622; discussion 1623.

5. Clark, W. M., Lessov, N. S., Dixon, M. P., and Eckenstein, F. (1997) Monofilament intraluminal middle cerebral artery occlusion in the mouse. *Neurol Res.* **19**, 641–648.

6. Schmid-Elsaesser, R., Zausinger, S., Hungerhuber, E., Baethmann, A., and Reulen, H. J. (1998) A critical reevaluation of the intraluminal thread model of focal cerebral ischemia: evidence of inadvertent premature reperfusion and subarachnoid hemorrhage in rats by laser-Doppler flowmetry. *Stroke.* **29**, 2162–2170.

7. Tsuchiya, D., Hong, S., Kayama, T., Panter, S. S., and Weinstein, P. R. (2003) Effect of suture size and carotid clip application upon blood flow and infarct volume after permanent and temporary middle cerebral artery occlusion in mice. *Brain Res.* **970**, 131–139.

8. Hata, R., Mies, G., Wiessner, C., Fritze, K., Hesselbarth, D., Brinker, G., and Hossmann, K. A. (1998) A reproducible model of middle cerebral artery occlusion in mice: hemodynamic, biochemical, and magnetic resonance imaging. *J Cereb Blood Flow Metab.* **18**, 367–375.

9. Tureyen, K., Vemuganti, R., Sailor, K. A., and Dempsey, R. J. (2005) Ideal suture diameter is critical for consistent middle cerebral artery occlusion in mice. *Neurosurgery.* **56**, 196–200; discussion 196–200.

10. Bederson, J. B., Pitts, L. H., Tsuji, M., Nishimura, M. C., Davis, R. L., and Bartkowski, H. (1986) Rat middle cerebral artery occlusion: evaluation of the model and development of a neurologic examination. *Stroke.* **17**, 472–476.

11. Freitag, S., Schachner, M., and Morellini, F. (2003) Behavioral alterations in mice deficient for the extracellular matrix glycoprotein tenascin-R. *Behav Brain Res.* **145**, 189–207.

12. Ikegami, S., Harada, A., and Hirokawa, N. (2000) Muscle weakness, hyperactivity, and impairment in fear conditioning in tau-deficient mice. *Neurosci Lett.* **279**, 129–132.

13. Li, X., Blizzard, K. K., Zeng, Z., DeVries, A. C., Hurn, P. D., and McCullough, L. D. (2004) Chronic behavioral testing after focal ischemia in the mouse: functional recovery and the effects of gender. *Exp Neurol.* **187**, 94–104.

14. Zhang, L., Schallert, T., Zhang, Z. G., Jiang, Q., Arniego, P., Li, Q., Lu, M., and Chopp, M. (2002) A test for detecting long-term sensorimotor dysfunction in the mouse after focal cerebral ischemia. *J Neurosci Methods.* **117**, 207–214.

15. Lin, T. N., He, Y. Y., Wu, G., Khan, M., and Hsu, C. Y. (1993) Effect of brain edema on infarct volume in a focal cerebral ischemia model in rats. *Stroke.* **24**, 117–121.

16. Swanson, R. A., Morton, M. T., Tsao-Wu, G., Savalos, R. A., Davidson, C., and Sharp, F. R. (1990) A semiautomated method for measuring brain infarct volume. *J Cereb Blood Flow Metab.* **10**, 290–293.

17. Laing, R. J., Jakubowski, J., and Laing, R. W. (1993) Middle cerebral artery occlusion without craniectomy in rats. Which method works best? *Stroke.* **24**, 294–297; discussion 297–298.

18. Cho, S., Park, E. M., Febbraio, M., Anrather, J., Park, L., Racchumi, G., Silverstein, R. L., and Iadecola, C. (2005) The class B scavenger receptor CD36 mediates free radical production and tissue injury in cerebral ischemia. *J Neurosci.* **25**, 2504–2512.

19. Cho, S., Park, E. M., Zhou, P., Frys, K., Ross, M. E., and Iadecola, C. (2005) Obligatory role of inducible nitric oxide synthase in ischemic preconditioning. *J Cereb Blood Flow Metab.* **25**, 493–501.

20. Kunz, A., Abe, T., Hochrainer, K., Shimamura, M., Anrather, J., Racchumi, G., Zhou, P., and Iadecola, C. (2008) Nuclear factor-kappaB activation and postischemic inflammation are suppressed in CD36-null mice after middle cerebral artery occlusion. *J Neurosci.* **28**, 1649–1658.

21. Kunz, A., Anrather, J., Zhou, P., Orio, M., and Iadecola, C. (2007) Cyclooxygenase-2 does

not contribute to postischemic production of reactive oxygen species. *J Cereb Blood Flow Metab.* **27**, 545–551.

22. Kunz, A., Park, L., Abe, T., Gallo, E. F., Anrather, J., Zhou, P., and Iadecola, C. (2007) Neurovascular protection by ischemic tolerance: role of nitric oxide and reactive oxygen species. *J Neurosci.* **27**, 7083–7093.

23. Park, E. M., Cho, S., Frys, K., Racchumi, G., Zhou, P., Anrather, J., and Iadecola, C. (2004) Interaction between inducible nitric oxide synthase and poly(ADP-ribose) polymerase in focal ischemic brain injury. *Stroke.* **35**, 2896–2901.

24. Park, E. M., Cho, S., Frys, K. A., Glickstein, S. B., Zhou, P., Anrather, J., Ross, M. E., and Iadecola, C. (2006) Inducible nitric oxide synthase contributes to gender differences in ischemic brain injury. *J Cereb Blood Flow Metab.* **26**, 392–401.

25. Abe, T., Shimamura, M., Jackman, K., Kurinami, H., Anrather, J., Zhou, P., and Iadecola, C. Key role of CD36 in Toll-like receptor 2 signaling in cerebral ischemia. *Stroke.* **41**, 898–904.

26. Crack, P. J., Taylor, J. M., Flentjar, N. J., de Haan, J., Hertzog, P., Iannello, R. C., and Kola, I. (2001) Increased infarct size and exacerbated apoptosis in the glutathione peroxidase-1 (Gpx-1) knockout mouse brain in response to ischemia/reperfusion injury. *J Neurochem.* **78**, 1389–1399.

27. Ducruet, A. F., Hassid, B. G., Mack, W. J., Sosunov, S. A., Otten, M. L., Fusco, D. J., Hickman, Z. L., Kim, G. H., Komotar, R. J., Mocco, J., and Connolly, E. S. (2008) C3a receptor modulation of granulocyte infiltration after murine focal cerebral ischemia is reperfusion dependent. *J Cereb Blood Flow Metab.* **28**, 1048–1058.

28. Krieger, D. W., and Yenari, M. A. (2004) Therapeutic hypothermia for acute ischemic stroke: what do laboratory studies teach us? *Stroke.* **35**, 1482–1489.

29. McIlvoy, L. H. (2005) The effect of hypothermia and hyperthermia on acute brain injury. *AACN Clin Issues.* **16**, 488–500.

30. Hasegawa, Y., Latour, L. L., Sotak, C. H., Dardzinski, B. J., and Fisher, M. (1994) Temperature dependent change of apparent diffusion coefficient of water in normal and ischemic brain of rats. *J Cereb Blood Flow Metab.* **14**, 383–390.

31. Ergul, A., Li, W., Elgebaly, M. M., Bruno, A., and Fagan, S. C. (2009) Hyperglycemia, diabetes and stroke: focus on the cerebrovasculature. *Vascul Pharmacol.* **51**, 44–49.

32. Tureyen, K., Vemuganti, R., Sailor, K. A., and Dempsey, R. J. (2004) Infarct volume quantification in mouse focal cerebral ischemia: a comparison of triphenyltetrazolium chloride and cresyl violet staining techniques. *J Neurosci Methods.* **139**, 203–207.

Chapter 14

Pharmacological Models of Parkinson's Disease in Rodents

Peter Klivenyi and Laszlo Vecsei

Abstract

Parkinson's disease (PD) is one of the most common neurodegenerative disorders. Despite the substantial progress that has been achieved, the precise mechanisms involved in the development of this disease are still not fully understood. The most common concepts relate to the genetic background and environmental/ toxic effects. A number of model systems have been introduced, which mimic the human disease to varying extents. In this chapter, we introduce some of the most widely accepted protocols of the pharmacological models of Parkinson's disease.

Key words: Parkinson's disease, Toxin models, Genetic models

1. Introduction

Parkinson's disease (PD) is one of the most common neurodegenerative disorders. The prevalence is increasing, and a recent study estimated that by 2030 more than two million patients with PD can be expected in the EU and the USA (1). The annual economic impact of this disease in the USA alone is estimated at $10.8 billion, including both the direct and the indirect medical costs (2). The medical, social, and economic impacts have naturally generated the need for a better understanding of the pathomechanism of PD, with the consequent aim of better treatment. A considerable number of PD models have been proposed in recent decades; unfortunately, none of them faithfully reflects all of the features of the human disease. Accordingly, the search for ever-better models is continuing.

Various toxins have been used to produce selective neuronal damage mimicking different neurodegenerative disorders. Most of these toxins act on the mitochondria, resulting in a respiratory

Giovanni Manfredi and Hibiki Kawamata (eds.), *Neurodegeneration: Methods and Protocols,*
Methods in Molecular Biology, vol. 793, DOI 10.1007/978-1-61779-328-8_14, © Springer Science+Business Media, LLC 2011

dysfunction, an energy deficit, oxidative stress, and ultimately more or less selective neuronal degeneration (for a review, see ref. 3). Neuronal cell death can be verified by measurement of the striatal dopamine level by means of HPLC, histological analysis (cell count estimation and unbiased stereology). Numerous studies have been published in recent years in which the neuroprotective effects of certain drugs have been demonstrated. However, most of them failed to produce the same efficacy in human clinical trials as seen in animal models. One explanation for these failures is the methodological problems originating from inadequate models and methods. The establishment of reliable unified models, methods, and interpretations would clearly make drug screening more efficacious and obviously cheaper. In this chapter, we describe the most widely accepted and used protocols, methods, and readouts.

2. Materials

2.1. Toxins

The concentration of toxins varies depending on the method of administration and regimen for each model, and is indicated in Subheading 3.

6-Hydroxydopamine

1. 6-Hydroxydopamine (6-OHDA) dissolved in cold 0.9% saline containing 1 mM ascorbic acid.

2. Sterilize by passage through a 0.22-μm filter.

3. Made freshly before each experiment.

1-Methyl-4-Phenyl-1,2,3,6-Tetrahydropyridine Hydrochloride

1. Phosphate-buffered saline (PBS) 10× stock solution: 1.37 M NaCl, 27 mM KCl, 100 mM Na_2HPO_4, and 18 mM KH_2PO_4 (adjusted to pH 7.4). Store at room temperature. The working solution is prepared by dilution of one part with nine parts of water.

2. 1-methyl-4-phenyl-1,2,3,6-tetrahydropyridine (MPTP) hydrochloride dissolved in 0.1 M PBS or saline: The solution can be stored overnight at 4°C or prepared fresh just prior to each experiment.

3. Chronic administration can be achieved by using subcutaneously implanted osmotic minipumps.

4. MPTP can be coadministered with probenecid, which can be dissolved in 0.1 N NaOH to 250 mg/kg.

Rotenone

1. Rotenone dissolved in DMSO or carboxymethyl cellulose.

2. Alternatively, rotenone can be dissolved in sunflower oil.

3. Prepared freshly just prior to each experiment.

Paraquat

Paraquat dichloride hydrate is dissolved in saline.

Lipopolysaccharide

Lipopolysaccharide (LPS) – *Escherichia coli* Serotype 026:B6 or
O111:B4 dissolved in saline at a dose of 250 µg/kg.

Manganese

1. $MnCl_2$ can be administered as a subcutaneous injection.

2. As an alternative way of administration, a mixture of 0.04 M $MnCl_2$ and 0.02 M $Mn(OAc)_3$ can be used for inhalation experiments (see Note 1).

3. Inhalation has been performed in closed acrylic boxes (35 cm wide, 44 cm long, and 20 cm high) connected to an ultranebulizer.

2.2. Procedures

Intraperitoneal Injections

1-ml syringe with 30G hypodermic needle for intraperitoneal injections.

Transcardial Perfusion

1. 4% (w/v) paraformaldehyde, prepared in PBS. The solution needs to be heated carefully. A stirring hot plate in a fume hood is used to facilitate dissolution, and the solution then cooled to room temperature before use.

2. Scissors, forceps, and scalpel may be needed.

3. 20- or 50-ml syringe and a microperfusion set or electronic minipump for perfusion.

Stereotaxic Surgery

1. Isoflurane.

2. Urethane (1,000–1,200 mg/kg), ketamine (50–200 mg/kg i.p.), ketamine and xylazine mixture (100 mg/kg:10 mg/kg), or chloral hydrate (400–600 mg/kg).

3. Equithesin mixture: Dissolve 2.13 g chloral hydrate and 1.0 g magnesium sulfate in 0.9% saline, add 8.1 ml sodium pentobarbitone solution (60 mg/ml), 5.0 ml 100% ethanol, and 19.8 ml propylene glycol, and adjust the final volume to 50.0 ml with saline.

4. The stereotaxic apparatus.

5. A 5–10-µl 30 G Hamilton syringe.

Rotational Test

1. 0.05–1.0 mg/kg apomorphine hydrochloride dissolved in 0.1% ascorbinated saline and prepared freshly.

2. A rotometer is a cylindrical container with a diameter of 33 cm and a height of 35 cm.

3. A quiet isolated room

Detection of Dopamine and Its Metabolites by Means of HPLC

1. An HPLC system, such as Series 1100 Agilent Technologies, Santa Clara, CA, USA or ESA, Inc. Chelmsford, MA, USA.

2. An electrochemical detector (e.g., Model 105 Precision Instruments, Marseille, France) or a coulometric detector (e.g., the Coulochem II detector ESA, Inc. Chelmsford, MA, USA).

3. A reversed-phase column (HR-80 C18, 80×4.6 mm, 3-μm particle size; ESA Biosciences, Chelmsford, MA, USA).

4. A pre-column (Hypersil ODS, 20×2.1 mm, 5-μm particle size; Agilent Technologies, Santa Clara, CA, USA).

5. A dual-channel interface (35900E Agilent Technologies).

6. Evaluation software (ChemStation Rev.1.10.02, Agilent Technologies).

7. Bio-Rad protein analysis protocol (e.g., Bio-Rad Laboratories, Hercules, CA USA).

8. Homogenization buffer: 70% w/w perchloric acid (7.2 μl), 0.1 M sodium metabisulfite (1 μl), 0.1 M disodium ethylene-diaminetetraacetate (1.25 μl), and distilled water (240.55 μl). Use 250 μl of ice-cooled solution.

9. A mobile phase: 75 mM NaH_2PO_4, 2.8 mM sodium octylsulfate, and 50–100 μM disodium ethylenediaminetetraacetate, supplemented with 7–10% v/v acetonitrile and the pH is adjusted to 2.8–3.0 with 85% w/w H_3PO_4. Alternatively, another mobile phase containing 0.1 M LiH_2PO_4, 0.85 mM 1-octanesulfonic acid, and 10% (v/v) methanol may also be used.

10. For MPP^+ meauserments, use 90% 0.1 M acetic acid and 75 mM triethylamine HCl, pH 2.3, adjusted with formic acid and 10% acetonitrile.

Immunohistochemistry

1. 0.1% Triton X-100, diluted in PBS

2. 10% sucrose or glycerol, diluted in 0.1 M phosphate buffer (pH 7.4)

3. 0.05 M Tris-buffered saline (TBS), pH7.4

4. 1% (w/w) H_2O_2 in 0.05 M TBS.

5. ABC kit for immunostaining (Santa Cruz Biotechology, USA) or alternatively biotinylated IgG (1:200, diluted in PBS/0.5% BSA; Vector Laboratories, Burlingame, CA, USA) and avidin–biotin peroxidase complex (1:200 in PBS) and 3,3′-diaminobenzidine tetrahydrochloride dihydrate (DAB) (Vector Laboratories,).

6. Mouse or rabbit anti-TH affinity-purified monoclonal antibody (1:1,000–1:5,000 in TBS) (Chemicon, Temecula, CA, USA; or R&D Systems, Minneapolis MN, USA).

7. 2–10% Normal horse serum (Vector Laboratories Inc.) diluted in TBS.

8. Microscope coverslips ($22 \times 40 \times 0.15$ mm).

Open-Field Test

1. A rectangular ($48 \times 48 \times 40$ cm) or a circular box.

2. An LED sensor system (Experimentia Ltd., Budapest, Hungary) or a video-associated detection system (e.g., SMART video-tracking system Panlab, S.L. Technology for Bioscience; Barcelona, Spain).

Stereology

1. A Nikon Eclipse E600 photomicroscope equipped with an LEP motorized X–Y mechanical stage.

2. StereoInvestigator software (Microbrightfield, Burlington, VT, USA).

3. Methods

3.1. Intraperitoneal and Subcutaneus Injections

1. For intraperitoneal injections, the mouse is grabbed by the back and held tightly so that the animal is not able to move at all (see Note 2).

2. For rats, thick gloves should be used and the animal is pressed gently to the grid of the cage to prevent movements.

3. The desired solution is injected slowly in the middle of the midline. Injection of at most a volume of 0.3 ml is advisable in mice and of 0.6 ml in rats (Fig. 1).

Risk for liver injury

Risk for injecting stomach

Suggested area for i.p. injections

Risk for injecting bladder

Fig. 1. Intraperitoneal injection.

4. The control animals receive an equal volume of vehicle at the same time.

5. For subcutaneous injections, the mouse or rat is grabbed by the back and gently pressed to the grid. The solution should be injected into the neck region. Injection of at most 0.2 ml is advisable in mice and at most 0.3 ml in rats.

6. The animal is then released in the home cage (see Notes 3–5).

3.2. Stereotaxic Surgery

1. All surgical procedures should be performed at the same time of the day, typically between 9.00 and 11.00 h (4, 5).

2. Rats can be anesthetized with an intraperitoneal injection of either urethane, ketamin, xylazine, equithensin, or chloralhydrate.

3. In mice, isoflurane can be used with an adequate apparatus.

4. The anesthetized animal is immobilized on a stereotaxic frame in flat skull position.

5. Body temperature is maintained at 37–38°C with a heating pad placed beneath the animal.

6. The fur is shaved off if necessary.

7. A skin incision is made on the scalp and cleaned with H_2O_2.

8. A hole is drilled after localization of the desired coordinates (see Subheading 3.3).

9. The infusion cannula is advanced slowly through the hole until the tip is located at the desired position.

10. The solution is infused slowly at a rate of 0.5–1 μl/min.

11. Following toxin administration, the cannula is allowed to remain in place for 5 min and then slowly withdrawn (see Note 6).

12. The skin incision is closed with stainless steel wound clips or with sutures (see Note 7).

13. The animals may be given ibuprofen (15–60 mg/kg) and penicillin (100.000 IU/kg) in their drinking water for 24 h to alleviate potential postsurgical discomfort and to prevent infection.

3.3. Coordinates

1. Several parts of the brain can be injected.

2. In rats, striatum, substantia nigra (SN), globus pallidus, ventral tegmental area, and medial forebrain bundle are the most common targets.

3. In mice, intrastriatal and intraventricular injections are used commonly.

4. The typical coordinates are summarized in Table 1.

3.4. 6-OHDA Model

1. 6-OHDA, at a dose of 4 μg/μl (use 1–2 μl), can be administered stereotaxically into the substantia nigra, medial forebrain bundle, or striatum, as described in Subheading 3.3.

Table 1
Coordinates for stereotaxic surgery

Rat	AP (mm)	ML (mm)	DV (mm)	Note
Intrastriatal	1.0/−0.5	2.0/3.5/4.0	−5.5/−6.0/−6.5	Right side From Bregma
Intrastriatal	+9.2	−3	+4.5	From the center of the interaural line
Ventral tegmental area	5.0	1.0	7.8	
Substantia nigra	5.0	2.0	8.0	Right side
Right medial forebrain bundle	3.2	1.5	8.7	From Bregma Wistar 260–290 gr
Globus pallidus	1.0	3.0	6.9	From Bregma Right side Wistar 260–290 gr
Intraventricular	0.8/1.5	1.5	3.5/3.8	From Bregma

Mouse	AP (mm)	ML (mm)	DV (mm)	Note
Intrastriatal	0.5/1.1	1.5/2.1	3.5	C57Bl, 10–12 weeks of age, from Bregma
Intraventricular	1.0	1.0/1.6	3.0/2.0	From Bregma

2. It can also be injected intraperitoneally, at a dose of 5 µg/µl (use 5–10 µl), as described in Subheading 3.1.

3. Bilateral lesions involve difficulty because they induce adipsia and aphagia.

4. Thirty minute prior to 6-OHDA injection, 25 mg/kg desipramine may be administered i.p. in order to protect the noradrenergic cells, if this is desired (6, 7) (see Note 8).

3.5. Rotational Test

1. Two weeks after toxin administration, the rotating behavioral response to the apomorphine test can be examined in order to assess the severity of nigral lesions (6, 7).

2. The dose of apomorphine hydrochloride lies in the range between 0.05 and 0.5 mg/kg administrated subcutaneously.

3. One minute after apomorphine injection, full rotations are counted in a cylindrical container (a diameter of 33 cm and a height of 35 cm) at 10-min intervals for 60 min in a quiet isolated room. Alternatively, six consecutive 5-min periods (30 min total) may be applied. The net number of rotations is defined as the rotation opposite to the lesion minus rotation toward the lesion. Time really matters.

4. Animals performing at least 50 net rotations within 10 min after the apomorphine injection are considered to have significant lesions and may be included in the studies.

5. Three weeks after toxin injection (1 week after the first apomorphine test), the rotational test can be repeated with a lower dose (0.05 mg/kg).

3.6. MPTP Model

1. The serious toxicity to humans requires strict adherence to the special safety rules throughout the experiments (8, 9) (see Notes 9–14).

2. MPTP can be administered intraperitoneally or subcutaneously, at a dose of 10–30 mg/kg (5 ml/kg) (5, 6) (see Notes 15–21).

   (a) In the acute regimen, the toxin injections are given 3–5 times 1–2 h apart on a single day.

   (b) In the acute model, the mice move slowly between the injections, and the body temperature is decreased. The animals recover by the following day and show no obvious signs of intoxication.

   (c) The subacute regimen comprises a single systemic injection per day for several consecutive days (usually 5 days) and even for weeks in the chronic cases.

   With the subacute and chronic regimes, no obvious signs can be detected after MPTP administration.

   (d) Chronic administration can be achieved via subcutaneously implanted osmotic minipump filled with MPTP at 170 mg/ml, which delivers the toxin at a dose of 40 mg/kg body weight daily for approximately 28 days.

   (e) Probenecid as an adjuvant blocks the rapid clearance of the toxin and its metabolites. Coadministration with MPTP enhances the toxicity. This regime consists of the administration of MPTP (25 mg/kg, s.c.) and probenecid (250 mg/kg) administration at 3–4-day intervals over 5 weeks (10).

3.7. Rotenone Model

1. Rotenone dissolved in DMSO or carboxymethyl cellulose can be administered at a dose of 3–9 mg/kg/day subcutaneously or 30 mg/kg orally for 4 weeks.

2. Rotenone can be dissolved in sunflower oil and administered subcutaneously, at a dose of 2 mg/kg/day for 4 weeks (11).

3.8. Paraquat Model

1. Paraquat can be given subcutaneously at a dose of 5–15 mg/kg at 2-day intervals for a total of ten doses (12).

2. Alternatively, it can be injected intraperitoneally at a dose of 5–15 mg/kg twice per week for 6 weeks, for a total of 12 doses (see Note 22).

3.9. Manganese Model

1. For systemic administration, $MnCl_2$ can be given as a single subcutaneous injection (50 or 100 mg/kg) (single-dose regimen) or as multiple injections over 7 days (multiple-dose regimen) (13).

2. For inhalation, the animals are placed in closed acrylic boxes connected to an ultranebulizer, with a continuous flux of 10 l/min. The ultranebulizer is designed to produce droplets in the range of 0.5–5 μm.

3. A vapor trap is located in the opposite side containing a solution of $NaHCO_3$ to precipitate the residual metal.

4. The animals inhale a mixture of 0.04 M $MnCl_2$ and 0.02 M $Mn(OAc)_3$ for 1 h, twice a week for 5 months.

5. The control mice inhale the vehicle (deionized water) for the same period.

6. During exposure, the animals may be monitored for respiration rate, depth, and regularity (14).

7. The temperature, oxygen level, and Mn concentration of the system may also be monitored continuously.

3.10. LPS Model

1. Single injection model: A single stereotaxic injection of 2 μg of LPS into the substantia nigra or pallidum of Wistar, Fisher, or Sprague Dawley rats leads to a marked loss (50–85%) of the SNpc DA neurons (15). The procedure is performed as described in Subheading 3.2.

2. SN chronic infusion model: After stereotaxic surgery, nanograms of LPS in an osmotic minipump can be slowly delivered into the SN for a period of 2 weeks in rats. The loss of SNpc DA neurons begins after 2–4 weeks, but does not become significant until 6 weeks after the start of LPS infusion.

3. Systemic LPS injection model: Chronic microglial activation and progressive DA neurodegeneration can be observed following an intraperitoneal injection of 5 mg/kg of LPS in C57BL mice. The loss of nigral DA neurons has been reported to reach 23 and 43% at 7 and 10 months, respectively (16).

4. In utero LPS injection: LPS (10,000 endotoxin units) can be administered by an in utero injection to gravid Sprague Dawley rats on gestation day 10.5, a critical time point during embryonic DA neuron development. In 3-week-old pups from in utero LPS-exposed rat mothers, significant reductions in the number of SNpc DA neurons (27%) and striatal DA content (29%) were observed as compared with pups from saline-injected rat mothers (17).

3.11. Dopamine Measurement

1. The animals are sacrificed by rapid decapitation 1 week after MPTP administration.

2. Both striata are dissected rapidly on a chilled glass plate and transferred immediately into dry ice (see Note 23).

3. The samples can be stored at −80°C until measurement.

4. The samples are subsequently thawed in 0.25 ml of chilled 0.1 M perchloric acid (about 100 μl/mg tissue) and sonicated (see Note 24). Alternatively, the samples are manually homogenized in homogenizer solution for 1 min in a tube.

5. An aliquot is taken for protein quantification with measurement according to the Bio-Rad protein analysis protocol using Perkin Elmer Bio Assay Reader. Alternatively, the wet weight can be used.

6. Other aliquots (e.g., 250 μl) are centrifuged at $4,000 \times g$ at 4°C for 30 min, and dopamine, 3,4-dihydroxyphenylacetic acid (DOPAC), and homovanillic acid (HVA) are measured in the supernatants by HPLC with electrochemical detection.

7. The working potential of the detector is set at +750 mV, and a glassy carbon electrode and an Ag/AgCl reference electrode are used.

8. The flow of the mobile phase onto the reversed-phase column after passage through a precolumn is set to 1–1.2 ml/min at 40°C.

9. Ten-microliter aliquots are injected by the autosampler with the cooling module set at 4°C.

10. The signals captured by the electrochemical detector are converted by a dual-channel interface and the chromatograms are evaluated with software (see Note 25).

11. The concentrations of dopamine, DOPAC, and HVA are expressed as ng per mg protein or ng/mg wet weight (4, 5).

3.12. MPP+ Levels

1. To determine whether the MPTP uptake or metabolism is altered, MPTP is injected intraperitoneally twice at a dose of 20 mg/kg, 2 h apart.

2. The mice are sacrificed 2 h after the last dose. The precise timing is crucial in view of the rapid clearance of MPP+ (see Note 26).

3. Both striata are dissected rapidly on a chilled glass plate and transferred immediately into dry ice.

4. The samples can be stored at −80°C until measurement.

5. MPP+ levels can be quantified by HPLC with UV detection at 295 nm.

6. The samples are sonicated in 0.1 M perchloric acid (about 100 μl/mg tissue).

7. An aliquot of the supernatant is injected onto a Brownlee Aquapore X03–224 cation-exchange column and is eluted isocratically (4, 5).

3.13. Transcardial Perfusion

1. The animals are deeply anesthetized with the above-mentioned drugs.

2. The chest and the abdomen are opened with scissor.

3. A needle is inserted into the left ventricle and the right atrium is cut.

4. The animal is perfused with 10–200 ml normal saline, followed by 20–400 ml 4% paraformaldehyde in 0.1 mM PBS (see Notes 27–29).

5. The brain is then removed in toto.

6. The perfused brains are postfixed in paraformaldehyde for 2–4 h, and then placed in sucrose for 24 h.

7. In sucrose buffer, the brains may be stored further at 4°C (4, 5).

3.14. Immuno-histochemistry

1. The tissues are fixed for histological analysis in 4% paraformaldehyde in 0.1 M phosphate buffer (pH 7.4), and then cryoprotected in 10–30% sucrose or 10–30% glycerol overnight at 4°C. The storage solution may be supplemented with 0.05% (w/v) azide.

2. Blocks containing the whole SNpc are dissected.

3. Coronal brain sections 30–60 μm thick are cut and the sections are collected in a systematic fashion in three vials.

4. The sections are rinsed in PBS, twice for 5 min each.

5. They are then incubated in 10–30% (w/v) sucrose for cryoprotection.

6. If needed, these sections may be stored at −80°C in sealed polypropylene vials for a longer period of time.

7. The sections are rinsed in phosphate buffer, twice for 5 min each.

8. They are next incubated with 0.1% Triton X-100 for 30 min. Alternatively, sections can be frozen and thawed twice in liquid nitrogen.

9. Following rinsing in 0.05 M TBS, the free-floating sections are incubated in 1% (w/w) H_2O_2 in 0.05 M TBS for 30 min to suppress endogenous peroxidase activity.

10. After washing in 0.05 M TBS, the sections are incubated in normal horse serum (2% (v/v)), bovine serum albumin (2%), or human serum albumin (1%) diluted in 0.05 M TBS for 30 min.

11. The sections are incubated in solutions of appropriate primary antibodies (mouse, rabbit anti-TH affinity-purified monoclonal antibody or mouse anti-NeuN-purified monoclonal antibody) for 12–36 h at room temperature or 4°C, respectively.

12. The sections are rinsed with TBS, three times for 10 min each.

13. This is followed by incubation with appropriate biotinylated IgG (1:200, diluted in TBS/0.5% BSA) for 1–3 h at room temperature or at 4°C overnight.

14. The sections are rinsed with TBS, three times for 10 min each.

15. Sections are then incubated with the avidin–biotin peroxidase complex (1:200 in PBS) for 1–3 h.

16. The sections are rinsed with Tris buffer (pH 7.6) three times for a minimum of 10 min each.

17. The immunoreaction can be visualized by using 3,3′-diaminobenzidine tetrahydrochloride dihydrate and possibly with nickel intensification as the chromogen.

18. All incubations and rinses are performed on an orbital shaker.

19. The sections are mounted onto gelatin-coated slides, cleared in xylene, and coverslipped (see Note 30).

3.15. Stereological Cell Count

1. For stereological cell counts, serial coronal sections (50–60 μm) should be cut through the SN using a cryostat or vibratome.

2. Two to three sets of sections are prepared, each set consisting of four to six sections, 100–180 μm apart.

3. One set is stained with Nissl (cresyl violet) or NeuN, as described in Subheading 3.14.

4. The other set is stained for TH immunohistochemistry, as described in Subheading 3.14.

5. The numbers of Nissl-stained or TH-immunoreactive neurons in the SNpc are counted by using the optical fractionator method in the Stereo Investigator (v 4.35) software program.

6. Two to four sections per animal are analyzed.

7. The cell loss in the SNpc is determined by using the same software program.

8. With the optical fractionator method, a grid should be applied in order to produce a sufficient number of counting bricks of approximately $60–140 \times 60–140 \times 15$ μm. To avoid biased counting due to the uneven sectioning, a guard zone of 3–4 μm is set up.

9. The estimated cell count is calculated by interpolation of the cell numbers in the counting brick to the whole SNpc and expressed as the total estimated cell number (see Note 31).

3.16. Rotorod Apparatus

1. The rotorod apparatus for mouse generally consists of a wooden or plastic rod approximately 15 cm long and 7 cm in diameter, attached to a variable-speed motor.

2. The clockwise revolutions of the rotorod can be adjusted from 0 to 20 rpm. The rod is separated from the motor by a vertical 30×48-cm plane.

3. The time for which the mice are able to stay on the rod rotating at a constant rate or the maximum rpm is recorded.

4. The test can be performed 1–3 times per week on the same day of each week.

5. Two to three trials are recommended on the given day and the best performance is used (18).

6. The apparatus is cleaned with ethanol between sessions in order to avoid distractions due to smells (see Notes 32 and 33).

3.17. Open-Field Test

1. The spontaneous locomotor activity and the exploration activity can be detected.

2. The test is performed at the same time of the day and on the same day, at least 2 h after any intervention.

3. Each mouse is placed at the center of the open-field box and its behavior is recorded for 5 min with the aid of software (Fig. 2).

4. The ambulation distance, the mean velocity, the local time, and the number of rearings can be evaluated (19).

5. The apparatus is cleaned with ethanol between sessions in order to avoid distractions due to smells (see Notes 32 and 33).

3.18. Grid Test

1. This test is used to study coordination and rigidity (10).

2. The mice are placed at the center of a wire mesh grid.

3. The grid is rotated through 180° to suspend the mice upside down.

4. The mice are allowed to move freely on the grid.

5. Their movements are taped for a fixed period of time (e.g., 60 s).

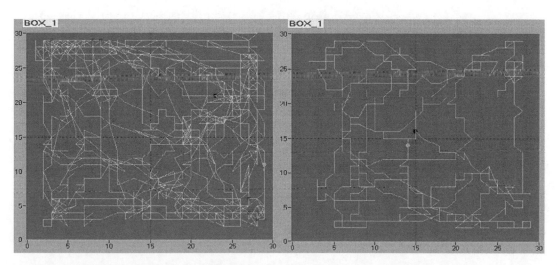

Fig. 2. Typical pattern of spontaneous movements of a mouse in an open-field test during 5 min. Left side normal locomotor activity, right side reduced activity (Conducta system).

6. Forepaw foot faults and total forepaw steps are recorded and the ratio is used as a characteristic of coordination and rigidity.

7. Three trials are performed for each mouse.

8. The best performance may be used or the data may be pooled.

3.19. Conclusions

An appreciable number of model systems are available for studies of the pathomechanism of PD and the neuroprotective or symptomatic effects of certain drugs. However, despite the enormous progress that has been made in this field, a model which exhibits all the features seen in patients is still missing. With the well-established models, the use of unified protocols, methods, and interpretations results in easier comprehension of the data obtained by different research groups, which facilitates pharmacological studies and ultimately improves the preclinical drug screening for clinical trials.

4. Notes

1. Inhalation of magnesium results more chronic model and may be related to pathomechanism of human intoxication. Use this way of administration if human toxicity is being studied.

2. All protocols involving animals must be carried out in strict accordance with the *Guide for the Care and Use of Laboratory Animals* and should be approved by the local institutional committee.

3. Irritating solutions should be avoided and the pH should be adjusted to neutral.

4. Injection into the bladder or liver should be also avoided.

5. For injections, the animals must be marked correctly (with ear tag or tail labeling) so as to avoid confusion if many animals are kept in one cage.

6. When stereotaxic surgery is performed, it is crucial to leave the needle in place for some time after the injection and then remove it very slowly, since toxin can be sucked out during the needle removal.

7. Stereotaxic surgery needs experience. Practice is demanded before any experiments are performed.

8. Taconic Inc., Hudson, NY, USA, offers 6-OHDA lesioned rodent model with positive apomorphin test for sale.

9. Use of these toxins involves a serious risk of human poisoning.

10. Only trained, experienced personnel can carry out MPTP experiments.

11. For MPTP preparation, a certified fume hood must be used.

12. As toxic metabolites of MPTP are excreted into the urine, the bedding is contaminated with toxic metabolites. A mask must be used and the bedding must be changed every day after MPTP administration.

13. Only an isolated room, exclusively designated for MPTP experiments, should be used.

14. If poisoning of personnel occurs, 5 mg selegiline must be taken orally immediately and notify the Safety Department.

15. Several mouse strains may be utilized. For more consistent results, sensitive strains, such as C57 Black or Swiss Webster mice, should be used.

16. MPTP itself is not toxic in rats since they do not possess the MAO-B enzyme necessary for active toxin metabolite formation. However, the stereotaxic administration of MPP^+ results in the same toxicity as seen in MPTP-intoxicated mice.

17. Female mice are more resistant than male mice to MPTP.

18. The sensitivity of animals to MPTP is age-, species-, and strain-dependent, with primates and aged animals the most sensitive.

19. To reach the level of a significant difference between groups, use of at least 8–10 mice per groups is advisable.

20. In the case of PBS solution, the pH is neutral. If saline is used as solvent, pH adjustment may be needed.

21. It is necessary to measure the exact body weight before each injection so as to deliver the desired dose.

22. It is the general experience that the paraquate model is not reliable, as the data can vary significantly between studies. Consistency may be increased by the use of minipumps.

23. After dissection of the brain, the samples must be put on dry ice immediately, as otherwise the levels of catecholamines may be decreased.

24. Sonication can be very noisy. A protective device should be used and consideration should be paid to other staff members.

25. For HPLC measurements, individual changes in the setup may be needed. Internal standards should be used for calibration.

26. MPP^+ measurements are highly dependent on the timing because of the rapid metabolism. A difference of 5–10 min between animals should be maintained to allow sufficient time for brain removal and dissection.

27. For transcardial perfusion, a sufficient volume of saline should be applied in order to remove all the blood from the brain so as to improve the quality of the immunohistochemistry.

28. PFA must be prepared under a hood, since its vapor is irritating.

29. For transcardial perfusion, adequate ventilation and drainage for PFA should be applied, since its inhalation can lead to lung damage. PFA cannot be poured into the sink and should be treated as a hazardous fluid.

30. Immunohistochemistry also requires practice. The most critical steps relate to the quality of the primary antibody and its dilution. For cell counting or immunoreactivity, only well-stained sections should be used.

31. Stereology is very time consuming and demands significant experience. The study should be designed well in order to avoid the unnecessary and invalid methods.

32. All behavioral experiments (open-field and rotarod tests) are very sensitive to external noise. Sudden noises can produce unexpected alterations in the behavior of animals. An isolated, quiet room should be used for such studies.

33. To avoid diurnal fluctuations in behavior, these experiments should always be performed on the same time of day.

References

1. Dorsey ER, Constantinescu R, Thompson JP, et al (2007) Projected number of people with Parkinson disease in the most populous nations, 2005 through 2030. *Neurology* **68**:384–386.

2. O'Brien JA, Ward A, Michels SL, et al (2009) Economic burden associated with Parkinson disease. *Drug Benefit Trend* **21**:179–190.

3. Henchcliffe C, Beal MF (2008) Mitochondrial biology and oxidative stress in Parkinson disease pathogenesis. *Nat Clin Pract Neurol* **4**:600–609.

4. Klivenyi P, Starkov AA, Calingasan NY, et al (2004) Mice deficient in dihydrolipoamide dehydrogenase show increased vulnerability to MPTP, malonate and 3-nitropropionic acid neurotoxicity. *J Neurochem* **88**:1352–1360.

5. Klivenyi P, Andreassen OA, Ferrante RJ, et al (2000) Mice deficient in cellular glutathione peroxidase show increased vulnerability to malonate, 3-nitropropionic acid, and 1-methyl-4-phenyl-1,2,5,6-tetrahydropyridine. *J Neurosci* **20**:1–7.

6. Przedborski S, Levivier M, Jiang H, et al (1995) Dose dependent lesions of the dopaminergic nigrostriatal pathway induced by intrastriatal injection of 6-hydroxydopamine. *Neuroscience* **67**:631–647.

7. Sauer H, Oertel WH (1994) Progressive degeneration of nigrostriatal dopamine neurons following intrastriatal terminal lesions with 6-hydroxydopamine: a combined retrograde tracing and immunocytochemical study in the rat. *Neuroscience* **59**:401–415.

8. Jackson-Lewis V, Przedborski S (2007) Protocol for the MPTP model of Parkinson's disease. *Nat Protoc* **2**:141–151.

9. Jackson-Lewis V, Jakowec M, Burke RE, Przedborski S (1995) Time course and morphology of dopaminergic neuronal death caused by the neurotoxin 1-methyl-4-phenyl-1,2,3,6-tetrahydropyridine. *Neurodegeneration* **4**:257–269.

10. Meredith GE, Totterdell S, Potashkin JA, Surmeier DJ (2008) Modeling PD pathogenesis in mice: advantages of a chronic MPTP protocol. *Parkinsonism Relat Disord* **14** Suppl 2:S112-115.

11. Fleming SM, Zhu C, Fernagut PO, Mehta A, et al (2004) Behavioral and immunohistochemical effects of chronic intravenous and subcutaneous infusions of varying doses of rotenone. *Exp Neurol* **187**:418–429.

12. Peng J, Stevenson FF, Doctrow SR, Andersen JK (2005) Superoxide dismutase/catalase mimetics are neuroprotective against selective paraquat-mediated dopaminergic neuron death in the substantia nigra: implications for Parkinson disease. *J Biol Chem* **280**:29194–29198.

13. Zhao F, Cai T, Liu M, et al (2009) Manganese induces dopaminergic neurodegeneration via microglial activation in a rat model of manganism. *Toxicol Sci* **107**:156–164.

14. Ordoñez-Librado JL, Anaya-Martinez V, Gutierrez-Valdez AL, et al (2010) L-DOPA treatment reverses the motor alterations induced by manganese exposure as a Parkinson disease experimental model. *Neurosci Lett* **471**:79–82

15. Castano A, Herrera AJ, Cano J, Machado A (1998) Lipopolysaccharide intranigral injection induces inflammatory reaction and damage in nigrostriatal dopaminergic system. *J Neurochem* **70**:1584–1592.

16. Qin L, Wu X, Block ML, et al (2007) Systemic LPS causes chronic neuroinflammation and progressive neurodegeneration. *Glia* **55**:453–462.

17. Ling Z, Gayle DA, Ma SY, et al (2002) In utero bacterial endotoxin exposure causes loss of tyrosine hydroxylase neurons in the postnatal rat midbrain. *Mov Disord* **17**: 116–124.

18. Klivenyi P, Ferrante RJ, Matthews RT, et al (1999) Neuroprotective effects of creatine in a transgenic animal model of amyotrophic lateral sclerosis. *Nat Med.* **5**:347–350.

19. Zádori D, Geisz A, Vámos E, et al (2009) Valproate ameliorates the survival and the motor performance in a transgenic mouse model of Huntington's disease. *Pharmacol Biochem Behav* **94**:148–153.

Chapter 15

Behavioral Phenotyping of Mouse Models of Neurodegeneration

Magali Dumont

Abstract

Neurodegenerative disorders, such as amyotrophic lateral sclerosis (ALS), Huntington's (HD), Parkinson's (PD) and Alzheimer's diseases (AD), are characterized by the loss of structure and function of specific neuronal circuitry in the brain. As a result of this loss, behavioral symptoms occur progressively.

Understanding the causes of neurodegeneration is fundamental for the development of new therapeutic targets. For this purpose, several animal models of neurodegenerative disorders have been generated and characterized. During the characterization, behavioral science plays a crucial role by identifying specific symptoms in these animal models of human disorders. Later on, it also allows scientists to verify the efficacy of new treatments.

This chapter describes some of the standard tests used to assess behavioral symptoms present in mouse models of neurodegenerative disorders. A list of procedures is provided to evaluate motor skills for the study of ALS, HD, and PD models, and to evaluate spatial learning and memory for the study of AD models.

Key words: Neurodegeneration, Mouse models, Alzheimer's disease, Parkinson's disease, Huntington's disease, Amyotrophic lateral sclerosis, Behavioral phenotype, Sensory skills, Motor skills, Spatial learning and memory

1. Introduction

Various mouse models are now available to study molecular aspects of neurodegeneration (1–3). The most commonly used models of neurodegeneration are transgenic mouse models of amyotrophic lateral sclerosis (ALS) (4), Huntington's (HD) (5, 6), Parkinson's (PD) (7, 8), and Alzheimer's diseases (AD) (9, 10). Studying the effect of targeted gene mutations on molecular aspects is fundamental in the field of neurodegeneration. However, this must be

Giovanni Manfredi and Hibiki Kawamata (eds.), *Neurodegeneration: Methods and Protocols*,
Methods in Molecular Biology, vol. 793, DOI 10.1007/978-1-61779-328-8_15, © Springer Science+Business Media, LLC 2011

complemented by studying the effect of targeted gene mutations on behavior (11). Behavior phenotyping provides a quantitative marker for human disease symptoms, and a preclinical tool to assess new treatments.

Motor dysfunction present in ALS, HD, and PD transgenic mice can be classified as locomotor deficit, impaired balance, or muscular weakness (12). Cognitive dysfunction present in AD transgenic mice can also be classified, such as spatial learning and memory deficits (13). For each symptom, one or more behavioral tests are available. This chapter describes some of the most standard procedures used to evaluate general health, sensory and motor skills, and spatial learning and memory in mouse models of neurodegeneration.

2. Materials

Behavioral testing on mice should be done with specific apparatus designed for mice, since the dimensions of the apparatus can influence the behavioral responses to the task.

2.1. General Health and Sensory Skills

1. Body weight: standard balance Ohaus (Scout II, VWR, West Chester, PA, USA).
2. Home cage activity: video camera and home cage from the animal facility.
3. Olfactory discrimination: extracts or oils such as vanilla and almond.
4. Visual placing test: metal grid or home cage lid.
5. Acoustic startle response: automated startle (SR-LAB, SD Instruments, San Diego, CA, USA).

2.2. Motor Skills

Locomotion

1. Openfield (Fig. 1a): square opaque plastic chamber (45 cm×45 cm, 20 cm height).
2. Circling: round transparent plastic chamber (30 cm diameter, 30 cm height).
3. Gait: plastic dark tunnel (10 cm width, 50 cm length, 10 cm height) with white paper on the bottom surface; black non-toxic and odor-free ink or paint.

Motor coordination and balance

1. Balance beam (Fig. 1b): plastic beam (2.5 cm diameter, 100 cm length, 35 cm height).
2. Rotarod (Fig. 1c): plastic apparatus (4 cm diameter, 8 cm width, 38 cm height) (Economex, Columbus Instruments, Columbus, OH, USA).

a Openfield test

b Balance beam test

c Rotarod test

Fig. 1. Behavioral tests evaluating motor skills. (**a**) Openfield test; (**b**) balance beam test; and (**c**) rotarod test.

Muscular strength

1. Hanging wire: metal grid or home cage lid (24 cm×20 cm, height: 30 cm).

2. Force: sensor and metal grid (Grip strength meter, Columbus Instruments, Columbus, OH, USA).

2.3. Spatial Learning and Memory

1. T-maze (Fig. 2a): opaque plastic maze with two arms (17 cm×7 cm, 10 cm height) perpendicular to a central stem (24 cm×7 cm, 10 cm height) (SD instruments, San Diego, CA, USA).

a T-maze test

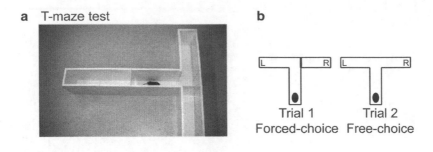

c   Morris water maze: hidden platform

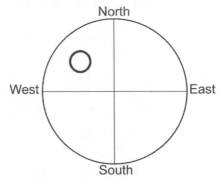

d   Ethovision 3.1. software   **e** Morris water maze: visible platform

Fig. 2. Behavioral tests evaluating spatial learning and memory. (**a**) T-maze test; (**b**) schematic view of the forced and free-choice trials in the T-maze; (**c**) schematic view of the Morris water maze for the acquisition period; (**d**) display from Ethovision 3.1. software; and (**e**) visible platform in the Morris water maze.

2. Morris water maze: round basin (120 cm diameter, 60 cm height) filled with water (20–25°C); nontoxic and odor-free white tempera paint; circular and transparent platform (10 cm diameter, 50 cm height).

3. Video tracking system (Ethovision 3.1, Noldus Technology, Attleborough, MA, USA).

3. Methods

Sources of stress must be controlled before and during the behavioral testing (see Note 1).

3.1. General Health and Sensory Skills

Animals must be familiarized with the experimenter prior to any behavioral testing (see Note 2). After that, examination of the general health and sensory skills is a crucial step in behavioral phenotyping. Compromise health or sensory impairments have detrimental consequences on behavioral performances (see Note 3). Therefore, this examination must be performed prior to any complex task.

1. Body weight: mice are placed in a plastic container to avoid any escape and are weighted on a scale at different stages of the disease.

2. Home cage activity (14): spontaneous activity (sleeping, grooming, nesting, feeding, drinking, etc.) is recorded in the home cage, inside the animal facility, by video camera for several 24 h periods. Scoring is done afterward by the experimenter.

3. Olfactory discrimination (15): odor extracts are painted into the side of a clean cage. Each mouse is placed in the cage and latency before the first sniff and cumulative time spent sniffing are recorded for 2–5 min with a stopwatch. Cages are cleaned with 70% ethanol in between animals.

4. Visual placing test (16): mice are suspended by the tail to a height of approximately 15 cm and lowered to a metal grid within seconds. Scoring is based on the distance of the animal's nose from the grid before extending the forelimbs toward it.

5. Acoustic startle response (17): mice are placed in the cylinder of the chamber and a white noise (70–75 dB) is delivered as background level. Brief tones from 75 to 120 dB are then delivered randomly. Amplitude of the whole body flinch in response to tones is recorded.

3.2. Motor Skills

Motor dysfunction is an important feature in ALS, Huntington (HD), and Parkinson's diseases (PD). Mouse models of these diseases also develop progressive motor impairments (18, 19), such as locomotor and balance deficits, as well as muscular weaknesses. Behavioral tests and procedures used to assess these deficits are describes in this section.

All tasks presented below evaluate motor skills and can be repeated for several days and at different stages of disease progression (see Note 4).

Locomotion

1. Openfield: mice are placed in one corner of the openfield for 5–30 min. Distance moved, number of rearing, time spent in the periphery or the center of the apparatus are recorded, using a video tracking system. The apparatus is cleaned with 70% ethanol in between animals.

2. Circling (20): mice are introduced in the chamber for 5 min. Time spent circling is recorded with a stopwatch.

3. Gait (21): mouse paws are dipped into black ink or paint. Each animal is placed at the entrance of the dark tunnel. By walking down the tunnel, each animal leaves black footprints on the white paper. Footprint patterns are analyzed, such as distance between stride, variability in stride length and around a linear axis.

Motor coordination and balance

1. Balance beam (22): the beam is divided into ten segments of 10 cm each with a black marker. Each mouse is placed in the center segment and is allowed to walk on the beam for 60 s. Number of segments crossed and latency to fall are recorded and averaged over four trials per day with 20 min intertrial intervals (see Note 5). Slips from the beam can also be counted.

2. Rotarod (23): mice are placed on the accelerating rod for 2–5 min, with acceleration from 4 to 40 rpm. The rotarod can also be used at constant speed for the same period of time. Latency to fall is recorded and averaged over 4–8 trials per day with an intertrial interval of 20 min.

Muscular strength

1. Hanging wire (24): mice are placed in the middle of a metal grid. The grid is gently shaken to allow animals to grip firmly, then turned upside down and placed at about 30–40 cm high (padding is necessary to avoid fall injury). Latency to fall is measured with a cutoff period of 60 s up to several minutes, using a stopwatch. One to four trials per day can be performed and averaged with 20 min intertrial intervals.

2. Force: mice are placed onto the metal bar to allow firm grip with the forelimbs or hindlimbs. The tail is gently pulled horizontally until the animal releases the bar. Maximum force is recorded by the sensor. One to four trials per day can be performed and averaged with 20 min intertrial intervals.

3.3. Spatial Learning and Memory

Cognitive dysfunction is an important feature in Alzheimer's disease. Both human patients and AD transgenic mice display learning and memory deficits (25), especially related to spatial aspects. Classically, mazes are used to assess this type of deficits in mice.

In this section, two examples are described: the T-maze (26) and the Morris water maze (27).

1. T-maze (Fig. 2b): mice are placed in the central stem of the T-maze with the right arm blocked by a plastic barrier (forced choice). Once they enter into the available arm, mice are kept in it for 60 s. Mice are then retrieved and the barrier removed. Mice are immediately placed back in the central stem for a free-choice trial in which both arms are open, with a cutoff period of 60 s. Number of alternations and latency before choosing during the free-choice trial are recorded over 5–10 days. The side of the blocked arm is changed every day (first day right, next day left) (26, 28). The delay between the forced-choice trial and the free-choice trial can be increased.

2. Morris water maze: the round basin is filled with water at 23°C (see Note 6). The water is opacified with nontoxic and odor-free white tempera paint. During the acquisition period, extra-maze visual cues, such as light fixtures and wall posters are arranged in the room. The hidden platform is submerged in the middle of the northwest (NW) quadrant of the pool, 1 cm beneath water level (Fig. 2c). Each day, mice are placed next to and facing the wall of the basin in four different starting positions: north, east, south, and west, respectively, corresponding to four successive trials. Latencies and total distances before reaching the platform are recorded for 5 days with a video tracking system (Fig. 2d). For 20 min after each trial, mice are placed in plastic holding cages filled with paper towels to keep them dry and warm. Whenever a mouse fails to reach the platform within the maximum allotted time of 60 s, it is manually placed on the platform for 5 s by the experimenter. The acquisition period can be extended from 5 to 10 days.

A probe trial is conducted 24 h after the acquisition period, in which the platform is removed from the pool. Mice are released on the north side of the maze for a single trial of 60 s, during which the percent time spent in each quadrant is measured for the first 15 s and for the entire 60 s (see Note 7).

Animals are also tested for four trials per day over 2 days in the visible platform version of the Morris water maze to ensure any visual or motor deficits (Fig. 2e) (see Note 8). Each trial lasts 60 s and is followed by a 20 min intertrial period. In this cued version, a 13 cm pole is fixed on the platform. Mice are released from the north side on the first day and from the south side of the pool on the second day. For each trial, the cued platform is moved to a different quadrant so that the mouse could only locate it visually. Latencies and total distances before reaching the visible platform are recorded.

4. Notes

1. Stress (noise, transportation, smells, other animals, etc.) can interfere with the behavioral response during the testing. For example, after transportation from the animal facility to the behavioral room, animals should be allowed to recover for at least 15–30 min before starting the testing. In addition, animals should be kept in a quiet room and handled gently. Behavioral testing should be organized to conduct stressful tasks at the end of the battery of tests (such as the Morris water maze).

2. Animals should be handled by the same experimenter for at least 1 week prior to starting any behavioral testing.

3. Sick animals should not be used for behavioral testing.

4. Motor tests can be repeated for several days. This repetition can provide evidences of motor learning and habituation. Repeating tests at various stages of the disease can also provide evidences of progressive motor impairment.

5. Stress and fatigue due to the testing can interfere with motor performances. For this reason, an intertrial interval must be given to the animals to recover between trials, typically 15–30 min.

6. Water temperature is very important during the Morris water maze testing. It should be maintained between 20 and 25°C, since hypothermia can interfere with behavioral performances.

7. During the probe trial, mice are allowed to search for the platform location for 60 s. Time spent in each quadrant is calculated over the 60 s. However, after several passages in the area of the missing platform, mice can quickly switch strategy and start an active search of the platform elsewhere. For this reason, measurements can also be calculated only for the first 15 s in the pool.

8. In order to exclude any visual and motor deficits that can affect performances during the acquisition period, a visible platform task is required. This cued platform version of the test can be conducted before or after the acquisition period.

References

1. Wong, P. C., Cai, H., Borchelt, D. R., and Price, D. L. (2002) Genetically engineered mouse models of neurodegenerative diseases. *Nat Neurosci.* **5**, 633–639.

2. Zoghbi, H. Y., and Botas, J. (2002) Mouse and fly models of neurodegeneration. *Trends Genet.* **18**, 463–471.

3. Langui, D., Lachapelle, F., and Duyckaerts, C. (2007) [Animal models of neurodegenerative diseases]. *Med Sci (Paris).* **23**, 180–186.

4. Jackson, M., Ganel, R., and Rothstein, J. D. (2002) Models of amyotrophic lateral sclerosis. *Curr Protoc Neurosci.* **Chapter 9**, Unit 9 13.

5. Ferrante, R. J. (2009) Mouse models of Huntington's disease and methodological considerations for therapeutic trials. *Biochim Biophys Acta*. **1792**, 506–520.

6. Heng, M. Y., Detloff, P. J., and Albin, R. L. (2008) Rodent genetic models of Huntington disease. *Neurobiol Dis*. **32**, 1–9.

7. Chesselet, M. F., Fleming, S., Mortazavi, F., and Meurers, B. (2008) Strengths and limitations of genetic mouse models of Parkinson's disease. *Parkinsonism Relat Disord*. **14 Suppl 2**, S84–87.

8. Hattori, N., and Sato, S. (2007) Animal models of Parkinson's disease: similarities and differences between the disease and models. *Neuropathology*. **27**, 479–483.

9. Morrissette, D. A., Parachikova, A., Green, K. N., and LaFerla, F. M. (2009) Relevance of transgenic mouse models to human Alzheimer disease. *J Biol Chem*. **284**, 6033–6037.

10. Gotz, J., and Ittner, L. M. (2008) Animal models of Alzheimer's disease and frontotemporal dementia. *Nat Rev Neurosci*. **9**, 532–544.

11. Crawley, J. N. (2008) Behavioral phenotyping strategies for mutant mice. *Neuron*. **57**, 809–818.

12. Fleming, S. M. (2009) Behavioral outcome measures for the assessment of sensorimotor function in animal models of movement disorders. *Int Rev Neurobiol*. **89**, 57–65.

13. Eriksen, J. L., and Janus, C. G. (2007) Plaques, tangles, and memory loss in mouse models of neurodegeneration. *Behav Genet*. **37**, 79–100.

14. Steele, A. D., Jackson, W. S., King, O. D., and Lindquist, S. (2007) The power of automated high-resolution behavior analysis revealed by its application to mouse models of Huntington's and prion diseases. *Proc Natl Acad Sci USA*. **104**, 1983–1988.

15. Yang, M., and Crawley, J. N. (2009) Simple behavioral assessment of mouse olfaction. *Curr Protoc Neurosci*. **Chapter 8**, Unit 8 24.

16. Pinto, L. H., and Enroth-Cugell, C. (2000) Tests of the mouse visual system. *Mamm Genome*. **11**, 531–536.

17. Logue, S. F., Owen, E. H., Rasmussen, D. L., and Wehner, J. M. (1997) Assessment of locomotor activity, acoustic and tactile startle, and prepulse inhibition of startle in inbred mouse strains and F1 hybrids: implications of genetic background for single gene and quantitative trait loci analyses. *Neuroscience*. **80**, 1075–1086.

18. Brooks, S. P., and Dunnett, S. B. (2009) Tests to assess motor phenotype in mice: a user's guide. *Nat Rev Neurosci*. **10**, 519–529.

19. Menalled, L., El-Khodor, B. F., Patry, M., Suarez-Farinas, M., Orenstein, S. J., Zahasky, B., Leahy, C., Wheeler, V., Yang, X. W., MacDonald, M., Morton, A. J., Bates, G., Leeds, J., Park, L., Howland, D., Signer, E., Tobin, A., and Brunner, D. (2009) Systematic behavioral evaluation of Huntington's disease transgenic and knock-in mouse models. *Neurobiol Dis*. **35**, 319–336.

20. Mura, A., Feldon, J., and Mintz, M. (1998) Reevaluation of the striatal role in the expression of turning behavior in the rat model of Parkinson's disease. *Brain Res*. **808**, 48–55.

21. Wooley, C. M., Sher, R. B., Kale, A., Frankel, W. N., Cox, G. A., and Seburn, K. L. (2005) Gait analysis detects early changes in transgenic SOD1(G93A) mice. *Muscle Nerve*. **32**, 43–50.

22. Lalonde, R., Hayzoun, K., Selimi, F., Mariani, J., and Strazielle, C. (2003) Motor coordination in mice with hotfoot, Lurcher, and double mutations of the Grid2 gene encoding the delta-2 excitatory amino acid receptor. *Physiol Behav*. **80**, 333–339.

23. Barlow, C., Hirotsune, S., Paylor, R., Liyanage, M., Eckhaus, M., Collins, F., Shiloh, Y., Crawley, J. N., Ried, T., Tagle, D., and Wynshaw-Boris, A. (1996) Atm-deficient mice: a paradigm of ataxia telangiectasia. *Cell*. **86**, 159–171.

24. Gallardo, G., Schluter, O. M., and Sudhof, T. C. (2008) A molecular pathway of neurodegeneration linking alpha-synuclein to ApoE and Abeta peptides. *Nat Neurosci*. **11**, 301–308.

25. Kobayashi, D. T., and Chen, K. S. (2005) Behavioral phenotypes of amyloid-based genetically modified mouse models of Alzheimer's disease. *Genes Brain Behav*. **4**, 173–196.

26. Lalonde, R. (2002) The neurobiological basis of spontaneous alternation. *Neurosci Biobehav Rev*. **26**, 91–104.

27. Morris, R. (1984) Developments of a water-maze procedure for studying spatial learning in the rat. *J Neurosci Methods*. **11**, 47–60.

28. Deacon, R. M., and Rawlins, J. N. (2006) T-maze alternation in the rodent. *Nat Protoc*. **1**, 7–12.

Part IV

Neurodegenerative Disease Mechanisms: Proteinopathies

Chapter 16

Characterization of Amyloid Deposits in Neurodegenerative Diseases

Ruben Vidal and Bernardino Ghetti

Abstract

The extracellular accumulation of insoluble fibrillar peptides in brain parenchyma and vessel walls as amyloid is the hallmark of neurodegenerative diseases, such as Alzheimer's disease and Prion diseases. Regardless their amino acid sequences, all amyloid peptides adopt an insoluble, highly ordered beta sheet structure when aggregated. Amyloid is homogeneous and eosinophilic and, common to most cross-beta-type structures; it is generally identified by apple-green birefringence when stained with Congo red and seen under polarized light. Amyloid can also be identified by an apple green color when stained with thioflavine-S and seen under a fluorescence microscope. By electron microscopy, the typical fibrillar ultrastructure of amyloid deposits is revealed. The biochemical nature of the amyloid subunits present in the deposits can be recognized by immunohistochemistry using specific antibodies or by amino acid sequencing analysis, western blot, and mass spectrometry after isolation of parenchymal or vascular amyloid proteins.

Key words: Amyloid, Fibrillar, Neurodegeneration, Dementia, Cerebral amyloid angiopathy

1. Introduction

Amyloid deposition in the parenchymal and vascular extracellular spaces is the main neuropathological finding in several neurodegenerative diseases, including Alzheimer's disease (AD), Prion diseases, sporadic cerebral amyloid angiopathy (CAA), familial British and Danish dementia (FBD and FDD), hereditary cerebral hemorrhage with amyloidosis-Dutch and Icelandic (HCHWA-D and HCHWA-I) types, hereditary leptomeningeal-type transthyretin amyloidosis, Finnish amyloidosis, and some forms of immunoglobulin light-chain deposition (Table 1) (1–12). Amyloid is an in vivo deposited material, which can be distinguished from non-amyloid deposits by characteristic fibrillar electron microscopic appearance and histological staining reactions, particularly the

Giovanni Manfredi and Hibiki Kawamata (eds.), *Neurodegeneration: Methods and Protocols,*
Methods in Molecular Biology, vol. 793, DOI 10.1007/978-1-61779-328-8_16, © Springer Science+Business Media, LLC 2011

Table 1
Amyloid proteins found in the central nervous system

Amyloid protein	Precursor	Disease
Aβ	Aβ protein precursor (AβPP)	AD, CAA, HCHWA-D
APrP	Prion protein	Spongiform encephalopathies
ATau	MAPT	AD, FTD-17, others
ACys	Cystatin C	HCHWA-I
ABri	ABriPP (BRI$_2$)	Familial dementia, British
ADan	ADanPP (BRI$_2$)	Familial dementia, Danish
ATTR	Transthyretin	Leptomeningeal amyloidosis
AL	Immunoglobulin light chain	Amyloidoma, others
AGel	Gelsolin	Finnish amyloidosis

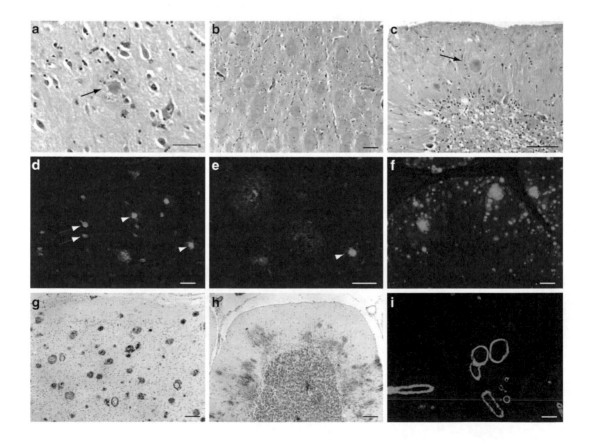

affinity for the dye Congo red with resulting green birefringence (1, 13) (Figs. 1 and 2). The increasing knowledge of the exact biochemical nature of amyloid disorders facilitated the neuro-pathological diagnosis of these disorders, development of animal models, and therapeutic approaches. However, in some instances, the identity of the amyloid subunit remains elusive and biochemical approaches are needed to identify the amyloid polypeptide and determine whether the disease represents a new entity (8, 9).

2. Materials

2.1. Fixative and Tissue Processing

1. Formalin solution: 100 ml formaldehyde (37–40%) (Fisher Scientific); 900 ml distilled water (dH_2O) (1:10 dilution).
2. Ethanol (100, 95, 80, 50% in dH_2O).
3. Paraplast brand paraffin.

2.2. Hematoxylin and Eosin

1. Stock solution of eosin:

   1% aqueous eosin-Y: 1 g eosin (Sigma); 100 ml dH_2O, mix to dissolve.

   1% aqueous phloxin B: 1 g phloxin B (Sigma); 100 ml dH_2O, mix to dissolve.
2. Working solution of eosin: 100 ml stock eosin-Y; 10 ml stock phloxin B; 780 ml 95% ethanol; 4 ml glacial acetic acid. Mix well.
3. Harris Hematoxylin (Thermo Shandon Inc.).
4. Working solution of saturated lithium carbonate: Add dH_2O to enough lithium carbonate to leave white sediment at the bottom of the bottle. No need to filter before use.
5. 0.25% acid alcohol: 2,578 ml 95% ethanol; 950 ml dH_2O; 9 ml HCL.

2.3. Congo Red

1. 0.5% Congo red in 50% alcohol: 0.5 g Congo red (Sigma); 100 ml 50% alcohol.
2. 1% sodium hydroxide: 1 g sodium hydroxide; 100 ml dH_2O.

Fig. 1. Photomicrographs of amyloid deposits from hippocampus (**a**), temporal cortex (**b, d, e, g**) and cerebellum (**c, f, h**) from a sporadic AD case (sAD) (**a, d**), a familial AD (FAD) case (*PSEN1 A431E* mutation) (**b, e, g, h**), and a Gerstmann–Sträussler–Scheinker syndrome (GSS) case (*PRNP F198S* mutation) (**f**). The *arrow* indicates the presence of a typical amyloid core (**a**). Numerous well-demarcated cotton wool plaques (CWPs) can be clearly observed by H&E in some cases of FAD (**b**). Amyloid cores (*arrowhead*) and neurofibrillary tangles (*arrows*) are fluorescent (**d**). CWPs are mildly fluorescent compared to amyloid cores (*arrowhead*) (**e**). APrP deposits are strongly fluorescent (**f**). Leptomeningeal vessel walls (**i**) are intensely fluorescent, indicating the presence of amyloid angiopathy in a case of sAD. Amyloid cores and CWPs can be identified using antibodies against the Aβ peptide, which also shows the presence of CAA (**g**). Extensive (*diffuse*) Aβ deposition in the cerebellum (**h**). (**a–c**), H&E stain, (**d–f, i**), thioflavin-S method, (**g, h**), immunohistochemistry using antibodies 10D5 (**g**) and 21F12 (**h**). Bar: 100 μm (**b, f, g, h, i**), 50 μm (**c, d**), and 20 μm (**a, e**).

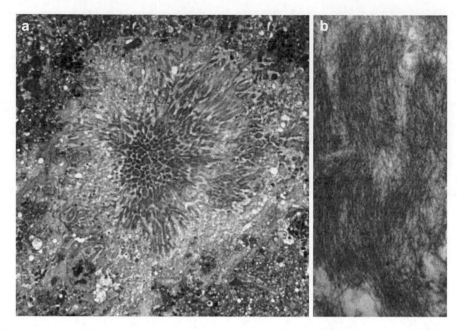

Fig. 2. TEM of an amyloid plaque (**a**) and amyloid fibrils in the plaque seen at higher magnification (**b**). Neocortex of a patient with FAD (*AβPP V717F* mutation).

3. Alkaline alcohol solution: 1 ml 1% sodium hydroxide; 100 ml 50% alcohol.

2.4. Thioflavine-S

1% Thioflavine-S (Th-S) solution: 0.5 gm Thioflavine-S (Sigma); 50.0 ml dH_2O. Make fresh, discard after use. Caution: Avoid contact and inhalation.

2.5. Immunohisto-chemistry

1. Xylene (to remove paraffin).

2. Ethanol (100, 95, 80, 70% in dH_2O).

3. 3% H_2O_2 in methanol: Add 3 ml of H_2O_2 to 97 ml of methanol.

4. Tris-buffered saline (TBS): 0.05 M Tris buffer, pH 7.4; NaCl (to 2.5% w/v). Dissolve 6.1 g tris (hydroxymethyl) methylamine in 50 ml of dH_2O. Add 37 ml of 1 M HCl. Add 25 g NaCl. Dilute to a total volume of 1 l with dH_2O. pH should be 7.4 at 25°C; adjust with 1 M HCl if necessary.

5. TBS plus 0.1% Triton X-100: 0.1 ml of Triton X-100 in 100 ml TBS.

   Milk buffer: 5% nonfat dry milk in TBS plus 0.1% Triton X-100.

6. Citrate buffer: 0.01 M citrate buffer; to 800 ml of dH_2O, add 2.94 g tri-sodium citrate (dihydrate). Adjust pH to 6.0 with 1N HCl; add dH_2O to 1 l. Store at 4°C.

   3,3'-Diaminobenzidine (DAB) solution: 6 mg DAB in 10 ml TBS.

2.6. Transmission Electron Microscopy

Fixatives:

1. 5.0% glutaraldehyde in 0.067 M phosphate buffer with 0.005% calcium chloride (pH 7.35). Best prepared and used fresh.

2. 2.5% glutaraldehyde in 0.1 M sodium cacodylate buffer (add 1 ml of 25% glutaraldehyde stock to 9 ml of buffer). Best prepared and used fresh.

3. Osmium tetroxide: 1% solution of osmium tetroxide in dH_2O.

Buffers (pH should be within the range 7.2–7.4, corrected with 0.1 M HCl):

1. 0.1 M sodium cacodylate: 10.7 g in 500 ml of dH_2O.

2. 0.2 M sodium cacodylate: 21.4 g in 500 ml of dH_2O.

3. 0.4 M sodium cacodylate: 42.8 g in 500 ml of dH_2O.

4. 10 mM HEPES (N-2-hydroxyethylpiperazine-N'-2-ethane-sulfonic acid): 0.59 g of HEPES in 250 ml of dH_2O. Adjust pH to 7.5 with potassium hydroxide (KOH) and store.

5. 1% ammonium acetate: 1.0 g in 100 ml of dH_2O.

Stains:

1. Toluidine blue: 1 part 5% toluidine blue; 1 part 2% sodium borate; 1 part dH_2O. Mix well and then filter. Store at room temperature (RT).

2. Uranyl acetate: Methanolic (saturated uranyl acetate in 50% methanol); aqueous (saturated uranyl acetate in dH_2O). (Store in a brown glass bottle).

3. Lead citrate: 1.33 g lead nitrate; 1.76 g sodium citrate; 30 ml dH_2O. Shake for 1 min. Allow to stand for 30 min, shaking the solution occasionally. Add 8 ml 1 M NaOH; mix. Dilute to 50 ml with dH_2O. Final pH should be pH 12.

Epoxy resin: Polybed 812 (Polysciences, Warrington, PA).

2.7. Rapid Isolation and Biochemical Analyses of Amyloid Deposits

1. Dounce glass homogenizer.

2. Tris-EDTA (TE): 0.1 M Tris, pH 7.4; 5 mM EDTA; 0.02% sodium azide. Dissolve 12.2 g tris (hydroxymethyl) methylamine in 600 ml of dH_2O. Add 1.86 g of EDTA disodium salt (Sigma) and 0.2 g of sodium azide. Add 1 M HCl to pH 7.4 at 25°C. Dilute to a total volume of 1 l with dH_2O.

3. Protease inhibitor (PI) cocktail: Complete with 1 mM Pepstatin, 100 mM TLCK-HCl, 200 mM TPCK, and 1 mM Leupeptin (all from Roche Applied Science) in TE buffer.

4. Ultracentrifuge: We use a benchtop Beckman ultracentrifuge (Beckman TLA 110 rotor) with a fixed-angle (28° angle) TLA 110 rotor.

5. 30% sodium dodecyl sulfate (SDS) solution in TE: 30 g of SDS; TE up to 100 ml.

6. 99% glass-distilled formic acid

7. Nitrogen gas tank with a nitrogen evaporator to evaporate formic acid and concentrate samples.

8. Collagenase/DNase solution: 0.3 mg/ml collagenase CLS-3; 10 μg/ml DNase I in TE buffer with 2 mM $CaCl_2$.

9. Mesh woven filter: 300, 100, 30-μm nylon mesh woven filter (Spectrum Laboratories, Inc) and filter holder.

10. Leica AS LMD laser micro dissection system and membrane coated slides (Leica Microsystems).

11. Paramagnetic beads: Paramagnetic Dynabeads M-450 coated with goat anti-mouse (rabbit) IgG (Dynal Biotech ASA, Oslo, Norway).

12. Phosphate-buffered saline (PBS): 10 mM phosphate, pH 7.4. To prepare, add 8.00 g NaCl; 0.20 g KCl; 1.44 g Na_2HPO_4; 0.24 g KH_2PO_4; dissolve in 800 ml of dH_2O and adjust the pH to 7.4 with HCl or NaOH. Add dH_2O to 1 l. The simplest way to prepare a PBS solution is to use PBS buffer tablets.

13. BSA solution: 0.1% bovine serum albumin (BSA): dissolve 0.1 g BSA in 100 ml PBS.

3. Methods

3.1. General Autopsy Procedures

All deceased patients undergoing autopsy must be treated as potentially infectious cases. The pathologist and assisting personnel must wear a gown, shoe covers, and eye protection. Work areas and spills must be decontaminated with chlorine bleach or chlorine bleach solution (1:10). Decontamination procedures must be carried out as described by the Committee on Health Care Issues of the American Neurological Association. If a Prion disease is suspected, all personnel are advised of the possible health hazard and instructed on the proper use of disinfectants. In addition, the skull is opened under the cover of a plastic bag barrier using a handsaw instead of an electric one. Tissues are considered contagious, even after prolonged formalin fixation, and all materials and instruments in contact with potentially infected material are incinerated. Nondisposable materials are decontaminated by soaking in 1.0N NaOH or 5% hypochlorite for at least 1 h. Blood, CSF, and other fluids are handled with the same precautions given to specimens with hepatitis. All personnel handling these fluids are advised that decontamination of materials with alcohol, formaldehyde, and ultraviolet sterilizing lamps is not appropriate. Brains are examined by sectioning and gross diagnosis. A gross description of the macroscopic findings is dictated the same day, along with a cassette summary. Cassettes are submitted on blocks of defined areas, including lesions, as well as areas relevant to the history and general autopsy findings.

3.2. Brain Harvesting

The brain is removed following standard procedures. Distortion of the specimen is carefully avoided. If allowed by the autopsy consent, the spinal cord and other pertinent tissues are removed. At the time of autopsy, the cerebrum, cerebellum, and brainstem are carefully inspected and weighed; the arteries of the Circle of Willis are measured and analyzed for the presence of atheromatous changes and other pathologies. The dissection of the brain specimen is performed as soon as possible using a carefully standardized protocol. Available clinical information is used in order to obtain the most representative material for neuropathologic and molecular studies. The dissection is carried out as follows: (a) separation of the cerebellum and brainstem from the cerebral hemispheres at the level of the upper midbrain; (b) separation of the cerebral hemispheres by a mid-sagittal incision; (c) both right and left hemispheres are weighed separately; (d) the left cerebral and cerebellar hemispheres and the brain stem are fixed (see Note 1); (e) the right cerebral and cerebellar hemispheres and the brain stem are dissected into a series of slices according to standard planes; (f) the serial coronal slices, including cortex, subcortical nuclei, cerebellum, and brainstem, are frozen in dry ice to avoid distortion, thus allowing detailed regional sampling for biochemical and molecular studies. For this purpose, 1-cm-thick sections are placed on a block of dry ice and allowed to freeze. After a few minutes, sections are turned over and placed between two blocks of dry ice. Each frozen section is sealed in plastic envelopes, labeled, and stored in a −70°C freezer. A 1 cm^3 sample of cerebellar cortex is routinely stored at −70°C for DNA isolation and genetic analyses.

3.3. Fixation and Tissue Process

1. Tissue is fixed at room temperature in formalin. Whole hemispheres are fixed for 2 weeks. Tissue may be cut for better fixation. 5-mm-thick slices may only require 24 h in fixative. Tissue is then prepared for submission (Table 2).

2. Tissue processing with automatic processor: Wash tissue with tap water for 30 min. Place in 50% ethanol for 2 h.

Step	Time (h)
80% ethanol	1.0
95% ethanol	2.0
95% ethanol	1.0
95% ethanol	1.0

(continued)

Step	Time (h)
100% ethanol	2.0
100% ethanol	1.5
100% ethanol	1.0
100% ethanol	2.5
Xylene (equal parts)	
Xylene	2.5
Xylene	3.0
Paraffin	3.0 (58–60°C)
Paraffin	4.0 (58–60°C with vacuum)

Embed in fresh paraplast paraffin. Store at RT in cool place.

Table 2
Tissue blocks that are routinely selected for study

1. Superior frontal and cingulate gyri
2. Middle frontal gyri
3. Caudate nucleus, putamen
4. Putamen, globus pallidus, substantia innominata
5. Amygdala
6. Superior and middle temporal gyri
7. Thalamus, subthalamic nucleus
8. Hippocampus, anterior
9. Hippocampus and entorhinal cortex
10. Superior parietal lobule
11. Calcarine cortex
12. Cerebellar cortex and dentate nucleus
13. Midbrain
14. Pons
15. Medulla

3.4. Stainings

H&E Staining

1. Dry sections in 60°C oven for at least 2 h.

2. Deparaffinize sections: Xylene, three changes, 5 min each.

3. Rehydrate in three changes of 100% alcohol, 5 min each.

  4. 95% alcohol for 5 min.

  5. Stain in Harris hematoxylin solution for 8 min. (Filter before each use to remove oxidized particles.)

  6. Rinse in dH_2O until water is clear.

  7. Wash in running tap water for 5 min.

  8. Differentiate in 1% acid alcohol for 30 s.

  9. Wash in running tap water for 5–10 min.

  10. Dilute saturated lithium carbonate 1:1 with dH_2O and dip slides two times.

  11. Wash in running tap water for 5–10 min.

  12. Place slides in 70% alcohol for 1 min.

  13. Counterstain in filtered eosin solution for 30 s.

  14. Dip two times in 70% alcohol.

  15. Dip five times in 95% alcohol.

  16. Dip five times in 100% alcohol, five times in second 100% alcohol, and 30 s in third 100% alcohol.

  17. Clear in three changes of xylene, 5 min each.

  18. Mount with xylene-based mounting medium (see Note 2).

Congo Red Staining

  1. Deparaffinize and hydrate sections to water as described above.

  2. Stain in Congo Red solution for 15–20 min.

  3. Rinse in dH_2O.

  4. Differentiate quickly (5–10 dips) in alkaline alcohol solution.

  5. Rinse in tap water for 1 min.

  6. Counterstain with hematoxylin for 30 s.

  7. Rinse in tap water for 2 min.

  8. Dehydrate through 95% alcohol, two changes of 100% alcohol, 3 min each.

  9. Clear in xylene or xylene substitute, two changes, 3 min each.

  10. Mount with resinous mounting medium.

  11. Observe under polarized light using fluorescence microscope equipped with a polarizer and an analyzer. Amyloid shows green birefringence (see Note 3).

Th-S Staining

  1. Deparaffinization and hydration of tissue sections:

     100% xylene – 5 min

     100% xylene – 5 min

     100% ethanol – 3 min

     95% ethanol – 3 min

70% ethanol – 3 min

50% ethanol – 3 min

Water – 2 × 3 min

2. Incubate in filtered 1% aqueous Th-S for 10 min at room temperature. Filter Th-S before each use.

3. Dip four times in 80% ethanol.

4. Dip two times in 70% ethanol.

5. Dip two times in 50% ethanol.

6. Wash with three exchanges of dH$_2$O.

7. Coverslip in aqueous mounting media and allow slides to dry in the dark. Seal coverslip with clear nail polish (see Note 4).

3.5. Immuno-histochemistry

Paraffin sections produce satisfactory results for the demonstration of majority of tissue antigens with the use of antigen-retrieval techniques. Since certain cell antigens do not survive routine fixation and paraffin embedding, the use of frozen sections still remains essential for the demonstration of many antigens. The disadvantages of frozen sections include poor morphology, poor resolution at higher magnifications, special storage needed, limited retrospective studies, and cutting difficulty over paraffin sections.

1. Deparaffinize and hydrate the tissue section: 6–8-μm paraffin sections should be incubated at 58–60°C overnight (recommended) or at 65°C for 1–2 h (for fast experiment). Cool down sections at RT before dipping the sections into four consecutive stain jars containing xylene to remove paraffin (3 × 5 min). Dip sections into 100% ethanol (3 × 3min) to remove xylene. If using a horseradish peroxidase (HRP) conjugate for detection, blocking of endogenous peroxidase can be performed here. Dip the sections in 3% H$_2$O$_2$ in methanol for 15 min. Rinse in dH$_2$O (2 × 5 min).

2. Antigen retrieval: Put sections in citrate buffer and heat in a microwave for 10~15 min (see Note 5). Cool down sections in gently running water.

3. Incubation with antibodies: Block in milk buffer for 15 min at room temperature. Drain slides for a few seconds (do not rinse) and wipe around the sections with tissue paper. Apply primary antibody diluted in milk buffer. Incubate for 1 h at RT. Rinse in dH$_2$O and 1 × 5 min TBS 0.1% Triton X-100 with gentle agitation. Apply enzyme-conjugated secondary antibody to the slide diluted to the concentration recommended by the manufacturer and incubate for 1 h at RT. Rinse 3 × 5 min TBS.

We use routinely the EnVision System (Dako), which is based on dextran polymer technology. This unique chemistry permits binding of a large number of enzyme molecules (HRP or alkaline phosphatase) to a secondary antibody via the dextran backbone. The benefits are many, including increased sensitivity, minimized nonspecific background staining, and a reduction in the total number of assay steps as compared to conventional techniques. The simple protocols are (1) application of primary antibody; (2) application of enzyme-labeled polymer; and (3) application of the substrate chromogen. EnVision+ was developed after EnVision to provide increased sensitivity.

Develop with chromogen DAB or commercial kit (Dako, recommended) for 10 min at RT. It produces a brown, permanent color. Rinse in running tap water for 5 min. Counterstain (if required). Some commonly used counterstains are hematoxylin (blue), nuclear fast red, or methyl green. Dehydrate, clear, and mount. Mount sections in a suitable organic mounting media. Sections mounted in organic mounting media have a better refractive index than those mounted in aqueous mounting media. This means that the image seen down the microscope is sharper and clearer if organic mounting media is used (see Note 6).

3.6. Transmission Electron Microscopy

For transmission electron microscopy (TEM), tissue is removed by dissection as rapidly as possible, and then cut into smaller pieces for fixation. The actual size of the piece depends on the nature of the fixative used and the density of the tissue. For primary fixation with aldehyde fixatives, the pieces should not be larger than 2 mm^3. If tissue is larger, then smaller cubes of 1 mm^3 can be cut before secondary fixation with osmium tetroxide. The fixative should be at least ten times greater in volume than the specimen, and tissue samples should never be allowed to dry.

TEM Processing Schedule: Epoxy Resin

1. Fix tissue in 2.5% glutaraldehyde in 0.1 M sodium cacodylate buffer at 4°C for a minimum of 4 h. Tissue should be cut into ~1 mm^3 cubes for fixing. This may be done in a drop of fix on a sheet of dental wax using a razor blade. Place in glass processing vials and close with plastic caps. (Tissue may be stored at this stage).

2. Wash in 0.1 M sodium cacodylate buffer – 2 × 1 h (or overnight at 4°C).

3. Postfix in osmium tetroxide in 0.2 M sodium cacodylate buffer – 1 h (mix equal quantities of 2% aqueous OsO$_4$ and 0.4 M sodium cacodylate buffer and use immediately) (see Note 7).

4. Rinse in 0.2 M sodium cacodylate buffer – 2 × 5 min.

5. Dehydrate in 70% ethanol – 2 × 20 min. (May be stored overnight at this stage if absolutely necessary).

6. Dehydrate in 90% ethanol – 2 × 10 min.

7. Dehydrate in 100% ethanol – 2 × 20 min.

8. Propylene oxide (1.2 epoxy propane) – 2 × 10 min.

9. Propylene oxide/epoxy resin mixture (1:1) – 1 h.

10. Epoxy resin – overnight – with caps removed from vials. (Allow any remaining propylene oxide to evaporate.)

11. Embed in labeled capsules with freshly prepared resin.

12. Polymerize at 60°C – 48 h.

Retrieval of tissue from histological blocks for TEM:
In some cases, specific areas of interest can be retrieved from the wax block. In addition, this method may be useful if tissue is not available for TEM processing.

1. Identify the area of interest on the microscope slide by circling it with a marker pen.

2. Match the area marked on the slide against the specimen in the wax block and cut around it with a razor blade.

3. Cut a few mm into the surface of the wax block all around the marked area. Lever out the piece of tissue carefully.

4. Cut the piece of tissue into suitable-sized blocks making sure that orientation can be recognized later by cutting so that one dimension is greater than the other two.

5. Place the tissue into a glass processing vial and fill it with a suitable wax solvent (Histo-Clear® or xylene) and leave for 24 h (preferably on a rotating mixer).

6. Place tissue into 100% ethanol for 2 × 1 h.

7. Place tissue into 90% ethanol for 2 × 30 min.

8. Place tissue into 70% ethanol for 2 × 30 min.

9. Place tissue into 0.1 M sodium cacodylate buffer for 2 × 30 min.

10. Continue with usual processing schedule for TEM specimens.

11. The flat (previously cut) surface is embedded facing the end of the embedding capsule so that the required area is accessible in the finished block.

Semithin sectioning and staining:
Several resin sections are cut at approximately 1 μm using glass knives and an ultramicrotome. The sections are dried onto a glass slide on a hot plate at 80°C and then heated over a flame for a few seconds to ensure adhesion. The sections are then stained with toluidine blue for 1 min at 80°C. The stain is rinsed off with dH$_2$O and the sections are dried and covered with a glass coverslip using a synthetic mounting medium. Sections can be viewed by light microscopy to allow selection of the appropriate tissue area before proceeding to TEM.

Thin sectioning and staining for TEM:
The sections are cut in the same way as for semithin sectioning, but using a diamond knife, with the ultramicrotome set to cut at around

100 nm. The sections are picked up onto 300-mesh (300 squares), thin-bar, copper grids unless they are for immunocytochemistry, in which case gold or nickel grids are used. Samples need to be suspended in distilled water or a suitable buffer, such as 10 mM HEPES or 1% ammonium acetate.

Timing for staining (epoxy resin):
Stain grids with uranyl acetate (10 min for methanolic and 20 min for aqueous) and lead citrate (5 min).

Immunoelectron microscopy:
We have utilized paraffin-embedded tissue to carry out successful pre-embedding immunogold EM. We use this technique when molecular and/or neuropathologic data obtained from tissue that has not been processed for EM merit ultrastructural analysis. In brief, 30-μm-thick paraffin sections are deparaffinized in graded ethanol to distilled water and washed in TBS. The sections are treated with 0.05% Triton X-100, washed in TBS, and incubated with 10% normal goat serum. The primary antibodies are used at a ten times higher concentration than used for light microscopy, incubated for 48 h at 4°C, and then washed in buffer. The sections are then incubated for 24 h in the secondary antibody conjugated with 10-nm gold particles, fixed in 2.5% glutaraldehyde with buffer, postfixed with 1% osmium tetroxide, dehydrated with graded ethanol, propylene oxide, 1:1 propylene oxide: Polybed 812, and stored in Polybed 812 overnight. The sections are embedded in fresh Polybed 812 between two coverslips, polymerized at 60°C overnight, placed on previously polymerized capsules, and polymerized 2 more days. Ultrathin sections are cut and stained as routinely performed for EM.

3.7. Isolation and Biochemical Analyses of Amyloid Deposits

Isolation of parenchymal amyloid deposits:
The histological study of the brain should guide the selection of the area(s) for dissection of frozen tissue. The final yield varies according to the extent of pathology (see Note 8). The cerebral cortex is dissected from the underlying white matter, free of any large vessel contamination (approximately 5 gm of frontal cortex is used). Amyloid peptides are isolated using a published procedure (14, 15) with some modifications.

1. The gray matter is homogenized in a Dounce homogenizer in 10 volumes of buffer TE containing PI on ice and centrifuged at 200,000×g for 1 h at 4°C. The supernatant represents the "soluble" fraction (S_{200}) and is stored at –70°C.

2. The 200,000×g pellet is rehomogenized in 10 volumes of TE buffer with PI.

3. Filter the homogenate through a series of nylon mesh of 300, 100, and 30 μm in order to remove small vessel contamination.

4. Add an equal volume of 30% SDS solution in TE buffer to obtain a final concentration of 15% SDS. Stir overnight at room temperature.

5. Centrifuge at $130,000 \times g$ for 1 h at 20°C (save pellet, see step 10).

6. Centrifuge the supernatant at $500,000 \times g$ for 2 h at 20°C.

7. Dissolve the resulting pellet (P_{500}) in 99% glass-distilled formic acid for 1 h.

8. Centrifuge for 30 min at $14,000 \times g$.

9. Collect the formic acid-soluble material and evaporate under a stream of nitrogen.

10. Resuspend the pellet from the $130,000 \times g$ centrifugation in buffer TE with PI and centrifuge at $2,000 \times g$ for 15 min (save pellet). Recentrifuge the supernatant from this $2,000 \times g$ centrifugation for 1 h at $130,000 \times g$ at 20°C and dissolve the resulting pellet (P_{130A}) in formic acid.

11. Centrifuge the pellet for 30 min at $14,000 \times g$ and collect the formic acid-soluble material. Evaporate under a stream of nitrogen.

12. Wash the pellet from the $2,000 \times g$ centrifugation in 0.1 M Tris, pH 8.0, four times and digest with collagenase and DNase I (0.3 mg/ml collagenase CLS-3 and 10 μg/ml of DNase I) for 16 h at 37°C.

13. Centrifuge at $130,000 \times g$ for 1 h at 4°C. Dissolve the resulting pellet (P_{130B}) in formic acid, and centrifuge for 30 min at $14,000 \times g$. Collect and evaporate the formic acid-soluble fraction.

Isolation of vascular amyloid deposits:
Amyloid can be isolated from cerebral cortical vessels or from leptomeningeal vessels. Here, we describe the procedure for rapid amyloid isolation from leptomeningeal vessels.

1. Remove the leptomeninges (3–6 g) carefully from the surface of the cerebral hemispheres.

2. Wash four times at 4°C with 200 ml of TE buffer to remove entrapped blood cells by centrifugation at $5,000 \times g$ for 10 min at 4°C.

3. Resuspend vessels in TE buffer. Cut down vessels and collect them by filtration through a 30-μm nylon mesh.

4. Wash three times with 100 ml of 1% SDS, 0.1 M Tris with PI.

5. Homogenize in 20 volumes of 0.1 M Tris with 2 mM of $CaCl_2$ and digest with collagenase/DNase I for 16 h at 37°C.

6. Centrifuge the digested material at $50,000 \times g$.

7. Dissolve the pellet (P_{50}) in formic acid, and centrifuge for 30 min at $14,000 \times g$. Evaporate the formic acid-soluble fraction.

Isolation of amyloid plaques by laser capture microdissection (LCM):
The use of LCM coupled with mass spectrometry-based proteomic profiling is a particularly useful approach to characterizing amyloid deposits (15–17). LCM is ideal for the isolation of amyloid plaques since it allows for isolation of individual structures, reducing the amount of contaminants while maintaining their morphology (18). A major limitation of LCM is that it requires the capture of large quantities of plaques since only small amounts of protein can be recovered from individual structures. We have used LCM on frozen as well as formalin-fixed, paraffin-embedded cerebral tissue.

1. For frozen tissue, tissue fixed in 75% ethanol is preferred since it preserves the morphology while maintaining the quality of the material for analysis. Eight-micrometer-thick histological sections are placed on membrane-coated slides.

2. For formalin-fixed material, tissue sections need to be dewaxed twice in xylene for 2 min each and passed through decreasing concentrations of ethanol (100, 90, and 70%) and water for 30 s each.

For the identification of amyloid lesions, we stain sections in eosin for 30 s, and then dehydrate in ethanol. Alternatively, sections can be stained with Th-S and viewed under fluorescence. Although structures could be visualized by immunohistochemistry (IHC), this adds additional proteins that may interfere with the proteomics analysis.

Laser dissection using a Leica AS LMD laser micro dissection system (Leica Microsystems): Amyloid deposits are selected and cut using a 63× objective. Once the region of interest is excised, it gently falls into the collection tube. No additional steps are necessary. Since the sample is not touched, there is no contamination. Nonamyloid-containing areas can be captured as controls. If frozen tissue is used, collected samples can be extracted with lysis buffer (PBS with 2% SDS, 10% glycerol, 10 mM dithiothreitol, 1 mM EDTA, and PI) at 65°C for 15 min. If formalin-fixed tissue is used, dissected plaques may need to be treated with 99% glass-distilled formic acid in a sealed siliconized tube for 6 h at 96 ± 1°C using an Eppendorf thermomixer. After incubation, samples are centrifuged for 30 min at $14,000 \times g$ and the formic acid-soluble material collected. The formic acid is evaporated under a stream of nitrogen.

Analysis of amyloid peptides:
For matrix-assisted laser desorption ionization time-of-flight mass spectrometry (MALDI-TOF-MS) analysis, samples are resuspended with 10 µl of isopropylic alcohol/water/formic acid (4:4:1) mixture and analyzed. For western blot analysis, samples were resuspended in 3× sample buffer and analyzed. In our experience, ~10,000 amyloid plaques need to be isolated for mass spectrometry analysis and ~2,500 amyloid plaques are enough for western blot analysis.

4. Notes

1. Use high-quality and sufficient amounts of fixative. Use sharp blades and fine instruments to handle all specimens. Treat tissues as gently as possible to minimize morphology artifacts. Do not let samples dry during preparation. Samples should always be covered by solutions.

2. After Hematoxylin and Eosin (H&E) staining, nuclei are stained blue and cytoplasm pink to red. Amyloid deposits are stained red. For less blue in the finished slides, dip 1–2 more times in the acid alcohol step. For less pink, either use less time in the eosin or leave in the alcohol longer. Rinsing well after the lithium carbonate is very important; if there is a crystal deposit on the finished slide, soak in acid alcohol to remove stain and restain in H&E.

3. After Congo red staining, amyloid, elastic fibers, and eosinophil granules are seen red and nuclei blue under light microscopy.

4. Protect Th-S from light, and protect the stained slides from light as much as possible. Th-S stain should be stored at 4°C.

5. Citrate buffer is commonly used in IHC and works perfectly with many antibodies. It gives very nice intense staining with very low background.

6. Tissue preparation is the cornerstone of IHC. To ensure the preservation of tissue architecture and cell morphology, prompt and adequate fixation is essential. However, inappropriate or prolonged fixation may significantly diminish the antibody-binding capability. There is no one universal fixative that is ideal for the demonstration of all antigens. However, in general, many antigens can be successfully demonstrated in formalin-fixed, paraffin-embedded tissue sections. Antigen-retrieval techniques may further enhance the use of formalin as routine fixative for IHC in many research laboratories. Some antigens do not survive even moderate amounts of aldehyde fixation. H_2O_2 suppresses endogenous peroxidase activity, and therefore reduces background staining. Most antibodies are used in IHC at a concentration between 0.5 and 10 µg/ml. Check specificity of secondary antibody before using. DAB is a suspected carcinogen. Wear the appropriate protective clothing. Deactivate it with chloros in a sealed container overnight (it produces noxious fumes when chloros is added) and dispose of it according to laboratory guidelines.

7. Osmium tetroxide, propylene oxide, and propylene oxide/resin waste should be collected in bottles for safe disposal. All steps must be performed in a fume cupboard and gloves should be worn throughout. For steps 2–10, the processing vials

should be on a rotating mixer. Buffers, such as PBS, may contaminate the grid with salt residues which have to be washed off leaving little contrast. All staining solutions should either be filtered through Millipore filters or centrifuged before use. Use filtered distilled water to wash between stains and 50% methanol and then distilled water to wash if using methanolic uranyl acetate. Care should be taken not to breathe on the lead citrate while staining as a precipitate of lead carbonate may form and contaminate the sections.

8. During isolation of amyloid peptides, please note that fractions may contain different proportions of amyloid peptide species based on their solubility. Further purification may be needed by chromatographic methods, including reverse-phase high-performance liquid chromatography (HPLC) or fast protein liquid chromatography (FPLC).

Acknowledgments

We are very grateful to all participants and their families who through their commitment make these studies possible. The authors are grateful to F. Epperson, B. Dupree, and R. Richardson for their technical assistance. This study was supported by grants from the National Institute of Health (NS050227, NS063056, AG10133), the Alzheimer's association (IIRG-05-14220), and by the American Health Assistance Foundation (A2008-304).

References

1. Pepys, M.B. (2006) Amyloidosis. *Annu. Rev. Med.* 57, 223–241.

2. Prelli, F., Castaño, E.M., Glenner, G.G., and Frangione, B. (1988) Differences between vascular and plaque core amyloid in Alzheimer's disease, *J. Neurochem.* 51, 648 651.

3. Vidal, R., Calero, M., Piccardo, P., Farlow, M.R., Unverzagt, F.W., Mendez, E., Jimenez-Huete, A., Beavis, R., Gallo, G., Gomez-Tortosa, E., Ghiso, J., Hyman, B.T., Frangione, B., and Ghetti, B. (2000) Senile dementia associated with amyloid beta protein angiopathy and tau perivascular pathology but not neuritic plaques in patients homozygous for the APOE-epsilon4 allele, *Acta Neuropathol. (Berl.)* 100, 1–12.

4. Vidal, R., Ghiso, J., and Frangione, B. (2000) New familial forms of cerebral amyloid and dementia. *Mol Psychiatry.* 5(6), 575–576.

5. Ghetti, B., Tagliavini, F., Masters, C.L., Beyreuther, K., Giaccone, G.,Verga, L., Farlow, M.R., Conneally, P.M., Dlouhy, S.R., Azzarelli,

B., and Bugiani ,O. (1989) Gerstmann-Sträussler-Scheinker disease. II. Neurofibrillary tangles and plaques with PrP-amyloid coexist in an affected family. *Neurology* 39,1453–1461.

6. Vidal, R., Frangione, B., Rostagno, A., Mead, S., Revesz, T., Plant, G., and Ghiso, J. (1999) A stop-codon mutation in the BRI gene associated with familial British dementia. *Nature* 399, 776–781.

7. Vidal, R., Revesz, T., Rostagno, A., Kim, E., Holton, J.L., Bek, T., Bojsen-Moller, M., Braendgaard, H., Plant, G., Ghiso, J., and Frangione, B. (2000) A decamer duplication in the 3' region of the BRI gene originates an amyloid peptide that is associated with dementia in a Danish kindred. *Proc Natl Acad Sci USA* 97, 4920–4925.

8. Vidal, R., Ghiso, J., and Frangione, B. (2000) New familial forms of cerebral amyloid and dementia. *Mol Psychiatry* 5, 575–576.

9. Vidal, R., Delisle, M.B., and Ghetti, B. (2004) Neurodegeneration caused by proteins with an aberrant carboxyl-terminus. *J Neuropathol Exp Neurol* **63**, 787–800.

10. Vidal, R., Garzuly, F., Budka, H., Lalowski, M., Linke, R.P., Brittig, F., Frangione, B., and Wisniewski, T. (1996) Meningocerebrovascular amyloidosis associated with a novel transthyretin mis-sense mutation at codon 18 (TTRD 18G) *Am J Pathol*. **148**, 361–366.

11. Vidal, R.G., Ghiso, J., Gallo, G., Cohen, M., Gambetti, P.L., and Frangione, B. (1992) Amyloidoma of the CNS. II. Immunohistochemical and biochemical study. *Neurology*. **42**, 2024–2028.

12. Westermark, P., Benson, M.D., Buxbaum, J.N., Cohen, A.S., Frangione, B., Ikeda, S., Masters, C.L., Merlini, G., Saraiva, M.J., and Sipe, J.D. (2007) A primer of amyloid nomenclature. *Amyloid*. **14**, 179–183.

13. Westermark, G.T., Johnson, K.H., and Westermark, P. (1999) Staining methods for identification of amyloid in tissue. *Methods Enzymol*. **309**, 3–25.

14. Roher, A.E., and Kuo, Y.M. (1999) Isolation of amyloid deposits from brain. *Methods Enzymol*. **309**, 58–67.

15. Miravalle, L., Calero, M., Takao, M., Roher, A.E., Ghetti, B., and Vidal, R. (2005) Amino-terminally truncated Aβ peptide species are the main component of cotton wool plaques. Biochemistry. **44**, 10810–10821.

16. Liao, L., Cheng, D., Wang, J., Duong, D.M., Losik, T.G., Gearing, M., Rees, H.D., Lah, J.J., Levey, A.I., and Peng, J. (2004) Proteomic characterization of postmortem amyloid plaques isolated by laser capture microdissection. *J Biol Chem*. **279**, 37061–37068.

17. Wittliff, J.L.,and Erlander, M.G. (2002) Laser capture microdissection and its applications in genomics and proteomics. *Methods Enzymol*.**356**, 12–25.

18. Cornea, A., Mungenast, A. (2002) Comparison of current equipment. *Methods Enzymol*. **356**, 3–12.

Chapter 17

Detection of Protein Aggregation in Neurodegenerative Diseases

Han-Xiang Deng, Eileen H. Bigio, and Teepu Siddique

Abstract

Protein aggregates/inclusions are pathological hallmarks of a wide spectrum of neurodegenerative diseases. These aggregates have different shapes, sizes, distribution, and protein composition, which are unique features used for pathological diagnosis. The aggregates per se are also used as molecular targets for designing therapeutic approaches. Detection of these aggregates is generally achieved by using immunostaining methods, most often by immunohistochemistry. In clinical and pathologic practice, the neurologic tissues to be examined are generally fixed with formalin and processed to paraffin-embedded tissue blocks. These treatments result in covalent cross-linking of the protein molecules and preserve the tissue morphology, but dramatically mask the antigens, making it often difficult to detect the aggregates. Therefore, removal of the cross-linking of antigens is a critical step for effective detection of these aggregates. In this chapter, we discuss and present immunostaining methods with a focus on the effectiveness of antigen-retrieval methods. In our experience, a treatment of tissues at 125°C for 20 min represents a relatively ideal antigen-retrieval method not only preserving the tissue morphology, but also providing efficient antigen retrieval. Using this method, we successfully detected some protein aggregates that escaped detection when other antigen-retrieval methods were employed.

Key words: Protein aggregation, Neurodegenerative diseases, Immunohistochemistry, Immunofluorescence, Antigen retrieval

1. Introduction

Protein aggregates have been observed in most of the common neurodegenerative diseases (1, 2). The shape, size, location, and protein composition of the aggregates are characteristic features in different diseases (Fig. 1). Examples include extracellular amyloid-beta plaques and intracellular tau neurofibrillary tangles in Alzheimer's disease (AD), alpha-synuclein-containing Lewy

Giovanni Manfredi and Hibiki Kawamata (eds.), *Neurodegeneration: Methods and Protocols*,
Methods in Molecular Biology, vol. 793, DOI 10.1007/978-1-61779-328-8_17, © Springer Science+Business Media, LLC 2011

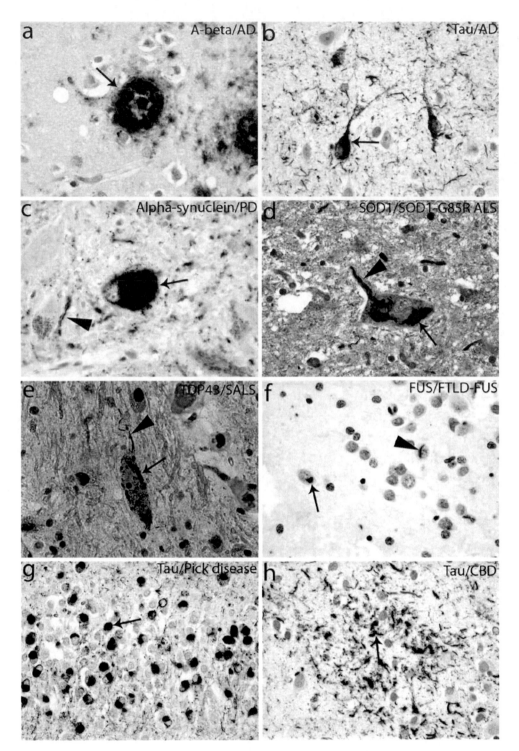

Fig. 1. Typical protein aggregates/inclusions in several neurodegenerative diseases. Aggregates shown here are extracellular amyloid-beta plaques (**a**) and intracellular tau neurofibrillary tangles (**b**) in Alzheimer's disease; alpha-synuclein-containing Lewy bodies (*arrow*) and Lewy neurites (*arrowhead*) in Parkinson's disease (**c**); cytoplasmic (*arrow*) and neuritic (*arrowhead*) SOD1-containing aggregates in spinal motor neurons in a patient with an SOD1 G85R mutation (**d**); cytoplasmic (*arrow*) and neuritic (*arrowhead*) TDP43-containing aggregates in spinal motor neurons in sporadic ALS (**e**); FUS-containing cytoplasmic (*arrow*) and nuclear (*arrowhead*) inclusions in frontotemporal lobar degeneration with FUS proteinopathy (FTLD-FUS) (**f**); tau-containing Pick bodies in Pick disease (FTLD-tau (PiD)) (**g**); tau-containing glial plaques in corticobasal degeneration (FTLD-tau (CBD)) (**h**).

bodies in Parkinson's disease (PD) and PD-related disorders (3), and ubiquitin, TDP-43, and FUS positive skein-like aggregates in motor neurons in amyotrophic lateral sclerosis (ALS) (4, 5). In some neurodegenerative disorders, the composition of the protein aggregates may differ in various subtypes of the same clinical entity, depending on the cause and pathogenic mechanisms. For example, the SOD1-immunoreactive aggregates are a prominent pathology in mutant SOD1-linked ALS, but not in non-SOD1 ALS (6). On the other hand, TDP43- and FUS-immunoreactive aggregates appear to be a pathologic hallmark in all the other types of ALS, but not in mutant SOD1 ALS (7–9). Therefore, detection of the specific aggregates may not only provide robust and objective lines of evidence for diagnosis and differential diagnosis, but also provides molecular insight into understanding the pathogenic mechanisms of the disease or subtypes of the same disease. Moreover, the proteins identified in the aggregates are now being used as molecular targets for designing rational therapies (10). The molecular mechanisms underlying aggregation are largely unknown for most of the proteins. But oxidation-mediated intermolecular disulfide cross-linking has been demonstrated as the molecular mechanism in the aggregation of SOD1 in the mutant SOD1 mouse models of ALS (11). Although in most cases the roles of aggregates in the pathogenesis of neurodegeneration remain to be further elucidated, it is generally accepted that these aggregates are pathogenic for most neurodegenerative diseases (1, 2, 12, 13).

The protein aggregates can be visualized under a light or fluorescence microscope after immunostaining. Immunohistochemistry (IHC) and immunofluorescence (IF) are the most commonly used immunostaining methods for detection of aggregates. IHC and IF involve the binding of an antibody to a cellular or tissue antigen of interest and visualization of the bound product by a detection system. IHC consists of the following steps: (1) primary antibody binds to specific antigen; (2) antibody–antigen complex is bound by a secondary, enzyme-conjugated antibody; and (3) in the presence of substrate and chromagen, the enzyme forms a colored deposit at the sites of antibody–antigen binding (Fig. 2). In IF, a fluorochrome can be directly conjugated with the secondary antibody so that the signal can be visualized under a fluorescence microscope (Fig. 2).

There are several immunostaining methods that can be employed depending on the type of specimen under study, the degree of sensitivity required, and the cost. The most frequently used technique is the avidin–biotin immunoperoxidase complex (ABC) method. In this chapter, we discuss the technical aspects of IHC with a focus on the critical issue of antigen retrieval.

Fig. 2. Basic diagram of immunohistochemistry (IHC) and immunofluorescence (IF).

1.1. Chemical Reactions During Formalin Fixation and Tissue Processing

To ensure the preservation of tissue architecture and cell morphology, prompt and adequate fixation is essential, especially for fragile tissues, such as brain and spinal cord. Formalin fixation, subsequent tissue processing in ethanol, and embedding in paraffin wax are routine procedures in pathology laboratories. Accordingly, the largest proportion of brain and spinal cord samples for IHC is formalin-fixed and paraffin-embedded. Some of the tissues may have been maintained in formalin for years. A number of chemical reactions take place during formalin fixation and tissue processing. These reactions result in tissue fixation, but also modify antigens, leading to impaired or lost immunoreactivity.

Formalin is a routinely used fixative solution containing 35–40% formaldehyde with about 10% methanol as a stabilizer. Some pathologists use 20% formalin and others use 10%. Basically, the fixation process has three stages. First, the formaldehyde dissolves in water, combining with it to form methylene hydrate (methylene glycol). Then, the methylene glycol can react with almost all end groups found in biological molecules, such as primary and secondary amines, amides, hydroxyls, sulfhydryls, reactive hydrogen atoms from aromatic amino acids, etc., resulting in the attachment of various hydroximethyl adducts at the original end groups. Finally, this attachment is followed by cross-linking with other end groups through the formation of a methylene bridge ($-CH_2-$) (14) (Fig. 3).

In addition to formaldehyde-induced cross-linking, subsequent use of ethanol in tissue processing may also result in additional cross-linking as shown in Fig. 3. Ethanol may lead to dehydration of the biological molecules. Acting on the hydroxymethyl adducts that have not cross-linked, ethanol can catalyze the formation of highly reactive imines by removing the hydrogen atom from the nitrogen of the original amine end group and the hydroxyl group from the formaldehyde adduct. The imine then reacts with another molecule of ethanol to form a variety of

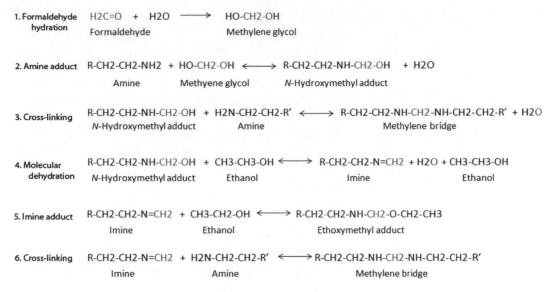

1. Formaldehyde   H2C=O + H2O ──────→   HO-CH2-OH
   hydration   Formaldehyde   Methylene glycol

2. Amine adduct   R-CH2-CH2-NH2 + HO-CH2-OH ⟷ R-CH2-CH2-NH-CH2-OH + H2O
                        Amine   Methyene glycol   N-Hydroxymethyl adduct

3. Cross-linking   R-CH2-CH2-NH-CH2-OH + H2N-CH2-CH2-R' ⟷ R-CH2-CH2-NH-CH2-NH-CH2-CH2-R' + H2O
                   N-Hydroxymethyl adduct   Amine   Methylene bridge

4. Molecular   R-CH2-CH2-NH-CH2-OH + CH3-CH3-OH ⟷ R-CH2-CH2-N=CH2 + H2O + CH3-CH3-OH
   dehydration   N-Hydroxymethyl adduct   Ethanol   Imine   Ethanol

5. Imine adduct   R-CH2-CH2-N=CH2 + CH3-CH2-OH ⟷ R-CH2-CH2-NH-CH2-O-CH2-CH3
                        Imine   Ethanol   Ethoxymethyl adduct

6. Cross-linking   R-CH2-CH2-N=CH2 + H2N-CH2-CH2-R' ⟷ R-CH2-CH2-NH-CH2-NH-CH2-CH2-R'
                        Imine   Amine   Methylene bridge

Fig. 3. Common reactions of formaldehyde or ethanol with protein side chains during formalin fixation and subsequent tissue processing in ethanol. Methylene bridges are formed through covalent bonds.

products, such as an ethoxymethyl adduct. Imines can also cross-link with neighboring end groups, such as amine, resulting in methylene-bridge cross-linking (Fig. 3).

1.2. Antigen Retrieval

As stated above, the methylene bridge-mediated covalent cross-linking is the major modification of antigens in the formalin-fixed, paraffin-embedded tissues. Such modifications may change the shape, charge, solubility, and accessibility of the antigens to the antibodies, resulting in antigen masking. Therefore, an antibody raised against a natural antigen may not be able to recognize the modified antigen or may recognize it with less affinity. Moreover, the cross-linking can prevent antibodies from penetration into the tissues to react with the antigen. However, the antigens are rarely lost. An important premise is that the formaldehyde- and ethanol-mediated modifications are reversible. Demodification of the cross-linked antigens retrieves some of the native states of the antigens in the fixed tissues.

Two basic methods are available for antigen retrieval: enzymatic digestion and heat-induced epitope retrieval (HIER). The latter has many advantages and is now most commonly used (15).

HIER involves energy transfer from the heat that is applied during antigen retrieval, leading to breakage of the cross-linking. Bonded atoms are held together by energy and different bonding pairs have different bond energies. Carbon–carbon bonds are strong, meaning that they have high energy and are not easily broken. Weaker bonds, such as carbon–nitrogen and carbon–oxygen pairs, have less bond energy, and thus are broken more readily (14).

All molecules vibrate as a function of the energy content of the system. Temperature is a measure of that energy level. With HIER, heat or microwave energy applied to the system causes molecules to vibrate at increasing speed. Bonds between atoms break when vibrational energy exceeds bond energy. As expected, weaker bonds break first while stronger bonds require more externally applied energy before breaking. The most important issue in immunostaining is to break enough cross-linking bonds so that specific antibodies can interact with previously masked or modified sites while maintaining structural integrity and morphology. Theoretically, energy (temperature) applied to the antigen retrieval system is the most important factor for effective antigen retrieval. But the system may take some time for equilibration of the applied energy. Therefore, time may also be an important factor for antigen retrieval. Because different antigens have different amino acid composition and charge, the optimum pH of the antigen retrieval system varies for different antigens.

Different laboratories may have their own preferred HIER methods. The commonly used methods are boiling and especially microwaving. The maximum sustained temperature for either boiling or microwaving method is not more than 100°C. Although boiling and microwaving methods have been successfully used in detection of nearly all of the reported protein aggregates in neurodegenerative diseases, it is quite possible that some protein aggregates are missed due to insufficient antigen retrieval, especially when a protein is not abundant in the aggregates and/or an antibody is not sensitive enough. Theoretically, increasing the energy applied to the antigen retrieval system should improve the sensitivity of IHC and IF. To test this possibility, we compared three antigen-retrieval methods: boiling, microwaving, and a high-pressure decloaking chamber. The high-pressure decloaking chamber can maintain a temperature of 125°C. We observed that the high-pressure decloaking chamber consistently yielded much stronger signals compared to the other two methods, when other conditions remained the same (9). Using the high-pressure decloaking chamber method, we observed skein-like FUS aggregates in the spinal motor neurons of all the ALS patients, except those with SOD1 mutations (9). These results demonstrate that FUS aggregates are a common component in the characteristic skein-like aggregates in all types of ALS, except for SOD1-linked ALS, implying a shared pathogenic pathway involving FUS in sporadic ALS and non-SOD1 ALS. However, these FUS aggregates could not be detected using the boiling and microwaving methods (9) (Fig. 4). In the sections from the same patient, even though ubiquitin, p62, and TDP43 can be detected using all three methods, the signals are much stronger when antigens are retrieved with the high-pressure decloaking chamber than with boiling and microwaving methods (Fig. 4).

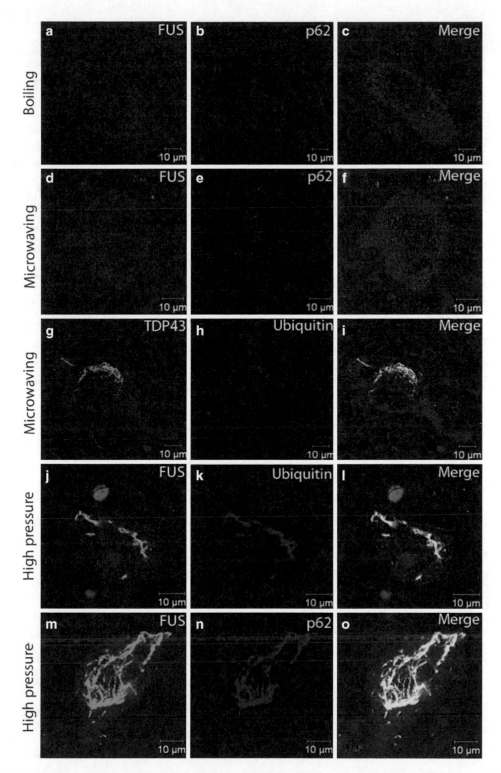

Fig. 4. Effect of antigen-retrieval protocols on detection of protein aggregates. Antigens in spinal cord sections of the same patient with sporadic amyotrophic lateral sclerosis were treated with different antigen-retrieval protocols and detected by immunofluorescent staining. P62, ubiquitin, and TDP43, but not FUS, in the skein-like aggregates could be detected using boiling (**a–c**) or microwaving protocol (**d–i**), whereas FUS aggregates were only shown when the sections were treated with high-pressure decloaking protocol for antigen retrieval (**j–o**). The staining intensity of the p62 and ubiquitin is much stronger in high-pressure decloaking protocol (**j–o**) than in boiling and microwaving protocols (**a–i**).

Powerful antigen retrieval is not necessarily better. Maintaining the integrity and morphology of the tissues is also important for IHC and IF. We examined the optimal time for high-pressure decloaking chamber method. We observed that a treatment of 2–5 min in the high-pressure decloaking chamber is sufficient for most of the antigens tested. Meanwhile, we also noticed that the morphology of the cells is satisfactorily maintained even when the sections are treated up to 20 min.

Described below are the basic protocols for IHC using peroxidase-conjugated avidin as a detection system (9). These protocols can be applied to other detection systems with minor modifications.

2. Materials

2.1. Deparaffinization and Rehydration

1. Fourteen glass containers with lids to be prepared as follows:
    (a) Xylene in containers 1–3.
    (b) Absolute ethanol in containers 4–6.
    (c) 95% Ethanol in containers 7–9.
    (d) 75% Ethanol in container 10.
    (e) 50% Ethanol in container 11.
    (f) Distilled water in containers 12–14.
2. Stainless steel slide stand for holding slides.
3. Phosphate-buffered saline (PBS, pH 7.4).

2.2. Antigen Retrieval

1. 10 mM citrate buffer at pH 6.0.
2. Stainless steel slide stand for holding slides.
3. Glass container larger than stainless steel slide stand.
4. Decloaking chamber (Biocare Medical, Walnut Creek, CA).

2.3. Blocking

1. Kimwipes tissue pad.
2. Plastic transfer pipettes.
3. Humid chamber lined with wet paper towel.
4. 2% Hydrogen peroxide.
5. 1% Bovine serum albumin (BSA) or 5% nonfat milk in 1× PBS (pH 7.4).

2.4. Antibody Binding

1. Humid chamber lined with wet paper towel.
2. Plastic transfer pipettes.
3. 1× PBS (pH 7.4).

4. PAP pen (Electron Microscopy Science, Hatfield, PA).

5. Primary antibody with appropriate dilution.

6. Biotinylated secondary antibody solution.

7. Working solution of peroxidase-conjugated avidin or streptavidin.

8. Working substrate solution containing hydrogen peroxide and 3-amino-9-ethyl-carbazole (AEC, fresh preparation).

2.5. Counterstaining and Coverslipping

1. Container with hematoxylin.

2. Aqueous mounting media.

3. Coverslips.

4. Kimwipes tissue pad.

5. Plastic transfer pipettes.

3. Methods

3.1. Deparaffinization and Rehydration

1. Prepare 6-μm thick sections from formalin-fixed tissue (see Note 1).

2. Allow tissue to affix overnight at 37°C or 1 h at 57°C.

3. Transfer fixed tissue slides to fresh xylene bath for 10 min. Repeat twice.

4. Transfer slides to fresh absolute ethanol for 5 min. Repeat twice.

5. Transfer slides to 95% ethanol for 3 min. Repeat twice.

6. Transfer slides to 75% ethanol for 3 min.

7. Transfer slides to 50% ethanol for 3 min.

8. Transfer slides to distilled water for 3 min. Repeat twice.

3.2. Antigen Retrieval

1. Transfer slides to a stainless steel slide stand and put the stand in a glass beaker containing 10 mM sodium citrate buffer (see Note 2).

2. Put the glass beaker in the decloaker chamber.

3. Set temperature at 125°C and time at 20 min.

4. Allow to cool at room temperature for 30 min.

5. Rinse the slides with distilled water for 5 min.

3.3. Blocking

1. Wipe off excess water from the slide (see Note 3).

2. Incubate sections with 2% hydrogen peroxide for 20 min to block potential endogenous peroxidase activities (see Note 4).

3. Rinse the slides in distilled water twice for 5 min each.

4. Incubate sections with 1% BSA or 5% nonfat milk in 1× PBS (pH 7.4) for 20 min at room temperature (see Note 5).

3.4. Antibody Binding

1. Remove blocking solution. Do not rinse.

2. Wipe off excess solution from the periphery of the tissue.

3. Draw a water-repellent circle around slide-mounted tissue with a PAP pen (see Note 6).

4. Apply enough primary antibody at appropriate concentration to cover the tissue (see Notes 7–9).

5. Incubate at room temperature for 1 h.

6. Remove excessive primary antibody.

7. Rinse slide in PBS for 3 min. Repeat twice.

8. Apply working biotinylated secondary antibody at appropriate concentration.

9. Incubate at room temperature for 30 min.

10. Remove excessive secondary antibody.

11. Rinse in PBS for 5 min. Repeat twice.

12. Incubate peroxidase-conjugated avidin or streptavidin at appropriate concentration in PBS at room temperature for 30 min.

13. Remove excessive peroxidase-conjugated avidin.

14. Rinse in PBS for 5 min. Repeat twice.

15. During rinsing, make freshly prepared working substrate solution containing hydrogen peroxide and AEC for color development. A number of manufacturers provide premade hydrogen peroxide solution and AEC solution. You may use these premade solutions by mixing them according to manufacturer's instruction immediately before use (see Notes 10 and 11).

16. Apply freshly prepared working substrate solution to cover the tissue.

17. Incubate the tissue sections with the substrate solution at room temperature. You may observe the color development under a light microscope until suitable staining intensity develops. Development time may significantly vary for different antibodies and their dilutions. Therefore, it should be individually determined. Generally, it takes 5–15 min.

18. Remove the substrate solution.

19. Immerse the slide in distilled water and rinse the slide for 3 min. Repeat twice.

3.5. Counterstaining and Coverslipping

1. Transfer slides to a jar containing hematoxylin solution.
2. Incubate for 1–3 min.
3. Rinse slide with distilled water for 3 min. Repeat twice.
4. Place an appropriate amount of an aqueous mounting media onto the tissue section.
5. Place a coverslip of appropriate size onto the section.
6. Press down the coverslip to allow air bubbles to float to outer edges of coverslip.
7. Turn slide over onto a Kimwipe pad and press to absorb excess mounting media.
8. Air dry.

3.6. Interpretation

Cells and cellular structures bearing the antigen recognized by the antibody display brown–red staining materials in addition to the blue counterstaining. Cells which do not bear the antigen display only the counterstaining (see Notes 3, 12–14).

4. Notes

1. Tissue sections often may detach from the slide during antigen retrieval and other procedures, especially for larger tissue sections. The use of positive-charged glass slides can avoid this problem.

2. The optimal pH of the sodium citrate buffer may vary from 2.0 to 10.0, depending on the antigen to be retrieved. In general, a pH 6.0 appears appropriate for most of the antigens.

3. Do not allow slides to dry at any time during the entire immunostaining procedures. Dried sections result in nonspecific binding of the antibodies and other reagents, leading to nonspecific signals.

4. Described above are the basic protocols for IHC using a peroxidase detection system as an example. In addition to peroxidase, alkaline phosphatases (APs) are also commonly used for IHC. When an AP is used, peroxidase blocking procedure can be omitted. The AP activity can be quenched at high temperature during antigen retrieval. These basic protocols can also be used for an IF detection system with minor modifications.

5. The blocking procedure significantly reduces nonspecific signals. Instead of BSA or nonfat milk, an ideal blocking solution is diluted normal serum of the hosts in which the antibody is raised (i.e., rabbits for rabbit primary antibodies).

6. Drawing a water-repellent circle around the tissue on the slides not only prevents the waste of valuable reagents (such as antibodies), but also maintains sufficient amount of reagents for the tissue by keeping the liquid pooled in a single droplet and preventing the tissue from drying.

7. In the IF detection system, two or more than two antigens can be simultaneously detected using fluorescence confocal microscopy. The primary antibodies should be derived from different hosts. Those from closely related hosts should not be used simultaneously (such as sheep and goats or rats and mice) due to the similarity of their immunoglobulin. The secondary antibodies may be directly conjugated with different fluorochromes.

8. Manufacturers generally recommend dilution ranges of their reagents or offer prediluted reagents ready for use. Due to variations between different protocols, optimal working dilutions, especially the dilution for the primary antibody, need to be determined by titration while the other conditions are fixed.

9. When a new antibody is tested, a wide range of dilutions (4×) should be examined for optimal conditions. A very high concentration of a primary antibody may yield no detectable specific signals in IHC. In this case, the entire section may appear uniformly light yellowish.

10. Any reagent stored at 4°C should be prewarmed to room temperature before use, except for the antibodies.

11. Storage and reuse of the dilute solutions of the enzyme conjugates, antibodies, and substrate solutions should be avoided, unless they can be used within 2 h.

12. The presence of lipofuscin in neurons is a common phenomenon in human brain and spinal cord samples. Lipofuscin is most prominent in large neurons, such as motor neurons and dopaminergic neurons, and its abundance increases with aging. Lipofuscin appears immunoreactive with most antibodies. The use of monoclonal or highly diluted antibodies may reduce the intensity of lipofuscin staining. Lipofuscin is a native autofluorescent material that persists even in the unstained sections. In IF, the presence of autofluorescence from lipofuscin and other native autofluorescent materials in the sections interferes with the observation of specific signals. The autofluorescence can be significantly quenched by treating the slides with a 0.3% solution of Sudan Black after application of the fluorochrome-conjugated secondary antibodies.

13. Differences in the quality of the primary antibodies strongly influence the quality of immunostaining. Each antibody, even a monoclonal antibody, may have various affinities and

cross-reactivities to nonspecific targets. This issue is more serious for polyclonal antibodies, which represent a mixture of various species (clones). Therefore, a positive stain may not be proof of the presence of a target in the section. On the other hand, different antibodies have different binding sites and have various sensitivities for IHC, and antigen retrieval is certainly not absolute. Therefore, a negative result from one antibody may not completely exclude the possibility that one specific protein or antigen is present in the section. Generally, the use of multiple antibodies to different epitopes of the same protein may be helpful. Moreover, the amount of an antigen may need to reach a threshold so that sufficient signals can be visualized through a detection system. Improvement of the sensitivity of the detection system may be beneficial, such as the use of signal amplification system.

14. For quality assurance, it is always a good idea to include positive and negative controls if possible. An ideal positive control should contain not just the antigen of interest, but also in small amounts that would be expected to correspond to the level of antigen in the studied tissue. A negative control is a tissue known to completely lack the antigen. An alternative negative control may include replacing the primary antibody, with either nonimmune serum or any other antibody of irrelevant specificity, or incubating without the primary antibody. Preabsorbed antibody is a good choice for testing the specificity of an antibody if possible.

References

1. Taylor JP, Hardy J, Fischbeck KH. (2002) Toxic proteins in neurodegenerative disease. *Science* **296**, 1991–5.

2. Aguzzi A, O'Connor T. (2010) Protein aggregation diseases: pathogenicity and therapeutic perspectives. *Nat Rev Drug Discov.* **9**, 237–48.

3. Lansbury PT, Lashuel HA. (2006) A century-old debate on protein aggregation and neurodegeneration enters the clinic. *Nature* **443**, 774–9.

4. Leigh PN, Whitwell H, Garofalo O, Buller J, Swash M, Martin JE, et al. (1991) Ubiquitin-immunoreactive intraneuronal inclusions in amyotrophic lateral sclerosis. Morphology, distribution, and specificity. *Brain* **114** (Pt 2), 775–88.

5. Lowe J. (1994) New pathological findings in amyotrophic lateral sclerosis. *J Neurol Sci.* **124 Suppl.**, 38–51.

6. Shibata N, Hirano A, Kobayashi M, Siddique T, Deng HX, Hung WY, et al. (1996) Intense superoxide dismutase-1 immunoreactivity in intracytoplasmic hyaline inclusions of familial amyotrophic lateral sclerosis with posterior column involvement. *J Neuropathol Exp Neurol.* **55**, 481–90.

7. Mackenzie IR, Bigio EH, Ince PG, Geser F, Neumann M, Cairns NJ, et al. (2007) Pathological TDP-43 distinguishes sporadic amyotrophic lateral sclerosis from amyotrophic lateral sclerosis with SOD1 mutations. *Ann Neurol.* **61**, 427–34.

8. Neumann M, Sampathu DM, Kwong LK, Truax AC, Micsenyi MC, Chou TT, et al. (2006) Ubiquitinated TDP-43 in frontotemporal lobar degeneration and amyotrophic lateral sclerosis. *Science* **314**, 130–3.

9. Deng HX, Zhai H, Bigio EH, Yan J, Fecto F, Ajroud K, et al. (2010) FUS-immunoreactive inclusions are a common feature in sporadic and non-SOD1 familial amyotrophic lateral sclerosis. *Ann Neurol.* **67**, 739–48.

10. Dohm CP, Kermer P, Bahr M. (2008) Aggregopathy in neurodegenerative diseases: mechanisms and therapeutic implication. *Neurodegener Dis.* **5**, 321–38.

11. Deng HX, Shi Y, Furukawa Y, Zhai H, Fu R, Liu E, et al. (2006) Conversion to the amyotrophic lateral sclerosis phenotype is associated with intermolecular linked insoluble aggregates of SOD1 in mitochondria. *Proc Natl Acad Sci USA.* **103**, 7142–7.

12. Krainc D. (2010) Clearance of mutant proteins as a therapeutic target in neurodegenerative diseases. *Arch Neurol.* **67**, 388–92.

13. Irvine GB, El-Agnaf OM, Shankar GM, Walsh DM. (2008) Protein aggregation in the brain: the molecular basis for Alzheimer's and Parkinson's diseases. *Mol Med.* **14**, 451–64.

14. Dapson RW. (2007) Macromolecular changes caused by formalin fixation and antigen retrieval. *Biotech Histochem.* **82**, 133–40.

15. Shi SR, Key ME, Kalra KL. (1991) Antigen retrieval in formalin-fixed, paraffin-embedded tissues: an enhancement method for immunohistochemical staining based on microwave oven heating of tissue sections. *J Histochem Cytochem.* **39**, 741–8.

Chapter 18

Assessment of Proteasome Impairment and Accumulation/Aggregation of Ubiquitinated Proteins in Neuronal Cultures

Natura Myeku, Maria Jose Metcalfe, Qian Huang, and Maria Figueiredo-Pereira

Abstract

The ubiquitin/proteasome pathway (UPP) is the major proteolytic quality control system in cells and involves tightly regulated removal of unwanted proteins and retention of those that are essential. In addition to its function in normal protein degradation, the UPP plays a critical role in the quality control process by degrading mutated or abnormally folded proteins. The proteolytic component of the UPP is a multiprotein complex known as the proteasome. Many factors, including the aging process, can cause proteasome impairment leading to formation of abnormal ubiquitin-protein aggregates that are found in most progressive neurodegenerative diseases, including Alzheimer's and Parkinson's diseases. In this chapter, we describe protocols to measure proteasome activity, evaluate its state of assembly, and assess the accumulation and aggregation of ubiquitinated proteins in two types of neuronal cultures: human neuroblastoma cells and rat primary cortical cultures. These protocols can be used with different types of neuronal cultures to estimate proteasome activity and the levels and aggregation of ubiquitinated proteins. In addition, they can be used to identify compounds potentially capable of preventing a decline in proteasome activity and formation of ubiquitin-protein aggregates associated with neurodegeneration.

Key words: Proteasome activity, Proteasome assembly, Ubiquitinated proteins, Protein aggregates, Protein turnover, Neurodegeneration, Neuronal cell cultures

1. Introduction

In a wide variety of neurodegenerative disorders, such as Alzheimer's, Parkinson's, and Huntington's diseases as well as amyotrophic lateral sclerosis, aggregates of ubiquitinated proteins are detected in neuronal inclusions (reviewed in ref. (1)) indicating that the

Giovanni Manfredi and Hibiki Kawamata (eds.), *Neurodegeneration: Methods and Protocols*,
Methods in Molecular Biology, vol. 793, DOI 10.1007/978-1-61779-328-8_18, © Springer Science+Business Media, LLC 2011

ubiquitin/proteasome pathway (UPP) may be deficient. The accumulation of ubiquitinated proteins in inclusions is thought to reflect a failure in proteasome activity although the mechanisms leading to inclusion formation remain unclear. There is a rising interest in the proteasome as a therapeutic target to prevent protein accumulation/aggregation and therefore delay/treat neurodegeneration in these progressive disorders (reviewed in ref. (2)).

In this chapter, we describe three different methods to measure proteasome activity in two types of neuronal cultures: (a) human neuroblastoma SK-N-SH cells and (b) rat E18 primary cortical neuronal cultures. First, the "in-gel" assay distinguishes three forms of the proteasome: the 26S with two and one cap as well as the 20S proteasome. It semiquantitatively assesses the chymotrypsin-like activity, the levels of each of the three proteasome forms and proteasome assembly. Second, the glycerol gradient centrifugation provides a means to quantitatively measure the individual activities of 26S and 20S proteasomes. Finally, the total cell lysate assay is a quick but somewhat unspecific way to measure proteasome activity without differentiating between 26S and 20S proteasomes or other enzymes that cleave the substrates. Besides the three assays to measure proteasome activity, we describe how to assess the accumulation of ubiquitinated proteins by western blot analysis, and how to detect ubiquitin-protein aggregates using the filter trap assay and immunofluorescence. To promote the accumulation of ubiquitinated proteins, we treated neuronal cultures with the irreversible proteasome inhibitor epoxomicin (3) or with the neurotoxic product of inflammation prostaglandin J2 (PGJ2) (4, 5). All of these protocols can be used to evaluate changes in proteasome activity and levels as well as accumulation/aggregation of ubiquitinated proteins under conditions that lead to neurodegeneration or prevent it.

2. Materials

2.1. Common Reagents/Equipment

1. Phosphate-buffered saline (PBS): Prepare 1× PBS using PBS tablets dissolved in ultrapure water. Store at 4°C.
2. Bradford assay for protein concentration.
3. Bicinchoninic acid (BCA) assay for protein concentration.
4. Bromophenol blue solution: 250 mg bromophenol blue dissolved in 10 mL buffer A (refer to Subheading 2.3 below). Centrifuge in a microcentrifuge at maximum speed for 5 min. Aliquot supernatant and store at 4°C.
5. 30% acrylamide/Bis solution, 29:1. Store at 4°C.
6. Transfer buffer: 25 mM Trizma base; 192 mM glycine, pH 8.3; 15% (v/v) methanol. Store at room temperature.

7. Antibody dilution buffer: SuperBlock blocking buffer (Thermo Scientific, Rockford, IL).

8. ECL stock solutions: 250 mM luminol in dimethyl sulfoxide (DMSO), store at −20°C; 90 mM p-Coumaric acid in DMSO, store at −20°C. ECL solution A: 100 mM Tris–HCl, pH 8.5; 2.5 mM luminol; 0.4 mM p-Coumaric acid. ECL solution B: 100 mM Tris–HCl, pH 8.5; 0.02% H_2O_2. ECL reagent: Mix solution A and B (1:1) just prior to use.

9. Repeater plus pipettor.

10. X-ray film and film developer.

11. Heidolph type RZR 50 benchtop homogenizer.

12. Spectrophotometer.

2.2. Cell Culture

1. SK-N-SH cells: These cells are an established human neuroblastoma cell line, which has a neuronal phenotype, synthesizes catecholamines, and expresses dopamine-β-hydroxylase. The cell line is derived from peripheral tissue (6). The cells are maintained as described in (7) and were obtained from ATCC. Culture media: Minimal essential media (MEM) with Eagle's salts containing 2 mM L-glutamine; 1 mM sodium pyruvate; 0.4% MEM vitamins; 0.4% MEM nonessential amino acids; 100 units/mL penicillin; 100 μg/mL streptomycin; and 5% normal fetal bovine serum (FBS) (see Note 1).

2. Rat E18 cortical neuronal cultures: These cultures are prepared from E18 embryos obtained from pregnant Sprague Dawley females following the methods described in (8). Culture media: Neurobasal media supplemented with B27 and 0.5 mM L-glutamax. Plating: Precoat 100-mm dishes with 50 μg/mL poly-D-lysine; add 6 million cells per dish.

3. Treatment: Treat SK-N-SH cells or neurons (DIC 8) with DMSO or PGJ2 (15 or 20 μM, PGJ2) for 16 or 24 h. PGJ2 stock: 3 or 4 mM in DMSO, respectively (see Notes 2 and 3).

2.3. In-Gel Assay

2.3.1. Cell Harvesting and Homogenization

1. Buffer A: 50 mM Tris–HCl, pH 7.4; 5 mM $MgCl_2$; 10% glycerol (v/v). Store at 4°C.

2. ATP and DTT stock: 200 mM ATP (made in buffer A); 200 mM DTT (made in ultrapure water). Place on ice (see Note 4).

3. Buffer for harvesting cells and protein standards: Buffer A; 5 mM ATP; 1 mM DTT. Place on ice (see Note 4).

2.3.2. Native PAGE

1. Buffer B: 0.18 M boric acid; 0.18 M Trizma base; 5 mM $MgCl_2$. Store at 4°C.

2. Gel buffer: Buffer B; 1 mM ATP; 1 mM DTT. For two minigels: 35 mL buffer B, 175 μL of 200 mM ATP, and 175 μL of 200 mM DTT. ATP and DTT prepared from stock (see Note 4).

3. Running buffer: Same as "gel buffer" and kept at 4°C until running gel.

4. Rhinohide polyacrylamide gel strengthener. Store at 4°C.

2.3.3. Proteasome Activity

1. Proteasome substrate stock solution: Succinyl-leucine-leucine-valine-tyrosine-(7-aminomethyl)-coumarine (Suc-LLVY-AMC). Make 20 mM stock reconstituted in DMSO, aliquot, and store at −80°C. This substrate is used to measure the chymotrypsin-like activity.

2. Proteasome activity mix: Buffer A; 1 mM ATP; 1 mM DTT. Just before measuring proteasome activity, add 300 μM of the proteasome substrate.

3. To make 15 mL of the proteasome activity mix: 14,625 μL buffer A; 75 μL 200 mM ATP; 75 μL 200 mM DTT; 225 μL 20 mM Suc-LLVY-AMC (see Note 4).

2.3.4. Proteasome Levels and Assembly

1. Blocking buffer: SuperBlock Blocking Buffer (Thermo Scientific, Rockford, IL).

2. Washing buffer: 0.5× PBS; 0.1% Tween 20.

3. PVDF membrane: Immobilon-P membrane (0.45 μm).

4. Primary antibodies: Rabbit anti-β5 (1:2,000); mouse anti-Rpt6 (1:2,000).

5. Secondary antibodies: Goat anti-rabbit IgG conjugated to horse radish peroxidase (1:10,000); goat anti-mouse IgG conjugated to horse radish peroxidase (1:10,000).

2.4. Glycerol Gradient Centrifugation

2.4.1. Cell Harvesting and Homogenization

1. Stock buffer: 0.1 M Tris–HCl, pH 7.5; 0.2 M ATP; 0.1 M DTT.

2. Harvesting buffer: 25 mM Tris–HCl, pH 7.5; 2 mM ATP; 1 mM DTT.

3. Epoxomicin reconstituted in DMSO to 1 mM. Aliquot and store at −80°C.

2.4.2. Glycerol Gradient Centrifugation and Fractionation

1. 40% glycerol buffer: 25 mM Tris–HCl, pH 7.5; 2 mM ATP; 1 mM DTT; 40% glycerol (v/v).

2. 10% glycerol buffer: 25 mM Tris–HCl, pH 7.5; 2 mM ATP; 1 mM DTT; 10% glycerol (v/v).

3. Gradient maker.

4. Beckman ultra-clear centrifuge tubes (14×89 mm).

5. Ultracentrifuge with SW41-rotor. Store the SW41-rotor and the Beckman tube buckets at 4°C.

6. Fraction recovery system for puncturing.

7. Fraction collector.

2.4.3. Chymotrypsin-Like Activity

1. Substrate: Suc-LLVY-AMC. Make 10 mM stock reconstituted in DMSO, aliquot, and store at −80°C.

2. Assay buffer: 0.05 M Tris–HCl, pH 7.5.

3. 10% (w/v) trichloroacetic acid (TCA).

4. 0.1% (w/v) sodium nitrite. Store in dark bottle.

5. 0.5% (w/v) ammonium sulfamate.

6. 0.05% (w/v) N-(1-naphthyl)ethylenediamine 2HCl (N-NEDA). Dissolve in 95% ethanol and store in dark bottle.

2.5. Peptidase Activities in Total Cell Lysate

2.5.1. Cell Harvesting and Homogenization

1. Stock buffer: 100 mM Tris–EDTA, pH 7.5; EDTA (see Note 5). Store at 4°C.

2. Harvesting buffer: 10 mM Tris–EDTA, pH 7.5. Store at 4°C.

3. Epoxomicin reconstituted in DMSO to 1 mM. Aliquot and store at −80°C.

2.5.2. Peptidase Activities

1. Peptidase substrates: Chymotrypsin-like activity, Suc-LLVY-AMC; trypsin-like activity, benzyloxycarbonyl-glycine-glycine-arginine-β-naphthylamide (Z-GGR-βNA); caspase-like activity, benzyloxycarbonyl-leucine-leucine-glutamic acid-β-naphthyl-amide (Z-LLE-βNA). Make 10 mM stock solutions in DMSO, aliquot, and store at −20°C.

2. Assay buffers: Chymotrypsin-like activity, 0.05 M Tris–HCl, pH 7.5; trypsin- and caspase-like activities, 0.05 M Tris–HCl, pH 8.0.

3. 10% (w/v) TCA.

4. 0.1% (w/v) sodium nitrite. Store in dark bottle.

5. 0.5% (w/v) ammonium sulfamate.

6. 0.05% (w/v) N-NEDA. Dissolve in 95% ethanol and store in dark bottle.

2.6. SDS-PAGE and Western Blotting for Proteasome Subunits α4, β5, and Rpt6, for β-Actin, and for Ubiquitinated Proteins

2.6.1. SDS-PAGE

1. 1% SDS lysis buffer: 10 mM Tris–EDTA, pH 7.5; 1% SDS. Store at room temperature.

2. Loading buffer: 50% glycerol; 50% β-mercaptoethanol; bromophenol blue. Store at room temperature.

3. Resolving buffer: 3 M Tris–HCl, pH 8.8. Store at 4°C.

4. Stacking buffer: 0.5 M Tris–HCl, pH 6.8. Store at 4°C.

5. Running buffer: 25 mM Trizma base; 192 mM glycine, pH 8.3; 0.1% (w/v) SDS. Store at room temperature.

6. Sonicator cell disruptor.

2.6.2. Western Blotting

1. Blocking buffer: 10 mM Tris–HCl, pH 7.3; 5% (w/v) nonfat dry milk; 10 mM NaCl; 0.1% (v/v) Tween 20 (see Note 4).

2. Washing buffer: 0.5× PBS; 0.1% Tween 20. Store at room temperature.

3. PVDF membrane: Immobilon-P membrane (0.45 μm).

4. Primary antibodies: Mouse monoclonal anti-α4 (1:500), rabbit polyclonal anti-β5 (1:2,000), and mouse anti-Rpt6 (1:2,000); mouse anti-β-actin (1:10,000); rabbit anti-ubiquitin antibody (1:1,500).

5. Secondary antibodies: Goat anti-rabbit IgG conjugated to horse radish peroxidase (1:10,000); goat anti-mouse IgG conjugated to horse radish peroxidase (1:10,000).

2.7. Filter Trap Assay for Ubiquitinated Protein Aggregates

2.7.1. Cell Harvesting and Homogenization

1. 2× lysis buffer base: 40 mM Tris–HCl, pH 7.5; 274 mM NaCl; 2 mM EGTA; 20% glycerol (v/v). Store at 4°C.

2. Lysis (RIPA) buffer (5 mL): 2.5 mL of 2× lysis buffer base; 100 μL protease inhibitor cocktail; 1 mM sodium orthovanadate; 1 mM phenylmethylsulfonyl fluoride; 1 mM β-glycerophosphate; 2.5 mM sodium pyrophosphate; 50 mM sodium fluoride; 1% Nonidet P40 (see Note 4).

3. Normalizing buffer: 2% SDS; 10 mM Tris–EDTA, pH 7.5. Store at 4°C.

4. Disposable cell scrapers.

2.7.2. Filter Trap

1. Washing buffer: 0.1% SDS; 10 mM Tris–EDTA, pH 7.5. Store at 4°C.

2. Trans-Blot nitrocellulose membrane, 0.2 μm.

3. Qualitative filter paper.

4. 96-well minifold Dot-blot system.

5. Vacuum source.

2.7.3. Western Blotting for Ubiquitinated Proteins

1. Blocking buffer: 10 mM Tris–HCl, pH 7.3; 5% (w/v) nonfat dry milk; 10 mM NaCl; 0.1% (v/v) Tween 20 (see Note 4).

2. Washing buffer: 0.5× PBS; 0.1% Tween 20. Store at room temperature.

3. Primary antibody: Rabbit anti-ubiquitin antibody (1:1,500).

4. Secondary antibody: Goat anti-rabbit IgG conjugated to horse radish peroxidase (1:10,000).

2.8. Immuno fluorescence for Detecting Ubiquitinated Protein Aggregates

2.8.1. Cell Culture and Fixation

1. Cells are plated onto eight-well chamber slides precoated with poly-d-lysine at a density of 50,000 cells/mL.

2. Paraformaldehyde stock: 16% paraformaldehyde solution, EM grade.

3. Paraformaldehyde (4%): Dilute paraformaldehyde stock with 1× PBS. Store at 4°C.

4. Ice-cold methanol.

2.8.2. Immunofluorescence

1. Dilution buffer: 25 mM Tris–HCl, pH 7.2; 300 mM NaCl; 0.3% v/v Triton X-100; 0.5 mg/mL bovine serum albumin; 0.01% w/v thimerosal. Store at 4°C.

2. Blocking buffer: 5% v/v normal goat serum in dilution buffer (see Note 4).

3. Primary antibodies: Rabbit polyclonal anti-ubiquitin antibody (1:150); mouse monoclonal anti-β III tubulin (1:1,000). Dilute both in the same microtube in blocking buffer.

4. Secondary antibodies: Alexa fluor 488 goat anti-rabbit (1:500); Alexa fluor 546 goat anti-mouse (1:500). Prepare both in the same microtube in dilution buffer.

5. Mounting medium hard set with DAPI.

3. Methods

3.1. In-Gel Assay

3.1.1. Cell Harvesting and Homogenization

1. Prepare harvesting buffer: For 4 mL harvesting buffer: 100 μL of 200 mM ATP; 20 μL of 200 mM DTT; 3,880 μL of buffer A. Place on ice.

2. When ready to harvest the cells, place dishes on ice and carefully aspirate the media from the dishes.

3. Wash cells once with 5 mL ice-cold 1× PBS (see Note 6).

4. Add 100 μL of harvesting buffer to one dish and collect cells by scraping (see Note 7).

5. Homogenize each sample on ice for 1 min with a benchtop homogenizer at maximum speed.

6. Centrifuge for 15 min at $19,000 \times g$ (4°C). Transfer supernatant to new chilled tubes and save the pellets for future SDS-PAGE analysis [see Subheading 3.4 (steps 1 and 2) Samples from "in-gel" assay – for SDS-PAGE and western blotting].

7. Determine protein concentration with the Bradford assay.

8. Normalize samples to equal protein concentration with harvesting buffer. Add to each sample 2 μL of bromophenol blue solution. Mix. Samples should remain on ice at all times.

3.1.2. Native PAGE (9)

1. Prepare three gel mixes (3, 4, and 5%) to make a 1.5-mm thick native gradient minigel. The components and respective volumes of each mix are shown in Table 1.

2. Add the 5% mix first to make a 2-cm layer on the bottom of the minigel cassette and overlay it with water. Once the gel layer polymerizes, carefully remove the water, dry with filter paper by gently tapping, and add the 4% mix to make a 3-cm layer above the 5% layer. Add water, and follow the same steps as above. Finally, add the 3% mix to completely fill the cassette and quickly insert the comb. Once the gel polymerizes, carefully remove the comb and with a 3-mL Falcon transfer pipette wash the wells with ultrapure water followed by running buffer.

Table 1
Native gel solutions

Gel components	3% (top)	4% (middle)	5% (bottom)
Layer height (cassette 8 × 8 cm)	3 cm	3 cm	2 cm
30% acrylamide/Bis solution, 29:1	1,260 μL	1,680 μL	2,100 μL
Rhinohide polyacrylamide gel strengthener	240 μL	320 μL	400 μL
Ammonium persulfate (1.5%, w/v)	600 μL	600 μL	600 μL
Gel buffer (complete with ATP and DTT)	9.888 mL	9.388 mL	8.888 mL
TEMED	12 μL	12 μL	12 μL
Total volume	12 mL	12 mL	12 mL

3. Prepare the running buffer. For example: 396 mL buffer B; 2 mL of 200 mM ATP; 2 mL of 200 mM DTT. Set up the gel apparatus to run the minigel and load the samples (40 μg of protein per lane) at 4°C.

4. Run the nondenaturing native minigel at 4°C with 150 V for 2 h.

5. Following electrophoresis, transfer the gel into a small plastic box containing 15 mL gel buffer without substrate; make sure that the entire gel is covered with gel buffer. Wash the gel on a rocker at room temperature for 1 min.

3.1.3. Proteasome Activity

1. Remove the gel buffer and add newly prepared proteasome activity mix (15 mL buffer A; 1 mM ATP; 1 mM DTT; 300 μM Suc-LLVY-AMC; see Note 4) into the box making sure that the entire gel is covered. Incubate the gel on a rocker for 5–10 min at 37°C.

2. Remove the gel from the box and place it on a UV transilluminator with the proteasome activity mix. Three proteasome bands, including 26S with two caps, 26S with one cap, and 20S, should be visible (see Fig. 1a).

3. Photograph the gel with a NIKON Cool Pix 8700 camera with a 3-4219 fluorescent green filter. For a better image, remove all bubbles on top of the gel.

4. Semiquantification of proteasome bands is done by image analysis with the ImageJ program (Rasband, W.S., ImageJ, U.S. NIH, Maryland, http://rsb.info.nih.gov/ij/, 1997–2006).

3.1.4. Proteasome Levels and Assembly

1. Following proteasome activity, gently wash gel with transfer buffer.

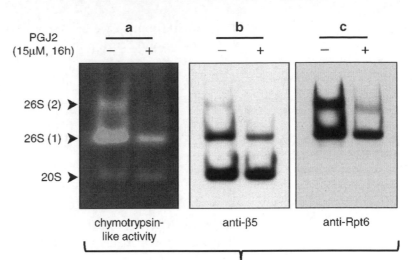

Fig. 1. In-gel assay for proteasome activity and assembly in rat E18 primary cortical neuronal cultures. Crude extracts were prepared from control cultures (–) or cultures treated with 15 μM PGJ2 for 16 h (+). Cleared lysates (40 μg/sample) were subjected to nondenaturing gel electrophoresis. In (**a**) the proteasomal chymotrypsin-like activity was measured with Suc-LLVY-AMC by the in-gel assay. In (**b** and **c**) 26S and 20S proteasomes were detected by immunoblotting with an anti-β5 antibody (**b**), a subunit of the core proteasome particle (20S), and with the anti-Rpt6 antibody (**c**), an ATPase subunit of the 19S regulatory particle. Proteasomal 26S (two caps and one cap) and 20S forms are indicated by on the left.

2. Transfer proteins from the minigel onto a PVDF membrane for 2 h at 110 mA per gel at 4°C.

3. Block the membrane with blocking buffer for 30 min at 37°C.

4. Incubate the membrane with a primary antibody in blocking buffer overnight at 4°C. The anti-β5 antibody reacts with a subunit of the 20S core particle, thus detecting both the 26S and 20S proteasomes (see Fig. 1b). The anti-Rpt6 antibody reacts with a subunit of the regulatory particle, thus detecting 26S proteasomes only (see Fig. 1c).

5. The next day, remove the primary antibody (can be reused) and wash the membrane three times (10 min each) with washing buffer on a rocker at room temperature.

6. Incubate the membrane with the appropriate secondary antibody in blocking buffer on a rocker at room temperature for 45 min.

7. Wash the membrane again three times (10 min each) with washing buffer on a rocker at room temperature.

8. Develop the blot by a chemiluminescent horseradish peroxidase method with the ECL reagent. Wrap the blot in a thin plastic bag and with tape attach it to an X-ray film-developing cassette.

9. In the darkroom, place an X-ray film in the cassette for the desired time and develop the film in an X-ray developer.

10. Three proteasome bands, including 26S with two caps, 26S with one cap, and 20S, should be visible depending on which antibody was used to probe the blots (see Fig. 1b, c).

11. Semiquantification of proteasome bands is done by image analysis with the ImageJ program.

3.2. Glycerol Gradient Centrifugation

3.2.1. Cell Harvesting and Homogenization

1. SK-N-SH cells are treated at 75–80% confluence with 25 nM epoxomicin (irreversible proteasome inhibitor) or DMSO (control) for 24 h (see Note 8).

2. When ready to harvest the cells, place dishes on ice and carefully aspirate the media from the dishes.

3. Wash cells twice with 5 mL ice-cold 1× PBS (see Note 6).

4. Add 100 μL of harvesting buffer to each dish and collect cells by scraping. Transfer to 1.5-mL chilled microtubes.

5. Centrifuge for 1 min at $19,000 \times g$ (4°C).

6. Homogenize each sample on ice for 1 min with a benchtop homogenizer at maximum speed.

7. Centrifuge for 10 min at $19,000 \times g$ (4°C). Transfer supernatant to new chilled microtubes and discard the pellet.

8. Determine protein concentration with the Bradford assay. Adjust protein concentration to 2 mg/sample in 500 μL total volume containing 10% glycerol. Samples should remain on ice at all times.

3.2.2. Glycerol Gradient Centrifugation and Fractionation (10, 11)

1. The gradient maker contains two chambers with a connecting valve centered between them that controls the generation of the gradient. Gradient flow rate is controlled by an outflow valve fitted on one of the chambers with attached tubing to deliver the gradient buffer (see Fig. 2).

2. Before starting to pour the gradient, both valves need to be closed and a magnetic spin bar should be placed into the chamber attached to the outflow valve (chamber 1).

3. Slowly dispense 6 mL of 40% glycerol buffer into chamber 1.

4. Slowly dispense 6 mL of 10% gradient buffer into chamber 2.

5. Turn on magnetic stirrer to gently mix the glycerol buffer.

6. Open the connecting valve first and then the outflow valve to allow pouring through gravity into each of the Beckman centrifuge tubes. The outflow rate should be slow.

7. Once all Beckman centrifuge tubes are filled, carefully load 500 μL of the cell lysate (2 mg of protein) onto the top of the gradient in each tube. Label each tube accordingly to the

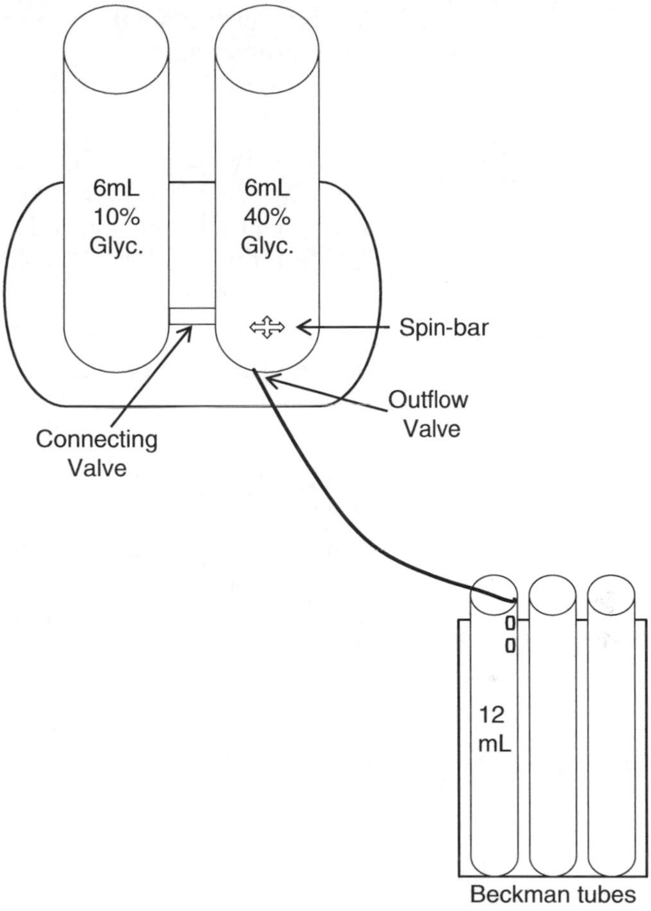

Fig. 2. Setup for making a glycerol gradient. For explanation, **see** Subheading 3.2, steps 1–8.

respective treatment. Transfer centrifuge tubes carefully into a chilled centrifuge bucket.

8. Equilibrate each centrifuge bucket/tube group by placing it in a small Erlenmeyer flask on a scale.

9. The weight difference between two opposing bucket/tube groups in the rotor should not be more than 0.1 g. Equilibrate the opposing bucket/tube groups by adding 10% glycerol buffer to the tube in the lighter group.

10. Carefully place the bucket/tube groups on the chilled rotor according to their numbers.

11. Centrifuge for 24 h at $83,000 \times g$ at 4°C in the Beckman ultracentrifuge.

12. The fraction collection in the following day should be done at 4°C (cold room).

13. Prepare 25 labeled 2.0-mL microtubes per sample (treatment) with caps removed (cut with scissors) to minimize any errors during fraction collection. Place tubes on the fraction collector.

14. Carefully place the first Beckman centrifuge tube on the fraction recovery system for puncturing. The fraction recovery system is attached to the fraction collector by a long tube (see Fig. 2). Puncture the bottom of the centrifuge tube and allow a few seconds for the flow to start.

15. Set the fraction collector to collect 500 µL of gradient mix per 2-mL microtube at 4°C. Monitor the flow and fraction collection throughout.

3.2.3. Chymotrypsin-Like Activity

1. Label new 1.5-mL microtubes (with caps) to correspond to each collected fraction.

2. Prepare the master mix to measure the chymotrypsin-like activity: 400 µM Suc-LLVY-AMC (substrate from 10 mM stock) in 0.05 M Tris–HCl, pH 7.5. Prepare enough master mix to add 50 µL per microtube (see Note 4). To save time, distribute the master mix with a repeater plus pipettor.

3. Add 50 µL of each fraction to the corresponding microtube and close the cap. Vortex.

4. Incubate at 37°C overnight on a rocker.

5. The next day, stop the reaction in each microtube with 100 µL of 10% TCA.

6. Vortex and to develop the color, add to each microtube in sequence 200 µL of 0.1% sodium nitrite; 200 µL of 0.5% ammonium sulfamate, and 400 µL 0.05% N-NEDA. Vortex before adding the next reagent. Upon addition of the last reagent, notice color development in the tubes where the substrate was cleaved.

7. Read absorbance in spectrophotometer at 560 nm.

8. Average the OD of every two consecutive tubes (example: average OD for tubes 1 and 2, then 3 and 4, and 5 and 6 to the end) to obtain the OD for a total of 13 fractions (tube #25 = fraction #13). Use the 13 averaged OD values (#1 the heaviest; #13 the lightest) to generate activity graphs for each condition (see Fig. 3a).

3.3. Peptidase Activities in Total Cell Lysate

3.3.1. Cell Harvesting and Homogenization

1. SK-N-SH cells are treated at 75–80% confluence for different time points with 25 nM epoxomicin (irreversible proteasome inhibitor) or DMSO (control) (see Note 9).

2. When ready to harvest the cells, place dishes on ice and carefully aspirate the media from the dishes.

3. Wash cells twice with 5 mL ice-cold 1× PBS (see Note 6).

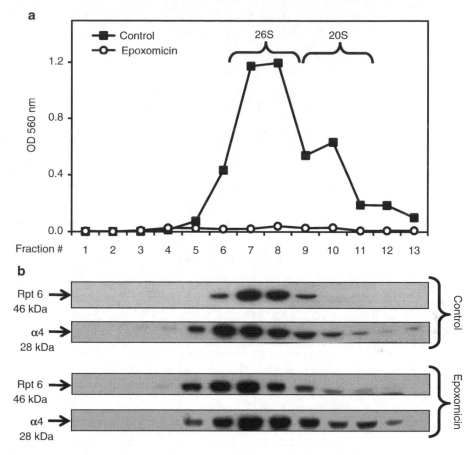

Fig. 3. Sedimentation velocity of proteasomes in SK-N-SH cells. Cells were treated for 24 h with DMSO (control, vehicle) or epoxo-micin (25 nM). Total lysates (2 mg protein/sample) were fractionated by glycerol density gradient centrifugation (10–40% glycerol corresponding to fractions 13 to 1). (**a**) Aliquots (50 μl) of each fraction obtained from control (*black squares*)- and epoxomicin (*white circles*)-treated cells were assayed for chymotrypsin-like activity with Suc-LLVY-AMC. (**b**) Immunoblot analyses of each fraction probed with antibodies that react with the proteasome (α4, core particle; Rpt6, 19S regulatory particle). Proteins were precipitated with acetone from 450 μl of each fraction. The fractions were obtained from control- and epoxomicin-treated cells.

4. Add 100 μL of harvesting buffer to each dish and collect cells by scraping. Transfer samples to 1.5-mL chilled microtubes.

5. Centrifuge for 1 min at 19,000 × g (4°C).

6. Homogenize each sample on ice for 1 min with a benchtop homogenizer at a maximum speed.

7. Centrifuge for 10 min at 19,000 × g (4°C). Transfer superna-tant to new chilled microtubes and discard the pellet.

8. Determine protein concentration with the BCA kit. Protein samples should remain on ice at all times.

9. Normalize samples to 2 μg/μL of protein, an optimal concen-tration to carry out this assay.

3.3.2. Peptidase Activities (12)

1. Label new microtubes in duplicate for each condition for three peptidase activities: chymotrypsin-, trypsin-, and caspase-like activities.

2. Prepare three master mixes for chymotrypsin-, trypsin-, and caspase-like activities (see Note 4):

(a) Chymotrypsin-like activity: 400 μM Suc-LLVY-AMC (substrate from stock 10 mM) in 0.05 M Tris–HCl, pH 7.5

(b) Tryspin-like activity: 400 μM of Z-GGR-βNA (substrate from stock 10 mM) in 0.05 M Tris–HCl, pH 8.0

(c) Caspase-like activity: 400 μM of Z-LLE-βNA (substrate from stock 10 mM) in 0.05 M Tris–HCl, pH 8.0

3. Prepare enough master mix for each activity to add 75 μL per microtube. To save time, distribute the master mix with a repeater plus pipettor.

4. Add 25 μL (50 μg) of sample per microtube and close caps. Vortex.

5. Incubate at 37°C overnight on a rocker.

6. The next day, stop the reaction in each microtube with 100 μL of 10% TCA.

7. Vortex and to develop the color, add to each microtube in sequence 200 μL of 0.1% sodium nitrite; 200 μL of 0.5% ammonium sulfamate, and 400 μL 0.05% N-NEDA. Vortex before adding the next reagent. Upon addition of the last reagent, notice color development in the tubes where the substrate was cleaved.

8. Read absorbance in spectrophotometer at 560 nm for chymotrypsin-like activity, and 580 nm for trypsin- and caspase-like activities.

9. Export data to Excel and generate activity graphs (see Fig. 4).

3.4. SDS-PAGE and Western Blotting for Proteasome Subunits α4, β5, and Rpt6, for β-Actin, and for Ubiquitinated Proteins

3.4.1. Samples from "In-Gel" Assay (Rat E18 Cortical Neuronal Cultures)

1. Supernatant remaining from "in-gel" assay (see Subheading 3.1 step 6): Adjust to 1% SDS lysis buffer.

2. Pellet remaining from "in-gel" assay (see Subheading 3.1 step 6): Resuspended in 1% SDS lysis buffer and sonicate until no particles are visible.

3. Heat all samples (supernatant and pellet) at 100°C for 5 min.

4. Determine protein concentration with BCA kit. Normalize samples with 1% SDS lysis buffer.

5. Add loading buffer to each sample (1:12.5, v/v).

6. Heat samples again at 100°C for 5 min. Samples are ready to load for SDS-PAGE and western blotting (see Fig. 5) or can be stored at –80°C until ready to load.

3.4.2. Samples from Glycerol Gradient Centrifugation (Human Neuroblastoma SK-N-SH Cells)

1. To the remaining 450 μL of each of the 25 original fractions collected in 2-mL microtubes (see Subheading 3.2 step 15), add 3 volumes (1,350 μL) of ice-cold acetone.

2. Mix well by vortexing and keep at –20°C overnight.

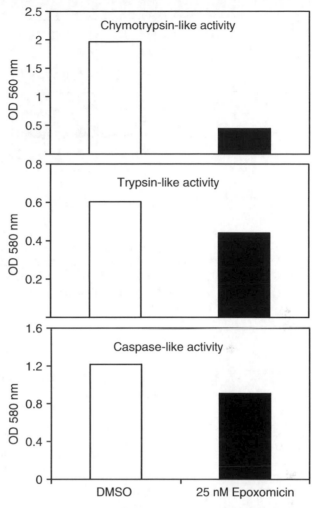

Fig. 4. Proteasome activities in SK-N-SH cells. Cells were treated for 24 h with DMSO (control, vehicle) or epoxomicin (25 nM). Proteasome activities were measured in cleared supernatants obtained from total cell homogenates (50 μg of protein/sample). Peptidase activities were assayed colorimetrically after a 24-h incubation at 37°C. The chymotrypsin-like activity was measured with Suc-LLVY-AMC, the trypsin-like activity with Z-GGR-βNA, and the caspase-like activity with Z-LLE-βNA.

3. Next day, spin the 2-mL microtubes for 15 min at 4°C at a maximum speed. Carefully discharge supernatant and retain the pellet.

4. Dry tubes by inversion on tissue paper (see Note 10).

5. Resuspend protein pellets in 25 μL of 1% SDS lysis buffer per sample.

6. Heat all samples at 100°C for 5 min.

7. Combine the samples of every two consecutive tubes (example: combine tubes 1 and 2, then 3 and 4, and 5 and 6 to the end)

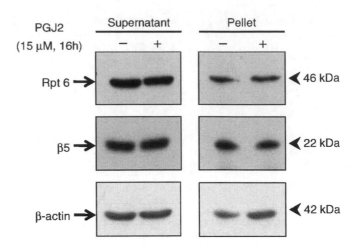

Fig. 5. Proteasome subunit levels in rat E18 primary cortical neuronal cultures. Aliquots of the supernatant (cleared lysate) and pellet fractions obtained from samples prepared for the "in-gel" assay (see Fig. 1) were run on SDS-PAGE (10% gel) followed by immunoblotting with the same antibodies listed in Fig. 1 as well as with anti-β-actin. 40 μg of protein were loaded per lane.

to obtain a total of 13 fractions (tube #25 = fraction #13). Fraction #1 is the heaviest and #13 the lightest. Each fraction now contains 50 μL of sample.

8. Add loading buffer to each fraction (1:12.5, v/v).

9. Heat samples again at 100°C for 5 min. Samples are ready to load (20 μL per lane) for SDS-PAGE and western blotting (see Fig. 3b) or can be stored at −80°C until ready to load.

3.4.3. Samples for Detecting Ubiquitinated Proteins (Rat E18 Cortical Neuronal Cultures)

1. Prepare hot 1% SDS lysis buffer in a 10-mL glass tube to harvest the cells (see Note 11).

2. When ready to harvest the cells, place dishes at room temperature and carefully aspirate the media from the dishes.

3. Wash the cells once with 5 mL PBS (room temperature, see Note 6).

4. Add 100 μL of hot 1% SDS lysis buffer to one dish and collect cells by scraping (see Note 7).

5. Heat samples at 100°C for 5 min.

6. Sonicate samples until no particles are visible.

7. Determine protein concentration with BCA kit. Normalize samples with 1% SDS lysis buffer.

8. Add loading buffer to each sample (1:12.5, v/v).

9. Heat samples again at 100°C for 5 min. Samples are ready to load for SDS-PAGE and western blotting (see Fig. 6) or can be stored at −80°C until ready to load.

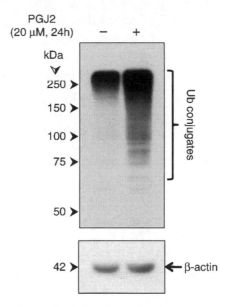

Fig. 6. Accumulation of ubiquitinated proteins in rat E18 primary cortical neuronal cultures. Cells were treated for 24 h with DMSO [control, vehicle (−)] or with 20 μM PGJ2 (+). Total cell extracts were subjected to western blot analysis (8% gel) to detect ubiquitinated proteins (40 μg of protein/lane). Equal protein loading was demonstrated by probing the immunoblots with the anti-β-actin antibody.

Table 2
SDS-polyacrylamide minigel solutions

Gel components	8%	10%	Stacking
30% acrylamide/Bis solution, 29:1	2.7 mL	3.4 mL	1.03 mL
Resolving buffer	1.3 mL	1.3 mL	None
Stacking buffer	None	None	2.07 mL
Ammonium persulfate (1.5%, w/v)	515 μL	515 μL	0.41 mL
Water	5.37 mL	4.67 mL	3.6 mL
SDS (10%, m/v)	105 μL	105 μL	0.83 mL
TEMED	10 μL	10 μL	6 μL
Total volume	10 mL	10 mL	8 mL

3.4.4. SDS-PAGE

1. Prepare gel mix (8% for ubiquitinated proteins; 10% for proteasome subunits and actin) to make a 1.0-mm thick SDS-polyacrylamide minigel. The components and respective volumes are shown in Table 2.

2. Pour the resolving gel leaving space for the stacking gel and overlay with water. The gel should polymerize in about 30 min.

3. Prepare the stacking gel mix as shown in Table 2.

4. Once the resolving gel polymerizes, carefully remove the water, dry with filter paper by gently tapping, add the stacking gel mix to completely fill the cassette, and quickly insert the comb.

5. Once the stacking gel polymerizes, carefully remove the comb and with a 3-mL Falcon transfer pipette wash the wells with ultrapure water followed by running buffer.

6. Set up the electrophoresis apparatus at room temperature. Add the running buffer into the upper and lower chambers of one gel unit. Load each sample (40 μg of protein per lane) and the molecular weight markers.

7. Run the SDS-polyacrylamide gel at room temperature with 150 V until the loading dye reaches the bottom of the gel.

3.4.5. Western Blotting

1. Remove the minigel from the cassette, place it in a small box, and wash it with transfer buffer.

2. Transfer proteins from the minigel onto a PVDF membrane for 2 h at 110 mA per gel at room temperature.

3. Block the membrane with blocking buffer for 30 min at 37°C on a rocker.

4. Incubate the membrane with primary antibody in SuperBlock blocking buffer overnight at 4°C.

5. The next day, remove the primary antibody (recycle) and wash the membrane three times (10 min each) with washing buffer on a rocker at room temperature.

6. Incubate the membrane with the appropriate secondary antibody in SuperBlock blocking buffer on a rocker at room temperature for 45 min.

7. Wash the membrane again three times (10 min each) with washing buffer on a rocker at room temperature.

8. Develop the blot by a chemiluminescent horseradish peroxidase method with the ECL reagent. Rap the blot in a thin plastic bag and with tape attach it to an X-ray film-developing cassette.

9. In the darkroom, place an X-ray film in the cassette for the desired time and develop the film in an X-ray developer.

10. After developing the film, wash the membrane for 10 min in washing buffer on the rocker. If desired, block the membrane again at 37°C for 30 min, and incubate it with another primary antibody overnight at 4°C for developing the next day. Follow the same steps as for the first primary antibody. Follow the same steps for incubating the membrane with yet another primary antibody and to develop the blot.

11. The molecular masses for Rpt6, α4, β5, and β-actin are approximately 46 kDa, 28 kDa, 22 kDa, and 42 kDa, respectively

(see Figs. 3 and 5). Ubiquitinated proteins are detected as a high-molecular-weight smear starting at the top of the blot (see Fig. 6).

12. Semiquantification of protein bands is done by image analysis with the ImageJ program.

3.5. Filter Trap Assay for Ubiquitinated Protein Aggregates

1. Prepare ice-cold lysis (RIPA) buffer to harvest the cells.

2. Label two 1.5-mL microtubes for each treatment and place them on ice.

3.5.1. Cell Harvesting and Homogenization

3. When ready to harvest the cells, place dishes on ice and carefully aspirate the media from the dishes.

4. Wash the cells once with 5 mL 1× PBS (ice cold, see Note 6).

5. Add 100 μL of lysis (RIPA) buffer to one dish and collect cells by scraping. Transfer cell lysate into a microtube.

6. Incubate samples at −80°C for 15 min to ensure total cell lysis.

7. Centrifuge for 10 min at $19,000 \times g$ (4°C). Transfer supernatant to a new chilled microtube. Discard pellet.

8. Determine protein concentration with the BCA kit. Protein samples should remain on ice at all times.

9. Normalize samples to 1 μg/μL of protein with normalizing buffer. Save remaining samples at −80°C.

3.5.2. Filter Trap Preparation

1. Cut the nitrocellulose membrane and Whatman filter paper to completely cover the chamber of the 96-well Dot-blot system.

2. Pre-wet the nitrocellulose membrane and Whatman filter paper with normalizing buffer for at least 5 min.

3. Set up the two bottom parts of the manifold.

4. Place two Whatman filter papers above the manifold. Place the nitrocellulose membrane over the top filter paper. Place the manifold top over the nitrocellulose membrane and seal tightly.

5. Establish a template for sample loading. Close tightly all the wells that are not going to be used with clear packing tape making sure to leave one empty row all around your samples. Do not load samples in wells at the edge of the apparatus; cover them with tape instead.

6. Set up vacuum, but DO NOT connect it to the manifold yet (see Note 12).

3.5.3. Filter Trap Running (13, 14)

1. Load 100 μg of each sample (in the same volume for all samples) into the respective wells.

2. Load 100 µl of washing buffer to the row of empty wells surrounding the samples.

3. Connect the vacuum line to the manifold and turn the vacuum on. All the samples should flow through the membrane and the wells are empty. Complete flow takes a few seconds.

4. Turn off the vacuum and disconnect its line from the manifold to release the vacuum.

5. Wash each untapped well two times with 100 µL of washing buffer. Connect the vacuum/turn it on each time and turn it off/disconnect it after each wash to release the vacuum.

6. Remove the top of the manifold. Mark the nitrocellulose membrane with a pencil to remember the sample loading order.

7. Place the membrane in a plastic box and proceed as for western blotting (see Subheading 3.4, steps 3–9), including blocking and probing the membrane with the anti-ubiquitin antibody. A positive signal appears as a dark dot (see Fig. 7b).

8. Semiquantification of protein dots is done by image analysis with the ImageJ program.

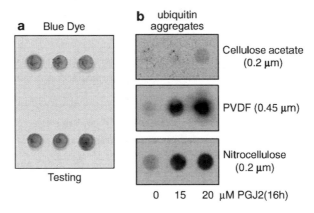

Fig. 7. Filter trap assay to measure ubiquitin-protein aggregates in rat E18 primary cortical neuronal cultures. (**a**) In a preliminary test, waterproof drawing ink (100 µl per well) was applied to the nitrocellulose membrane to assess flow through each well. If vacuum is properly equilibrated, dye dots corresponding to each well should appear round and neat on the nitrocellulose membrane. (**b**) Influence of membrane composition and porosity on detection of ubiquitin-protein aggregates. Cells were treated for 16 h with DMSO [control, vehicle (0)] or with 15 µM or 20 µM PGJ2. Cell extracts were subjected to the filter trap assay to detect ubiquitin-protein aggregates (100 µg of protein/sample). Nitrocellulose-yielded optimal assay sensitivity relative to polyvinylidene fluoride (PVDF) or cellulose acetate.

3.6. Immunoflu orescence for Detecting Ubiquitinated Protein Aggregates

3.6.1. Cell Culture and Fixation

1. Prepare the fixation solution (4% paraformaldehyde) in a fume hood.

2. When ready to immunostain, place slides at room temperature and carefully aspirate the media from each well.

3. Add 500 μL of 4% paraformaldehyde solution to each chamber and incubate for 15 min at room temperature in a fume hood (see Note 13).

4. Remove paraformaldehyde and discard it into a hazardous waste container (see Note 13). Repeat step 3.

5. Remove paraformaldehyde solution as in step 4. Wash wells three times with 1× PBS for 3 min each time at room temperature on a rocker.

6. Add 500 μL of ice-cold methanol per well and incubate for 2 min at room temperature on a rocker. Remove methanol. Wash wells three times with 1× PBS for 3 min each time at room temperature on a rocker.

3.6.2. Immunofluorescence

1. Block with blocking buffer for 1 h at room temperature on a rocker.

2. Remove blocking buffer. Add primary antibodies in blocking buffer.

3. Incubate overnight with both primary antibodies prepared in the same microtube. Place slides in a humid chamber (to avoid evaporation) on a rocker at 4°C.

4. Remove primary antibodies. Wash wells three times with 1× PBS for 3 min each time at room temperature on a rocker.

5. Prepare fresh secondary antibodies in dilution buffer in the same tube. Incubate for 1 h in the dark at room temperature.

6. Remove secondary antibodies. Wash wells three times with 1× PBS for 3 min each time at room temperature on a rocker.

7. Remove the wells, gasket, and biocompatible adhesive from the slide (a razor blade may be useful).

8. Add a few drops of mounting medium to the slide surface and quickly place the coverslip on top making sure to remove all bubbles.

9. The slide can be viewed immediately or can be stored in the dark at 4°C for several months.

10. View slides with a fluorescence microscope. Ubiquitin staining – green fluorescence; β-III tubulin – red fluorescence; DAPI – blue fluorescence. See Fig. 8.

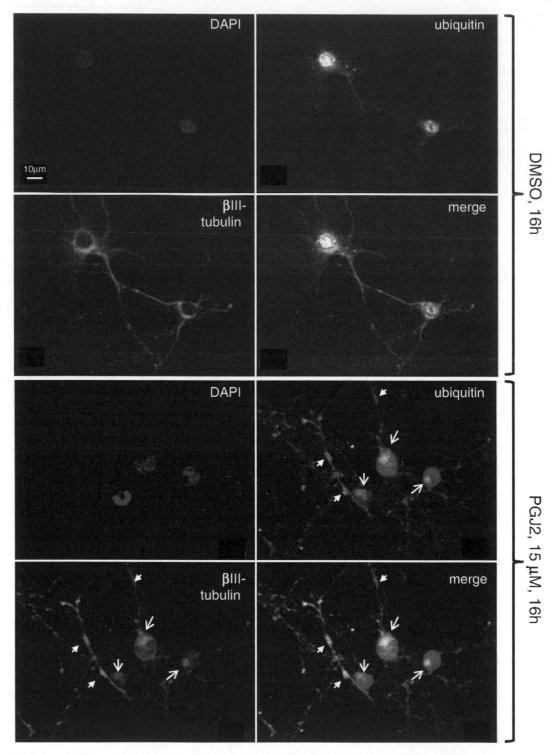

Fig. 8. Immunofluorescence detection of ubiquitin-protein aggregates in rat E18 primary cortical neuronal cultures. Ubiquitinated proteins and βIII-tubulin immunofluorescence staining of cortical cultures treated with DMSO (top 4 panels) or 15 μM PGJ2 for 16 h (bottom 4 panels). Nuclei are stained with DAPI. Large *arrows* point to protein aggregates and small *arrows* to dystrophic neurites. Scale bar = 10 μm.

4. Notes

1. Do not heat-inactivate the normal FBS.

2. The final DMSO concentration should be 0.5%.

3. Both cell cultures as well as other types of neuronal cultures can be used for all the assays described here.

4. Prepare fresh each time. Do not freeze.

5. Make sure that the EDTA is the acid form with no salts. Sodium and potassium inhibit proteasome activity. Use the acid EDTA (powder) to adjust the pH of the Tris solution.

6. While aspirating upon washing, tilt dishes slightly to make sure that all of the PBS is removed.

7. To increase protein concentration per treatment, transfer cell lysate to another dish with the same treatment and harvest cells the same way. Combine all of the cell lysates corresponding to the same treatment into one 1.5-mL chilled microtube and place on ice.

8. It is recommended to use four 100-mm dishes per treatment since a large amount (2 mg) of protein is needed.

9. It is recommended to use two 100-mm dishes per treatment.

10. Pellet may be difficult to see in heavier fractions.

11. Keep the glass tube in boiling water for at least 5 min to make sure that the buffer is hot.

12. To test if the setup is properly sealed, add 100 μL of any dye to the wells selected for sample loading. Remove the top part of the manifold. The dye should form round dots on the nitrocellulose membrane. Smeared, not perfectly round dye dots indicate that the setup is not properly sealed (see Fig. 7a). Redo the sealing and test again until the dye dots are perfectly round.

13. Paraformaldehyde is a hazardous chemical. Respiratory, skin, and eye protection should be used when working with it. Work within a fume hood or with local exhaust ventilation. Absorb incidental spills with damp absorbent pads. Collect and submit for waste disposal. Request hazardous waste pickup service for disposal.

Acknowledgments

Supported by NIH: [NIA-AG028847 to M.F.-P.; NINDS-NS41073 (SNRP) to M.F.-P. (head of subproject); NCRR-RR03037 (infrastructure) to Hunter College, City University of New York].

References

1. Alves-Rodrigues A, Gregori L, Figueiredo-Pereira ME (1998) Ubiquitin, cellular inclusions and their role in neurodegeneration. *Trends Neurosci* 21:516–520

2. Huang Q, Figueiredo-Pereira ME (2010) Ubiquitin/proteasome pathway impairment in neurodegeneration: therapeutic implications. *Apoptosis* 15(11):1292–1311

3. Meng L, Mohan R, Kwok BH, Elofsson M, Sin N, Crews CM (1999) Epoxomicin, a potent and selective proteasome inhibitor, exhibits in vivo antiinflammatory activity. *Proc Natl Acad Sci USA* 96:10403–10408

4. Li Z, Melandri F, Berdo I, Jansen M, Hunter L, Wright S, Valbrun D, Figueiredo-Pereira ME (2004) delta12-Prostaglandin J2 inhibits the ubiquitin hydrolase UCH-L1 and elicits ubiquitin-protein aggregation without proteasome inhibition. *Biochem Biophys Res Commun* 319:1171–1180

5. Wang Z, Aris VM, Ogburn KD, Soteropoulos P, Figueiredo-Pereira ME (2006) Prostaglandin J2 alters pro-survival and pro-death gene expression patterns and 26S proteasome assembly in human neuroblastoma cells. *J Biol Chem* 281:21377–21386

6. Biedler JL, Roffler-Tarlov S, Schachner M, Freedman LS (1978) Multiple neurotransmitter synthesis by human neuroblastoma cell lines and clones. *Cancer Res* 38:3751–3757

7. Li Z, Jansen M, Ogburn K, Salvatierra L, Hunter L, Mathew S, Figueiredo-Pereira ME

(2004) Neurotoxic prostaglandin J2 enhances cyclooxygenase-2 expression in neuronal cells through the p38MAPK pathway: a death wish? *J Neurosci Res* 78:824–836

8. Biederer T, Scheiffele P (2007) Mixed-culture assays for analyzing neuronal synapse formation. *Nat Protoc* 2:670–676

9. Elsasser S, Schmidt M, Finley D (2005) Characterization of the proteasome using native gel electrophoresis. *Methods Enzymol* 398:353–363

10. Eytan E, Ganoth D, Armon T, Hershko A (1989) ATP-dependent incorporation of 20S protease into the 26S complex that degrades proteins conjugated to ubiquitin. *Proc Natl Acad Sci USA* 86:7751–7755

11. Hirano Y, Murata S, Tanaka K (2005) Large- and small-scale purification of mammalian 26S proteasomes. *Methods Enzymol* 399:227–240

12. Wilk S, Orlowski M (1983) Evidence that pituitary cation-sensitive neutral endopeptidase is a multicatalytic protease complex. *J Neurochem* 40:842–849

13. Wanker EE, Scherzinger E, Heiser V, Sittler A, Eickhoff H, Lehrach H (1999) Membrane filter assay for detection of amyloid-like polyglutamine-containing protein aggregates. *Methods Enzymol* 309:375–386

14. Chang E, Kuret J (2008) Detection and quantification of tau aggregation using a membrane filter assay. *Anal Biochem* 373:330–336

Part V

Neurodegenerative Disease Mechanisms: Mitochondria

Chapter 19

Measurements of Threshold of Mitochondrial Permeability Transition Pore Opening in Intact and Permeabilized Cells by Flash Photolysis of Caged Calcium

Andrey Y. Abramov and Michael R. Duchen

Abstract

Changes in intracellular calcium concentration play a major role both in signal transduction and in cell death. In particular, mitochondrial Ca^{2+} overload is critically important as a determinant of irreversible cell injury. When accumulated above a threshold, matrix Ca^{2+} triggers opening of the mitochondrial permeability transition pore (mPTP), initiating ATP depletion and cell death via necrosis or by promoting cytochrome c release and initiating the apoptotic cascade. Measurement of mitochondrial Ca^{2+} uptake capacity (or the threshold for mPTP opening) is, therefore, important for understanding the mechanisms of pathophysiology in a variety of disease models and also for testing neuro- or cardioprotective drugs. We have, therefore, devised an approach that delivers Ca^{2+} directly to the matrix of mitochondria independently of uptake and therefore independently of potential ($\Delta\psi_m$) that allows direct study both of the Ca^{2+} efflux pathway and of the specific sensitivity of mPTP to Ca^{2+}. This is achieved using the photolytic release of Ca^{2+} by flash photolysis of caged Ca^{2+} using compounds, such as o-nitrophenyl EGTA, introduced into the cell as the acetoxymethyl (AM) ester (NP-EGTA, AM). This method can be used in both intact and permeabilized cells.

Key words: Caged Ca^{2+}, Flash photolysis, Mitochondria, Permeability transition pore

1. Introduction

Changes in intracellular Ca^{2+} concentration $[Ca^{2+}]_c$ play a major role both in signal transduction and in cell death. In particular, mitochondrial Ca^{2+} overload is critically important as a determinant of irreversible cell injury. This principle has emerged as a common theme recapitulated in a number of widely different pathological states, perhaps best established as a mechanism of ischaemic reperfusion injury in the heart and in the CNS (1, 2), but also in widely

Giovanni Manfredi and Hibiki Kawamata (eds.), *Neurodegeneration: Methods and Protocols*,
Methods in Molecular Biology, vol. 793, DOI 10.1007/978-1-61779-328-8_19, © Springer Science+Business Media, LLC 2011

different disease states, including the collagen VI deficiencies in Ullrich and Bethlem myopathies (3) and in acute pancreatitis (4). Therefore, measurement and analysis of mitochondrial responses to Ca^{2+} overload is a useful part of the experimental repertoire in trying to address mechanisms of cell injury in many model systems, including age-related neurodegenerative diseases, such as Alzheimer's and Parkinson's diseases (5).

When $[Ca^{2+}]_c$ rises close to mitochondria, they accumulate Ca^{2+}, which moves down its electrochemical gradient through an electrogenic uniporter that facilitates Ca^{2+} transport across the inner mitochondrial membrane into the matrix. The electrochemical gradient favours Ca^{2+} uptake as the matrix $[Ca^{2+}]_m$ is kept low by the activity of Ca^{2+}/nNa^+ and/or $Ca^{2+}/2H^+$ exchangers (6, 7) and by the negative mitochondrial membrane potential ($\Delta\psi_m$). Net Ca^{2+} accumulation occurs when the rate of influx exceeds the capacity of the exchangers to remove Ca^{2+}. This seems to occur at a cytosolic $[Ca^{2+}]$ of around 500 nM, often referred to as the set point for Ca^{2+} accumulation, but it also depends on the mitochondrial membrane potential and the cytosolic Na^+ concentration.

Under physiological conditions, raised intramitochondrial $[Ca^{2+}]_m$ stimulates the enzyme activity of the tricarboxylic acid (TCA) cycle, thereby increasing mitochondrial oxidative phosphorylation and energy production (8). This is thought to represent a valuable mechanism that matches energy supply to demand. Accumulated above a certain threshold, matrix Ca^{2+} can trigger opening of the mitochondrial permeability transition pore (mPTP), especially if a rise in $[Ca^{2+}]_c$ is associated with oxidative or nitrosative stress (9). The PTP opening usually triggers cell death, either by precipitating rapid ATP depletion and necrosis or – in cells with a high glycolytic capacity – by causing mitochondrial cytochrome c release and initiating an apoptotic cascade. In either case, this is a catastrophic event that signals cell death.

Measurement of mitochondrial uptake Ca^{2+} capacity (or the threshold for mPTP opening) is, therefore, important for understanding mechanisms of the pathology and also for testing neuro- or cardioprotective drugs. Many approaches use isolated mitochondria or bulk measurements from permeabilized cells and titration of the Ca^{2+} capacity by the progressive application of external Ca^{2+} (10, 11). In intact cells, the threshold of mPTP opening can be estimated by using different concentrations of electrogenic Ca^{2+} uniporters or by causing mitochondrial oxidative stress through illumination of photosensitising mitochondrially localized fluorescent dyes (12, 13). However, we have devised an approach that allows these measurements to be made at the level of a single cell, where this approach is a requirement of the experimental system. Thus, the experimental system demands that these measurements are made at the level of single cells when cell numbers are small, the population is heterogeneous, or when a modest proportion of cells are transfected and express a protein of interest.

Furthermore, it is important to bear in mind that, as mitochondrial Ca^{2+} uptake depends on $\Delta\psi_m$, comparative studies of mPTP threshold may be biased by differences in effective Ca^{2+} uptake in populations in which $\Delta\psi_m$ varies. We have, therefore, devised an approach that delivers Ca^{2+} directly to the matrix of mitochondria independently of uptake and therefore independently of $\Delta\psi_m$ that allows a direct study of the efflux pathway and the specific sensitivity of mPTP to Ca^{2+}. This is achieved by using flash photolysis of caged Ca^{2+} compounds, such as o-nitrophenyl EGTA, which are introduced into the cell as the acetoxymethyl (AM) ester (NP-EGTA, AM). This method can be used in intact and permeabilized cells, although results in intact cells can be misinterpreted because Ca^{2+} is released both in the cytosol and in the mitochondrial matrix. Permeabilization of cells with digitonin in a pseudo-intracellular solution localizes the caged Ca^{2+} signal to the matrix of mitochondria only.

The principle of the "caged" Ca^{2+} is that NP-EGTA is a photolabile Ca^{2+} chelator, whose affinity diminishes greatly following exposure to UV light (the K_d changes following the UV exposure from 80 nM to >1 mM). Thus, NP-EGTA acts as a chelator that binds Ca^{2+}avidly when loaded into cells, but releases Ca^{2+}as the affinity falls following a UV flash.

Measurement of mPTP opening can be achieved using cells co-loaded with a fluorescent Ca^{2+} indicator and with probes for mitochondrial membrane potential (fluo-4 and tetramethylrhodamine methylester (TMRM), respectively). The loss of mitochondrial membrane potential in response to a rise in $[Ca^{2+}]_c$ in intact or permeabilized cells can be attributed to mPTP opening, if it can be prevented by cyclosporine A (CsA), the archetypal inhibitor of the mPTP (14).

2. Materials

1. Fluo-4AM (Molecular Probes/Invitrogen) is supplied in aliquots. Dissolve an aliquot by addition of 50 μl dry DMSO to a tube before use, and store at −20°C for up to 1 month.

2. TMRM (Molecular Probes/Invitrogen) is water soluble. Prepare a 5 μM stock solution either in a physiological saline or in distilled water.

3. NP-EGTA, AM (Molecular Probes/Invitrogen) is also supplied in aliquots. Dissolve in DMSO to reach a final concentration of 1 mM in the tube before use, and store at −20°C.

4. Pluronic (Molecular Probes/Invitrogen) is soluble in DMSO. Prepare a 2% stock solution.

5. Use a dish that holds a glass coverslip in which cells are grown, and buffer while imaging cells or grow cells in a Petri dish with a glass coverslip base.

6. Confocal microscope equipped with HeNe (543- or 555-nm wavelength), Argon (488-nm wavelength), and UV lasers (351 and 360 nm) and ideally with inverted optics. Use a quartz UV-compatible objective lens and make sure that the UV laser is accurately focused on the sample by setting up the collimator correctly (see Note 1).

7. HEPES-buffered salt solution (HBSS): 156 mM NaCl, 3 mM KCl, 2 mM $MgSO_4$, 1.25 mM KH_2PO_4, 2 mM $CaCl_2$, 10 mM glucose, and 10 mM HEPES; pH adjusted to 7.35 with NaOH. This buffer is used to bathe cells in a static chamber. This avoids the need for continuous CO_2 equilibrated superperfusion. The chamber is ideally heated to 37°C.

8. "Pseudo-intracellular" solution: 135 mM KCl, 10 mM NaCl, 20 mM HEPES, 5 mM pyruvate, 5 mM malate, 0.5 mM KH_2PO_4, 1 mM $MgCl_2$, 5 mM EGTA, and 1.86 mM $CaCl_2$ (to yield a free $[Ca^{2+}]$ of ~100 nM), pH 7.1 + 20 µM digitonin + 40 nM TMRM.

3. Methods

TMRM partitions across the cell membrane and is accumulated by polarized mitochondria due to the delocalized positive charge of this organic lipophilic cation. According to Nernstian principles, at a mitochondrial potential of –180 mV, the intramitochondrial concentration of TMRM is about 1,000-fold greater than that in the cytosol. In most cells, TMRM takes about 30 min to reach full equilibration between compartments.

Fluo-4AM (5 µM) and NP-EGTA, AM (10 µM) cross the cell membrane, where the AM ester is cleaved by cellular esterases. Both NP-EGTA and fluo-4 as AM esters distribute evenly in the cytosol and cell organelles. Once the AM ester group is cleaved, the charged moiety remains trapped within a compartment. Loading the cells at room temperature may increase AM dye loading into the mitochondria, as lower temperatures reduce the esterase activity, allowing more time for accumulation of the dye into these organelles. NP-EGTA, AM is loaded into cells without Ca^{2+}. After intracellular cleavage of the ester group, NP-EGTA binds Ca^{2+} with a high affinity, forming a complex NP-EGTA:

Ca^{2+} both in the mitochondrial matrix and in the cytosol. It seems that the Ca^{2+} then re-equilibrates, as the intracellular homeostatic mechanisms maintain normal levels of free $[Ca^{2+}]$. Low temperatures reduce the rate of physiological processes, including mitochondrial metabolism, and limit Ca^{2+} "loading" of the chelator, and we have found that this reduces Ca^{2+} release following flash photolysis.

The optimum loading protocol depends on the cell type and has to be determined empirically. The K_d of NP-EGTA Ca^{2+} increases following UV illumination from 80 nM to >1 mM. UV "flashes" can be applied using a UV confocal and the intensity of the flash can be tuned to allow repeated uncaging, until mPTP opening is seen as the (CsA sensitive) release of fluo-4 and TMRM from the mitochondria. Each UV flash releases a set fraction of the caged Ca^{2+} (which depends on the flash intensity and efficiency of the optics), and so inevitably the absolute amount of Ca^{2+} released with successive flashes decreases. We have found that conditions can be established in several model systems such that wild-type cells maintain Ca^{2+} and membrane potential over many repeated flashes while cells expressing a mitochondrial pathology associated with neurodegeneration (e.g. in cells from the PINK1 knockout mouse; Figs. 1 and 2, and *see* also ref. 5) show mPTP opening after several flashes.

Caveat. In intact cells, a UV flash increases Ca^{2+} concentration in both the cytosol and the mitochondria inducing elevation of $[Ca^{2+}]_m$ both through direct release from the marix-localized caged compound and from uniporter-mediated uptake. Subsequent UV flashes in a short period of time (every 1 min) induced step-like increase of mitochondrial Ca^{2+} and mitochondrial depolarization (Fig. 1). This method allows detection of mPTP threshold, but also reveals the kinetics of mitochondrial Ca^{2+} efflux. Increases of $[Ca^{2+}]_c$ by flash photolysis in cells, in which the Na^+/Ca^{2+} exchanger is inhibited (either pharmacologically using the drug CGP37157 or, as we have found recently, in the case of PINK1-deficient neurons (5)), cause the appearance of a very bright fluo-4 signal co-localizing perfectly with the TMRM signal, reflecting high levels of $[Ca^{2+}]_m$ (Fig. 1).

In permeabilized cells, TMRM can be used as an indicator of $\Delta\psi_m$, but fluo-4 is generally useful only to indicate mitochondrial integrity, as fluo-4 is a relatively high-affinity Ca^{2+} indicator ($K_d = 345$ nM) and the signal tends to saturate (see Note 2). It is possible to use a low-affinity indicator for the same purpose (fluo-4 FF or Calcium Green 5N), although the Ca^{2+} threshold for mPTP opening in healthy mitochondria (it may be higher than 300 μM) is likely higher than the K_d of these indicators as well.

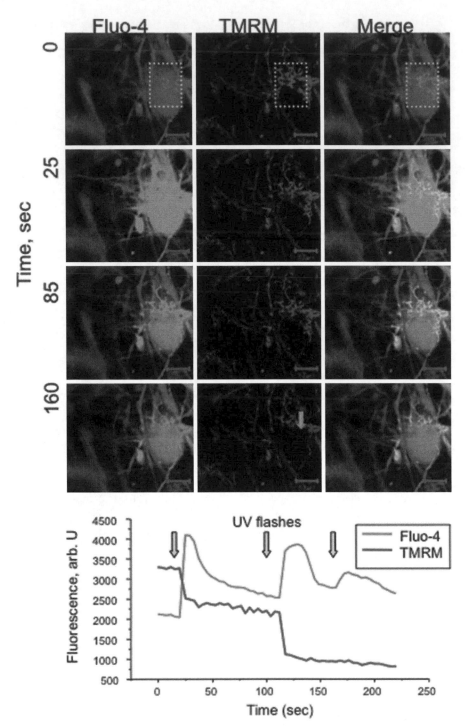

Fig. 1. Uncaging Ca^{2+} induces $[Ca^{2+}]_m$ overload and mitochondrial depolarization in a neuron. PINK1 knockout neurons demonstrate inhibition of the mitochondrial Na^+/Ca^{2+} exchanger and a reduced threshold for mPTP opening (5). The images show selected frames imaging cells loaded with *o*-nitrophenyl EGTA, fluo-4, and TMRM. Images selected from a time series showing the responses to flash photolysis of the caged Ca^{2+} in the area indicated by the *dashed line*. The trace below shows the continuous intensity measurements as a function of time at the area marked in the images with an *arrow*. The *arrows* on the trace indicate times of UV-induced flash photolysis. The photolysis-induced rise in $[Ca^{2+}]_c$ resulted in a dramatic increase in $[Ca^{2+}]_m$, as demonstrated by the fluo-4 signal in the mitochondrial area (note the co-localization of TMRM and the bright fluo-4 fluorescence in the images at 85″). This was rapidly followed by mitochondrial depolarization and the subsequent release of fluo-4 from the mitochondria. The calibration *bar* indicates 10 μm.

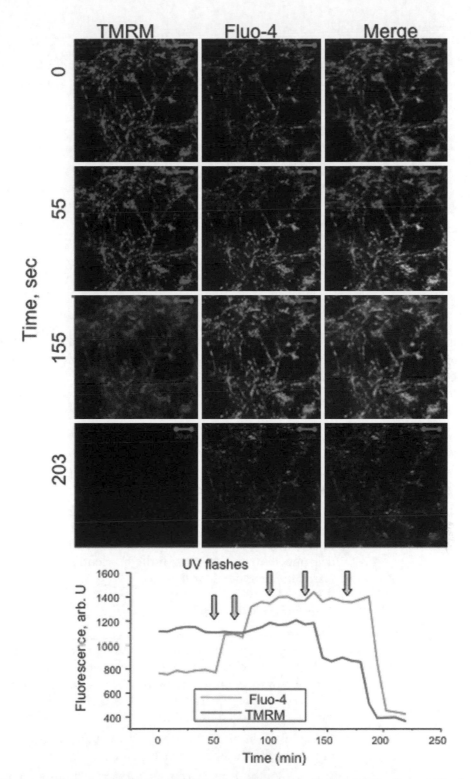

Fig. 2. Measurements of mitochondrial Ca^{2+} capacity in permeabilized cells. Flash photolysis of permeabilized neurons co-loaded with fluo-4 and TMRM demonstrated flash-induced increases in $[Ca^{2+}]_m$ followed by the concurrent release of Ca^{2+} and complete mitochondrial depolarization. The images show selected frames from a time series taken at the times indicated while the trace below shows the continuous changes in signal with time in a small area as indicated. In this experiment, the whole field of view shown was "flashed" with UV at time points indicated by the *blue arrows*. The calibration *red bar* indicates 20 μm.

With mPTP opening, the disappearance of individual mitochondria can be detected by the loss of both fluo-4 and TMRM fluorescence.

3.1. Preparing the Cells

1. Place glass coverslips at the bottom of 6-well tissue culture dishes (see Note 3).

2. Cells should be plated so that the density of the cells in the coverslip achieves ~70% confluence when they are imaged. In the mixed hippocampal cultures that we frequently use, containing a mixture of neurons and glia, this refers primarily to the astrocytes, as the neurons tend to grow unevenly over a coverslip.

3. Cells should be trypsinized and plated at least 24 h before the experiment to allow them to attach to the coverslip and recover.

3.2. Loading Fluo-4 AM, o-Nitrophenyl EGTA, AM, and TMRM into the Cells

1. In a 1.5-ml Eppendorf tube, add 5 µM fluo-4 AM, 10 µM NP-EGTA, AM, 25 nM TMRM + 0.05% pluronic, then add 1 ml HBSS buffer, and mix well.

2. Replace the buffer on the cells with HBSS buffer containing fluo-4 AM, o-nitrophenyl EGTA, AM, TMRM, and pluronic.

3. Leave the cells for 30 min at room temperature to equilibrate the dyes.

4. Place the cells in a tissue culture incubator at 37°C for 20 min for de-esterification of the dyes.

5. The buffer used for imaging should contain 25 nM TMRM for the whole duration of the experiment and in all solutions used prior to permeabilization, when the concentration is increased to compensate for the loss of the concentrating power of the plasma membrane potential.

3.3. Confocal Imaging of Fluo-4 and TMRM Fluorescence

1. Transfer the coverslip containing the cells to the imaging chamber.

2. Wipe any liquid from underneath the coverslip using a tissue and place the chamber on the objective of the microscope, adding a drop of oil to the objective, if necessary. Ensure that a UV-compatible quartz objective is used, as this is required for the UV-induced uncaging.

3. Using phase-contrast or transmitted light, adjust the focus until the cells are clearly visible.

4. In the confocal microscope software, choose appropriate imaging optics for imaging fluo-4, i.e. excitation using the 488-nm line of the laser, and use a band-pass filter between 505 and 550 nm to collect emitted light. This is most important – if a long-pass filter is used, light from TMRM (which is also

excited at 488 nm) will be included in the signal measured. For excitation of TMRM, use the 543-nm line of the laser while emitted light collected between 560 and 630 nm (or using a long-pass filter of >560 nm).

5. Keep the laser power to a minimum and the gain high as far as it is compatible with reasonable signal-to-noise ratios (we usually operate at ~1% or less) to avoid damaging the cells.

6. Start the continuous scan, and adjust the focus until mitochondria are clearly visible.

7. Open the pinhole setting in the software to give an optical section of approximately 3 μm and decrease the laser power and/or gain until a signal of ~50% intensity saturation is obtained. Use the lowest laser power compatible with a reasonable signal/noise to avoid phototoxicity.

8. Stop continuous scanning.

9. Increase image averaging (e.g. to average four frames) and/or decrease the scan speed to obtain a high-quality image to obtain clear resolution of mitochondria.

10. Most microscope control software have settings called "bleach" protocols. Using these settings, you are able to set up the requirements for the flash photolysis – we usually set the UV laser flash power at ~50%, but precise conditions have to be tuned according to your optical arrangement.

11. The software also allows you to select a region to be "bleached" – use this to choose the sector of the cell(s) to be exposed to UV.

12. Start the scan to obtain the images of fluo-4 and TMRM fluorescence in the cells and flash chosen regions of interest with UV.

3.4. Permeabilization of the Cells

1. Obtain the image of the cells in the selected areas of the coverslip.

2. Start a time series scanning images repeatedly.

3. During the scanning, replace HBSS in the chamber with "pseudo-intracellular" solution as described above (Subheading 2, Item 8). We do this simply using an Eppendorf pipette. Perfusion systems with these experiments are complicated by the continuous presence of the dye in the saline – perfusion requires large volumes of dye which become expensive and also stain all the tubing used.

4. When visualizing plasma membrane permeabilization – signalled by egress of fluo-4 from the cytosol, for example, and co-localization of the remaining fluo-4 signal with TMRM in the mitochondria – wash out the digitonin and replace the buffer with the "pseudo-intracellular" solution + 40 nM TMRM, but without digitonin.

5. Choose the sector of the cell(s) to be exposed to UV.

6. Start the scan to obtain the images of fluo-4 and TMRM fluorescence in the cells and flash chosen sectors with UV.

7. Analysis of the data is straightforward, as most software packages – ImageJ, Metamorph, The Zeiss, or Leica proprietary software – allow plotting of "Regions of Interest" intensities with time. As the dyes used are of single wavelength only, we would normally quantify the changes simply in terms of a "fold" change in signal.

4. Notes

1. It is useful to bear in mind that once mitochondria are loaded with fluo-4, the dye is trapped within the mitochondria and is released from mitochondria only in case of membrane destruction or mPTP opening.

2. In order to obtain images of sufficient resolution to distinguish mitochondria and in order to effectively release Ca^{2+} from the complex with NP-EGTA, it is essential to image the cells grown on a glass coverslip rather than imaging directly in the plastic tissue culture dish. An alternative is to grow cells on glass-bottomed tissue culture dishes or to use an upright microscope and image on plastic.

3. We do this by imaging NADH autofluorescence, which is excited at 351 nm and emits between 430 and 480 nm. If this is set up optimally, then the laser will be correctly aligned and will be optimal for flash photolysis.

References

1. Nicholls, D. G. , Budd, S. L. Mitochondria and neuronal survival. (2000) *Physiol Rev.* **80**, 315–360.

2. Duchen, M. R. Roles of Mitochondria in Health and Disease. (2004) *Diabetes* **53**, S96–S102.

3. Irwin, W. A., Bergamin, N., Sabatelli, P., Reggiani, C., Megighian, A., Merlini, L., Braghetta, P., Columbaro, M., Volpin, D., Bressan, G. M., Bernardi, P., , Bonaldo, P. Mitochondrial dysfunction and apoptosis in myopathic mice with collagen VI deficiency. (2003) *Nat Genet* **35**, 367–371.

4. Mukherjee, R., Criddle, D. N., Gukvoskaya, A., Pandol, S., Petersen, O. H., , Sutton, R. Mitochondrial injury in pancreatitis. (2008) *Cell Calcium* **44**, 14–23.

5. Gandhi, S., Wood-Kaczmar, A., Yao, Z., Plun-Favreau, H., Deas, E., Klupsch, K., Downward, J., Latchman, D. S., Tabrizi, S. J., Wood, N. W., Duchen, M. R., , Abramov, A. Y. PINK1-Associated Parkinson's Disease Is Caused by Neuronal Vulnerability to Calcium-induced Cell Death. (2009) *Molecular Cell* **33**, 627–638.

6. Crompton, M., Kunzi, M., , Carafoli, E. The calcium-induced and sodium-induced effluxes of calcium from heart mitochondria. Evidence for a sodium-calcium carrier. (1977) *Eur. J. Biochem.* **79**, 549–558.

7. Jurkowitz, M. S. , Brierley, G. P. H^{+}-dependent efflux of Ca^{2+} from heart mitochondria. (1982) *J Bioenerg. Biomembr.* **14**, 435–449.

8. Denton, R. M. , McCormack, J. G. On the role of the calcium transport cycle in heart and other

mammalian mitochondria. (1980) *FEBS Letters* **119**, 1–8.

9. Crompton, M. The mitochondrial permeability transition pore and its role in cell death. (1999) *Biochem. J.* **341 (Pt 2)**, 233–249.

10. Chalmers, S. , Nicholls, D. G. The Relationship between Free and Total Calcium Concentrations in the Matrix of Liver and Brain Mitochondria. (2003) *Journal of Biological Chemistry* **278**, 19062–19070.

11. Abramov, A. Y., Fraley, C., Diao, C. T., Winkfein, R., Colicos, M. A., Duchen, M. R., French, R. J., , Pavlov, E. Targeted polyphosphatase expression alters mitochondrial metabolism and inhibits calcium-dependent cell death. (2007) *Proc. Natl. Acad. Sci. USA*. **104**, 18091–18096.

12. Krasnikov, B. F. , Zorov, D. B. [Stimulation of mitochondrial respiration, induced by laser irradiation in the presence of rhodamine dyes]. (1996) *Biokhimiia*. **61**, 1793–1799.

13. Abramov, A. Y. , Duchen, M. R. Actions of ionomycin, 4-BrA23187 and a novel electrogenic Ca^{2+} ionophore on mitochondria in intact cells. (2003) *Cell Calcium* **33**, 101–112.

14. Broekemeier, K. M., Dempsey, M. E., , Pfeiffer, D. R. Cyclosporin A is a potent inhibitor of the inner membrane permeability transition in liver mitochondria. (1989) *Journal of Biological Chemistry* **264**, 7826–7830.

Chapter 20

Isolation and Functional Assessment of Mitochondria from Small Amounts of Mouse Brain Tissue

Christos Chinopoulos, Steven F. Zhang, Bobby Thomas, Vadim Ten, and Anatoly A. Starkov

Abstract

Recent discoveries have brought mitochondria functions in focus of the neuroscience research community and greatly stimulated the demand for approaches to study mitochondria dysfunction in neurodegenerative diseases. Many mouse disease models have been generated, but studying mitochondria isolated from individual mouse brain regions is a challenge because of small amount of the available brain tissue. Conventional techniques for isolation and purification of mitochondria from mouse brain subregions, such as ventral midbrain, hippocampus, or striatum, require pooling brain tissue from six to nine animals for a single mitochondrial preparation. Working with pooled tissue significantly decreases the quality of data because of the time required to dissect several brains. It also greatly increases the labor intensity and the cost of experiments as several animals are required per single data point.

We describe a method for isolation of brain mitochondria from mouse striata or other 7–12 mg brain samples. The method utilizes a refrigerated table-top microtube centrifuge, and produces research grade quality mitochondria in amounts sufficient for performing multiple enzymatic and functional assays, thereby eliminating the necessity for pooling mouse brain tissue. We also include a method of measuring ADP-ATP exchange rate as a function of mitochondrial membrane potential ($\Delta\Psi$m) in small amounts of isolated mitochondria, adapted to a plate reader format.

Key words: Adenine nucleotide translocase, Adenine nucleotide carrier, Systems biology

1. Introduction

The availability of a multitude of genetical mouse models of neurodegenerative diseases greatly stimulated the demand for in vivo and in vitro approaches to study mouse brain mitochondria functions. Although the in vivo methodology to study mitochondrial functions is not yet well developed, a wealth of in vitro approaches is readily available for in-depth studies with isolated

Giovanni Manfredi and Hibiki Kawamata (eds.), *Neurodegeneration: Methods and Protocols*,
Methods in Molecular Biology, vol. 793, DOI 10.1007/978-1-61779-328-8_20, © Springer Science+Business Media, LLC 2011

mitochondria. However, such studies still represent a challenge because of small amount of the available brain tissue, especially when a particular brain region is of interest. For example, dissected mouse striata weight only about ~12 mg (wet weight). If using conventional techniques for isolation, purification and functional assessment of striatal mitochondria, one has to pool brain tissue from six to nine animals for a single mitochondrial preparation. Tissue pooling has long been a routine practice in the field (e.g., where striata from ten mice had to be pooled per single experiment) (1). A disadvantage of this experimental approach is that *working with pooled tissue significantly decreases the quality of the data* because of the time required to dissect several brains; it also greatly increases the labor intensity and the cost of experiments as several animals are required per single data point. To overcome these limitations, we developed a method for isolation of brain mitochondria that allows us to isolate a workable amount of mitochondria from 7 to 12 mg brain samples, which eliminates the necessity for pooling mouse brain tissue.

One of most useful and frequently assessed mitochondria functional characteristics is their oxidative phosphorylation. Classically, it is evaluated by respirometry as the ratio of the amount of phosphorylated ADP to the amount of the oxygen consumed (ADP:O ratio), which characterizes the efficiency of oxidative phosphorylation and a degree of intactness of mitochondria. The amount of produced ATP is not measured in these experiments; however, it can be evaluated separately by enzymatic assays. Recently, we have developed a kinetic assay to measure the efflux of ATP from mitochondria (2). This method allows one to estimate ADP-ATP exchange rate mediated by the ANT in mitochondria, which is influenced by many parameters (3), most steeply by mitochondrial membrane potential ($\Delta\Psi$m) (2, 4–7). We describe this methodology adapted to a small volume assay and using commonly available instrumentation, such as a plate reader.

2. Materials

2.1. Materials and Equipment

1. *Blunt-blunt tip straight Student Surgical Scissors* (*1*, Fig. 1).
2. *Sharp-sharp tip straight Extra Fine Bonn Scissors* (*2*, Fig. 1).
3. Razor blade (3, Fig.1).
4. *Curved tip Dumont #7 Medical Forceps – Inox Standard Tip* (*4*, Fig. 1).
5. Straight tip Dumont #5 Mirror Finish Forceps (5, Fig. 1).
6. *Curved serrated tip Standard Pattern Forceps* (6, Fig. 1).
7. Kontes glass homogenizer, small clearance pestle "B" (*7*, Fig. 1).
8. Bucket filled with ice.

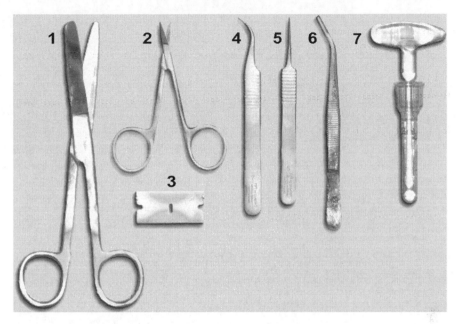

Fig. 1. Tools. (1) Blunt-blunt tip straight student surgical scissors; (2) Sharp-sharp tip straight Extra Fine Bonn Scissors; (3) Razor blade; (4) Curved tip Dumont #7 Medical Forceps – Inox Standard Tip; (5) Curved serrated tip standard pattern forceps; (6) Dounce homogenizer.

9. Glass or plastic dissection support surface (e.g., a Petri dish) and a piece of Whatman paper.

10. Wash bottle (~50 ml) filled with MSEGTA buffer (see Subheading 2.2, item 1).

11. A refrigerated microtube centrifuge (see Note 1).

12. A vacuum aspirator (see Note 2).

13. 10–12 1.7 ml Eppendorf® tubes (see Note 3).

14. 1 ml pipettor and tips.

15. Plate reader capable of measuring multiwavelength fluorescence and absorbance (see Note 4).

16. White opaque 96-well plates with flat-bottom wells and general lab equipment (e.g., pipettors, tips, etc.).

2.2. Reagents

1. MSEGTA buffer: 225 mM mannitol, 75 mM sucrose, 5 mM HEPES (pH 7.4), 1 mM EGTA, dissolved in water. Store at +0°C to +4°C (see Note 5).

2. MSEGTA-BSA: MSEGTA buffer supplemented with 0.2 mg/ml bovine serum albumin (BSA) essentially fatty acid-free. Store at +0°C to +4°C.

3. 100% Percoll™-MSEGTA buffer: 225 mM mannitol, 75 mM sucrose, 5 mM HEPES (pH 7.4), 1 mM EGTA, dissolved in 100% Percoll™. Store at +0°C to +4°C.

4. Experimental buffer: 8 mM KCl, 110 mM potassium gluconate, 10 mM NaCl, 10 mM Hepes (acid), 10 mM KH_2PO_4,

0.005 mM EGTA, 10 mM mannitol, 1.5 mM $MgCl_2$, 0.5 mg/ml BSA essentially fatty acid-free, pH 7.25 (adjusted with KOH). Store at +4°C (see Note 6).

5. 50 mM diadenosine pentaphosphate, (Ap5A), dissolved in water. Store at –20°C (see Note 7).

6. 2 mM magnesium green (MgG) 5 K^+ salt (Invitrogen) dissolved in water. Store in 30 µl aliquots at –20°C.

7. 1 M K^+ glutamate, pH 7.0 and 0.5 M K^+ malate dissolved in water, pH 7.0 (adjusted with KOH). Store at –20°C in 1 ml aliquots (see Note 8).

8. 200 mM K^+ ADP, pH 6.3–6.9. Store at –20°C (or preferably at –80°C) in 0.2 ml aliquots (see Note 9).

9. 200 mM K^+ ATP, pH 6.3–6.9. Store at –20°C (or preferably at –80°C) in 0.2 ml aliquots (see Note 9).

10. Solution of 0.5 M EDTA.

11. Solution of 1 M $MgCl_2$.

12. 1 mM safranin O dissolved in water. Make at least 20 ml and store at room temperature in the dark for at least 5 days before the experiment. Stable for a year if stored as described (see Note 10).

13. 1 mM SF 6847 ("Tyrphostin 9," 3,5-di-tert-butyl-4-hydroxy-benzylidenemalononitrile) in ethanol. Store at –20°C.

3. Methods

3.1. Dissection of Mouse Striata for Mitochondrial Preparation (see Fig. 2)

1. Sacrifice the mouse by a method approved at your Institution.

2. Decapitate with (1) *Blunt-blunt tip straight Student Surgical Scissors.*

3. Open the skull on the top with (2) *Sharp-sharp tip straight Extra Fine Bonn Scissors*, and carefully remove the brain from the skull. Take care not to damage the brain structure.

4. Place excised brain on the ice-chilled glass or plastic surface (e.g., a Petri dish) covered with wet Whatman paper; wash out the blood out of the brain surface (use mitochondria isolation buffer to wash and to wet the paper).

5. Using a razor blade (3), cut out the forebrain along the line *a* in Fig. 1, and cut it in two halves as shown in Fig. 2.

6. Flip the cut piece of forebrain on its side (see Fig. 3). Hold the block with (6) *Curved serrated tip Standard Pattern Forceps and* gently work out the space between hippocampus and striatum lengthwise using (4) *Curved tip Dumont #7 Medical Forceps – Inox Standard Tip.*

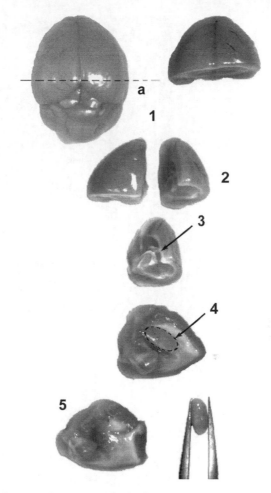

Fig. 2. Dissection procedure. see text for details.

7. Flip out tissue flap as shown in Fig. 4.

8. Using tips of (4) *Curved tip Dumont #7 Medical Forceps – Inox Standard Tip*, gently separate the striatum from surrounding tissue and lift ("scoop") it out.

9. Immediately place it in 1 ml of ice-cold MSEGTA-BSA buffer.

3.2. Isolation of Mitochondria

1. Homogenize brain tissue obtained in step 9 in Subheading 3.1 above in 1 ml MSEGTA-BSA with 2 ml Kontes™ Dounce homogenizer, tight pestle (7, see Fig. 1), with 30–35 strokes, transfer into two 1.7 ml Eppendorf® tubes;

2. Add MSEGTA-BSA to the tubes up to the groove at the top (~1.7 ml, total volume), mix by pipetting or inversion;

3. Centrifuge at ~500 × g × 5 min;

4. Transfer the supernatant into clean 1.7 ml Eppendorf® tubes, centrifuge at 14,000 × g × 10 min;

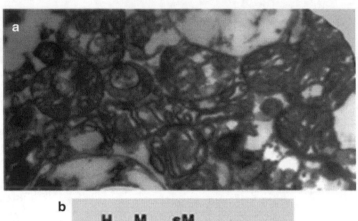

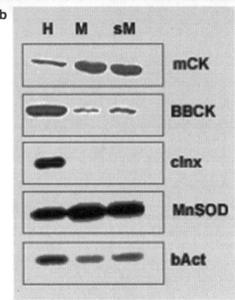

Fig. 3. Electron microphotograph (**a**) and immunoblot (**b**) of isolated mouse striatal mitochondria. Abbreviations: *H* whole brain homogenate, *M* whole brain mitochondria isolated by a conventional procedure described in (12), *sM* striatal mitochondria isolated as described in this manuscript, *mCK* mitochondrial isoform of creatine kinase, *BBCK* cytosolic isoform of creatine kinase, *clnx* calnexin, *MnSOD* mitochondrial superoxide dismutase, *bAct* beta-actin.

5. While centrifuging at step 4, fill 1.7 ml Eppendorf® tube with 1 ml of 24% Percoll™-(see Note 11) MSEGTA, and prepare 12% Percoll™-MSEGTA, 0.2 ml;

   (a) Take 0.1 ml of 24% Percoll™- MSEGTA and dilute with 0.1 ml MSEGTA;

   (b) Keep both 24% and 12% Percoll™-MSEGTA-filled tubes in ice;

6. Aspirate off the supernatant, resuspend both pellets (combine them) in 0.2 ml of 12% Percoll™-MSEGTA;

7. Carefully layer the suspension over the 24% Percoll™-MSEGTA solution, by holding the tube at about 30 degrees, placing the tip of the pipettor into the tube's groove and slowly releasing the suspension. Close the tube, turn it upright and place in ice before proceeding to the next step. Fill another 1.7 ml

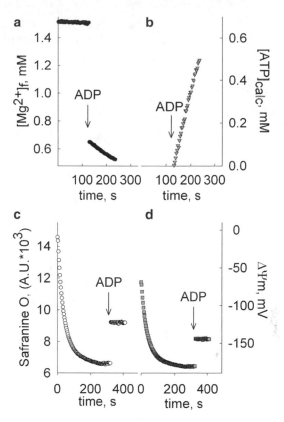

Fig. 4. Estimation of ATP appearing in the medium and $\Delta\Psi$m upon ADP addition in isolated mitochondria. (**a**) Time course of $[Mg^{2+}]_{free}$. (**b**) Time course of $[ATP]_{calc}$ appearing in the medium, calculated from $[Mg^{2+}]_{free}$. (**c**) Reconstructed time course of safranin O fluorescence. (**d**) Reconstructed time course of $\Delta\Psi$m in mV.

Eppendorf® tube with 1.2 ml of MSEGTA to use as a counterbalance in the centrifuge.

8. Centrifuge the prepared sample from step 7 at $18,000 \times g \times 15$ min. After the centrifugation is completed, check the appearance of the sample. There should be almost transparent band approximately in the middle of the tube, with cloudy top and bottom portions of the sample.

9. Aspirate off 0.7 ml of the top portion of the sample.

10. Add 1.2 ml MSEGTA, mix by inversion, and centrifuge at $18,000 \times g \times 5$ min;

11. Aspirate 1.5 ml of the supernatant, resuspend the pellet in the remaining 0.2 ml of the isolation buffer, add 1.5 ml MSEGTA, mix by inversion, centrifuge at $14,000 \times g \times 5$ min.

12. If the pellet is fuzzy, repeat previous step 11;

13. If the pellet appears solid with well-defined shape, aspirate off the supernatant completely (place 20 µl tip over the vacuum aspirator nozzle) and resuspend the pellet in 0.1 ml MSEGTA. This is purified mitochondrial fraction.

3.3. [Mg²⁺]ᶠ Determination from Magnesium Green Fluorescence in the Extramitochondrial Volume of Isolated Mitochondria

1. (See Note 4) Set the acquisition rate of plate reader at 0.33 Hz (one acquisition every 2 s plus 1 s for mixing in between each acquisition) and set 505 excitation and 535 nm emission wavelengths. Turn on the incubator and set it to 37°C.

2. Perform all experiments at 37°C. Prewarm the plate and the experimental buffer to this temperature.

3. Use white opaque 96-well plates with flat-bottom wells.

4. Add 0.2 ml of the experimental buffer minus the volume of the mitochondrial suspension that is intended to be added, 2 μM of magnesium green 5 K⁺ salt, 5 mM glutamate, 5 mM malate and 50 μM Ap5A to a well (see Notes 7 and 12).

5. Add 20–40 μg of mitochondria (see Note 13).

6. Record baseline until steady (usually a couple of minutes).

7. Add 2 mM of K⁺ ADP. Record the first abrupt and second progressive drop in magnesium green fluorescence for at least 2 min. (If de-energized mitochondria are to be used, i.e., in the absence of substrates, use 2 mM K⁺ ATP in lieu of ADP. For the remaining section, it will be assumed that mitochondria are energized by substrates).

8. At the end of the experiment, record minimum fluorescence (F_{min}) by adding 10 mM EDTA, followed by recording of maximum fluorescence (F_{max}) elicited by addition of 20 mM $MgCl_2$.

9. Free Mg^{2+} concentration, (Mg^{2+}_f), is calculated from the equation: $Mg^{2+}_f = (K_d(F - F_{min})/(F_{max} - F)) - 0.068$ mM, assuming a K_d of 0.9 mM for the MgG–Mg^{2+} complex (7), (see Fig. 4a) (see Note 6). The correction term −0.068 mM is empirical, and possibly reflects chelation of other ions by EDTA that have an affinity for MgG, and alter its fluorescence.

3.4. Conversion of [Mg²⁺]ᶠ to ADP-ATP Exchange Rate

ADP-ATP exchange rate is estimated using the recently described method (2) based on a concept developed previously (8, 9), exploiting the differential affinity of ADP and ATP to Mg^{2+}. The rate of ATP appearing in the medium following addition of ADP to energized mitochondria (or vice versa in case of de-energized mitochondria), is calculated from the measured rate of change in free extramitochondrial [Mg^{2+}] using the following equation:

$$[ATP]_t = \left(\frac{[Mg^{2+}]_t}{[Mg^{2+}]_f} - 1 - \frac{[ADP]_t(t=0) + [ATP]_t(t=0)}{K_{ADP} + [Mg^{2+}]_f} \right)$$

$$/ \left(\frac{1}{K_{ATP}[Mg^{2+}]_f} - \frac{1}{K_{ADP} + [Mg^{2+}]_f} \right) \quad (1)$$

Here, $[ADP]_t$ and $[ATP]_t$ are the total concentrations of ADP and ATP, respectively, in the medium, and $[ADP]_t(t=0)$ and $[ATP]_t(t=0)$ are $[ADP]_t$ and $[ATP]_t$ in the medium at time zero. For the calculation of [ATP] or [ADP] from free [Mg^{2+}], the apparent K_d values are identical to those in (2) due to identical experimental conditions ($K_{ADP} = 0.906 \pm 0.023$ mM, and $K_{ATP} = 0.114 \pm 0.005$ mM), (see Note 6). As a comparison, the program Winmaxc (http://www.stanford.edu/~cpatton/maxc.html), using constants from (10) predicts apparent K_d values of $K_{ADP} = 0.739$ mM, and $K_{ATP} = 0.060$ mM – in reasonable agreement with our measured values. Equation 1 is available for download as an executable file at: http://tinyurl.com/ANT-calculator. ADP-ATP exchange rate can be calculated by performing a linear regression on the values shown in Fig. 4b.

3.5. Mitochondrial Membrane Potential Determination

1. (See Note 4) Set the acquisition rate at 0.33 Hz (one acquisition every 2 s plus 1 s for mixing in between each acquisition); set 495 excitation and 585 nm emission wavelengths. Turn on the incubator and set it to 37°C.

2. Perform all experiments at 37°C. Prewarm the plate and the experimental buffer to this temperature.

3. Use white opaque 96-well plates with flat-bottom wells.

4. Add 0.2 ml of the experimental buffer minus the volume of the mitochondrial suspension that is intended to be added, 5 μM of safranin O, 5 mM glutamate, 5 mM malate, and 50 μM Ap5A to a well (see Notes 7 and 12).

5. Add 20–40 μg of mitochondria.

6. Record baseline until steady (usually a couple of minutes).

7. Add 2 mM of K^+ ADP. Record for at least 2 min (see Fig. 4c).

8. At the end of the experiment, record maximum fluorescence elicited by addition of 1 μM SF 6847.

3.6. Conversion of Safranin O Fluorescence to mV

Safranin O fluorescence can be "converted" to approximate values of $\Delta\Psi$m expressed in mV using the following three assumptions: (1) well-coupled mitochondria respiring on glutamate plus malate, at pH 7.25, exhibit $\Delta\Psi$m equal to –180 mV (2, 3) (see Note 14); (2) in the same conditions, addition of ADP clamps $\Delta\Psi$m to –145 mV for as long as there is ADP in the well (2, 3); and (3) addition of 1 μM SF 6847 collapses $\Delta\Psi$m to 0 mV. From these three points, the power function calibration can be constructed: $f = y_0 + a \times x^b$, in which safranin O fluorescence can be converted to mV. The "y_0" constant of the power function represents the residual safranin O fluorescence after mitochondria are fully charged; "x" is for safranin O raw fluorescence, "a" and "b" are constants and "f" represents calculated $\Delta\Psi$m in mV (see Fig. 4d).

3.7. Results

To assess the quality of mitochondria, we measured the respiration rates of the isolated striatal brain mitochondria (bSm). The oxygen consumption was measured with Hansatech oxytherm respirometry system ("Hansatech," UK) in 0.25 ml of incubation buffer (125 mM KCl, 2 mM KH_2PO_4, 20 mM HEPES (pH 7.2), 1 mM Mg^{2+}, 0.2 mg/ml BSA essentially fatty acid-free, 0.2 mM EGTA; 5 mM α-ketoglutarate, 1 mM Na^+-glutamate, 5 mM Na^+-pyruvate, 1 mM Na^+-malate). Mitochondria were added at 0.1 mg/ml concentration. We found that mitochondria were well coupled and exhibited excellent phosphorylation efficiency (Table 1. Note that ADP:O ratios >3 is due to the choice of oxidative substrates that activate the so-called substrate phosphorylation carried out by the succinyl thiokinase in the matrix of mitochondria, e.g., *see* (11)).

Electron microphotography revealed normal appearance of isolated striatal mitochondria and relatively low level of contaminating nonmitochondrial membranes (see Fig. 3a). The purity of the mitochondrial preparation was further assessed by immunoblotting (see Fig. 3b). The contamination by cytosol and unbroken nerve terminals (creatine kinase, cytosolic isoform BB "BBCK," marker) and that by endoplasmic reticulum (calnexin, marker) was

Table 1
The yield of the striatal (bSm) mitochondria and their respiratory characteristics

Parameter	bSm	Units
Tissue weight	12.5 ± 0.3	mg
Protein yield	110 ± 14	μg
State 4	32 ± 4	nmol O_2/min/mg
State 3	193 ± 15	nmol O_2/min/mg
R.C.I.	6.1 ± 0.5	State 3/State 4 ratio
A.C.I.	12 ± 1.1	State 3/V(cAtr) ratio
ADP:O	3.19 ± 0.07	ADP(nmol)/O(natoms) ratio

State 4 is the resting respiration rate when no active oxidative phosphorylation occurs; State 3 is the active respiration rate when mitochondria phosphorylate added ADP; RCI is the respiratory control index which reflect the quality of the mitochondrial preparation; ACI is the acceptor control index which under our experimental conditions characterizes the permeability of the inner membrane to protons, thereby reflecting the "true coupling" of the mitochondrial preparations; the V(cAtr) is the rate of respiration in the presence of carboxyatractylate, a specific inhibitor of ADP/ATP exchange across the mitochondrial membrane. ($n=6$)

very low and comparable with that in a conventional whole brain mitochondria preparation obtained by isopycnic centrifugation according to (12).

3.8. Conclusions

Overall, the method produces mitochondria of research grade quality in amounts sufficient for performing multiple enzymatic and functional assays, thereby eliminating the necessity for pooling mouse brain tissue.

4. Notes

1. We use a "Microfuge 22R centrifuge" (Beckman-Coulter, USA) refrigerated table top centrifuge for the procedure described above.

2. The vacuum aspirator is optional though preferable, but a pipettor can be used instead.

3. Eppendorf® tubes can be substituted with any other brand of small polypropylene tubes of similar volume, with a conical bottom.

4. These instructions assume the use of the monochromator-based Spectramax M5 plate reader (Molecular Devices, Sunnyvale, CA 94089, USA). If a monochromator-based plate reader is not available and a filter-based plate reader is to be used instead, a 505/535 nm excitation/emission filter is required for MgG, and a 495/585 nm excitation/emission filter for safranin O.

5. All water solutions and reagents are prepared using distilled (>17.4 MΩ resistance) water, such as e.g., Milli-Q.

6. If an alternative buffer is used, K_d values of ADP and ATP for Mg^{2+} must be determined as described in (2). K_d values of ADP and ATP for Mg^{2+} are temperature-dependent, so all experiments must be performed at the same temperature. Also, it is important to use no less than 1.5 mM $MgCl_2$ if 2 mM ADP is to be subsequently added because $[Mg^{2+}]_f$ must be sufficiently high to fall within the dynamic range of MgG used in the plate reader. It is encouraged to calculate own "empirical correction" values, pertaining to the experimental conditions. During the conversion of MgG fluorescence to $[Mg^{2+}]_f$, an additional point emerges that serves as a quality-control for the validity of the experiments: assuming that the total amount of $MgCl_2$ present in the buffer is 1.5 mM, and the K_d value of ADP for Mg^{2+} for the experimental conditions is 0.906 mM, addition of 2 mM ADP into the medium will decrease $[Mg^{2+}]_f$

from 1.5 mM to 0.658 mM. This value was calculated by the following standard binding equation:

$$[Mg^{2+}]_f = \cfrac{}{0.5 \cdot \left([Mg^{2+}]_t - K_d - [L]_t + \sqrt{\left([Mg^{2+}]_t - K_d - [L]_t\right)^2 + 4K_d[Mg^{2+}]_t}\right)} \quad (2)$$

where $[Mg^{2+}]_t$ is the total $[Mg^{2+}]$, $[L]_t$ is the total concentration of added ADP, and K_d is the affinity constant of ADP for Mg^{2+} under these experimental conditions, i.e., 0.906 mM. If after conversion of MgG fluorescence to $[Mg^{2+}]_f$, a value different than 0.658 mM is recorded, that probably means that either the K_d of ADP for Mg^{2+} has not been correctly estimated, or the ADP stock has not been properly made/stored.

7. Diadenosine pentaphosphate (Ap5A) stock, the inhibitor for adenylate kinase, does not require pH adjustment and can withstand at least ten freeze-thaw cycles. It is generally a good practice to supplement a 10 ml aliquot of the buffer with Ap5A, the dye (MgG or safranin O cannot be simultaneously added due to absorption overlap), and the mitochondrial substrates on the day of the experiment, and add only the mitochondria and the adenine nucleotides to the well, during the experiment. The purpose of this is to keep the increase of the experimental volume in the well due to additions of the materials to less than 5%.

8. The frozen stocks of the mitochondrial substrates are stable for at least 1 year. It is difficult to prepare K^+ malate stocks at a concentration higher than 0.5 M.

9. ADP and ATP stocks must be carefully made from the highest purity available materials, and the exact concentrations known, otherwise results obtained from Eq. 1 are invalid. During their preparation, ADP and ATP unbuffered stocks are highly acidic, and require pH adjustment to a pH range 6.3–6.9.

10. Upon long-term storage, the safranin O solution tends to form a dark precipitate. Do not shake the safranin O containing bottles, and always take from the top of the solution. If stored in the dark, a 1 mM safranin O stock solution is stable for at least 1 year.

11. 24% Percoll™-MSEGTA is prepared from 100% Percoll™-MSEGTA by diluting it with MSEGTA. Can be stored for 2 weeks at +4°C or frozen for longer storage.

12. The experimental buffer contains BSA, and therefore it tends to foam. While pipetting materials in the well, care must be taken to avoid extra foaming, that may affect the fluorescence signal.

13. For mouse brain, 20 µg of energized mitochondria in 0.2 ml volume exhibit sufficiently high ADP-ATP exchange rates to produce changes in $[Mg^{2+}]_f$ with an adequate signal-to-noise ratio. If, however, mitochondria are partially de-energized, i.e., uncoupled for any reason, or alternative type of mitochondria are used with low ADP-ATP exchange rates, it is advisable to increase the amount of mitochondria added in the well in order to increase the signal-to-noise ratio of the MgG fluorescence measurements. For the mitochondria protein determination, bicinchoninic acid (BCA) method was used (BCA kit is currently available from Thermo Scientific company).

14. If the mitochondria are uncoupled, either because of poor preparation, or for whatever reason they were impaired to begin with (experimental tissue damage, genetic/pharmacological manipulation, etc.), they will exhibit a lower membrane potential. In this case, the assumption of a resting membrane potential of −180 mV must be abandoned, and a two-point (ADP effect and SF 6847 effect) calibration of safranin O to mV conversion is advised. A two-point calibration is not as accurate as the three-point calibration elaborated above.

Acknowledgments

This work was supported by the Országos Tudományos Kutatási Alapprogram-Nemzeti Kutatási és Technológiai Hivatal (OTKA-NKTH) grant NF68294 and OTKA NNF78905 grant and Egeszsegügyi Tudományos Tanács (ETT) grant to C.C., the NIH grant 5POAG014930-11 to A.A.S. and NIH grants NS060885 and NS062165 to B.T.

References

1. Brustovetsky, N., LaFrance, R., Purl, K. J., Brustovetsky, T., Keene, C. D., Low, W. C., and Dubinsky, J. M. (2005) Age-dependent changes in the calcium sensitivity of striatal mitochondria in mouse models of Huntington's Disease. *J Neurochem.* **93**, 1361–1370.

2. Chinopoulos, C., Vajda, S., Csanady, L., Mandi, M., Mathe, K., and Adam-Vizi, V. (2009) A novel kinetic assay of mitochondrial ATP-ADP exchange rate mediated by the ANT. *Biophys J.* **96**, 2490–2504.

3. Metelkin, E., Demin, O., Kovacs, Z., and Chinopoulos, C. (2009) Modeling of ATP-ADP steady-state exchange rate mediated by the adenine nucleotide translocase in isolated mitochondria. *Febs J.* **276**, 6942–6955.

4. Klingenberg, M. (1980) The ADP-ATP translocation in mitochondria, a membrane potential controlled transport. *J Membr Biol.* **56**, 97–105.

5. Klingenberg, M. (2008) The ADP and ATP transport in mitochondria and its carrier. *Biochim Biophys Acta.* **1778**, 1978–2021.

6. Kramer, R., and Klingenberg, M. (1980) Modulation of the reconstituted adenine nucleotide exchange by membrane potential. *Biochemistry.* **19**, 556–560.

7. Metelkin, E., Goryanin, I., and Demin, O. (2006) Mathematical modeling of mitochondrial adenine nucleotide translocase. *Biophys J.* **90**, 423–432.

8. Silverman, H. S., Di Lisa, F., Hui, R. C., Miyata, H., Sollott, S. J., Hanford, R. G., Lakatta, E. G., and Stern, M. D. (1994) Regulation of intracellular free Mg^{2+} and contraction in single adult mammalian cardiac myocytes. *Am J Physiol.* **266**, C222–233.

9. Leyssens, A., Nowicky, A. V., Patterson, L., Crompton, M., and Duchen, M. R. (1996) The relationship between mitochondrial state, ATP hydrolysis, $[Mg^{2+}]i$ and $[Ca^{2+}]i$ studied in isolated rat cardiomyocytes. *J Physiol.* **496 (Pt 1)**, 111–128.

10. Fabiato, A., and Fabiato, F. (1979) Calculator programs for computing the composition of the solutions containing multiple metals and ligands used for experiments in skinned muscle cells. *J Physiol (Paris).* **75**, 463–505.

11. Chinopoulos, C., Gerencser, A. A., Mandi, M., Mathe, K., Torocsik, B., Doczi, J., Turiak, L., Kiss, G., Konrad, C., Vajda, S., Vereczki, V., Oh, R. J., and Adam-Vizi, V. Forward operation of adenine nucleotide translocase during F0F1-ATPase reversal: critical role of matrix substrate-level phosphorylation. *Faseb J.* **24**, 2405–2416.

12. Sims, N. R. (1990) Rapid isolation of metabolically active mitochondria from rat brain and subregions using Percoll density gradient centrifugation. *J Neurochem.* **55**, 698–707.

Chapter 21

Monitoring Mitophagy in Neuronal Cell Cultures

Jianhui Zhu, Ruben K. Dagda, and Charleen T. Chu

Abstract

Proper control of mitochondrial turnover is critical for maintenance of cellular energetics under basal and stressed conditions, and for prevention of endogenous oxidative stress. Whole organelle turnover is mediated through macroautophagy, a process by which autophagosomes deliver mitochondria to the lysosome for hydrolytic degradation. While mitochondrial autophagy can occur as part of a nonselective upregulation of autophagy, selective degradation of damaged or unneeded mitochondria (mitophagy) is a rapidly growing area in development, cancer, and neurodegeneration, particularly with regard to Parkinson's disease. Due to its dynamic nature, and the potential for regulatory perturbation by disease processes, no single technique is sufficient to evaluate mitophagy. Here, we describe several complementary techniques that include electron microscopy, single cell analysis of LC3 fluorescent puncta, and Western blot, each used in conjunction with a flux inhibitor to trap newly formed autophagosomes in order to monitor mitophagy in neuronal cells.

Key words: Autophagy, Mitophagy, Electron microscopy, Western Blot, RFP-LC3, GFP-LC3, Immunofluorescence

1. Introduction

Mitochondria are the power generators of the cell, converting oxygen and nutrients into adenosine triphosphate (ATP). Mitochondria also play an important role in maintaining calcium homeostasis, regulating lipid metabolism and heme biogenesis (1). However, excessive or damaged mitochondria may generate ROS and release cytochrome c, AIF, or other proapoptotic proteins to promote cell death (2–4). During nutrient deprivation or chronic hypoxia, mitochondria could be recycled to provide more urgently needed molecules. In pathological situations, timely elimination of dysfunctional mitochondria represents a cytoprotective response (5, 6) while global mitochondrial elimination has been implicated as part of a regulated cell death program (7).

Giovanni Manfredi and Hibiki Kawamata (eds.), *Neurodegeneration: Methods and Protocols*,
Methods in Molecular Biology, vol. 793, DOI 10.1007/978-1-61779-328-8_21, © Springer Science+Business Media, LLC 2011

Mitochondrial quality control is essential for cells to maintain mitochondrial integrity and normal function (8). Even under basal conditions, continuous cycles of mitochondrial fusion and fission and of biogenesis and degradation serve to produce daughter mitochondria and remove dysfunctional or effete mitochondria. The predominant mechanism for whole organelle turnover is autophagic sequestration and delivery to the lysosome for hydrolytic degradation. Selective autophagy of mitochondria is termed mitophagy (1, 9, 10), although mitochondria are also degraded during nonselective bulk cytoplasmic degradation. Mitochondrial dynamics and mitophagy are believed to play a key role in neurodegenerative diseases and the aging process itself (10, 11).

Recent studies indicate that the efficiency of mitophagy may be genetically controlled. Two proteins whose mutations are associated with autosomal recessive parkinsonism have been implicated in mitophagy. Parkin, an E3 ubiquitin ligase, is recruited to depolarized mitochondria to promote mitophagy in cancer cells (12). While loss of PTEN-induced kinase 1 (PINK1) results in increased mitophagy (6), overexpression of full-length PINK1 enhances Parkin recruitment to depolarized mitochondria (13–15), and mutations in either protein may impair mitochondrial depolarization-induced mitophagy. Nix, a BH3-only member of the Bcl-2 family, is a mitochondrial outer membrane protein that is essential for autophagic elimination of mitochondria during red blood cell development (16). The mitochondrial protein Atg32 interacts with the autophagy machinery in yeast (17), and other yeast proteins essential for mitophagy include the phosphatase AUP1P (18) and outer membrane-localized UTH1p (19). Mitophagy is also regulated by serine/threonine kinases, such as ERK2 (20), and mitochondrial fission and fusion-related proteins, such as Drp1 and mitofusin, may play a permissive role (10). Drp1 knockdown or expression of OPA1 or dominant negative Drp1 suppresses mitophagy (6, 21), which is of course dependent upon general autophagy machinery proteins Atg5, Atg7, Ulk1, and the Atg8 homolog microtubule-associated protein 1 light chain 3 (LC3) (22–24).

To better understand the role and mechanism of mitophagy during disease processes, several protocols to monitor and visualize mitochondrial autophagy have been reported. Each has caveats and limitations; therefore, multiple complementary techniques are necessary to establish induction of mitophagy by a particular cellular condition. As with any other dynamic process in cell biology, steady-state levels of a given organelle, such as a mitochondria-containing autophagosome ("mitophagosome"), are regulated by multiple potential factors, including rate of autophagosome formation, rate of maturation and lysosomal degradation, efficacy of cargo targeting to the forming autophagosomes, and the stability of the cargo and/or its detection method to the local environment inside maturing autophagic vacuoles (AVs) and autolysosomes.

Ultrastructural evaluation of mitophagy is a direct method to provide confirmation of mitochondrial autophagy or clearance. However, maturation and lysosomal fusion is a relatively rapid process, and it may be difficult to recognize mitochondrial morphology in autophagosomes. Preventing autophagosome maturation with short-term pulses of bafilomycin A1 traps newly formed autophagosomes, facilitating identification of cargo as well as providing comparative information on rates of autophagosome formation (8).

Disappearance of fluorescent signal for individual proteins, such as translocase of the outer membrane (TOM20), cytochrome c, HtrA2/Omi, or mitochondrially targeted fluorescent proteins, are frequently used as assays for mitophagy. However, intermembrane space proteins are lost with permeability transition in damaged mitochondria, and mitochondrially targeted proteins, such as TOM20 and Mito-GFP, are degraded by the proteasome under conditions that disrupt mitochondrial import (25). GFP itself shows pH-sensitive fluorescence and can be degraded by the lysosome (26). The possibility that observed loss of an overexpressed cytosolic protein reflects nonselective bulk autophagy should also be considered. Thus, we recommend that these types of studies, which are very useful for demonstrating potential mitophagic flux, are performed in conjunction with assays that (1) directly demonstrate association of mitochondria with autophagosomes and (2) confirm the ability to reverse the loss of mitochondrial signal using specific inhibitors of the autophagolysosomal pathway.

Decreases in expression levels of mitochondrial proteins is another useful method to confirm loss of the organelle, although electron microscopy allows for a direct assessment that the entire mitochondrion is absent as opposed to more selective degradation of certain protein components by other mechanisms. Ideally, several different proteins from different mitochondrial subcompartments to include the matrix and inner membrane are monitored for parallel decreases. Changes in protein levels due to alterations in biosynthesis, epitope destruction without loss of the entire protein, or alternative degradation pathways, such as proteasomal and intramitochondrial proteases, are important factors to consider (8).

In general, inferences of autophagy that depend solely upon loss of a signal are less optimal than assays that produce a positive signal confirming delivery to autophagosomes and lysosomes. On the other hand, steady-state increases in "mitophagosomes" may reflect increased sequestration or decreased maturation, requiring specific attention to issues of flux. Tracking multiple proteins in different subcompartments of the mitochondria and reversal studies using specific methods to inhibit autophagy or proteasomal degradation should be employed to evaluate relative contributions of these major degradative pathways. Herein, we describe methods employing electron microscopy, single cell analysis of LC3 puncta and mitochondrial cofluorescence, and Western blot experiments to evaluate mitophagy and mitophagic flux in neuronal cells.

2. Materials

2.1. Cell Culture

1. Cell lines: SH-SY5Y, a human neuroblastoma cell line (American Type Culture Collection, Rockville, MD).

2. Culture medium for SH-SY5Y cells (Dulbecco's modified Eagle's medium, DMEM): Antibiotic-free DMEM with 4.5 g/l D-glucose (BioWhittaker, Walkerville, MD); 10% heat-inactivated fetal bovine serum (Gibco/Invitrogen, Carlsbad, CA); 10 mmol/l HEPES; 2 mmol/l glutamine.

3. 10 mM Retinoic acid (1,000× stock solution) (Sigma, St. Louis, MO, USA), dissolved in DMSO and stored at –20°C in 20–50 µl aliquots.

4. Poly-D-lysine (10 mg/ml stock) (Sigma).

5. Culture medium for primary neurons (NB/B27): Serum-free neurobasal medium supplemented with l-gluta Max™ (2.0 mM) and B27™ (Invitrogen, Carlsbad, CA).

6. 6-hydroxydopamine (6-OHDA; Sigma), freshly prepared in distilled water or media.

7. 1-methyl-4-phenylpyridinium (MPP+; Sigma), prepared in distilled water, sterile filtered, and stored at –20°C.

8. 10 mM E64-D (Calbiochem, San Diego, CA): Stock made in DMSO, stored at –20°C; use at 10 µM for SH-SY5Y cells.

9. 25 mM Pepstatin-A (Calbiochem): Dissolved in methanol or DMSO; use at 25 µM for SH-SY5Y cells.

10. 10 µM Bafilomycin A1 (Sigma), dissolved in DMSO as stock solution.

11. 10 mM MG132 (Sigma), dissolved in methanol as stock solution; stored at –20°C.

12. Chambered Lab-tek II cover glasses (#1.5 German borosilicate; Nalge Nunc International, Naperville, IL, USA).

2.2. Electron Microscopy

1. 0.1 M PBS, pH 7.4.

2. 2.5% glutaraldchyde in 0.1 M PBS, pH 7.4.

3. 1% osmium in 0.1 M PBS.

4. 2% uranyl acetate solution, prepared in distilled water. Store at 4°C.

5. 1% lead citrate solution: Prepared in distilled water. Store for 3–6 months at 4°C.

6. Polybed 812 epoxy resin (Polysciences, Warrington, PA).

7. JEOL JEM 1210 transmission electron microscope (JEOL, Peabody, MA).

2.3. Image (Fluoresence)-Based Analysis of Mitophagy

1. 1 mM MitoTracker Red dye 580 (MTR; Molecular Probes, Eugene CA), dissolved in DMSO and stored at –20°C.

2. 1 mM MitoTracker Green FM dye (Molecular Probes), dissolved in DMSO and stored at –20°C.

3. 1 mM LysoTracker Red DND-99 (Molecular Probes), dissolved in DMSO and stored at –20°C.

4. FluoView 1000 (Olympus America) or Zeiss LSM 510 Meta laser-scanning confocal microscope (Carl Zeiss MicroImaging, Thornwood, NY).

2.4. Western Blot Analysis of Mitophagy

1. Lysis buffer: 25 mM HEPES, pH 7.5; 150 mM NaCl; 1% Triton X-100; 10% glycerol with freshly added proteinase and phosphatase inhibitors, including 100 µM E64, 1 mM sodium orthovanadate, 2 mM sodium pyrophosphate, 2 mM PMSF.

2. 5–15% polyacrylamide gradient gel.

3. Immobilon-P membranes (Millipore, Bedford, MA, USA).

4. Blocking solution: 5% nonfat dry milk in 20 mM potassium phosphate, 150 mM potassium chloride, pH 7.4, containing 0.3% (w/v) Tween-20 (PBST).

5. Mouse anti-pyruvate dehydrogenase antibody (1:1000, Molecular Probes).

6. Mouse anti-inner mitochondrial membrane protein human mitochondrial antigen of 60 kDa antibody (1:1000, Biogenex, San Ramon, CA).

7. Rabbit anti-outer mitochondrial membrane protein TOM20 antibody (1:10,000, Santa Cruz Biotechnologies, Santa Cruz, CA).

3. Methods

3.1. Ultrastructural Assay of Mitophagy

1. SH-SY5Y cells are maintained in antibiotic-free media and used at passages 30–45. Cells are plated at a density of $3 \times 10^4/cm^2$ in six-well culture plates. Parallel cultures of the experimental condition being tested (such as molecular or pharmacologic manipulation) are treated with 5–10 nM bafilomycin A1, an inhibitor of vacuolar-type H(+)-ATPase and autophagosome–lysosome fusion, or its vehicle for 2–4 h prior to fixation (see Note 1).

2. Cells are rinsed three times with 0.1 M PBS, fixed for at least 60 min in 2.5% glutaraldehyde at room temperature, or overnight at 4°C.

3. After fixation, cell monolayers are washed three times in PBS, and then postfixed in aqueous 1% OsO_4 and 1% K3Fe(CN)6 for 1 h.

4. After three PBS washes, the cultures are dehydrated through a graded series of 30–100% ethanol solutions, infiltrated, and then embedded in Polybed 812 epoxy resin.

5. Ultrathin (60 nm) sections are collected on copper grids and stained with 2% uranyl acetate in 50% methanol for 10 min, followed by an incubation in 1% lead citrate for 7 min.

6. Sections are photographed using a JEOL JEM 1210 transmission electron microscope at 80 kV.

As an example of ultrastructural analysis of mitophagy, Fig. 1 shows a comparison of early autophagosomes (AVis) trapped by bafilomycin A1 treatment of control SH-SY5Y cells (Fig. 1a) versus the PINK1 shRNA line A14 (Fig. 1b). Because of the low rate of autophagy induction in control cells, a 24 h treatment with bafilomycin A1 was used to derive sufficient numbers of AVis for analysis of their content, although such a lengthy pulse is not recommended for stressed cells or for flux analysis, due to saturation. For comparative analysis of sequestration rates, pulses of 2–6 h are ideal (8) (see Note 2). While the majority of AVis trapped by bafilomycin A1 in control cells contain a variety of organellar and membranous structures (Fig. 1a, arrowheads) with only rare mitochondria identified amid other structures (Fig. 1a, arrow), multiple AVis containing mitochondrial profiles (Fig. 1b, arrows) are identified in PINK1 shRNA cells. Stress and injury can change the appearance of mitochondria in cells. Comparison of putative mitochondrial profiles within the AVi with nearby mitochondria in the cytoplasm (Fig. 1, asterisks) facilitates their identification.

3.2. Fluorescence Methods for Analyzing Mitophagy

Confocal or epifluorescence microscopy using GFP-LC3 as a marker of autophagy can be used to monitor mitophagy induction in primary neurons or SH-SY5Y cells.

1. SH-SY5Y cells are seeded in complete DMEM medium on uncoated chambered Lab-tek II cover glasses. SH-SY5Y cells can be studied in their undifferentiated state or following retinoic acid treatment (10 µM for at least 3 days as judged by the extension of neurites of at least two soma lengths) to induce neuronal differentiation.

    Chambered cover glasses (four well) are coated with 100 µg/ml per well of poly-D-lysine for at least 4 h prior to seeding with primary cortical or midbrain neurons (100,000 neurons per well) in NB/B27.

2. Cells are transiently transfected with GFP-LC3 (1 µg DNA diluted in OPTIMEM in 0.10% Lipofectamine 2000). Four to six hours following transfection, one volume of complete DMEM medium or NB/B27 medium is added to the SH-SY5Y

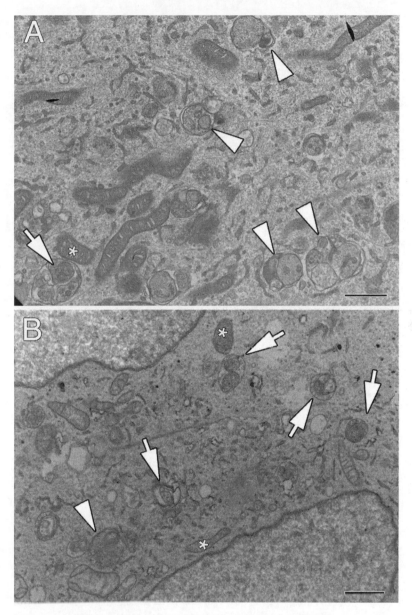

Fig. 1. Analysis of AVi content by electron microscopy using an inhibitor of autophagosome–lysosome fusion. (**a**) Control SH-SY5Y cells were treated with 5 nM bafilomycin A1 × 24 h prior to fixation and analysis by electron microscopy. Note the accumulation of several AVis containing heterogeneous cytoplasmic material (*arrowheads*). An occasional AVi exhibits a mitochondrial profile (*arrow*) similar to an adjacent mitochondrion in the cytoplasm (*asterisk*). (**b**) The PINK1 shRNA line A14 was treated with 10 nM bafilomycin A1 × 2 h prior to fixation and analysis by electron microscopy. Mitochondrial profiles are readily identified in the bafilomycin-trapped AVis (*arrows*), although generalized cargo was also observed in some (*arrowhead*). Note heterogeneity in electron density of mitochondria within AVis that are similar to heterogeneity observed among free mitochondria in the cytoplasm (*asterisks*). In the absence of bafilomycin A1, cargo is typically altered to the extent that it can no longer be recognized (see, for example, published images of the A14 cell line in (6, 8)). Scale bars: 1 μm.

or primary neurons, respectively, and the cultures incubated overnight in a 5% CO_2 incubator at 37°C. The day following transfection, approximately two-thirds of the media is replenished with fresh media.

3. To visualize the "colocalization" of mitochondria with autophagosomes, GFP-LC3-expressing cultures are loaded with
   100 nM MTR dye 580 (diluted from 1 mM DMSO stocks
   directly into media overlying cells). This loading should be
   done prior to experimental treatments that may affect mitochondrial membrane potential (see Note 3). The cultures are
   washed once with warm media to remove unbound/cytosolic
   background staining prior to performing image analysis using
   an epifluorescence or a laser confocal microscope.

In order to achieve statistical significance for measuring
mitophagy, it is necessary to analyze high-quality images captured
at a high magnification (60× or 90× if a 10× amplifier is used),
imaging at least 25–30 cells per condition (27). We use an inverted
FluoView 1000 or Zeiss LSM 510 Meta laser-scanning confocal
microscope (excitation/emission filter, 488/510 nm; 561/592 nm)
to image for mitophagy. Optimization experiments to compare the
results of live imaging with imaging after fixation in a positive control condition are advised for each cell type, as fewer GFP-LC3
puncta are detected in fixed cells under basal conditions, but the
background is also decreased with formaldehyde fixation (27).
Glutaraldehyde fixation may produce autofluorescent puncta (possibly related to lipofuscin) as an artifact of fixation, and use of this
fixative is not recommended for fluorescence studies.

There are two patterns of fluorescence association reflective of
mitophagy. First, a straightforward colocalization of LC3 and
mitochondrial signals in a punctate form (Fig. 2, arrows) is most
often observed. Notably, in merged images, colocalizing LC3 and
mitochondria do not always appear yellow due to differences in
intensity of staining. For example, a strong GFP-LC3 signal may
overwhelm a weaker MTR signal exhibited by marginally polarized
mitochondria or a dim mito-RFP signal in low-expressing cells (see
Note 4). Conversely, yellow pixels due to overlay of a GFP-LC3
puncta over a larger region of MTR signal most likely do not represent mitophagy. The second pattern that can be observed is of
ring-like GFP-LC3 structures surrounding discrete mitochondrial
fragments (Fig. 2, bottom row, inset); while there is technically no
colocalization of the fluorescent pixels, this is counted as a
"mitophagosome."

In addition to analyzing for mitophagy (see Note 5), changes
in the general level of macroautophagy can be analyzed within the
same experiment by quantifying the average number of GFP-LC3
puncta per cell in the green channel (27).

3.3. Flux Analysis

For each of the image-based methods of analyzing mitochondria
contained in autophagosomes (EM and dual fluorescence), an estimate

GFP-LC3   MTR

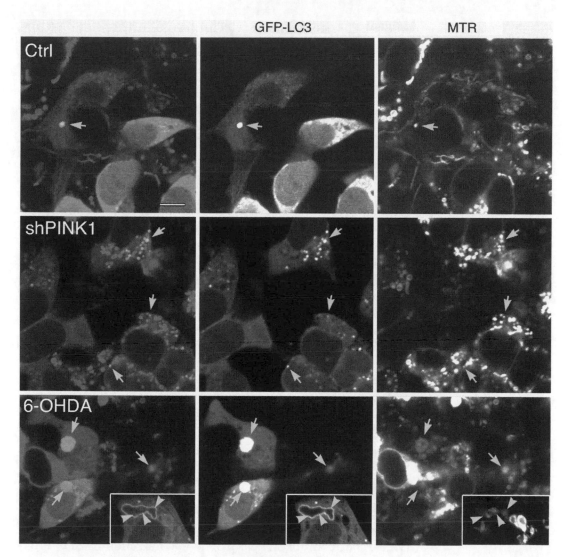

Fig. 2. Analysis of mitophagic sequestration by dual fluorescence. Stable cell lines that express an empty vector (Ctrl) or that stably knock down endogenous PINK1 (shPINK1) were transfected with GFP-LC3 for 2 days. Mitochondria were labeled by loading cells with 100 nM MTR 580 dye for 30 min at 37°C prior to treating some wells with 6-hydroxydopamine (6-OHDA, 120 μM × 4 h). Confocal images show a significant increase in GFP-LC3 puncta that colocalize with mitochondria in shPINK1 or 6-OHDA-treated cells (*arrows*). Note that 6-OHDA increases the average size of AVs and causes mitochondrial swelling, often associated with decreased MTR staining intensity, while shPINK1 results in fragmentation of the mitochondrial network with smaller autophagosomes consistent with the EM image of Fig. 21.1b. The inset shows a large, irregular GFP-LC3 ring encircling three distinct mitochondrial profiles (*arrowheads*); typically, the GFP-LC3 rings are rounder, encircling only one mitochondrial profile. Scale bar: 10 μm.

of flux can be derived from comparisons in the presence and absence of bafilomycin A1 (see Note 1) using the following formula:

Mitophagic Flux = (mean mitophagosome #/ cell with Baf

−mean mitophagosome #/ cell without Baf)/ time in Baf.

Alternatively, a qualitative assessment can be made through monitoring lysosomal delivery of mitochondria. Primary cortical or SH-SY5Y neuroblastoma cells are colabeled with 250 nM MitoTracker Green FM dye, which is relatively independent of mitochondrial membrane potential in many cell types, and loaded with LysoTracker Red DND-99 (100 nM) to label lysosomes, followed by a wash with warmed complete DME media (6) (see Note 6). RFP-LC3, an AV reporter construct that is more stable in the acidic environment of lysosomes compared to GFP-LC3 (26), can also be used to study both early and late AVs (28). The cells can be live-cell imaged immediately or they can be fixed in 4% paraformaldehyde (15 min at room temperature).

Finally, disappearance of mitochondrial fluorescence relative to diffuse GFP fluorescence in transfected cells (percent cellular pixels occupied by mitochondria) can be used to quantify mitochondrial loss, particularly at later time points (see Note 7).

3.4. Western Blot Analysis of Mitophagy

Cells are treated with a possible inducer of mitophagy in the presence or absence of 10 nM bafilomycin A1 (see Note 1) for at least 4 h, which blocks the fusion of AVs with lysosomes, or with the cell-permeable lysosomal protease inhibitors E64-D and pepstatin-A. Alternatively, RNAi against autophagy-related genes Atg7 and Atg8 can be employed 2–3 days prior (20, 29). Western blots (see Note 8) are probed for the total cellular levels of multiple mitochondrial proteins (e.g., manganese superoxide dismutase, porin, adenine nucleotide translocase, pyruvate dehydrogenase, or the complex IV human mitochondrial antigen of 60 kDa) (see Notes 9 and 10). Western blots are also probed for nonmitochondrial proteins, such as cytochrome P450 of the endoplasmic reticulum, to determine selectivity for mitochondria.

Reprobing the Western blot membranes for LC3, and for β-actin as a normalizing protein, is used to verify efficacy of the autophagy inhibitors (bafilomycin A1 or RNAi). The lipidated form of LC3 (LC3-II) exhibits a faster electrophoretic migration by SDS-PAGE compared to LC3-I, and the LC3-II/β-actin ratio is a well-accepted measure of autophagosome content (reviewed in (30)). For example, an increase in the LC3-II-to-β-actin ratio in cells pretreated with bafilomycin A1 (4 h at 10 nM for SH-SY5Y cells) is indicative of successful pharmacological suppression of autophagosome maturation. Likewise, a significant decrease in LC3-II induced in the presence of RNAi targeting essential Atg proteins is indicative of successful suppression of autophagy induction (20, 29, 31).

3.5. Conclusion

To summarize, no single technique is sufficient to establish an effect of a given experimental manipulation on mitophagy. Employing several complementary techniques that include (1) direct electron microscopy visualization of "mitophagosomes," (2) flux analysis of fluorescent puncta exhibiting colocalization of cargo and markers of early and late AVs, and (3) carefully controlled Western blot experiments should be sufficient for evaluating effects of a particular, gene product, treatment, or small compound on mitophagy (6, 20, 29).

4. Notes

1. The optimal bafilomycin A1 concentrations and treatment times are likely to be cell type-dependent. The ideal concentration and time course should be optimized to minimize toxicity and prevent saturation of the observed LC3 II accumulation. LysoTracker Red staining should be titrated to establish the minimum dose that is effective in increasing lysosomal pH for a particular cell type (32). In our hands, a concentration of bafilomycin A1 of 10 nM for more than 6 h leads to saturation of AV content in SH-SY5Y cells while a 20 nM concentration leads to cytotoxicity. Even though 10 nM does not cause toxicity in this time frame, there are ultrastructural effects on mitochondria (pallor, cristae separation) in control cells; thus, 5 nM may be optimal for ultrastructural studies.

2. At the EM level, the decrease of mitochondria and increase of autophagy/mitophagy could be quantitatively analyzed. Comparisons in the presence or absence of bafilomycin A1 provide information about whether neurotoxin or other experimental treatments increase or decrease lysosome-dependent flux using the formula indicated above.

3. The disappearance of MTR or tetramethylrhodamine methyl ester (TMRM) signal from cells by flow cytometry or fluorescence microscopy has been used to infer autophagic clearance of mitochondria. However, the loading and accumulation of MTR and TMRM in mitochondria are dependent on transmembrane potential. Loss of TMRM signal indicates a membrane potential change that could initiate mitophagy, but does not necessarily reflect mitophagy (33). Even the advertised membrane potential independent dye Mitotracker Green may show cell type- and concentration-dependent sensitivity to oxidative and depolarizing treatments (34), necessitating optimization experiments for each cellular context. In short, these probes are ideally loaded into cells prior to treatments that may dissipate mitochondrial membrane potential.

4. Cells may be cotransfected with monomeric RFP (mito-RFP) targeted to mitochondria via the leader sequences of various cytochrome c oxidase (COX) subunits to visualize colocalization of mitochondria with GFP-LC3 (6, 27). A word of caution is that transient overexpression of RFP targeted into mitochondria by COX subunit presequences may cause increased basal autophagy/mitophagy in neurons for several

days after transfection. Also, proper mitochondrial import is dependent upon healthy mitochondrial membrane potentials, and an overexpressed protein may compete with other proteins at the level of the outer mitochondrial membrane translocase system (Dagda & Chu, unpublished observations).

5. An increase in the percentage of GFP-LC3 puncta, that colocalize with MTR or mito-RFP-labeled mitochondria, or the percentage of RFP-LC3 puncta that colocalize with mitochondria labeled with mito-GFP suggests induction of mitophagy, rather than simply increased bulk autophagy.

6. Bafilomycin A1 cannot be used with LysoTracker Red, as it quenches lysosomal-specific fluorescence by disrupting pH (27, 35).

7. Given that GFP-LC3 and RFP-LC3 show dual cytosolic and punctate distributions, the percentage of the cellular area occupied by mitochondria in GFP-LC3 or RFP-LC3 transfected cells can be measured using NIH Image J (Bethesda, MD) to quantify the extent of mitochondrial loss induced by a specific treatment (27, 29). A low threshold level is set for the green channel of the RGB image containing the GFP-LC3 fluorescence in order to detect faint diffuse LC3 labeling that highlights the entire cell area (as opposed to detecting only bright spots >1.5 standard deviations above the background for detection of autophagic puncta). A decrease in the percentage of cellular area occupied by mitochondria, induced by a specific treatment compared to untreated cells, suggests mitochondrial degradation. However, reversing this mitochondrial loss by cotreating cells with bafilomycin A1, lysosomal protease inhibitors, or Atg7/Atg8 RNAi is needed to confirm induction of mitophagy (6). Rates of mitochondrial biogenesis is another factor to be considered.

8. Depending upon the mitochondrial protein of interest, a zwitterionic detergent, such as CHAPS, may be added to the lysis buffer for better extraction of membrane proteins.

9. Interpretation of immunoblots for apoptogenic factors of the intermembrane space, which can be released by different toxic insults, can be difficult (2). Complex I proteins frequently show disproportionate decreases compared to other mitochondrial proteins, which may reflect mechanisms other than whole organelle turnover (selective degradation or impaired biosynthesis). It is important to keep in mind that mitochondrially targeted proteins can undergo proteasomal degradation under conditions that impair potential-dependent import (25).

10. To rule out that the effects of a particular treatment on mitochondrial protein levels are not contributed by (a) the ubiquitin proteasome system, (b) localized degradation of mitochondrial proteins (8), or (c) decreased efficiency of biosynthesis,

import, or assembly of complex proteins, additional studies are required. In addition to proteasome inhibitors (i.e., MG132 and clasto-lactacystin-lactone), quantitative RT-PCR or reprobing the membrane for TFAM, a nuclear-encoded mitochondrial-specific transcription factor, may be used to implicate alterations in biogenesis (36, 37).

Acknowledgments

This work was supported in part by the National Institutes of Health (AG026389, NS065789). CTC is recipient of an AFAR/Ellison Medical Foundation Julie Martin Mid-Career Award in Aging Research.

References

1. Bueler, H., (2010) Mitochondrial dynamics, cell death and the pathogenesis of Parkinson's disease. *Apoptosis* **15**: 1336–1353.

2. Chu, C. T., J. H. Zhu, G. Cao, A. Signore, S. Wang & J. Chen, (2005) Apoptosis inducing factor mediates caspase-independent 1-methyl-4-phenylpyridinium toxicity in dopaminergic cells. *Journal of neurochemistry* **94**: 1685–1695.

3. Jennings, J. J., Jr., J. H. Zhu, Y. Rbaibi, X. Luo, C. T. Chu & K. Kiselyov, (2006) Mitochondrial aberrations in mucolipidosis Type IV. The Journal of biological chemistry **281**: 39041–39050.

4. Gomez-Lazaro, M., N. A. Bonekamp, M. F. Galindo, J. Jordan & M. Schrader, (2008) 6-Hydroxydopamine (6-OHDA) induces Drp1-dependent mitochondrial fragmentation in SH-SY5Y cells. *Free radical biology & medicine* **44**: 1960–1969.

5. Suen, D. F., D. P. Narendra, A. Tanaka, G. Manfredi & R. J. Youle, Parkin overexpression selects against a deleterious mtDNA mutation in heteroplasmic cybrid cells. *Proceedings of the National Academy of Sciences of the United States of America* **107**: 11835–11840.

6. Dagda, R. K., S. J. Cherra, 3rd, S. M. Kulich, A. Tandon, D. Park & C. T. Chu, (2009) Loss of PINK1 function promotes mitophagy through effects on oxidative stress and mitochondrial fission. *The Journal of biological chemistry* **284**: 13843–13855.

7. Tolkovsky, A. M., L. Xue, G. C. Fletcher & V. Borutaite, (2002) Mitochondrial disappearance from cells: a clue to the role of autophagy in programmed cell death and disease? *Biochimie* **84**: 233–240.

8. Chu, C. T., (2010) A pivotal role for PINK1 and autophagy in mitochondrial quality control: implications for Parkinson disease. *Human molecular genetics* **19**: R28-37.

9. Knott, A. B. & E. Bossy-Wetzel, (2008) Impairing the mitochondrial fission and fusion balance: a new mechanism of neurodegeneration. *Annals of the New York Academy of Sciences* **1147**: 283–292.

10. Chen, H. & D. C. Chan, (2009) Mitochondrial dynamics--fusion, fission, movement, and mitophagy--in neurodegenerative diseases. *Human molecular genetics* **18**: R169-176.

11. Weber, T. A. & A. S. Reichert, (2010) Impaired quality control of mitochondria: aging from a new perspective. *Experimental gerontology* **45**: 503–511.

12. Narendra, D., A. Tanaka, D. F. Suen & R. J. Youle, (2008) Parkin is recruited selectively to impaired mitochondria and promotes their autophagy. *The Journal of cell biology* **183**: 795–803.

13. Narendra, D. P., S. M. Jin, A. Tanaka, D. F. Suen, C. A. Gautier, J. Shen, M. R. Cookson & R. J. Youle, (2010) PINK1 Is Selectively Stabilized on Impaired Mitochondria to Activate Parkin. *PLoS biology* **8**: e1000298.

14. Ziviani, E., R. N. Tao & A. J. Whitworth, (2010) Drosophila parkin requires PINK1 for mitochondrial translocation and ubiquitinates mitofusin. *Proceedings of the National Academy of Sciences of the United States of America* **107**: 5018–5023.

15. Kawajiri, S., S. Saiki, S. Sato, F. Sato, T. Hatano, H. Eguchi & N. Hattori, (2010) PINK1 is recruited to mitochondria with parkin and associates with LC3 in mitophagy. *FEBS letters* **584**: 1073–1079.

16. Novak, I., V. Kirkin, D. G. McEwan, J. Zhang, P. Wild, A. Rozenknop, V. Rogov, F. Lohr, D. Popovic, A. Occhipinti, A. S. Reichert, J. Terzic, V. Dotsch, P. A. Ney & I. Dikic, (2010) Nix is a selective autophagy receptor for mitochondrial clearance. *EMBO reports* **11**: 45–51.

17. Kanki, T., K. Wang, Y. Cao, M. Baba & D. J. Klionsky, (2009) Atg32 is a mitochondrial protein that confers selectivity during mitophagy. *Developmental cell* **17**: 98–109.

18. Tal, R., G. Winter, N. Ecker, D. J. Klionsky & H. Abeliovich, (2007) Aup1p, a yeast mitochondrial protein phosphatase homolog, is required for efficient stationary phase mitophagy and cell survival. *The Journal of biological chemistry* **282**: 5617–5624.

19. Kissova, I., M. Deffieu, S. Manon & N. Camougrand, (2004) Uth1p is involved in the autophagic degradation of mitochondria. *The Journal of biological chemistry* **279**: 39068–39074.

20. Zhu, J. H., C. Horbinski, F. Guo, S. Watkins, Y. Uchiyama & C. T. Chu, (2007) Regulation of autophagy by extracellular signal-regulated protein kinases during 1-methyl-4-phenylpyridinium-induced cell death. *The American journal of pathology* **170**: 75–86.

21. Twig, G., A. Elorza, A. J. Molina, H. Mohamed, J. D. Wikstrom, G. Walzer, L. Stiles, S. E. Haigh, S. Katz, G. Las, J. Alroy, M. Wu, B. F. Py, J. Yuan, J. T. Deeney, B. E. Corkey & O. S. Shirihai, (2008) Fission and selective fusion govern mitochondrial segregation and elimination by autophagy. *The EMBO journal* **27**: 433–446.

22. Kundu, M., T. Lindsten, C. Y. Yang, J. Wu, F. Zhao, J. Zhang, M. A. Selak, P. A. Ney & C. B. Thompson, (2008) Ulk1 plays a critical role in the autophagic clearance of mitochondria and ribosomes during reticulocyte maturation. *Blood* **112**: 1493–1502.

23. Chu, C. T., J. Zhu & R. Dagda, (2007) Beclin 1-independent pathway of damage-induced mitophagy and autophagic stress: implications for neurodegeneration and cell death. *Autophagy* **3**: 663–666.

24. Zhang, J., M. S. Randall, M. R. Loyd, F. C. Dorsey, M. Kundu, J. L. Cleveland & P. A. Ney, (2009) Mitochondrial clearance is regulated by Atg7-dependent and -independent mechanisms during reticulocyte maturation. *Blood* **114**: 157–164.

25. Wright, G., K. Terada, M. Yano, I. Sergeev & M. Mori, (2001) Oxidative stress inhibits the mitochondrial import of preproteins and leads to their degradation. *Experimental cell research* **263**: 107–117.

26. Kimura, S., T. Noda & T. Yoshimori, (2007) Dissection of the autophagosome maturation process by a novel reporter protein, tandem fluorescent-tagged LC3. *Autophagy* **3**: 452–460.

27. Chu, C. T., E. D. Plowey, R. K. Dagda, R. W. Hickey, S. J. Cherra, 3rd & R. S. Clark, (2009) Autophagy in neurite injury and neurodegeneration: in vitro and in vivo models. *Methods in enzymology* **453**: 217–249.

28. Cherra, S. J., S.M. Kulich, G. Uechi, M. Balasubramani, J. Mountzouris, B.W. Day & C.T. Chu, (2010) Regulation of the autophagy protein LC3 by phosphorylation. *J. Cell. Biol.* **190**: 533–539.

29. Dagda, R. K., J. Zhu, S. M. Kulich & C. T. Chu, (2008) Mitochondrially localized ERK2 regulates mitophagy and autophagic cell stress: implications for Parkinson's disease. *Autophagy* **4**: 770–782.

30. Klionsky, D. J., H. Abeliovich, P. Agostinis, D. K. Agrawal, G. Aliev, D. S. Askew, M. Baba, E. H. Baehrecke, B. A. Bahr, A. Ballabio, B. A. Bamber, D. C. Bassham, E. Bergamini, X. Bi, M. Biard-Piechaczyk, J. S. Blum, D. E. Bredesen, J. L. Brodsky, J. H. Brumell, U. T. Brunk, W. Bursch, N. Camougrand, E. Cebollero, F. Cecconi, Y. Chen, L. S. Chin, A. Choi, C. T. Chu, J. Chung, P. G. Clarke, R. S. Clark, S. G. Clarke, C. Clave, J. L. Cleveland, P. Codogno, M. I. Colombo, A. Coto-Montes, J. M. Cregg, A. M. Cuervo, J. Debnath, F. Demarchi, P. B. Dennis, P. A. Dennis, V. Deretic, R. J. Devenish, F. Di Sano, J. F. Dice, M. Difiglia, S. Dinesh-Kumar, C. W. Distelhorst, M. Djavaheri-Mergny, F. C. Dorsey, W. Droge, M. Dron, W. A. Dunn, Jr., M. Duszenko, N. T. Eissa, Z. Elazar, A. Esclatine, E. L. Eskelinen, L. Fesus, K. D. Finley, J. M. Fuentes, J. Fueyo, K. Fujisaki, B. Galliot, F. B. Gao, D. A. Gewirtz, S. B. Gibson, A. Gohla, A. L. Goldberg, R. Gonzalez, C. Gonzalez-Estevez, S. Gorski, R. A. Gottlieb, D. Haussinger, Y. W. He, K. Heidenreich, J. A. Hill, M. Hoyer-Hansen, X. Hu, W. P. Huang, A. Iwasaki, M. Jaattela, W. T. Jackson, X. Jiang, S. Jin, T. Johansen, J. U. Jung, M. Kadowaki, C. Kang, A. Kelekar, D. H. Kessel, J. A. Kiel, H. P. Kim, A. Kimchi, T. J. Kinsella, K. Kiselyov, K. Kitamoto, E. Knecht, et al., (2008) Guidelines for the use and interpretation of assays for monitoring autophagy in higher eukaryotes. *Autophagy* **4**: 151–175.

31. Plowey, E. D., S. J. Cherra, 3rd, Y. J. Liu & C. T. Chu, (2008) Role of autophagy in G2019S-LRRK2-associated neurite shortening in differentiated SH-SY5Y cells. *Journal of neurochemistry* **105**: 1048–1056.

32. Shacka, J. J., B. J. Klocke, M. Shibata, Y. Uchiyama, G. Datta, R. E. Schmidt & K. A. Roth, (2006) Bafilomycin A1 inhibits chloro-

quine-induced death of cerebellar granule neurons. *Molecular pharmacology* **69**: 1125–1136.

33. Tolkovsky, A. M., (2009) Mitophagy. *Biochimica et biophysica acta* **1793**: 1508–1515.

34. Buckman, J. F., H. Hernandez, G. J. Kress, T. V. Votyakova, S. Pal & I. J. Reynolds, (2001) MitoTracker labeling in primary neuronal and astrocytic cultures: influence of mitochondrial membrane potential and oxidants. *Journal of neuroscience methods* **104**: 165–176.

35. Lang-Rollin, I. C., H. J. Rideout, M. Noticewala & L. Stefanis, (2003) Mechanisms of caspase-independent neuronal death: energy depletion and free radical generation. *J Neurosci* **23**: 11015–11025.

36. Rantanen, A., M. Jansson, A. Oldfors & N. G. Larsson, (2001) Downregulation of Tfam and mtDNA copy number during mammalian spermatogenesis. *Mamm Genome* **12**: 787–792.

37. Fisher, R. P., J. N. Topper & D. A. Clayton, (1987) Promoter selection in human mitochondria involves binding of a transcription factor to orientation-independent upstream regulatory elements. *Cell* **50**: 247–258.

Part VI

Neurodegenerative Disease Mechanisms: Biological Functions

Chapter 22

Analysis of Vesicular Trafficking in Primary Neurons by Live Imaging

Davide Tampellini and Gunnar K. Gouras

Abstract

Alzheimer's disease, the most common neurodegenerative disease, is characterized by a progressive loss of synapses and accumulation of amyloid-beta (Aβ) peptides in the brain. Previous studies demonstrated that acute increase in synaptic activity in cultured hippocampal slices and mouse brains (Cirrito et al. Neuron 48: 913–922, 2005; Kamenetz et al. Neuron 37: 925–937, 2003) enhanced secretion of Aβ. Since synaptic activity promotes Aβ secretion, it could also affect the trafficking and processing of its precursor, the amyloid precursor protein (APP). Here, we describe a method to investigate the effect of acute synaptic activation on APP trafficking within dendrites.

Key words: Amyloid precursor protein, Dendrites, Kymograph, Transport, Neurons

1. Introduction

Recent studies have shown that synaptic activity promotes secretion of amyloid-beta (Aβ) in hippocampal slice cultures and in mouse brain (1, 2). It is important to understand the underlying biology of how synaptic activity alters Aβ. In order to investigate the mechanism for this modulation of Aβ, we analyzed in cultured neurons the effects of synaptic activity on its precursor, amyloid precursor protein (APP). Altered trafficking, transport, and aberrant intracellular protein accumulation are common features of neurodegenerative diseases, including Alzheimer's disease (AD). We demonstrated that synaptic activation induced anterograde transport of APP to synapses, where APP is internalized and its amyloidogenic processing is enhanced (3). The methods used for these studies can be broadly applied to quantify changes in vesicular transport in neurons.

Giovanni Manfredi and Hibiki Kawamata (eds.), *Neurodegeneration: Methods and Protocols,*
Methods in Molecular Biology, vol. 793, DOI 10.1007/978-1-61779-328-8_22, © Springer Science+Business Media, LLC 2011

In our laboratory, we use primary neuronal cultures derived from embryonic day 15 (E15) Tg2576 AD transgenic mice. Tg2576 mice overexpress the Swedish 670/671-mutation human APP and are a well-established model of AD-like β-amyloidosis. Tg2576 primary neurons release elevated amounts of Aβ into the culture medium and accumulate Aβ42 intracellularly with time in culture (4). We utilized this primary neuron model to investigate whether synaptic stimulation is able to affect APP trafficking, processing, and transport. Depending on its location in the cell, APP can undergo either nonamyloidogenic or amyloidogenic processing. At the cell surface, APP is thought to be preferentially processed in the nonamyloidogenic pathway by α-secretase. When endocytosed, APP has been shown to be preferentially cleaved by β- and then γ-secretases to generate Aβ (3, 5). In order to induce synaptic stimulation in cultured neurons that mirror, as best as possible, what happens in the brain, we used an established protocol for chemically induced long-term potentiation (LTP) in 12 days in vitro (DIV) Tg2576 neurons. This protocol consists of specific activation of synaptic NMDA receptors using a previously described glycine treatment method (6, 7). As a control to demonstrate that this glycine-induced LTP-like synaptic activation is effective, elevated levels of phospho-CaMKII are demonstrated following a 1-h chase (3).

2. Materials

2.1. Cell Culture and Treatments

1. Neurobasal medium (Invitrogen, CA) supplemented with 2% B27 (Invitrogen), 1% penicillin/streptomycin, and 0.5% glutamine (Invitrogen).

2. Petri dishes with glass bottoms allow for live-imaging inverted microcopy with an oil-immersion objective. 35 mm live imaging Petri dishes (MatTek Corporation) are used. Dishes are precoated with 0.1 mg/ml poly-d-lysine (MW > 40,000; Sigma).

3. LTP buffer: 140 mM NaCl, 1.3 mM $CaCl_2$, 10 mM KCl, 25 mM HEPES, 33 mM glucose, 0.5 μM tetrodotoxin (TTX), 1 μM strychnine, 20 μM bicuculline.

4. 200 mM (1,000×) glycine stock solution. The final concentration of glycine, to activate neurons in the dish, is 200 μM. The LTP and glycine solution are prepared with Mg-free water to prevent blockage of NMDA receptors.

2.2. Neuronal Transfection

1. Lipofectamine 2000 (Invitrogen).

2. Opti-MEM 31985 reduced-serum medium (Invitrogen).

3. APP–YFP plasmid construct (1 µg/µl) (8): This construct generated in the laboratory of Carlos Dotti, Catholic University of Leuven, contains YFP at the C-terminus of full-length human APP695.

2.3. Fluorescent Microscopy

Olympus IX-70 inverted microscope equipped with a Hamamatsu ORCA-ER CCD digital camera and a 40×, 1.4 NA plan apochromat objective. Live-imaging Petri dishes containing transfected neurons are placed in a heated chamber (open perfusion micro incubator, model PDMI-2, Harvard Apparatus Inc.) at 37°C mounted on the microscope stage. Images are taken every 10 s using Metamorph software 7.5 (Molecular Devices). Comparable imaging equipment can be used.

3. Methods

Neurons are grown on poly-d-lysine-coated glass bottom microwell dishes. Neurons are transfected with human APP–YFP for 12 h (see Note 1) and then imaged in 37°C LTP solution at steady state or upon activation with glycine.

3.1. Cell Culture

Primary neuronal cultures are prepared from Tg2576 E15 embryos using standard procedures (9). Neurons are used at 8–10 DIV for these APP transport studies.

3.2. Transfection of Primary Neurons with Lipofectamine 2000

1. To transfect neurons in a live-imaging 35-mm plate (~2 ml of neurobasal medium), mix 25 µl of Opti-MEM with 1 µl of Lipofectamine 2000. Let the mixture sit for 5 min at room temperature.

2. Mix 25 µl of Opti-MEM with 1–1.5 µg of cDNA.

3. Combine the two solutions from steps 1 and 2, mix *gently* with the pipette, and let it sit for 20 min.

4. Add the final volume of the solution to the neurons in the 35-mm plate.

3.3. Treatment for Synaptic Activation

1. For chemical LTP, neurons are first treated for 1 h with 1 µM TTX in the culture medium to reduce endogenous synaptic activity and also to prevent phosphorylation of proteins involved in synaptic activation (6).

2. After TTX incubation, neurons are washed once with LTP buffer (6, 7).

3. Neurons are then incubated for 15 min in LTP buffer with or without 200 µM glycine (see Note 2).

4. After 15 min, neurons are washed once with LTP solution and imaged live in fresh LTP solution without glycine.

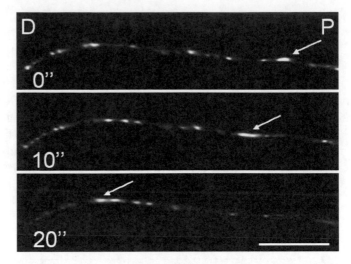

Fig. 1. Live imaging of APP–YFP. Representative Tg2576 neuron (8 DIV) transfected overnight with human APP–YFP. *Arrows* show the anterograde movement of an APP–YFP containing vesicle along a dendrite. Frames were automatically and sequentially acquired every 10 s with a YFP filter using Metamorph. *D* distal, *P* proximal, Scale bar = 10 μm.

3.4. Live-Cell Imaging

1. Neurons are imaged in a 37°C chamber using an Olympus Optical IX-70 microscope and a 40×, 1.4 NA plan apochromat oil-immersion objective. Frames are automatically and sequentially acquired every 10 s for 5 min with a specific YFP filter (Olympus 49003 ET-EYFP, 26 mm) using Metamorph (see Fig. 1).

2. One or two coverslips from each culture are analyzed (one to two neurons per coverslip). For each neuron, one to two dendritic segments 30–40-μm long are selected from areas, where single fluorescent vesicles can be outlined (see Notes 3–6).

3. Preparation of kymographs and kymographic analyses are performed using Metamorph (see Fig. 2). The total number of vesicles detected in one neurite (100%) are counted and the percent of vesicles that were stationary, moving anterogradely, or moving retrogradely are quantified (9).

3.5. Kymograph Analysis

Once the kymograph has been generated, it is possible to calculate the distance and the speed of the vesicles (see Fig. 3, Table 1). The track of each vesicle can be divided arbitrarily into segments, and the software (Methamorph in this case) auto matically calculates the distance, the interval of time, and the speed of the vesicle analyzed for each segment. The total distance traveled and the average speed for each vesicle can then be calculated.

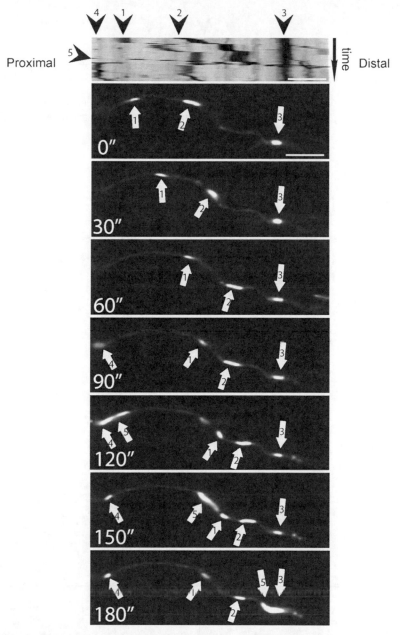

Fig. 2. Anterograde transport of APP is enhanced with synaptic activity. *Upper panel:* Kymograph of APP–YFP moving vesi-cles in neurons activated by glycine. Images were acquired at 10-s intervals, but only frames acquired every 30 s are shown here. *Black arrowheads* indicate movement of five APP–YFP-containing vesicles in a dendritic segment. *Black arrowhead #3* shows a stationary vesicle. The track of vesicle #4 starts at 90 s when it first appears in the proximal part of the analyzed neurite. The track of vesicle #5 starts at 120 s when this fast-moving vesicle first appears proximally and quickly transverses the neurite. *Lower panels:* Each *white arrow* indicates an APP–YFP-containing vesicle and its number corresponds to the respective kymograph generated by its trafficking. Vesicle #3 is stationary and its track is, therefore, vertical on the kymograph indicating no retrograde or anterograde movement. Vesicles #1 and #2 move progressively toward the distal portion of the neurite. Vesicle #4 appears at 90 s and remains stationary. Its track in the kymograph shows that it starts to move in the anterograde direction when the recording time is almost over. Vesicle #5 appears at 120 s and moves rapidly to the distal portion of the neurite. The corresponding horizontal track of the fast-moving vesicle #5 is perpendicular and contrasts the track of the nonmoving vesicle #3 in the kymograph. Scale bar = 10 μm.

Fig. 3. Quantitative analyses of vesicles in a kymograph. *Upper panel:* Tracks of vesicles #1, 3, and 5 were divided into segments (*red*, *yellow*, and *green*, respectively).

Table 1
Kymograph analysis

Segment	Distance (pixel)	Time (min:sec)	Velocity (pixel/second)
Vesicle 1			
1	0	00:00.0	0
2	20.00001	01:30.6	0.22084329
3	7.333337	00:50.3	0.145751421
4	23.333345	00:17.0	0.772960049
5	0.666667	02:00.8	0.005521006
6	7.333337	01:50.7	0.066254716
7	4.666669	00:50.3	0.092749061
8	4.666669	00:40.3	0.115930566
	68.000034		**0.202858587**
	17.5 μm	**480″**	**0.04 μm/sec**
Vesicle 3			
1	0	00:00.0	0
2	0	08:00.0	0
3	2.000001		**0.033126311**
	0.5 μm	**480″**	**0.00 μm/sec**
Vesicle 5			
1	0	00:00.0	0
2	41.333354	00:00.0	0
3	19.333343	00:10.1	1.920657957
4	22.666678	00:10.1	2.252701053
5	21.333344	00:20.1	1.060094613
6	14.666674	00:20.1	0.728742621
7	18.000009	00:10.1	1.789621098
8	10.000005	00:20.1	0.496820598
9	21.333344	00:00.0	0
10	15.333341	00:10.1	1.523280449
	184.000092		**1.221489799**
	47.4 μm	**100.7″**	**0.47 μm/sec**

Values of distance, time, and speed were calculated for each segment for vesicle #1, 3, and 5. Vesicle #3 (stationary) shows a movement of only 0.5 μm in a time of 480 s with a speed close to 0 μm/s. Note that vesicle #5 (very fast) moves approximately ten times faster than vesicle #1. Scale bar = 10 μm

4. Notes

1. The transfection time required to have a good expression of fluorescence-tagged protein is usually in the range of 4–24 h, and is also dependent on the construct. Typically, overnight transfection is sufficient to have good expression of APP–YFP.

2. Glycine is an NMDA receptor coactivator acting in concert with glutamate. Because glutamate release occurs in stochastic events only at synapses, the NMDA receptors that are activated by glycine are only those present at synapses and not those at extrasynaptic sites; glycine alone is not sufficient at extrasynaptic sites. Activation of NMDA receptors at synapses leads to LTP. It is known that binding of glycine to its receptor induces hyperpolarization with a consequent inhibitory action on neurons; but the buffer used to induce chemical LTP contains strychnine (1 µM), which prevents activation of inhibitory glycine receptors.

3. After imaging, neurons can be washed three times with PBS and then fixed with 4% paraformaldehyde in a PBS solution containing 4% sucrose for subsequent immunofluorescence.

4. Each set of cultures derived from a single embryo and transfected with APP–YFP is considered as $n = 1$.

5. Always know and mark in your notes what side of the imaged neurite is proximal and what side is distal. It is important to be clear about what direction the vesicles are moving.

6. Try to focus on a well-defined neurite. It is recommended not to include a transfected soma (including the one from the neurite of interest) in the imaging field because it is usually too bright and blocks the signal from the smaller moving vesicles.

Acknowledgments

Supported by an Alzheimer's Association New Investigator Award (DT) and Zenith Award (GKG) and National Institute of Health grants AG027140 and AG028174 (GKG). We thank Dr. Carlos Dotti, Catholic University of Leuven, Belgium, for providing the APP–YFP construct.

References

1. Cirrito JR, Yamada KA, Finn MB, Sloviter RS, Bales KR, May PC, Schoepp DD, Paul SM, Mennerick S, Holtzman DM. (2005) Synaptic activity regulates interstitial fluid amyloid-beta levels in vivo. *Neuron*, **48**(6):913–922.

2. Kamenetz F, Tomita T, Hsieh H, Seabrook G, Borchelt D, Iwatsubo T, Sisodia S, Malinow R. (2003) APP processing and synaptic function. *Neuron*, **37**(6):925–937.

3. Tampellini D, Rahman N, Gallo EF, Huang Z, Dumont M, Capetillo-Zarate E, Ma T, Zheng R, Lu B, Nanus DM *et al.* (2009) Synaptic activity reduces intraneuronal Abeta, promotes APP transport to synapses, and protects against Abeta-related synaptic alterations. *J Neurosci*, **29**(31):9704–9713.

4. Takahashi RH, Almeida CG, Kearney PF, Yu F, Lin MT, Milner TA, Gouras GK. (2004) Oligomerization of Alzheimer's beta-amyloid within processes and synapses of cultured neurons and brain. *J Neurosci*, **24**(14):3592–3599.

5. Koo EH, Squazzo SL. (1994) Evidence that production and release of amyloid beta-protein involves the endocytic pathway. *J Biol Chem*, **269**(26):17386–17389.

6. Ehlers MD. (2003) Activity level controls post-synaptic composition and signaling via the ubiquitin-proteasome system. *Nat Neurosci*, **6**(3):231–242.

7. Lu W, Man H, Ju W, Trimble WS, MacDonald JF, Wang YT. (2001) Activation of synaptic NMDA receptors induces membrane insertion of new AMPA receptors and LTP in cultured hippocampal neurons. *Neuron*, **29**(1):243–254.

8. Kaether C, Skehel P, Dotti CG. (2000) Axonal membrane proteins are transported in distinct carriers: a two-color video microscopy study in cultured hippocampal neurons. *Mol Biol Cell*, **11**(4):1213–1224.

9. Tampellini D, Magrane J, Takahashi RH, Li F, Lin MT, Almeida CG, Gouras GK. (2007) Internalized antibodies to the Abeta domain of APP reduce neuronal Abeta and protect against synaptic alterations. *J Biol Chem*, **282**(26): 18895–18906.

Chapter 23

Surface Trafficking of Sodium Channels in Cells and in Hippocampal Slices

Doo Yeon Kim and Dora M. Kovacs

Abstract

The voltage-gated sodium channel (Na_v1) plays an important role in initiating and propagating action potentials in neuronal cells. We and others have recently found that the Alzheimer's disease-related secretases BACE1 and presenilin (PS)/γ-secretase regulate Na_v1 function by cleaving auxiliary subunits of the channel complex. We have also shown that elevated BACE1 activity significantly decreases sodium current densities in neuroblastoma cells and acutely dissociated adult hippocampal neurons. For detailed molecular studies of sodium channel regulation, biochemical methods are now complementing classical electrophysiology. To understand how BACE1 regulates sodium current densities in our studies, we setup conditions to analyze surface levels of the pore-forming Na_v1 α-subunits. By using a cell surface biotinylation protocol, we found that elevated BACE1 activity significantly decreases surface Na_v1 α-subunit levels in both neuroblastoma cells and acutely prepared hippocampal slices. This finding would explain the decreased sodium currents shown by standard electrophysiological methods. The biochemical methods used in our studies would be applicable to analyses of surface expression levels of other ion channels as well as Na_v1 in cells and adult hippocampal neurons.

Key words: Voltage-gated sodium channel, BACE1, Presenilin, γ-Secretase, Cell surface biotinylation, Neuroblastoma, Hippocampal neurons, Adult hippocampal slices

1. Introduction

The voltage-gated sodium channel (Na_v1) is one of the major ion channels regulating electrical signaling in excitable cells (1). In neuronal cells, Na_v1 plays an important role in generating and propagating the action potential and therefore, regulates neuronal excitability (1, 2). While standard electrophysiological methods are used to study Na_v1 activity, biochemical studies are required to understand the detailed molecular mechanisms of channel regulation. Na_v1 is composed of a single pore-forming α-subunit and one

Giovanni Manfredi and Hibiki Kawamata (eds.), *Neurodegeneration: Methods and Protocols,*
Methods in Molecular Biology, vol. 793, DOI 10.1007/978-1-61779-328-8_23, © Springer Science+Business Media, LLC 2011

or two auxiliary β-subunits (3, 4). Ten α- and four β-subunits are currently known (4, 5). The β-subunits are type I membrane proteins that modulate the total/surface expression and inactivation of the α-subunits by direct interaction (2, 6–11). In neuronal cells, the majority of mature Na_v1 α-subunits is retained inside the cells while only a small portion gets to the cell surface and mediates sodium currents (6, 12). Therefore, it is very important to study how the surface trafficking of Na_v1 α-subunits is regulated.

We and others have recently reported that two specific proteases, BACE1 and presenilin (PS)/γ-secretase, cleave auxiliary subunits of the Na_v1 and thereby regulate Na_v1 function (13–16). Both proteases also mediate the generation of the Alzheimer's disease-related amyloid β-peptide (Aβ) and elevated BACE1 activity is associated with sporadic forms of the disease (17–21). We found that elevated BACE1 activity in neuroblastoma cells and acutely dissociated adult hippocampal neurons significantly decreases sodium current densities (15). To study the molecular mechanism underlying BACE1-mediated sodium current regulation, we set up conditions to analyze the surface expression levels of Na_v1 α-subunits, the major pore-forming Na_v1 subunits in neuronal cells. To analyze surface Na_v1 α-subunit levels in conditions similar to those used for sodium current measurements, we modified a protocol developed by Thomas-Crusells et al. (22). By applying this protocol, we were able to detect a significant decrease of surface Na_v1 α-subunit levels in both neuroblastoma cells and acutely prepared hippocampal slices from BACE1 transgenic animals. This would explain the decreased sodium currents found by standard electrophysiological methods.

2. Materials

2.1. Cell Culture

1. B104 neuroblastoma medium: Dulbecco's modified Eagle's medium (DMEM, Lonza/BioWhittaker, Walkersville, MD, USA) supplemented with 10% fetal bovine serum (FBS, HyClone, Logan, Utah, USA), penicillin/streptomycin (Lonza/BioWhittaker), and 2 mM L-glutamine (Lonza/BioWhittaker).

2. Dulbecco's phosphate buffered saline (DPBS, Lonza/BioWhittaker).

3. Trypsin-EDTA solution (Lonza/BioWhittaker).

4. 100 and 150 mm cell culture plates.

2.2. Cell Surface Biotinylation of Cultured Neuroblastoma Cells

1. 10 and 50 mL Falcon tubes.

2. Hank's balanced salt solution (HBSS) w/Ca^{2+}/Mg^{2+} buffer (pH 7.4): To make 1 L buffer, mix 100 mL 10×HBSS (Invitrogen/GIBCO, Grand Island, NY, USA), 0.47 mL 7.5%

sodium bicarbonate (Invitrogen/GIBCO), and 899.5 mL HPLC water. Make it ice-cold before use (see Note 1).

3. Biotinylation solution: Dissolve 7.5 mg of sulfo-NHS-LC-biotin (Thermo/Pierce, Rockford, IL, USA) in 0.75 mL HPLC water and mix well (10 mg/mL, 20× biotinylation stock); add 0.75 mL of 20× biotinylation stock into 14.25 mL HBSS with Ca^{2+}/Mg^{2+} buffer (see Note 2).

4. DPBS without Ca^{2+}/Mg^{2+} (Invitrogen/GIBCO).

5. Biotinylation stopping solution: HBSS with 200 mM lysine (Sigma, St. Louis, MO, USA).

6. BCA protein assay kit (Thermo/Pierce).

7. Disposable cell scraper (Corning incorporated, Corning, NY, USA).

2.3. Preparation of Acute Brain Slices and Cell Surface Biotinylation

1. 6-well cell culture plates for washing and other general purposes.

2. Tissue culture filter insert (Millipore, Billerica, MA, USA).

3. Superglue (Elmer's products Inc, Columbus, OH, USA).

4. Vibratome (model DTK-1500E, Ted Pella Inc, Redding, CA, USA).

5. Slice incubation chamber (*for a picture,* see Fig. 1a).

6. Carbogen gas (95% O_2 and 5% CO_2).

7. Dissection buffer (pH 7.4): 15 mM HEPES, 250 mM sucrose, 2.5 mM KCl, 0.1 mM $CaCl_2$, 1 mM Na_2HPO_4, 4 mM $MgSO_4\cdot7H_2O$, and 180 mM glucose (see Note 3).

8. Low calcium buffer (pH 7.4): 15 mM HEPES, 140 mM Isethionate, 2 mM KCl, 0.1 mM $CaCl_2$, 2 mM $MgCl_2$, 23 mM d-glucose (see Note 3).

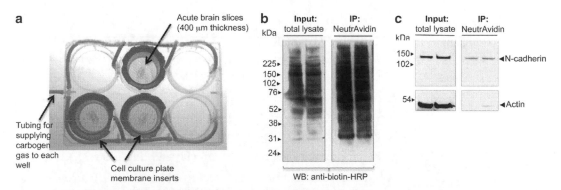

Fig. 1. Surface-biotinylation analysis of adult brain slices acutely prepared from wild-type control mice. (a) The slice incubation chamber. (b) Total lysates and NeutrAvidin-captured biotinylated proteins were stained with anti-biotin-horseradish peroxidase (HRP) to measure the efficiency of capturing biotinylated proteins. (c) Total lysates (*left panel*) and NeutrAvidin-captured biotinylated proteins (*right panel*) were analyzed by Western blot using antibodies against N-cadherin as a membrane protein control, and Actin for a control intracellular protein.

9. NaHCO$_3$-buffered Earl's balanced salt solution (EBSS), Sigma (see Note 3).

10. 2× biotinylation solution: Dissolve 3.6 mg of sulfo-NHS-LC-biotin (Thermo/Pierce) in 0.6 mL HPLC water and mix well (6 mg/mL, 20× biotin stock); add 0.6 mL biotin stock into 5.4 mL NaHCO$_3$-buffered EBSS (see Note 2).

11. Biotinylation stopping solution: EBSS with 200 mM lysine.

2.4. Affinity-Purification of Biotinylated Proteins

1. Complete protease inhibitor cocktail tablet (Roche, Indianapolis, IN, USA).

2. Extraction buffer (pH 7.4): 10 mM Tris–HCl, pH 6.8, 1 mM EDTA, 150 mM NaCl, 0.2% SDS, 0.5% sodium deoxycholate (Sigma), 1% Triton X-100 (Roche), complete inhibitor cocktail tablet (1 tablet per 10 mL), 1 mM PMSF (Sigma), and 1 mM 1,10-phenanthroline (PNT, ICN Biochemicals Inc, Aurora, OH, USA).

3. Sonicator (Ultasonic cleaner, Branson Ultrasonic Corp, Danbury, CT, USA).

4. NeutrAvidin (Thermo/Pierce): Wash three times with extraction buffer before use.

2.5. SDS-Polyacrylamide Gel Electrophoresis

1. 4× LDS sample loading buffer (Invitrogen) with 8% (v/v) β-mercaptoethanol.

2. Tris-Acetate gel running buffer (1×): Dilute the 4× Tris-acetate running buffer (Invitrogen) with three volumes of deionized water.

3. NuPage 3–8% Tris-Acetate gel (1.0 mm thickness, 10 well, Invitrogen).

4. Prestained molecular weight markers: Rainbow full-range markers (GE healthcare, UK).

2.6. Western Blotting Analysis for Sodium Channel α-Subunits and Control Proteins

1. Transfer buffer (1×): Dilute 1 volume of 20× NuPAGE transfer buffer (Invitrogen) with 17 volumes of deionized water and 2 volumes of HPLC grade methanol (Merck, Fair Lawn, NJ, USA) and store in a close capped plastic bottle at 4°C.

2. Immuno-Blot PVDF membranes (for protein blotting, 0.2 μm, Bio-Rad, Hercules, CA, USA).

3. Extra thick filter paper (Bio-Rad).

4. Hoefer semidry gel transfer unit (Hoefer, Holliston, MA, USA).

5. Tris-buffered saline with Tween-20 (1× TBST): Prepare 20× stock with 1.37 M NaCl, 27 mM KCl, 250 mM Tris–HCl, pH 7.4, 0.3% (v/v) Tween-20; dilute 100 mL with 900 mL water before use.

6. Blocking buffer: 5% (w/v) nonfat dry milk in TBST.

7. Primary antibody dilution buffer: TBST supplemented with 1% (w/v) bovine serum albumen (BSA, Sigma).

8. Primary antibody solutions used in our study: Anti-pan α-subunit antibody (1:500, Sigma), anti-Na$_v$1.1 α-subunit antibody (1:500, Sigma), anti-Na$_v$1.2 α-subunit antibody (1:500, Alomone, Isael), anti-GAPDH antibody (BD bioscience), anti-N-cadherin antibody (BD bioscience/transduction laboratory).

9. Secondary antibody solution: Anti-mouse IgG conjugated to horseradish peroxidase (HRP) (1:6,000) in TBST with 1% skim milk or anti-rabbit IgG conjugated to HRP in TBST (1:6,000, GE healthcare).

10. Enhanced chemiluminescent (ECL) reagents (GE healthcare) and SuperSignal (Thermo/Pierce).

11. VersaDoc imaging system (Biorad).

12. Bio-Max ML film (Kodak, Rochester, NY).

13. Stripping buffer: 60 mM Tris–HCl (pH 6.8) buffer with 2% (w/v) SDS.

3. Methods

We made several changes to the surface biotinylation protocol originally developed by Thomas-Crusells et al. (22) to optimally analyze Na$_v$1 α-subunits. First, we used strong ionic detergents, such as sodium deoxycholate and SDS together with a nonionic detergent to extract biotinylated proteins. Since Na$_v$1 α-subunits are extremely hydrophobic membrane proteins with high molecular weight, strong detergent conditions are required to efficiently solubilize them from the plasma membrane. In the same detergent conditions, we confirmed that NeutrAvidin effectively captures the biotinylated proteins by using a total biotin staining protocol (see Fig. 1b). Second, we tested and optimized a combination of protease inhibitors that can effectively block the nonspecific degradation of Na$_v$1 α-subunits, even with 4°C overnight incubation. Third, for adult hippocampal slice preparation, we used the exact same conditions and buffer systems that had been used to measure the sodium currents in our previous electrophysiological studies (15). These conditions allow for direct comparison of the electrophysiology and surface expression data. To analyze Na$_v$1 α-subunits in neuroblastoma cells, here we used B104 cells stably expressing human Na$_v$1 β2-subunit. Since most neuroblastoma cells lack endogenous expression of the β2-subunit, we made this cell line to study Na$_v$1 regulation specifically by the β2-subunit. The methods described here may also be applicable to analyze the surface expression of hydrophobic α-subunits of other ion channels in addition to Na$_v$1.

3.1. Cell Surface Biotinylation of Cultured Neuroblastoma Cells

1. B104 rat neuroblastoma cells stably expressing human Na_v1 β2-subunit are maintained in B104 neuroblastoma medium supplemented with 0.2 mg/mL zeocin (Invitrogen).

2. Detach the cells from the plates with the trypsin-EDTA solution when the cells are approaching full confluence.

3. Plate the trypsinized cells into 100 or 150 mm cell culture dishes in 1:3 dilutions.

4. When the plated cells approach to 50–70% confluency (it generally takes 1–2 days), wash cells three times with 15 mL of ice-cold HBSS w/Ca^{2+} and Mg^{2+} buffer.

5. Place the cell plates on ice and add 15 mL of ice-cold biotinylation solution.

6. Incubate the cells for 1 h on ice in a dark place with gentle rocking (see Notes 4 and 5).

7. Wash the cells two times with ice-cold biotinylation stopping solution (5 min incubation each).

8. Wash the cells two times with 15 mL ice-cold HBSS w/Ca^{2+} and Mg^{2+} buffer followed by a DPBS (without Ca^{2+}/Mg^{2+}) washing.

9. Immediately add 400 μL extraction buffer (see Note 6).

10. Incubate on ice for 5 min.

11. Scrape the cells from the plates with disposable cell scraper and transfer the cell extracts to 1.5 mL tubes on ice.

12. Disrupt the cell clumps with gentle pipetting.

13. Incubate on ice for 10 min.

14. Centrifuge the cell extracts at $12,000 \times g$ (tabletop centrifuge), 4°C for 15 min.

15. Transfer the supernatant to new tubes on ice (see Note 7).

16. Measure protein concentration by BCA protein assay kit.

3.2. Cell Surface Biotinylation of Acute Adult Brain Slices

1. Saturate the dissection buffer, low calcium buffer, and EBSS buffer with carbogen gas (95% O_2/5% CO_2) by bubbling the buffers with the gas for more than 30 min.

2. Add 100 mL dissection buffer into glass beaker and store at −80°C for 15–20 min until the buffer reaches 50% frozen ice/water status.

3. Sacrifice the mice by deep isoflurane anesthesia/cervical dislocation and quickly decapitate.

4. Rapidly open the scull and pour 5 mL ice-cold dissection buffer directly into the scull, remove the brain, and transfer the brains to ice-cold dissection buffer (see Note 8).

5. Trim the top and bottom of the brain and fix it with superglue to the cooled metal stage.

6. After fixing the brains with the cooled metal stage, cut the brain slices (300–400 μm) by using the vibratome (see Note 9).

7. Transfer the slices to ice-cold low-calcium buffer with wide-open glass pipettes and wash three times with the same buffer.

8. Transfer the washed slices onto the top of the tissue culture filter insert in a slice incubation chamber that is submerged in EBSS, continuously bubbled with carbogen gas (95% CO_2/5% CO_2).

9. Incubate the slices for more than 1 h at room temperature.

10. Briefly wash the slices three times with ice-cold low-calcium buffer.

11. Transfer the slices to another slice incubation chamber with 3 mL ice-cold EBSS in each well.

12. Incubate the slices on ice for 10 min.

13. Add 3 mL of 2× biotin solution (0.3 mg/mL final) to the wells of the slice incubation chamber.

14. Incubate on ice for 45 min (see Note 5).

15. Briefly wash the slices three times with ice-cold low-calcium buffer.

16. Transfer the slices to another incubation chamber containing biotinylation stopping solution and incubate for 5 min.

17. Repeat step 16 two times.

18. Briefly wash the slices three times with ice-cold low-calcium buffer

19. Dissect the individual sections under a surgical microscope and transfer hippocampal, cortical, and striatal regions to dry-ice frozen 1.5 mL tubes (see Note 10).

20. For extracting the hippocampal sections, 100 μL of extraction buffer is added per hippocampal section.

21. Disrupt the hippocampal slices by pipetting, followed by 5 min sonication on a water-bath type sonicator with maximum energy.

22. Incubate samples on ice for additional 10 min.

23. Centrifuge the slice extracts at 12,000×g (tabletop centrifuge), 4°C for 15 min.

24. Transfer the supernatant to new tubes on ice.

25. Measure protein concentration by the BCA method.

3.3. Affinity Purification of Biotinylated Proteins

1. Dilute the extracted samples with the extraction buffer for a final total protein concentration of 1 μg/μL.

2. For total protein analysis, 25–30 μg samples are mixed with 10 μL 4× LDS sample loading buffer with 8% (v/v) β-mercaptoethanol, quickly frozen on dry ice, and stored at

−20°C until SDS-polyacrylamide gel electrophoresis (SDS-PAGE). Frozen samples will be heated at 95°C for 5 min before gel loading.

3. To capture the biotinylated proteins, 100–300 μg of samples (100 μg total for hippocampal slices; 300 μg for neuroblastoma cells) are mixed with 40 μL of NeutrAvidin beads (see Note 11).

4. Incubate for 3 h or overnight at 4°C with rocking (for hippocampal section, 3 h incubation is recommended to avoid a potential protein degradation).

5. Wash four times with the 500 μL extraction buffer for each 10 min.

6. Add 50 μL of 1× sample loading buffer (4× LDS, BME, and Tris-Acetate buffer) and heat at 95°C for 5 min.

7. Briefly centrifuge the beads and collect the supernatant.

3.4. SDS-Polyacrylamide Gel Electrophoresis

1. 3–8% NuPage Tris-Acetate gels are used to resolve Na_v1 α-subunits with high molecular weight (~260 kDa). The rainbow marker is used as a protein size marker.

2. After loading total and biotin-captured proteins samples, run the gels at 170 V for approximately 60 min.

3. At the end, the gel units are disassembled and the gels are transferred to clean plastic trays containing 1×NuPAGE transfer buffer.

3.5. Western Blot Analysis of Total and Cell Surface Na_v1 α-Subunits and Control Proteins

1. Hoefer semidry gel transfer unit is used to transfer proteins in the NuPAGE gels to PVDF membranes.

2. Once the transfer is complete, PVDF membrane blots are marked and transferred to the blocking buffer.

3. After 1 h incubation at room temperature with gentle rocking, discard the blocking buffer and wash the blots three times for 5 min with TBST.

4. Add primary antibody solution and incubate at 4°C for overnight with gentle rocking.

5. Collect the primary antibody solution and wash three times with TBST for 10 min.

6. After the final wash, place the blots in a clear plastic plate and add 2 mL of ECL plus or SuperSignal.

7. Incubate the blots for 1–3 min and discard the solution.

8. Cover the blots with clean wrap or clear film.

9. The chemiluminescent signal can be captured digitally with an instrument, such as VersaDoc imaging system. Quantification of data is done by attached Quantityone software version 4.6.5 (see Note 12). Example results are shown in Figs. 1c and 2.

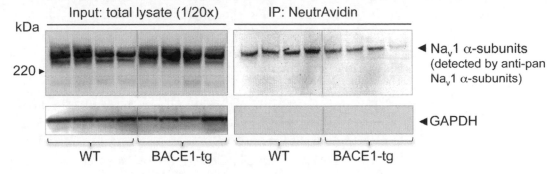

Fig. 2. BACE1 overexpression reduces the surface levels of Na_v1 a-subunits in adult hippocampal slices as we previously reported (15). Four matching slice pairs from similar locations of the brain were selected from BACE1-transgenic (BACE1-tg) and age-matched wild-type control mice (WT), respectively. Total lysate (1/20th of total input lysate, *left panel*) and NeutrAvidin captured surface biotinylated proteins (*right panel*) were analyzed by Western blot (WB) using antibodies against Na_v1 α-subunits (pan) and control GAPDH.

3.6. Stripping and Reprobing Blots for Control Proteins

1. After getting the optimal exposure, the blots are stripped and reprobed with other primary antibody solutions, including antibodies for isoform-specific sodium channel antibodies and control proteins, including $Na_v1.1$, 1.2, 1.6, *N*-cadherin, and GAPDH.

2. For stripping, the blots are incubated with stripping buffer for 15 min at 65°C.

3. After washing three times with TBST, reincubate the blots with blocking buffer for 1 h at room temperature.

4. After washing three times again, add another primary antibody solution and incubate for 1 h at room temperature or overnight at 4°C.

4. Notes

1. Alternatively, DPBS with Ca^{2+} and Mg^{2+} (Invitrogen/GIBCO, Cat # 14040–133) can be used instead of HBSS.

2. Prepared the biotin solution fresh just before the experiments.

3. All the solutions for brains slices must be cooled and saturated with carbogen gas (95% O_2 and 5% CO_2) by bubbling for more than 30 min before use.

4. It is very important to use a gentle rocker to avoid stressing the cells. Alternatively, we found that the biotinylation works well without rocking if the plates are leveled to ensure even distribution of the biotinylation buffer.

5. Biotinylation should be done in dark conditions. We generally cover each whole plate with aluminum foil.

6. It is important to add the extraction buffer drop by drop to cover the entire cell culture plate.

7. After the protein assay, the extracted samples are directly used for affinity purification of biotinylated proteins or can also be frozen and used later.

8. The speed is the key to the successful experiment. Brains should be removed and kept on ice-cold dissection buffer within 5 min after sacrificing the mice.

9. We used 12.5° cutting angle and the lowest speed for making coronal sections. We generally obtained five to six brain slices containing the hippocampal region. Dissection buffers in the vibratome should be continuously bubbled with carbogen gas (95% O_2/5% CO_2).

10. The dissected slices can be stored at $-80°C$ until the extraction.

11. To avoid potential pipetting errors, we vortex the beads every time just before transferring NeutrAvidin beads to the samples. In addition, it is strongly recommended to cut the end of pipette tips to avoid disrupting the beads.

12. Alternatively, Bio-Max ML film can be used to visualize chemiluminescence signals.

Acknowledgments

We would like to thank Mr. Manuel T. Gersbacher and Ms. Ateka Ahmed for their technical help. This work is supported by grants from the Cure Alzheimer's Fund and the NIH/NIA to DMK and DYK.

References

1. Lai, H., and Jan, L. (2006) The distribution and targeting of neuronal voltage-gated ion channels. *Nat Rev Neurosci.* 7, 548–562.

2. Catterall, W. (2000) From ionic currents to molecular mechanisms: the structure and function of voltage-gated sodium channels. *Neuron.* 26, 13–25.

3. Isom, L., De Jongh, K., and Catterall, W. (1994) Auxiliary subunits of voltage-gated ion channels. *Neuron.* 12, 1183–1194.

4. Isom, L., and Catterall, W. (1996) Na⁺ channel subunits and Ig domains. *Nature.* 383, 307–308.

5. Brackenbury, W. J., Djamgoz, M. B. A., and Isom, L. L. (2008) An emerging role for voltage-gated Na⁺ channels in cellular migration: regulation of central nervous system development and potentiation of invasive cancers. *Neuroscientist.* 14, 571–583.

6. Schmidt, J., and Catterall, W. (1986) Biosynthesis and processing of the alpha subunit of the voltage-sensitive sodium channel in rat brain neurons. *Cell.* 46, 437–444.

7. Isom, L. (2001) Sodium channel beta subunits: anything but auxiliary. *Neuroscientist.* 7, 42–54.

8. Catterall, W. (2002) Molecular mechanisms of gating and drug block of sodium channels. *Novartis Found Symp.* 241, 206–218; discussion 218–232.

9. Isom, L., Ragsdale, D., De Jongh, K., Westenbroek, R., Reber, B., Scheuer, T., and Catterall, W. (1995) Structure and function of the beta 2 subunit of brain sodium channels, a transmembrane glycoprotein with a CAM motif. *Cell.* **83**, 433–442.

10. Chen, C., Bharucha, V., Chen, Y., Westenbroek, R., Brown, A., Malhotra, J., Jones, D., Avery, C., Gillespie, P., Kazen-Gillespie, K., Kazarinova-Noyes, K., Shrager, P., Saunders, T., Macdonald, R., Ransom, B., Scheuer, T., Catterall, W., and Isom, L. (2002) Reduced sodium channel density, altered voltage dependence of inactivation, and increased susceptibility to seizures in mice lacking sodium channel beta 2-subunits. *Proc Natl Acad Sci USA.* **99**, 17072–17077.

11. Lopez-Santiago, L., Pertin, M., Morisod, X., Chen, C., Hong, S., Wiley, J., Decosterd, I., and Isom, L. (2006) Sodium channel beta2 subunits regulate tetrodotoxin-sensitive sodium channels in small dorsal root ganglion neurons and modulate the response to pain. *J Neurosci.* **26**, 7984–7994.

12. Schmidt, J., Rossie, S., and Catterall, W. (1985) A large intracellular pool of inactive Na channel alpha subunits in developing rat brain. *Proc Natl Acad Sci USA.* **82**, 4847–4851.

13. Wong, H., Sakurai, T., Oyama, F., Kaneko, K., Wada, K., Miyazaki, H., Kurosawa, M., De Strooper, B., Saftig, P., and Nukina, N. (2005) {beta} Subunits of Voltage-gated Sodium Channels Are Novel Substrates of {beta}-Site Amyloid Precursor Protein-cleaving Enzyme (BACE1) and {gamma}-Secretase. *J Biol Chem.* **280**, 23009–23017.

14. Kim, D., Ingano, L., Carey, B., Pettingell, W., and Kovacs, D. (2005) Presenilin/{gamma}-Secretase-mediated Cleavage of the Voltage-gated Sodium Channel {beta}2-Subunit Regulates Cell Adhesion and Migration. *J Biol Chem.* **280**, 23251–23261.

15. Kim, D., Carey, B., Wang, H., Ingano, L., Binshtok, A., Wertz, M., Pettingell, W., He, P., Lee, V., Woolf, C., and Kovacs, D. (2007) BACE1 regulates voltage-gated sodium channels and neuronal activity. *Nat Cell Biol.* **9**, 755–764.

16. Vassar, R., Kovacs, D. M., Yan, R., and Wong, P. C. (2009) The beta-secretase enzyme BACE in health and Alzheimer's disease: regulation, cell biology, function, and therapeutic potential. *J Neurosci.* **29**, 12787–12794.

17. Fukumoto, H., Cheung, B., Hyman, B., and Irizarry, M. (2002) beta-Secretase Protein and Activity Are Increased in the Neocortex in Alzheimer Disease. *Arch Neurol.* **59**, 1381–1389.

18. Tyler, S., Dawbarn, D., Wilcock, G., and Allen, S. (2002) alpha- and beta-secretase: profound changes in Alzheimer's disease. *Biochem Biophys Res Commun.* **299**, 373–376.

19. Yang, L., Lindholm, K., Yan, R., Citron, M., Xia, W., Yang, X., Beach, T., Sue, L., Wong, P., Price, D., Li, R., and Shen, Y. (2003) Elevated beta-secretase expression and enzymatic activity detected in sporadic Alzheimer disease. *Nat Med.* **9**, 3 4.

20. Li, R., Lindholm, K., Yang, L., Yue, X., Citron, M., Yan, R., Beach, T., Sue, L., Sabbagh, M., Cai, H., Wong, P., Price, D., and Shen, Y. (2004) Amyloid beta peptide load is correlated with increased beta-secretase activity in sporadic Alzheimer's disease patients. *Proc Natl Acad Sci USA.* **101**, 3632–3637.

21. Zhao, J., Fu, Y., Yasvoina, M., Shao, P., Hitt, B., O'Connor, T., Logan, S., Maus, E., Citron, M., Berry, R., Binder, L., and Vassar, R. (2007) Beta-site amyloid precursor protein cleaving enzyme 1 levels become elevated in neurons around amyloid plaques: implications for Alzheimer's disease pathogenesis. *J Neurosci.* **27**, 3639–3649.

22. Thomas-Crusells, J., Vieira, A., Saarma, M., and Rivera, C. (2003) A novel method for monitoring surface membrane trafficking on hippocampal acute slice preparation. *J Neurosci Methods.* **125**, 159–166.

Chapter 24

Imaging Presynaptic Exocytosis in Corticostriatal Slices

Minerva Y. Wong, David Sulzer, and Nigel S. Bamford

Abstract

Optical imaging is a valuable tool for investigating alterations in membrane turnover and vesicle trafficking. Established techniques can easily be adapted to study the mechanisms of synaptic dysfunction in models of neuropsychiatric disorders and neurodegenerative diseases, such as drug addiction, Parkinsonism, and Huntington's disease. Fluorescent endocytic tracers, including FM1-43, have been used to optically monitor synaptic vesicle fusion and measure synaptic function in various preparations, including chromaffin cells, dissociated cell cultures, and brain slices. In this chapter, we describe a technique that provides a direct measure of pathway-specific exocytosis from glutamatergic corticostriatal terminals.

Key words: Optical, Imaging, FM1-43, ADVASEP-7, Multiphoton, Microscopy, Corticostriatal, Mouse, Electrophysiology

1. Introduction

Neurotransmitter release and reuptake from recycling synaptic terminals is tightly regulated and alterations in vesicular turnover or in the availability of neuromodulators that act presynaptically can be features of neurodegenerative conditions (1). Optical tracers that label individual axon terminals in an activity-dependent manner have become useful tools in neurobiology and are responsible for improving our understanding about membrane trafficking and synaptic activity. When combined with standard electrophysiological techniques, optical recordings enable synapse modeling by showing how neuromodulators select subsets of presynaptic terminals, leading to changes in postsynaptic activation.

Perhaps the first optical tracer to monitor neurotransmitter release used horseradish peroxidase, which in the presence of appropriate substrates can produce a colored or electron-dense

Giovanni Manfredi and Hibiki Kawamata (eds.), *Neurodegeneration: Methods and Protocols*,
Methods in Molecular Biology, vol. 793, DOI 10.1007/978-1-61779-328-8_24, © Springer Science+Business Media, LLC 2011

reaction product (2). Eric Holtzman showed that synaptic vesicles in lobster muscle accumulate and then release horseradish peroxidase upon stimulation (3), providing initial evidence that these organelles recycle and undergo multiple rounds of fusion with the plasma membrane.

Fluorophores were then adapted for endocytic labeling of synaptic vesicles, initially using Lucifer yellow in photoreceptors (4). Shortly after Jeff Lichtman and colleagues (5) used rhodamine, fluorescein, and 8-hydroxypyrene (which emit in red, green, and blue wavelengths, respectively) and stimulated multiple nerves innervating a neuromuscular junction to identify which axon terminals were associated with a particular nerve. Quantum dots have also been adapted for similar purpose (6). Additional classes of optical probes that have been used to study synaptic vesicle fusion are mutants of vesicle lumen proteins that possess pH sensitive mutant forms of green fluorescent protein (7). These "pHluorin" proteins have relatively low levels of emission in the typically acidic environment of synaptic vesicles, and become far brighter when exposed to the neutral extracellular milieu following fusion. Mutations of synaptic vesicle transporters are sometimes used as such tracers (8). Finally, fluorescent false neurotransmitters are accumulated by synaptic vesicle transporters, providing both a means by which destaining can be measured during exocytosis and a means to estimate vesicle accumulation under a range of circumstances (9).

To date, the most successful approach for studying the activity of corticostriatal terminals has been to use fluorescent styryl dyes, particularly FM1-43. This class of probes offers distinct advantages, including reversible staining, tissue impermeability and activated fluorescence when dissolved into the lipid bilayer membrane (10). These specialized qualities are bestowed by a positively charged head, which prevents membrane permeability, and a lipophilic tail of varying length that defines the dye's spectral properties and washout kinetics (11). The compounds can also, like horseradish peroxidase, be used to form an electron dense reaction product (12) and a fixable derivative is available (13).

The FM dyes differ from other fluorescent tracers in that they are "amphiphilic" and can partition into membrane or aqueous media, but are most fluorescent when associated with lipid. Thus, the probe initially brightly labels the extracellular membrane. When neuronal stimulation induces endocytosis, the vesicle membrane becomes exposed to the synaptic cleft and FM1-43 is internalized within the recycling synaptic vesicle. When the synaptic vesicle membrane is endocytosed, FM1-43 fluorescence is visualized as bright fluorescent puncta. Restimulation produces exocytosis of recycling dye-loaded vesicles, FM1-43 is released into the extracellular medium or is dispersed in the surrounding membrane (14) and the fluorescence intensity of the terminal declines as it is desorbed from the membrane.

Preliminary evidence for the success of FM1-43 was demonstrated by William Betz and colleagues in frog, rat, and mouse motor nerve terminals (10). More recently, FM1-43 has been used to study secretory activity in chromaffin cells (11), cultured neurons (15, 16), as well as in hippocampal (17) and corticostriatal brain slices (18–21). Destaining of FM1-43 typically follows first to second order kinetics (15, 21) and is calcium dependent (18, 21), consistent with regulated neurotransmitter release.

An early identified impediment to using FM1-43 in brain slices was nonspecific "adventitious" binding of the dye that resisted washing and obscured the clear visualization of synaptic terminals. This problem was minimized following the introduction by Alan Kay of a carrier molecule, a sulfobutylated derivative of β-cyclodextrin, named ADVASEP-7 (22). This carrier was found to have a higher affinity for FM1-43 than plasma membranes and extracellular molecules, allowing for an efficient removal of FM1-43 from the extracellular space, and thus significantly reduced background staining. Additionally, two-photon laser-scanning microscopy (TPLSM) of FM1-43-labeled presynaptic terminals was found to be a superior method for visualizing dynamics of vesicle release and uptake (17, 23). Compared to confocal laser microscopy, TPLSM preserves tissue health by requiring less energy (at twice the wavelength), suppresses background fluorescence by membrane-bound FM1-43, increases the depth of excitation and leads to better imaging resolution of individual presynaptic boutons deep within a slice (24). Thus, we have found that optical imaging of presynaptic release in brain slices with FM1-43 is optimized by using TPLSM along with the external application of ADVASEP-7.

Here, we describe a technique that provides a direct measure of pathway-specific exocytosis from glutamatergic corticostriatal terminals. We provide instructions for loading FM1-43 into striatal terminals via nonspecific potassium-driven endocytosis and by pathway specific cortical stimulation. The instructions are tailored toward investigators with some experience in electrophysiology and knowledge of TPLSM function and tuning.

2. Materials

2.1. Media

1. Artificial cerebrospinal fluid solution (ACSF): 109 mM NaCl, 5 mM KCl, 35 mM $NaHCO_3$, 1.25 mM NaH_2PO_4, 20 mM HEPES, 1 mM $MgCl_2$, 2 mM $CaCl_2$, and 10 mM glucose (pH 7.3–7.4, 295–305 mOsm; see Note 1).

2. High potassium (K^+) ACSF: 74 mM NaCl, 40 mM KCl, 35 mM $NaHCO_3$, 1.25 mM NaH_2PO_4, 20 mM HEPES, 1 mM $MgCl_2$, 2 mM $CaCl_2$, and 10 mM glucose (pH 7.3–7.4, 295–305 mOsm; see Note 1).

2.2. Equipment for Two-Photon Excitation Time-Lapse Imaging

1. Zeiss LSM 510 NLO TPLSM equipped with a titanium-sapphire laser (excitation 810 nm/emission 650 nm; Carl Zeiss, Germany; see Notes 2 and 3).

2. Plan Fluor ×40 oil-immersion objective (NA 1.3; Zeiss).

3. IBM computer.

4. Software.

   (a) LSM 510 Browser software (Zeiss).

   (b) Image J (Wayne Rosband, National Institutes of Health, Rockville, MD).

   (c) SigmaPlot (SPSS, Chicago, IL), Excel (Microsoft, WA) or another graphics software package.

2.3. Chemicals and Supplies

1. FM1-43 (8 μM; N-[3-(triethylammonio)propyl]-4-(4-dibutyl-aminostyryl) pyridinium dibromide; Invitrogen, Carlsbad, CA). Protect from light (see Note 3).

2. ADVASEP-7 (CyDex, Overland Park, KS).

3. NBQX (10 μM; 2,3-dihydroxy-6-nitro-7-sulphamoylbenzo(f)-quinoxaline-2,3-dione; A.G. Scientific, San Diego, CA).

4. APV (50 μM; D-2-amino-5-phosphonovaleric acid; Sigma, St. Louis, MO).

5. CPCCOEt (40 μM; 7-(hydroxyimino) cyclopropa(b)chromen-1a-carboxylate ethyl ester; Tocris Bioscience, Ellisville, MO).

6. RC-27 L slice incubation chamber (56 μL/mm; Warner Instruments, Hamden, CT).

7. Custom slice anchor (see Note 4).

8. Twisted tungsten bipolar electrode (22 mm; Plastics One, Roanoke, VA).

9. Tektronix R564B wave generator (Tektronics, Gaithersburg, MD).

10. Stimulation isolator (AMPI, Jerusalem, Israel).

11. S88 storage oscilloscope (Grass-Telefactor, West Warwick, RI).

12. Carbogen gas (95% O_2, 5% CO_2).

3. Methods

Optical imaging of release from corticostriatal terminals is measured using the endocytic tracer FM1-43, which reports synaptic vesicle fusion activity at individual terminals, using techniques introduced previously (18–21). In brain slice preparations, TPLSM, combined with the judicious use of the adventitious quencher ADVASEP-7, provides a high signal to noise ratio and excellent three-dimensional spatial resolution with minimal photo bleaching and photo damage (24).

3.1. Preparation of the Corticostriatal Slice

1. Prepare 1 L of ACSF and perfuse with carbogen gas. Check pH and osmolarity.

2. Isolate 50 cc of the ACSF, bubble in carbogen and cool in an ice bath.

3. Anesthetize mice with nembutal 200 mg/kg i.p. or other agent prior to sacrifice.

4. Decapitate, rapidly remove the brain and sever the cerebellum.

5. Affix the caudal portion of the forebrain to a vibratome tissue holder using a few drops of cyanoacrylate glue and fill the holder with the prepared ice-cold ACSF.

6. Cut tissue using a vibratome into four to six 250 μm-thick coronal sections, containing the cortex and striatum between bregma +1.54 and +0.62 (25).

7. Bisect each coronal section at the midline creating two half-sections, each containing a hemisphere and striatum (see Fig. 1); remove to a slice holding chamber filled with carbogenated ACSF at room temperature.

8. Allow slices to recover for at least 1 h prior to use.

9. Prepare the following solutions, store in foil-lined plastic containers and bubble with carbogen for a few min just prior to using.

   (a) 8 μM FM1-43, 10 cc in High K$^+$ ACSF.

   (b) 1 mM ADVASEP 7, 10 cc in ACSF.

   (c) 100 μM ADVASEP-7, 10 cc in High K$^+$ ACSF.

   (d) 8 μM FM1-43, 10 cc in ACSF.

   (e) 100 μM ADVASEP-7, 500 cc in ACSF.

3.2. Preparation for Optical Imaging with FM1-43

1. Activate and tune the TPLSM (see Subheading 2.2).

2. Place a single corticostriatal slice in the imaging chamber with the cortex facing the electrodes (see Fig. 1a). Place the slice anchor over the slice so that harp strings run parallel with the corticostriatal fibers. Perfuse with warmed (35°C) carbogenated ACSF at 2–3 mL/min.

3.3. Loading and Unloading FM1-43 Using High K$^+$

1. Load FM1-43 nonspecifically into striatal terminals with high K$^+$.

2. Stop ACSF perfusion, close the suction, and remove most of the ACSF from the imaging chamber. Slowly add 2 ml of FM1-43 in high K$^+$ ACSF (see Step 9a in Subheading 3.1) to the

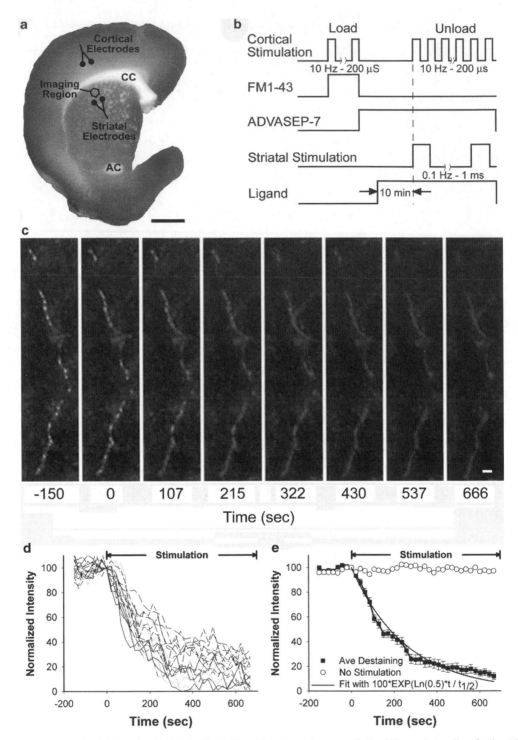

Fig. 1. (a) A corticostriatal slice stained with 3,3'-diaminobenzidine shows the areas of stimulation and recording. Corticostriatal terminals are loaded with FM1-43 by electrical stimulation with bipolar electrodes placed over cortical layers V–VI (Cortical Electrodes), located 1.5–2.0 mm from the imaging site. Multiphoton images of corticostriatal terminals are obtained from the corresponding motor striatum. Dopamine can be released by amphetamine or by local striatal stimulation (Striatal Electrode). *CC* corpus callosum, *AC* anterior commissure. Bar, 1 mm. (b) Protocol for loading and destaining corticostriatal terminals with FM1-43. (c) Multiphoton images of corticostriatal terminals captured every 21.5 s from the dorsal striatum revealed an *en passant* array of corticostriatal terminals. Restimulation at $t=0$ with 10 Hz pulses shows activity-dependent destaining of fluorescent puncta. Bar, 2 μm. (d) Time-intensity analysis of FM1-43 release from individual puncta ($n=18$), shown in panel C, demonstrate diverse terminal kinetics. Stimulation began at $t=0$ s. (e) Mean fluorescence intensity of puncta over time shown in Panel **c** and **d** is compared to a first-order exponential curve. The plateau line represents fluorescence measurements from a nondestaining punctum (reproduced from see ref. 21 with permission from the Society of Neuroscience).

imaging chamber with a pipette. Bubble FM1-43 gently with carbogen applied locally using a micropipette (see Note 5).

3. Allow at least 2 min for the FM1-43 to load into synaptic terminals. Remove FM1-43 in high K^+ ACSF and add 2 cc of 1 mM ADVASEP-7 (see Step 9b in Subheading 3.1) with a pipette. Bubble ADVASEP-7 gently with carbogen for 15 min. This will help to remove adventitious staining.

4. Focus the TPLSM on the striatum and capture images of fluorescent puncta (see Subheading 3.5 and Fig. 2).

5. After 9 baseline frames, continue TPLSM scanning and unload FM1-43 by superfusion with high K^+ ACSF with 100 μM ADVASEP-7 (see Step 9c in Subheading 3.1 and Note 6).

3.4. Loading and Unloading FM1-43 Using Cortical Stimulation

1. Load FM1-43 specifically into corticostriatal terminals.

2. Using a fresh brain slice, attach bipolar electrodes to a micromanipulator and gently place the electrodes over cortical layers V–VI (see Note 7 and Fig. 1).

3. Stop ACSF perfusion, close the suction and carefully add 2 ml of FM1-43 (see Step 9d in Subheading 3.1) into the imaging chamber with a pipette. Bubble FM1-43 gently with carbogen (see Note 5).

4. Allow 10 min for the FM1-43 to penetrate the slice.

5. Load FM1-43 into corticostriatal presynaptic terminals using a 10 min train of 200 μs, 400 μA pulses at 10 Hz (see Figs. 1 and 2).

6. Stop stimulation. Without bumping the electrodes, carefully remove FM1-43 and add 1 mM ADVASEP-7 (see Step 9b in Subheading 3.1) with a pipette. Bubble ADVASEP-7 gently with carbogen, applied locally using a micropipette, for 2 min to remove adventitious staining.

7. Open suction drain and perfuse slices in ACSF containing 100 μM ADVASEP-7 (see Step 9e in Subheading 3.1 and Note 6).

8. To prevent feedback synaptic transmission, expose sections to the pharmacological agents NBQX (AMPA receptor blocker), APV (NMDA receptor blocker), and CPCCOEt (to block metabotropic glutamate receptors).

9. Dopamine can be released by amphetamine or by local striatal stimulation. Likewise, receptor ligands can be bath-applied to detect alterations in FM1-43 destaining (see Fig. 1 and Note 8).

10. Focus the TPLSM on the striatum and capture images of fluorescent puncta (see Subheading 3.5 and Figs. 1 and 2).

11. After 9 baseline frames, continue TPLSM scanning and deliver 200 μs, 400 μA pulse trains to the cortex at 10 Hz for 10 min.

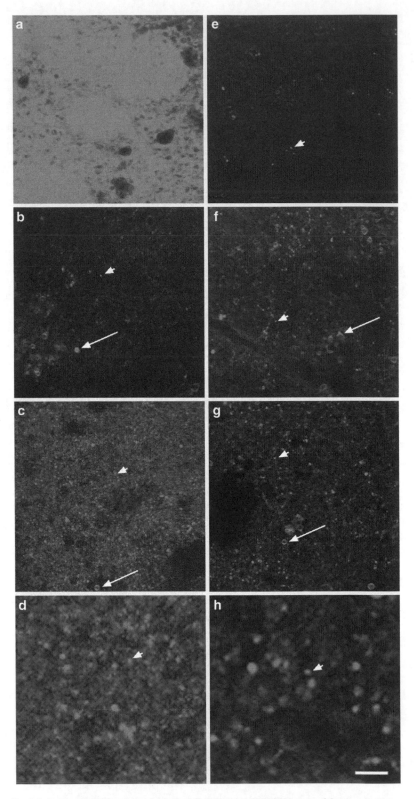

Fig. 2. Labeling striatal synaptic terminals with FM1-43. (**a**) TPLSM image of the dorsal striatum shows dramatic adventitious staining following incubation of a striatal slice in FM1-43 (8 μM) for 10 min, followed by a 5 min wash in ACSF. The more darkly stained ovals are bundles of

3.5. Two-Photon Image Acquisition

1. Fluorescent corticostriatal terminals are visualized using a Zeiss TPLSM system. Images captured in 8-bit, 123×123 μm regions of interest (ROI) at 512×512 pixel resolution and acquired at 21.5 second time intervals (see Notes 2 and 3).

2. During each time interval, obtain a z-series of five images, separated by 1 μm in the z-plane. This will compensate for any minor z-axis shift.

3. Collect image stacks for 40 time intervals (3,600 s; 200 images). Perfuse with high K+ ACSF (see Step 5 in Subheading 3.3) or begin cortical stimulation at the start of the tenth time interval (see Step 11 in Subheading 3.4).

3.6. Data Analysis

1. Condense the image stack. Using the LSM 510 Browser software, align images in each z-series, condense the z-series stack with maximum transparency and export the 40 condensed images to a folder in tiff format (see Note 9).

2. Prepare the image stack for analysis (see Note 10).

3. Identify fluorescent puncta measuring 0.5–1.5 μm in diameter (see Note 11).

4. Measure average FM1-43 fluorescence intensity of each punctum over the course of the time series (see Note 12).

5. Subtract the background fluorescence. Identify an area of tissue with no fluorescent puncta. Measure a sample of the slice background fluorescence for each time interval and subtract from the fluorescence intensity of each individual punctum.

6. Normalize the fluorescence intensity. For each punctum, normalize the time-dependent fluorescence intensity by the maximal puncta fluorescence just prior to the application of destaining stimulation (time frame 9).

7. Graph the normalized, background subtracted, intensity of each individual punctum using SigmaPlot software or another graphics software package. Active terminals are identified by the exponential decay of fluorescent intensity following stimulation (see Fig. 1 and Note 13). Reject any punctum that does

---

Fig. 2 (continued) corticofugal axons that course through the striatum. (**b**) When ADVASEP-7 (1 mM) was applied to the slice, most of the background staining was removed, leaving a few solitary puncta (*Arrow head*) and residual staining of larger diameter tube-like structures (*Arrow*), likely representing myelinated fibers or blood vessels. (**c**) FM1-43 in high K+ ACSF followed by ADVASEP-7 produced multiple fluorescent puncta. (**d**) Panel C was magnified 4×, revealing puncta with diameters of ~0.5–2 μm. (**e**) FM1-43 was loaded into synaptic terminals using cortical bipolar stimulation at 10 Hz for 2 min. A progressively greater number of fluorescent puncta were seen after cortical stimulation for (**f**) 5 and (**g**) 10 min. (**h**) Magnifying panel G by 4×, revealed puncta with diameters of ~0.5–2 μm. Bar, 20 μM for panels **a–c** and **e–g** and 5 μM for panels **d** and **h**.

not begin to destain following stimulation. Determine the halftime of fluorescence intensity decay during destaining ($t_{1/2}$). Average the fluorescence intensity for all punctum across the time interval using Excel software or other software (see Fig. 1).

8. The fractional release of dye for each stimulation pulse can be determined using SigmaPlot software (see Note 14).

---

4. Notes

1. Mix solutions fresh each day. Alternatively, a 10× solution of salt (NaCl, KCl, $NaHCO_3$, NaH_2PO_4, and HEPES) and a 10× solution of $MgCl_2$ and $CaCl_2$ can be prepared ahead of time and stored at 4°C for several weeks. The two 10× solutions are combined together with dH_2O on the day of experiment and glucose is added. HEPES is used to help reduce slice edema and secondary movement artifact (26).

2. Several alternative TPLSMs can be used. We also use a Prairie Ultima TPLSM equipped with a titanium-sapphire laser (excitation 900 nm/emission 625 nm; Prairie, Middleton, WI), a Plan Fluor ×60 water-immersion objective (NA 1.2; Olympus) and an IBM computer with Prairie View software (Prairie). When using a Prairie TPLSM, images are captured in 16-bit, 75.2×75.2 μm ROI at 512×512 pixel resolution and acquired at 35-second intervals. An upright or inverted TPLSM can be used. In our experience, an upright microscope provides the best resolution (as the beam does not travel through a cover slip) while an inverted microscope affords a reduction in movement artifact. Whether inverted or upright, the TPLSM requires a stage platform and at least one electrode-holding micromanipulator.

3. FM1-43 dye has a single photon excitation maximum at 479 nm. The ideal laser excitation wavelength using two-photon imaging in these experiments should be determined to achieve adequate fluorescence signal yet maintain minimal laser power usage to reduce photo-bleaching of FM1-43. Optimal laser excitation wavelength and power (generally around ~10%) may vary between different set-ups. The pixel size (typically 0.22 μm) may vary, depending on chosen parameters. To further prevent photo-bleaching of the dye, minimize the intensity of mercury epifluorescence and bright field light and limit the duration of light source use in the presence of FM1-43.

4. Slice anchors are constructed from platinum wire and lycra threads with 1 mm spacing. An appropriately designed anchor will hold the brain slice in place while in the chamber bath and will

reduce the amount of slice movement during imaging. Fabricated slice anchors are also available from Warner Instruments.

5. Carbogen is applied locally using a small (1–10 µL) plastic micropipette. Attach a gas line to the micropipette. Place the tip of the pipette into the imaging bath, directing the flow of gas away from the brain slice. Minimize the gas flow so that the slice is not agitated.

6. ADVASEP-7 will help prevent recurrent endocytosis of FM1-43 dye into the recycling synaptic vesicles.

7. Robust placement of the stimulating electrode and imaging region are critical. Follow the referenced figure closely.

8. Amphetamine (10 µM) can be bath applied to release dopamine from nigrostriatal terminals (18). Dopamine (with other neurotransmitters) can also be released using a separate set of bipolar electrodes placed over the striatum, just adjacent to the imaging region. Striatal stimulation at 0.1 Hz produces little change in FM1-43 release while providing pulsatile dopamine efflux (18). To ensure equilibrium, the brain slice is generally exposed to pharmacological agents for 10 min before stimulation-mediated unloading (see Fig. 1).

9. Start the Zeiss LSM Image Browser and import the imaging file: name.lsm. To condense the z-stack, create a projection image: Process<Projections. The Projection window will open. Set the following parameters: Transparency = Maximum, Turning Axis = X, First Angle = 0, Number of Projections = 1, Difference Angle = 0. Uncheck Single Time Index option. Choose "Apply". The projection image with 40 slices will open in a new window. Export the condensed images to tiff files. File>Export>Set: Save as type "Tagged Image File 16-bit". Label the files xxx-0.tif through xxx-39.tif.

10. Open the condensed image series in ImageJ software (v. 1.43u).

    (a) Start ImageJ and import the condensed image files. File>Import>Image Sequence. Choose the first image of the stack "xxx-0.tif". The Sequence Options dialog box opens; leave at the default settings. Choose "OK".

    (b) Set properties of the images acquired on the TPLSM. Image>Properties>Type "microns" as "Unit of Length" and assign the appropriate pixel width, height and depth (0.22 µm; see Note 3). Choose "OK".

    (c) Identify puncta by setting the threshold range of fluorescence intensity. Image>Adjust>Threshold. The Threshold window will open. The "Auto" function generally gives a good result. Select "Default", "B&W", and "Dark Background". Choose "Apply". If the "Convert to Mask" window opens check "Black Background" and OK. Close the Threshold window.

11. In ImageJ, select Analyze>Analyze Particles. Set size to 0.25–2.25. As nerve terminals are generally 0.5–1.5 μm in diameter, the software requires the square of this range of values or 0.25–2.25. Adjust the circularity of desired objects to be identified, if desired. Select "Outlines" under the "Show" option. Check "Display Results", "Clear Results", "Add to Manager", and "Exclude on Edges". Choose "OK". Choose "No" when asked to process all 40 images. A Drawing window opens. This shows the location of fluorescent puncta on the first image of the sequence.

12. Identify puncta over the remaining time points of the image sequence.

    (a) In the "Results" window, Select Edit<Set Measurements. Select "Area", "Mean Gray Value", "Display Label", and then "OK".

    (b) To ensure analysis of the time series, reselect the original image stack.

    (c) Within the "ROI Manager" window, select all of the listed codes. Click the first code, shift bar then click on the last code. Select "More" and "Options". Within the Options dialog box, select "Associate 'show all' ROIs with Slices" and then "OK".

    (d) Within the "ROI Manager" window, select "More" and click "Multi-Measure". Check the option to "Measure All Slices" only. Choose "OK".

    (e) Save data as an excel spreadsheet. Results<File<Save As: name.xls and organize the data within Excel.

13. Nearness of fit to first-order kinetics is determined by comparing FM1-43 destaining with $A=100*EXP\ (ln(0.5)*t/t_{1/2})$ (an integrated form of the first-order kinetics equation, $-d[A]/dt=k[A]$), using the square of the correlation coefficient (R^2; see Fig. 1e).

14. Fractional release (f) provides a measure of the change in fluorescent intensity over time. Calculate f for each punctum from $ln\ (F1/F2)/\Delta STIM$ (21, 27), where ln is the natural logarithm, F_1 and F_2 are the fluorescent intensities at t_1 and t_2, respectively, and $\Delta STIM$ is the number of stimuli delivered during that period.

Acknowledgments

This work was supported by DA07418, Picower and Parkinson's Disease Foundations (DS) and by NS052536, NS060803, HD02274, University of Washington Vision Research Center and Children's Hospital, Seattle, WA (NSB).

References

1. Bamford, N. S., and Cepeda, C. (2009) The Corticostriatal Pathway in Parkinson's Disease, in *Cortico-Subcortical Dynamics in Parkinson's Disease* (Tseng, K. Y., Ed.), pp 87–104, Humana Press, New York.

2. Graham, R. C., Jr., and Karnovsky, M. J. (1965) The histochemical demonstration of uricase activity, *J Histochem Cytochem 13*, 448–453.

3. Holtzman, E., Freeman, A. R., and Kashner, L. A. (1971) Stimulation-dependent alterations in peroxidase uptake at lobster neuromuscular junctions, *Science 173*, 733–736.

4. Wilcox, M., and Franceschini, N. (1984) Illumination induces dye incorporation in photoreceptor cells, *Science 225*, 851–854.

5. Lichtman, J. W., Wilkinson, R. S., and Rich, M. M. (1985) Multiple innervation of tonic endplates revealed by activity-dependent uptake of fluorescent probes, *Nature 314*, 357–359.

6. Zhang, Q., Cao, Y. Q., and Tsien, R. W. (2007) Quantum dots provide an optical signal specific to full collapse fusion of synaptic vesicles, *Proc Natl Acad Sci USA 104*, 17843–17848.

7. Miesenbock, G., De Angelis, D. A., and Rothman, J. E. (1998) Visualizing secretion and synaptic transmission with pH-sensitive green fluorescent proteins, *Nature 394*, 192–195.

8. Voglmaier, S. M., Kam, K., Yang, H., Fortin, D. L., Hua, Z., Nicoll, R. A., and Edwards, R. H. (2006) Distinct endocytic pathways control the rate and extent of synaptic vesicle protein recycling, *Neuron 51*, 71–84.

9. Gubernator, N. G., Zhang, H., Staal, R. G., Mosharov, E. V., Pereira, D. B., Yue, M., Balsanek, V., Vadola, P. A., Mukherjee, B., Edwards, R. H., Sulzer, D., and Sames, D. (2009) Fluorescent false neurotransmitters visualize dopamine release from individual presynaptic terminals, *Science 324*, 1441–1444.

10. Betz, W. J., and Bewick, G. S. (1992) Optical analysis of synaptic vesicle recycling at the frog neuromuscular junction, *Science 255*, 200–203.

11. Betz, W. J., Mao, F., and Smith, C. B. (1996) Imaging exocytosis and endocytosis, *Curr Opin Neurobiol 6*, 365–371.

12. Harata, N., Ryan, T. A., Smith, S. J., Buchanan, J., and Tsien, R. W. (2001) Visualizing recycling synaptic vesicles in hippocampal neurons by FM 1–43 photoconversion, *Proc Natl Acad Sci USA 98*, 12748–12753.

13. Rhee, M., and Davis, P. (2006) Mechanism of uptake of C105Y, a novel cell-penetrating peptide, *J Biol Chem 281*, 1233–1240.

14. Ryan, T. A., Smith, S. J., and Reuter, H. (1996) The timing of synaptic vesicle endocytosis, *Proc Natl Acad Sci USA 93*, 5567–5571.

15. Stevens, C. F., and Tsujimoto, T. (1995) Estimates for the pool size of releasable quanta at a single central synapse and for the time required to refill the pool, *Proc Natl Acad Sci USA 92*, 846–849.

16. Pothos, E. N., Davila, V., and Sulzer, D. (1998) Presynaptic recording of quanta from midbrain dopamine neurons and modulation of the quantal size, *J Neurosci 18*, 4106–4118.

17. Zakharenko, S. S., Zablow, L., and Siegelbaum, S. A. (2001) Visualization of changes in presynaptic function during long-term synaptic plasticity, *Nat Neurosci 4*, 711–717.

18. Bamford, N. S., Zhang, H., Schmitz, Y., Wu, N. P., Cepeda, C., Levine, M. S., Schmauss, C., Zakharenko, S. S., Zablow, L., and Sulzer, D. (2004) Heterosynaptic dopamine neurotransmission selects sets of corticostriatal terminals, *Neuron 42*, 653–663.

19. Bamford, N. S., Zhang, H., Joyce, J. A., Scarlis, C. A., Hanan, W., Wu, N. P., Andre, V. M., Cohen, R., Cepeda, C., Levine, M. S., Harleton, E., and Sulzer, D. (2008) Repeated exposure to methamphetamine causes long-lasting presynaptic corticostriatal depression that is renormalized with drug readministration, *Neuron 58*, 89–103.

20. Bamford, N. S., Robinson, S., Palmiter, R. D., Joyce, J. A., Moore, C., and Meshul, C. K. (2004) Dopamine modulates release from corticostriatal terminals, *J Neurosci 24*, 9541–9552.

21. Joshi, P. R., Wu, N. P., Andre, V. M., Cummings, D. M., Cepeda, C., Joyce, J. A., Carroll, J. B., Leavitt, B. R., Hayden, M. R., Levine, M. S., and Bamford, N. S. (2009) Age-dependent alterations of corticostriatal activity in the YAC128 mouse model of Huntington disease, *J Neurosci 29*, 2414–2427.

22. Kay, A. R., Alfonso, A., Alford, S., Cline, H. T., Holgado, A. M., Sakmann, B., Snitsarev, V. A., Stricker, T. P., Takahashi, M., and Wu, L. G. (1999) Imaging synaptic activity in intact brain and slices with FM1-43 in C. elegans, lamprey, and rat, *Neuron 24*, 809–817.

23. Winterer, J., Stanton, P. K., and Muller, W. (2006) Direct monitoring of vesicular release and uptake in brain slices by multiphoton excitation of the styryl FM 1–43, *Biotechniques 40*, 343–351.

24. Mainen, Z. F., Maletic-Savatic, M., Shi, S. H., Hayashi, Y., Malinow, R., and Svoboda, K. (1999) Two-photon imaging in living brain slices, *Methods 18*, 231–239, 181.

25. Franklin, K. B. J., and Paxinos, G. (1997) *The Mouse Brain in Stereotaxic Coordinates*, Academic Press, San Diego.

26. MacGregor, D. G., Chesler, M., and Rice, M. E. (2001) HEPES prevents edema in rat brain slices, *Neurosci Lett 303*, 141–144.

27. Isaacson, J. S., and Hille, B. (1997) GABA(B)-mediated presynaptic inhibition of excitatory transmission and synaptic vesicle dynamics in cultured hippocampal neurons, *Neuron 18*, 143–152.

Chapter 25

Microscopic Imaging of Intracellular Calcium in Live Cells Using Lifetime-Based Ratiometric Measurements of Oregon Green BAPTA-1

Carli Lattarulo, Diana Thyssen, Kishore V. Kuchibholta, Bradley T. Hyman, and Brian J. Bacskaiq

Abstract

Calcium is a ubiquitous intracellular messenger that has important functions in normal neuronal function. The pathology of Alzheimer's disease has been shown to alter calcium homeostasis in neurons and astrocytes. Several calcium dye indicators are available to measure intracellular calcium within cells, including Oregon Green BAPTA-1 (OGB-1). Using fluorescence lifetime imaging microscopy, we adapted this single wavelength calcium dye into a ratiometric dye to allow quantitative imaging of cellular calcium. We used this approach for in vitro calibrations, single-cell microscopy, high-throughput imaging in automated plate readers, and in single cells in the intact living brain. While OGB is a commonly used fluorescent dye for imaging calcium qualitatively, there are distinct advantages to using a ratiometric approach, which allows quantitative determinations of calcium that are independent of dye concentration. Taking advantage of the distinct lifetime contrast of the calcium-free and calcium-bound forms of OGB, we used time-domain lifetime measurements to generate calibration curves for OGB lifetime ratios as a function of calcium concentration. In summary, we demonstrate approaches using commercially available tools to measure calcium concentrations in live cells at multiple scales using lifetime contrast. These approaches are broadly applicable to other fluorescent readouts that exhibit lifetime contrast and serve as powerful alternatives to spectral or intensity readouts in multiplexing experiments.

Key words: Alzheimer's disease, Intracellular calcium, Oregon Green BAPTA-1, Fluorescence lifetime imaging microscopy, Lifetime plate reader

1. Introduction

Calcium is an important intracellular messenger implicated in numerous cell functions as well as neuronal development (1) and synaptic plasticity (2). Intracellular calcium regulation is altered in neurodegenerative diseases, including models of Alzheimer's

Giovanni Manfredi and Hibiki Kawamata (eds.), *Neurodegeneration: Methods and Protocols*,
Methods in Molecular Biology, vol. 793, DOI 10.1007/978-1-61779-328-8_25, © Springer Science+Business Media, LLC 2011

disease (AD). The addition of amyloid beta (Aβ) triggers a fast and constant elevation in calcium concentrations in cultured cells (3–6). Several reports have demonstrated the involvement of aberrant calcium regulation resulting from mutations in presenilin 1 (PS1) mediated through intracellular stores (7–11). It has also been shown in vivo that neuritic calcium homeostasis is severely disrupted in the presence of senile plaques, a hallmark of AD, that results in both structural and functional changes of neuronal networks (12). In additional work, it was shown that resting calcium and calcium signaling in astrocytes was affected by the presence of plaques (13). To investigate the mechanisms of calcium dysregulation in single cells, we established protocols for imaging intracellular calcium using the commonly used single wavelength fluorescent calcium indicator Oregon Green BAPTA-1 (OGB-1) in combination with lifetime contrast using fluorescence lifetime imaging microscopy (FLIM). This approach allows ratiometric imaging in live cells to permit determinations of absolute intracellular calcium levels.

Calcium dye indicators, such as OGB, are used to monitor dynamic changes in intracellular calcium. OGB reports changes in $[Ca^{2+}]$ with a change in both intensity and lifetime, but it is impossible to discriminate $[Ca^{2+}]$ from dye concentration with intensity measures alone. FLIM, however, produces images based on the differences in decay rates from a fluorescent sample instead of looking at brightness alone. By using FLIM to measure fluorescence lifetimes, absolute calcium concentration can be determined independently of variations in dye concentration by comparing bound and unbound calcium decay curves, which exhibit different lifetimes. These measurements should be consistent across diagnostic platforms and should provide a single comparable metric across assays that test pathological alterations and/or therapeutic interventions. In this chapter, we outline protocols for fluorescence lifetime imaging of OGB across platforms, from single cells to high-throughput multiwell formats and finally to in vivo disease models, thereby allowing an integrative approach for evaluation of compounds that block or reduce elevated calcium in pathophysiological models of disease.

We demonstrate approaches to calibrate the known calcium reporter dye, OGB, with a ratiometric index based on lifetime measurements using FLIM. These calibrations were generated using both cell impermeant OGB-1 in calcium buffers and cell permeant OGB-1/AM loaded into live cells. We then investigated the feasibility of high-throughput measurements of absolute $[Ca^{2+}]$ using commercially available fluorescence lifetime plate readers. Finally, we provide in vivo evidence that OGB can be used to determine the resting calcium concentration in specific cell types in the living brain of Alzheimer's mouse models using FLIM.

2. Materials

2.1. OGB-1 Calibration in Microcuvettes

1. OGB-1, hexapotassium salt, cell impermeant (Molecular Probes, Carlsbad, CA, USA).
2. Calcium Calibration Kit #2 (Molecular Probes).
3. Glass capillary tubes, borosilicate glass with filament (O.D.: 1.5 mm, I.D.:0.86 mm, 10 cm), halved, (Sutter Instruments, Novato, CA, USA).
4. SPCM lifetime acquisition and SPCImage analysis software (Becker and Hickl, Berlin, Germany).
5. DCS 120 confocal scanning fluorescence lifetime imaging measurement light microscope (Becker and Hickl).

2.2. Cell Culture

1. OPTI-MEM I reduced serum medium 1×, (Gibco/Invitrogen, Grand Island, NY) enriched with 10% fetal bovine serum (FBS) (Sigma, St. Louis, MO, USA).
2. Dulbecco's phosphate buffered saline 1×, [–] calcium chloride, [–] magnesium chloride, (Gibco/Invitrogen) (see Note 3).
3. TrypLE express 1×, [+] phenol red (Gibco/Invitrogen).
4. 25 cm^2-flask, canted neck, nonpyrogenic, sterile, polystyrene (Corning Incorporated, Corning, NY, USA).

2.3. OGB-1/AM Calibration in Living Cells

1. OGB-1/AM, cell permeant, special packaging (Molecular Probes) (see Note 1).
2. Centrifugal filter units, Durapore PVDF 0.45 μm (Millipore Corporation, Billerica, MA, USA).
3. Ionomycin, calcium salt (Invitrogen).
4. Calcium Calibration Kit #2 (Molecular Probes).
5. Dulbecco's phosphate buffered saline 1×, [–] calcium chloride, [–] magnesium chloride, (see Note 3).
6. 35 mm glass bottom dishes, gamma irradiated, poly-D-lysine coated, (MatTek, Ashland, MA, USA).
7. 96-well optical CVG sterile glass bottom plate (Nalge Nunc International, Rochester, NY).

2.4. OGB-1/AM In Vivo

1. OGB-1/AM, cell permeant, special packaging, (Molecular Probes) (see Note 1).
2. Double transgenic APPswe/PS1dE9 mice and age-matched wild-type mice, 6–8 months old, dedicated investigator supply of B6C3-Tg(APPswe,PSEN1dE9)85Dbo/J (The Jackson Laboratory, Bar Harbor, ME, USA).
3. Sulforhodamine 101 (SR-101), (Invitrogen).

4. Pluronic® F-127, 20% solution in DMSO.

5. Artificial cerebral spinal fluid.

6. Patch pipette.

7. AM system Picospritzer 2000.

8. Olympus Fluoview 1000MPE with prechirp optics and an AOM mounted to an Olympus BX61WI upright microscope.

9. Mode-locked titanium/sapphire laser (Tsunami, Spectra-Physics, Fremont, CA).

10. Time-correlated single photon counting (TCSPC) acquisition card (SPC830, Becker and Hickl).

11. A 16-channel multispectral photomultiplier (PML-16) with a 300 nm tunable bandwidth.

3. Methods
Imaging Protocols

3.1. OGB-1 Calibration Using Visible Light Confocal FLIM

We constructed lifetime decay curves for OGB-1 in each of 11 calcium buffers using samples of free dye in glass capillary tubes at varying calcium concentrations. We used the DCS 120 visible light confocal FLIM system (Becker and Hickl) to collect FLIM images. In the absence of calcium, OGB-1 had a single, fast lifetime of ~650 ps. At saturating levels of calcium (39 μM), OGB-1 had a single, slow lifetime of ~3,700 ps (see Note 5). As the free calcium concentration changed between these extremes, the average lifetime (single exponential fit) varied, and the relative contribution of each of the two exponents measured with biexponential fits varied. We fixed the values of each individual lifetime, and determined the relative amplitude of each exponent. These amplitudes are independent of the concentration of OGB-1 (see Note 4). These amplitude values were used to generate the calibration curves (Fig. 1d). The K_d of OGB-1 was found to be 150 nM, which is consistent with the literature (14).

1. Reconstitute OGB-1 hexapotassium salt in double deionized water to 1 mM stock solution.

2. In 11 appropriately labeled Eppendorf tubes, combine each of the 11 calcium buffers, ranging from 0 μM to 39 μM free calcium, from Molecular Probes Calcium Calibration Kit #2 with 10 μM OGB-1.

3. Load about 3 μL of 0 μM calcium buffer and OGB-1 mixture into one clean glass capillary tube by submerging one end directly into the calcium buffer and sealing off the other end with wax.

4. On DCS 120 confocal scanning fluorescence lifetime imaging measurement (FLIM) light microscope (Becker and Hickl),

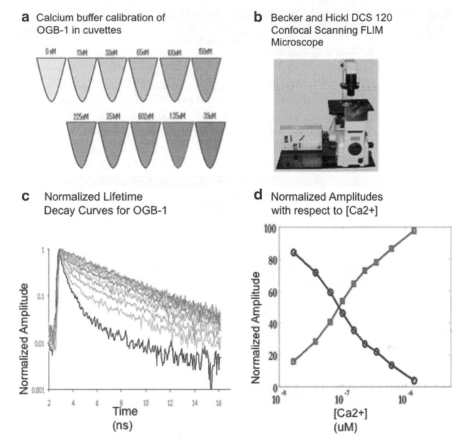

a Calcium buffer calibration of OGB-1 in cuvettes

b Becker and Hickl DCS 120 Confocal Scanning FLIM Microscope

c Normalized Lifetime Decay Curves for OGB-1

d Normalized Amplitudes with respect to [Ca2+]

Fig. 1. Calibration of OGB-1 using visible light confocal FLIM. (**a**) Eleven different calcium calibration buffers were used to obtain the calcium calibration curves. (**b**) Microcuvette calibration was performed using a DCS 120 confocal dual scanning FLIM system. (**c**) The raw decay curves of 11 different calcium concentrations. The high calcium condition shows a slow lifetime of ~3,700 ps. The low calcium condition shows a faster lifetime of ~650 ps. The intermediate calcium levels show a combination of these two exponential decays. (**d**) Using a biexponential fit with fixed lifetimes for the bound and unbound lifetimes, we measure the relative amplitudes of the exponential components to create a ratiometric calibration curve of OGB-1 based on lifetime imaging. The K_d of OGB-1 for calcium was determined to be 150 nM.

place glass capillary tube from step 3 on 20× objective at 1.8 zoom. Excite with 473 nm pulsed diode laser.

5. Using SPCM software (Becker and Hickl), scan fluorescent solution for 16 cycles at 30 s per cycle to collect FLIM image.

6. Analyze images in SPCImage. Adjust components to 2 in multiexponential decay. Calculate decay matrix to get pseudocolored image of OGB-1 lifetime(s) present in calcium buffer. Determine average and relative amplitude of each lifetime present.

7. Repeat steps 3–6 for the other ten calcium buffers with OGB-1.

8. Plot concentration of calcium versus relative amplitude of each lifetime for each buffer to create the calibration curve. Calculate the K_d value of OGB-1 from calibration curve (Fig. 1).

3.2. Multiphoton FLIM of Cultured Cells with OGB-1/AM

We next measured lifetimes of OGB-1/AM in live CHO cells using multiphoton microscopy and a multispectral FLIM detector. The cells were loaded with OGB-1/AM and equilibrated with buffers of known calcium concentration in the presence of ionomycin. Imaging was performed using a BioRad 1024 ES multiphoton microscope to obtain both intensity and FLIM images. We used the same approaches described above to determine the calcium-free and calcium-bound lifetimes of OGB-1/AM in live cells. The values were similar to those found in the capillary tube measurements with single photon excitation. The short lifetime of OGB-1/AM unbound to calcium was found to be ~540 ps while the long, bound lifetime was found to be ~3,000 ps (see Note 5). Fixing the exponents to these values, we determined the relative amplitudes of each component and plotted them versus the calcium concentration as shown in Fig. 4f. Examples of lifetime images are shown, where the images have been pseudocolored based on calcium concentration using this calibration curve (Fig. 3).

1. Split Chinese hamster ovary (CHO) cells into 35-mm plastic bottom sterile dishes in OPTI-MEM media with 10% FBS. Remove cells from T25 flask using TrypLE Express. Deactivate TrypLE and resuspend cells in OPTI-MEM media with FBS.

2. Add 5 μL of 80% DMSO, 20% pluronic acid (F127) to 50-μg tube of OGB-1/AM to dissolve dye powder. Transfer OGB-1/AM to microcentrifuge tube and mix with 45 μL of PBS for a total of 1 mM stock solution. Vortex solution and transfer to 0.45-μm filter unit. Microcentrifuge until dye solution passes through filter. Retain dye on ice until used (see Notes 1 and 2).

3. Add 20 μL 1 mM OGB-1/AM to 980 μL PBS.

4. Add 1 mL OGB-1/AM (from step 3 above) to each dish. Incubate dishes for 45 min at 37°C and 5% CO_2.

5. Aspirate off OGB-1/AM and wash cells three times for 5 min with calcium-free PBS (see Note 3).

6. Using the Molecular Probes Calcium Calibration Kit #2, add one of the 11 calcium buffers to each dish of CHO cells. Add ionomycin (calcium ionophore) to a final concentration of 20 μM. Incubate for 30 min prior to imaging.

7. Next, image the cells on a 1024 ES multiphoton microscope (Bio-Rad) using a 20× water objective (Olympus, numerical aperture = 0.95) at 2× zoom for 10 min at 0.33 Hz.

8. Switch to FLIM component of microscope using a TCSPC acquisition card (SPC 830, Becker and Hickl) and a PML-16 that allows simultaneous acquisition of 16 spectral bins across a tunable 300 nm bandwidth. 30 s acquisitions were repeated 6–12 times. Data was typically sampled at 128×128 pixels (0.42 pixels/μm) at 64 time bins per pulse period (12.5 ns/64).

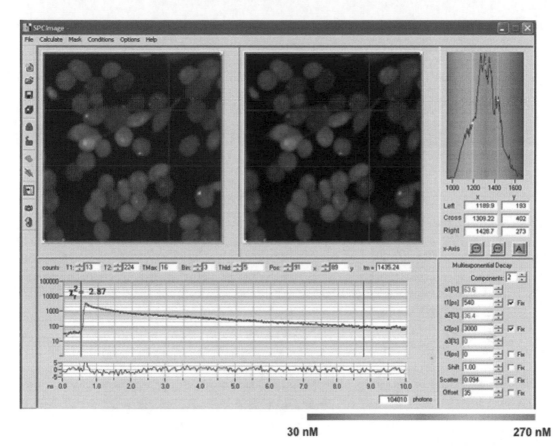

30 nM   **270 nM**

Fig. 2. SPCImage screenshot. OGB-1/AM loaded into CHO cells and corresponding FLIM image. The saved FLIM image is imported to SPCImage and analyzed using a two-component fit with short- and long lifetimes fixed at 540 ps and 3,000 ps, respectively. The image in the *top left* is the fluorescent intensity of the field of cells. To the *right* is the pseudocolored image based on the lifetime calculations. The bottom trace is from a single pixel, and shows the decay curve along with the curve fit.

9. Analyze lifetime images using SPCImage (Fig. 2) with biexponential fits (Fig. 3). Refer to Fig. 4 for sample data (see Notes 4 and 5).

3.3. High-Throughput Assays

We next translated these single-cell imaging approaches to high-throughput assays using commercially available lifetime plate readers. We used two different instruments that work with the same basic principle of using a pulsed diode laser source to excite fluorescence within a well, and time-domain photon counting to acquire fluorescence decays. Each well resulted in a fluorescent decay curve that could be fit in the same way described above, but without single-cell resolution. To increase signal to noise, we used suspensions of cells loaded with OGB-1/AM in both the UltraFLT (Tecan) and the Nanotaurus (Edinburgh Instruments) to create calibration curves of OGB-1/AM lifetime ratios with different concentrations of calcium buffers (Fig. 5b).

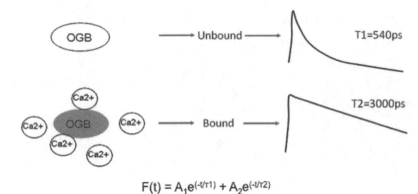

$$F(t) = A_1e^{(-t/\tau1)} + A_2e^{(-t/\tau2)}$$

Fig. 3. Short- and long lifetimes of OGB. OGB has two characteristic lifetimes depending on whether it is bound or unbound to calcium. When there is no free calcium in solution, then OGB exhibits a short lifetime of 540 ps, as determined on the 1024 ES multiphoton microscope (Bio-Rad). When calcium is bound to OGB, then the dye exhibits a long lifetime of 3,000 ps.

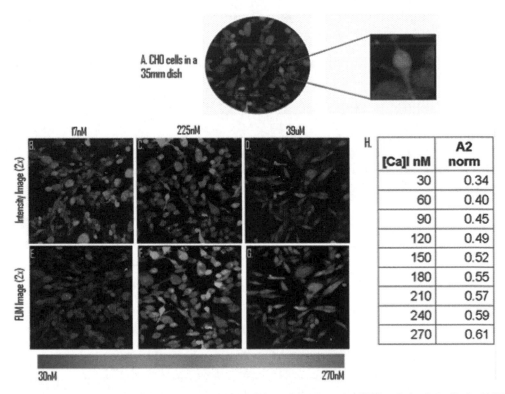

[Ca]I nM	A2 norm
30	0.34
60	0.40
90	0.45
120	0.49
150	0.52
180	0.55
210	0.57
240	0.59
270	0.61

Fig. 4. Multiphoton FLIM of OGB-1/AM in cultured cells. (a) Chinese hamster ovary (CHO) cells loaded with 8 µM OGB-1/AM for 1 h at 37°C and then imaged on a Bio-Rad 1024 ES multiphoton microscope using a 20× objective at 2× zoom. (b–d) Intensity images of CHO cells loaded with OGB-1/AM and incubated with different calcium buffers (a: 17 nM; b: 225 nM; c: 39 µM) and 20 µM ionomycin. (e) Representative intensity and FLIM images of cells at 17 nM (b, e), 225 nM (c, f), and 39 µM (d, g). (h) Calibration data showing [Ca^{2+}] in nM and the A2 norm values (relative amplitude of the long decay time) that correlate with calcium concentrations.

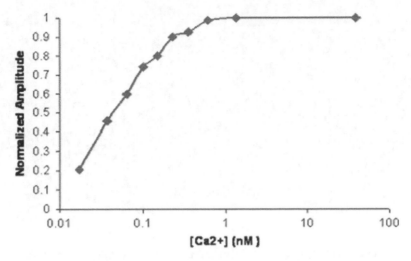

Fig. 5. High-throughput assay for OGB-1/AM. Normalized amplitude of the calcium bound component (*slow lifetime*), which was found on the Ultra FLT (Tecan Trading AG, Switzerland). The Nanotaurus (Edinburgh Instruments) produced similar lifetimes and relative amplitude curves allowing for quantitative analysis of [Ca^{2+}].

1. Remove CHO cells from T25 flask with TrypLE Express. Transfer suspended cells into 15-mL conical tube and spin down. Resuspend cells in fresh OPTI-MEM media.

2. Prepare OGB-1/AM as described above (see Notes 1 and 2).

3. Add OGB-1/AM to resuspended cells in 15-mL conical tube to a final concentration of 1 mM. Incubate cells for 45 min in 37°C and 5% CO_2.

4. Spin down cells again and aspirate off OGB-1/AM media. Resuspend cells in calcium-free PBS to wash them. Repeat two more times (see Note 3).

5. After final wash, divide cells evenly into 11 Eppendorf tubes and spin down. Resuspend cells in 1 of 11 calcium buffers, as above, and treat with 20 μM ionomycin for 20 min. Repeat each condition in triplicate.

6. Load cells treated with calcium buffers into a 96-well glass bottom plate.

7. Load plate into the UltraFLT (Tecan Trading AG, Switzerland) or the Nanotaurus (Edinburgh Instruments, UK) (Fig. 2). Using a 440 nm pulsed diode laser running at 40 MHz, excite OGB-1/AM and collect its emission at 544 nm for 5 s per well.

8. Analyze the raw data from each well using biexponential curve fitting. Create calibration curve for OGB-1/AM (Fig. 5) (see Notes 4 and 5).

3.4. OGB-1/AM In Vivo Imaging

Finally, the process of imaging OGB-1/AM ratios with FLIM was translated to an in vivo setting. We measured calcium concentration in astrocytes within the intact brains of APPswe/PS1dE9 mice

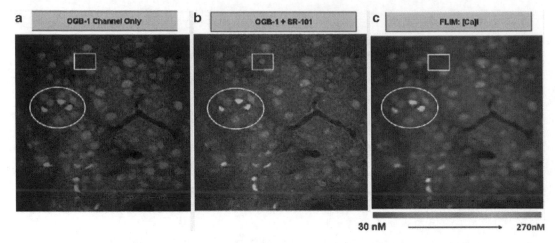

Fig. 6. In vivo image of cells loaded in the brain of APPswe/PS1dE9 mice using the multicell bolus loading technique. (**a**) The OGB-1/AM channel alone showing many loaded cells. The *white circle* highlights three astrocytes loaded with OGB-1/AM and the *white box* shows a cell that has not been loaded. (**b**) Combined OGB-1/AM and SR-101 channels, demonstrating the determination of astrocytes based on cell specific staining with SR-101. Astrocytes have been loaded with both dyes, as seen in the *white circle*. The *cell* in the *white box* was loaded only with SR-101. (**c**) FLIM image of the same field. Note that astrocytes have higher resting calcium than the surrounding neurons.

(at an age where amyloid plaques were detectable). The animals were anesthetized and a cranial window was implanted, as previously described (15, 16). The brain (neurons and astrocytes) was then loaded with OGB-1/AM using the multicell bolus loading technique (12, 17). The fluorescent dye SR-101 was included in the pipette to label astrocytes specifically (18). Intensity images were taken of OGB-1/AM (green) and SR-101 (red) using multiphoton microscopy, as previously described. Following this, FLIM images were acquired, and the biexponential decay curves within the images were fit with the lifetimes of the calcium-free and calcium-bound forms of OGB-1/AM fixed at their respective values. An overlay image was made to determine which cells were astrocytes and which cells were neurons. Overall, astrocytes exhibited higher resting calcium levels than neurons in these mice (Fig. 6).

1. Double transgenic APPswe/PS1dE9 hemizygous mice and age-matched wild-type mice were obtained at 6–8 months old from The Jackson Laboratory. At this age, significant plaque deposition is evident.

2. Animals were anesthetized using isoflurane at 0.5–1% as previously described (19) and placed into a stereotax to immobilize the head.

3. A craniotomy was done as previously described (15, 16) but the cover glass was not immediately implanted.

4. OGB-1/AM was prepared to 1 mM as outlined above (see Notes 1 and 2). SR-101 was added to OGB-1/AM mixture. These two dyes were combined for a total of 0.8nM OGB-1/

AM in 80% DMSO and 20% pluronic acid (F-127, Invitrogen) and mixed with ACSF, and 10 μM SR-101 (Invitrogen).

5. The mixture from step 4 was loaded into a patch pipette with tip diameter of 2–3 μm and inserted approximately 100 μm deep into the mouse cortex.

6. Next, the dyes are pressure ejected at five to ten PSI for 60–90 s using the AM system Picospritzer 2000 and the multicell bolus loading technique, as described elsewhere (17).

7. The cranial window was implanted after micropipette injection and the dyes were incubated in the brain for 1 h so that cortical neurons and astrocytes were loaded with OGB-1/AM, and astrocytes were selectively loaded with SR-101.

8. In vivo multiphoton imaging was used to image the dye-loaded cells using the Olympus Fluoview 1000MPE with prechirp optics and an AOM mounted to an Olympus BX61WI upright microscope. A mode-locked titanium/sapphire laser (Tsunami, Spectra-Physics, Fremont, CA) generated two-photon excitation at 860 nm, and detectors with photomultiplier tubes (Hamamatsu, Ichinocho, Japan) collected emitted light in the ranges of 500–540 and 560–650 nm for OGB-1/AM and SR-101, respectively.

9. FLIM images were acquired using a TCSPC acquisition card (SPC830, Becker and Hickl) and a PML-16 with a 300 nm tunable bandwidth. Thirty-second collection periods were repeated consecutively 6–12 times, resulting in acquisition of integrated lifetime images that required 3–6 min.

4. Notes

1. All OGB-1/AM dye should be made fresh before each experiment. Do not freeze and reuse.

2. Be sure to mix the OGB-1/AM well in the tube by repeatedly pipetting up and down several times before transferring solution to Eppendorf tube to ensure that all the dye powder is dissolved.

3. CHO cells should be washed with calcium- and magnesium-free PBS so that the only calcium surrounding the cells is from the calcium buffer. This ensures that the labeled calcium concentration of each buffer is the actual calcium concentration on the cells.

4. Other ratiometric approaches for calcium imaging, such as the genetically encoded yellow cameleon reporter (12) or the small-molecule fluorophores Indo-1 and Fura-2 (20), have been used to image $[Ca^{2+}]i$ quantitatively. The drawbacks to these

approaches include the need to transfect or transduce cells of interest with yellow cameleon, or to use UV excitation for Fura-2 or Indo-1, which can lead to significant phototoxicity. Ratiometric imaging eliminates confounds resulting from photobleaching, differences in dye concentrations from uneven loading, and variations in illumination intensity.

5. OGB should have only one lifetime in the 0 μM and 39 μM calcium buffers. These decay curves can be fit with a single exponential.

References

1. Wen Z, Guirland C, Ming GL, Zheng LQ (2004) A CaMKII/calcineurin switch controls the direction of Ca(2+)-dependent growth cone guidance. *Neuron* **43**, 835–846.

2. Malenka RC, Bear MF (2004) LTP and LTD: an embarrassment of riches. *Neuron* **44**, 5–21.

3. Demuro A, Mina E, Kayed R, Milton SC, Parker I, Glabe CG (2005) Calcium dysregulation and membrane disruption as a ubiquitous neurotoxic mechanism of soluble amyloid oligomers. *J Biol Chem* **280**, 17294–17300.

4. Guo Q, Sebastian L, Sopher BL, Miller MW, Ware CB, Martin GM, Mattson MP (1999) Increased vulnerability of hippocampal neurons from presenilin-1 mutant knock-in mice to amyloid beta-peptide toxicity: central roles of superoxide production and caspase activation. *J. Neurochem* **72**, 1019–1029.

5. Mattson MP, Cheng B, Davis D, Bryant K, Lieberburg I, Rydel RE (1992) Beta-Amyloid peptides destabilize calcium homeostasis and render human cortical neurons vulnerable to excitotoxicity. *J Neurosci* **12**, 376–389.

6. Mattson MP, Barger SW, Cheng B, Lieberburg I, Smith-Swintosky VL, Rydel RE (1993) Beta-Amyloid precursor protein metabolites and loss of neuronal Ca^{2+} homeostasis in Alzheimer's disease. *Trends Neurosci* **451**, 720–724.

7. Tu H, Nelson O, Bezprozvanny A, Wang Z, Lee SF, Hao, YH, Serneels L, deStrooper B, Yu G, Bezprozvanny I (2006) Presenilins form ER Ca^{2+} leak channels, a function disrupted by familial Alzheimer's disease-linked mutations. *Cell* **126**, 981–993.

8. Nelson O, Tu H, Lei T, Bentahir M, deStrooper B, Bezprozvanny I (2007) Familial Alzheimer disease-linked mutations specifically disrupt Ca^{2+} leak function of presenilin 1. *J Clin Invest*, 1230–1239.

9. Cheung KH, Shineman D, Müller M, Cárdenas C, Mei L, Yang J, Tomita T, Iwatsubo T, Lee VM, Foskett JK (2008) Mechanism of Ca^{2+} disruption in AD by presenilin regulation of InsP3 receptor channel gating. *Neuron* **58**, 871–883.

10. Stutzmann GE, Caccamo A, LaFerla FM, Parker I (2004) Dysregulated IP3 signaling in cortical neurons of knock-in mice expressing an Alzheimer's-linked mutation in presenilin1 results in exaggerated Ca^{2+} signals and altered membrane excitability. *J Neurosci* **24**, 508–513.

11. Stutzmann GE, Smith I, Caccamo A, Oddo S, LaFerla FM, Parker I (2006) Enhanced ryanodine receptor recruitment contributes to Ca^{2+} disruptions in young, adult, and aged Alzheimer's disease mice. *J Neurosci* **26**, 5180–5189.

12. Kuchibhotla KV, Goldman ST, Lattarulo CR, Wu H-Y, Hyman BT, Bacskai BJ (2008) Amyloid-beta plaques lead to aberrant regulation of calcium homeostasis *in vivo* resulting in structural and functional disruption of neuronal networks. *Neuron* **29**, 214–225.

13. Kuchibhotla KV, Lattarulo CR, Hyman BT, Bacskai BJ (2009) Synchronous hyperactivity and intercellular calcium waves in astrocytes in Alzheimer mice. *Science*, 1211–1215.

14. Hendel T, Mank M, Schnell B, Griesbeck O, Borst A, Reiff DF (2008) Fluorescence changes of genetic calcium indicators and OGB-1 correlated with neural activity and calcium *in vivo* and *in vitro*. *J Neurosci* **28**, 7399–7411.

15. Spires TL, et al. (2005) *J NeuroSci* **25**, 7278.

16. Skoch J, Hickey GA, Kajdasz ST, Hyman BT, Bacskai BJ (2005) In vivo imaging of amyloid-beta deposits in mouse brain with multiphoton microscopy. Methods Mol Biol **299**: 349–363.

17. Stosiek C, Garaschuk O, Holthoff K, Konnerth A (2003) *In vivo* two-photon calcium imaging of neuronal networks. *Proc Natl Acad Sci USA* **100**, 7319–7324.

18. Nimmerjahn A, Kirchhoff F, Kerr JN, Helmchen F (2004) Sulforhodamine 101 as a specific marker of astroglia in the neocortex *in vivo*. *Nat Methods* 1, 31–37.

19. Bacskai BJ, Klunk WE, Mathis CA, Hyman BT (2002) Imaging amyloid-beta deposits *in vivo*. *J Cereb Blood Flow Metab* **22**, 1035–41.

20. Grynkiewicz G, Poenie M, Tsien RY (1985) A new generation of Ca^{2+} indicators with greatly improved fluorescence properties. *J Biol Chem* **25**, 3440–3450.

Chapter 26

Electrophysiological Characterization of Neuromuscular Synaptic Dysfunction in Mice

Yoshie Sugiura, Fujun Chen, Yun Liu, and Weichun Lin

Abstract

Emerging evidence suggests that synaptic dysfunction occurs prior to neuronal loss in neurodegenerative diseases, such as amyotrophic lateral sclerosis (ALS). Therefore, monitoring synaptic activity during early stages of neurodegeneration may provide valuable information for the development of diagnostic and/or therapeutic strategies. Here, we describe an electrophysiological method routinely applied in our laboratory for investigating synaptic activity of the neuromuscular junction (NMJ), the synaptic connection between motoneurons and skeletal muscles. Using conventional intracellular sharp electrodes, both spontaneous synaptic activity (miniature end-plate potentials) and evoked synaptic activity (end-plate potentials) can be readily recorded in acutely isolated nerve–muscle preparations. This method can also be adapted to various simulation protocols for studying short-term plasticity of neuromuscular synapses.

Key words: Amyotrophic lateral sclerosis, Miniature end-plate potential, End-plate potential, Intracellular recording, Neuromuscular junction, Neurodegeneration

1. Introduction

Synaptic dysfunction/loss is among one of the earliest signs of neurodegeneration. Functional impairment of the neuromuscular junction (NMJ) occurs in amyotrophic lateral sclerosis (ALS) patients (1–3), and neuromuscular synapses degenerate prior to the loss of motoneurons (4–10). These observations lead to an emerging view that synaptic dysfunction/loss is likely the primary pathophysiological event, and therefore is an important target for developing diagnostic and/or therapeutic strategies (11–14).

Synaptic dysfunction of the NMJ can be conveniently assayed in acutely isolated nerve–muscle preparations. In this chapter, we describe methods routinely used in our laboratory for evaluating

Giovanni Manfredi and Hibiki Kawamata (eds.), *Neurodegeneration: Methods and Protocols*,
Methods in Molecular Biology, vol. 793, DOI 10.1007/978-1-61779-328-8_26, © Springer Science+Business Media, LLC 2011

neuromuscular synaptic transmission. To illustrate synaptic dysfunction of the NMJ, we describe electrophysiological analyses of the NMJ in mutant mice deficient in ubiquitin carboxyl-terminal hydrolase L1 (UCH-L1) (15). These mutant mice develop progressive age-dependent paralysis reminiscent of ALS.

We apply conventional intracellular recordings to record spontaneous miniature end-plate potentials (mEPPs) and evoked end-plate potentials (EPPs). We use a number of nerve–muscle preparations, including diaphragm muscles, triangularis sterni muscles (16), levator auris longus muscles (17), flexor digitorum brevis (FDB) muscles (18), soleus and extensor digitorum longus (EDL) muscles (19–21). Each muscle preparation has some advantages and disadvantages. For example, the diaphragm muscles are easy to dissect, but they are relatively large, which require a large volume of perfusion and thus large amount of drug, such as Mu-conotoxin. The triangularis sterni muscles are difficult to dissect but they are best suited for visual identification of individual motor end-plates (22). The levator auris longus muscles are particular useful for chronic injection of drugs and toxins, as these muscles are conveniently located just underneath the skin covering the dorsal aspect of head and neck (17). We prefer the lumbrical muscle (23–26) because they are small. The muscles are located in the hindpaw and innervated by a distal branch of the sciatic nerve, the plantar nerve. Their small size is particularly advantageous for morphological analyses, especially at electron microscopy level.

2. Materials

2.1. Solutions and Reagents

All solutions are prepared with purified water (mili-Q, Millipore Co. Bedford MA, USA). The equipments listed below are currently used in our laboratory.

1. Normal mouse Ringer's solution, based on Liley (27): 136.8 mM NaCl, 5 mM KCl, 12 mM $NaHCO_3$, 1 mM NaH_2PO_4, 1 mM $MgCl_2$, 2 mM $CaCl_2$, and 11 mM d-glucose, pH 7.3, stored at 4°C.

2. Mu-conotoxin GIIIB (Peptides International, Louisville, KY): Working concentration 1–2 μM. Prepare 100 μM stock solution, make aliquots, and store at –20°C (see Note 1).

3. Intracellular solution for filling recording micropipettes: A mixture of 2 M potassium citrate and 10 mM potassium chloride (see Note 2). This solution is stable and can be stored at room temperature.

4. Isoflurane (Webster Veterinary Supply Inc., Devens, MA, USA).

2.2. Dissection

1. Sylgard-lined dishes. The Sylgard 184 Silicone Elastomer kit (Dow Corning Co, Midland, MI, USA) is composed of two parts: Sylgard base and curing agent. Combine in a ratio of ten parts of base to one part of curing agent, by weight. Mix gently by inverting the tubes (avoid air bubbles) and pour into Petri dishes. Use a 100-mm dish (in diameter) for dissection, and a 35-mm dish for recording chamber. Keep Sylgard-lined dishes in a dust-free and flat area overnight at room temperature (see Note 3). More detailed information is described in manufacture instructions (http://www3.dowcorning.com/DataFiles/090007c8801d3f9e.pdf).

2. Minutien pins: Item No.26002-10 or No.26002-20 (Fine Science Tools, USA).

3. Fine forceps and scissors.

4. Stereomicroscope (Stemi 2000, Zeiss).

5. Oxygenated (95% O_2 and 5% CO_2) Ringer's solution.

2.3. Electrophysiology Equipments

1. Vibration isolation table and Farraday Cage.

2. Upright fixed stage microscope with water immersion objectives (Olympus BX51WI).

3. Micromanipulator: Model MP-225 (Sutter Instrument Co., CA, USA).

4. Stimulator: Grass-Telefactor SD9 (West Warwick, RI, USA).

5. Amplifier: AxoClamp-2B (Molecular Devices, Sunnyvale, CA, USA).

6. Computer and data acquisition software: Digidata 1332 (Molecular Devices), pClamp 9.0 (Molecular Devices) and Mini Analysis Program (Synaptosft, Inc., Decatur, GA, USA).

7. Micropipette puller: Model P-97 or model P-30 (Sutter Instrument Co., CA, USA).

8. Glass capillaries for recording micropipettes: Cat. No. TW100F-3 (1 mm OD with filament, World Precision Instruments Inc., Sarasota, FL, USA).

9. Glass capillaries for suction electrode: Cat. No. 628500 (A-M Systems Inc. Carlsborg, WA, USA).

10. Suction electrode Cat. No. 573000 (A-M Systems Inc. Carlsborg, WA, USA). This product includes tubing.

11. Audible baseline monitor: ABM (World Precision Instruments Inc. Sarasota, FL, USA).

3. Methods

3.1. Dissection of Plantar Nerve–Lumbrical Muscle

The lumbrical muscles are located between the digits of the hindpaw (see also ref. 24).

1. Anesthetize the mice with isoflurane and sacrifice the animal by cervical dislocation.

2. Remove one foot by cutting above the ankle.

3. Pin the foot in a Sylgard-lined dish with oxygenated Ringer's solution, ventral side up.

4. Make a medial incision to the skin from the ankle toward the toes to expose hypodermal surface.

5. Remove the surface muscles, including the FDB to expose the flexor digitorium longus (FDL) tendon.

6. Grasp the cut end of the FDL tendon and gently lift it from the proximal to distal (from the ankle to the toes) while carefully cutting the underlying connective tissues. Lumbrical muscles are attached to the tendon.

7. Isolate plantar nerves attached to the lumbrical muscles.

8. Cut the distal side of FDL tendons. Now, four lumbrical muscles attached to the FDL tendon can be removed as a whole.

9. Cut tendon and separate single nerve–muscle preparation. No need to remove the tendon attached to the proximal part of the muscle.

10. Place the preparation in oxygenated Ringer's.

3.2. Preparation of Suction Electrodes for Nerve Stimulation

We modify a commercially available suction electrode (from A-M Systems Inc.) by soldering a piece of silver wire (as a reference electrode) to the negative end of the BNC cable.

3.2.1. Preparation of Glass Pipettes for Suction Electrodes

1. Expose a glass capillary over the flame of a burner or alcohol lamp and pull manually (see Note 4).

2. Brake off the tip of the glass capillary.

3. Fire-polish the broken end of the pipette tips using an alcohol lamp (see Note 5).

4. Under a stereomicroscope, match the size of the pipette tip with the nerve so that a good seal can be achieved.

3.2.2. Assembly of Suction Electrodes

1. Cut a 30–40 cm length of tubing (see Subheading 2.3, item 10).

2. Connect one end of the tubing to the side port of the suction electrode holder.

3. Connect the other end of the tubing to a 3–5 ml syringe.

4. Connect a BNC connector of the suction electrode to a stimulator (SD9, see Subheading 2.3, item 4).

5. For a reference electrode, solder a thin silver wire to the outer conductor of the BNC cable, and then run it alongside of the suction electrode down to the length of the glass pipette.

6. Insert the glass pipette into the electrode holder. Be sure the silver wire of the holder is inside the glass pipette.

3.3. Preparation of Sharp Electrodes for Intracellular Recording

1. Adjust the heater and weights on the pipette puller till the tips of micropipettes are sharp and small, with resistances around 20–40 MΩ. Use thin-wall glass capillaries with filament to facilitate filling (see Subheading 2.3, item 8).

2. Fill the micropipette using a fine flexible plastic capillary connected to a 1-ml syringe (see Note 6).

3. Hold the micropipette vertically (tip down) and gently tap the pipette to remove air bubbles.

4. Connect the micropipette to a pipette holder of the amplifier (see Note 7).

3.4. Recording Procedures

1. Place the nerve–muscle preparation in a recording chamber with oxygenated Ringer's solution. It is advisable to measure the volume of the Ringer's solution, as the volume is used to determine the amount of Mu-conotoxin being added later (see Note 8).

2. Draw the nerve into the suction electrode by applying gentle suction. It is critical to obtain a good seal. Test if muscle responds to nerve stimulation by applying a single pulse of stimulation.

3. Add Mu-conotoxin to a final concentration 1 μM (see Note 9).

4. Locate the tip of the micropipette under the microscope by adjusting the X-Y-Z position using a micromanipulator. Start with a low-power objective (e.g., 4×, then 10×).

5. Use a high-power objective (e.g., 40×) to locate an end-plate area of the muscle.

6. Move the recording micropipette gently toward the end-plate area of the muscle. Use the audio baseline monitor (MBA) as a guide when approaching to the surface of the muscle membrane.

7. When muscle is impaled, the resting potential (RP) drops to around −70 to −80 mV.

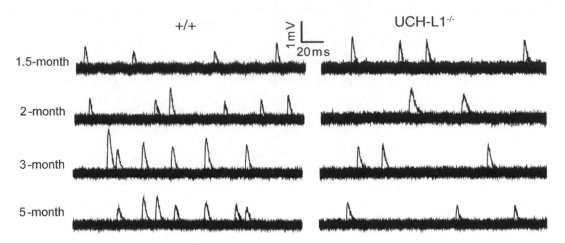

Fig. 1. Reduced spontaneous neuromuscular synaptic activity in *UCH-L1*^{-/-} mice. Sample traces of mEPPs from wild-type (WT, +/+) and *UCH-L1*^{-/-} mice at 1.5-, 2-, 3-, and 5 months of age: each trace is a set of 20 superimposed 200 ms-sweeps representing a 4-s continuous recording from an individual muscle fiber. MEPP frequencies are reduced in *UCH-L1*^{-/-} mice after 2 months of age, compared with age-matched WT mice (modified from ref. 15).

8. If the micropipette is near or at synaptic sites, spontaneous mEPPs with rise time less than 1.5 ms can be detected.

9. Record mEPPs (see Note 10). Examples of spontaneous mEPP traces are shown in Fig. 1.

10. Record nerve-evoked EPPs. Examples of nerve-evoked EPP traces are shown in Fig. 2.

11. You may apply various stimulation protocols (see Note 11) to obtain a variety of synaptic responses. For example, paired-pulse stimulation (see Fig. 3a, b) or repetitive stimulation (see Fig. 3c) can be applied to examine short-term plasticity of the NMJ.

4. Notes

1. To prevent muscle contraction, we use Mu-conotoxin GIIIB. It binds voltage-gated sodium channels in skeletal muscles and prevents the generation of muscle action potentials thus blocking muscle contraction (28, 29).

2. We prefer this solution (2 M potassium citrate and 10 mM potassium chloride) over 3 M potassium chloride because the latter often causes drifting of membrane potential, especially in small muscle fibers, such as the lumbrical muscles, due to the leakage of chloride channels (25).

3. The time required for the polymerization is dependent on the temperature (faster at higher temperature).

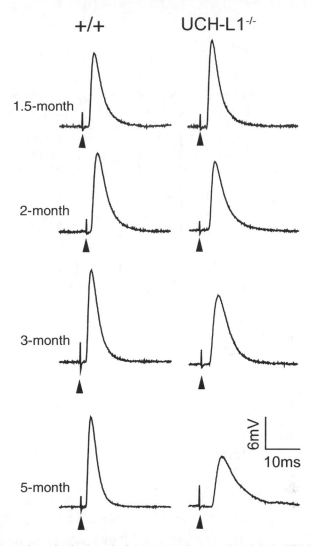

Fig. 2. Progressive reduction in evoked neuromuscular synaptic activity in *UCH-L1⁻/⁻* mice. Sample EPP traces from wild-type (WT, +/+) and *UCH-L1⁻/⁻* mice at 1.5-, 2-, 3-, and 5 months of age: EPP amplitudes are markedly decreased in *UCH-L1⁻/⁻* mice at 3- and 5 months of age, compared with age-matched WT mice. *Arrowheads* point to stimulus artifacts (modified from ref. 15).

4. Any capillary glass with outer diameters from 1.2 to 1.5 mm fits the electrode.

5. Fire polishing sometimes makes the tip completely close. Expose to the flame shorter time and observe the tip with a stereoscope.

6. Fine flexible plastic capillaries are made from yellow pipette tips. Expose a tip to flames and pull. It is also commercially available (MicroFil, World Precision Instruments Inc., Sarasota, FL, USA).

7. Glass micropipette can be prepared a day before an experiment and kept in a container.

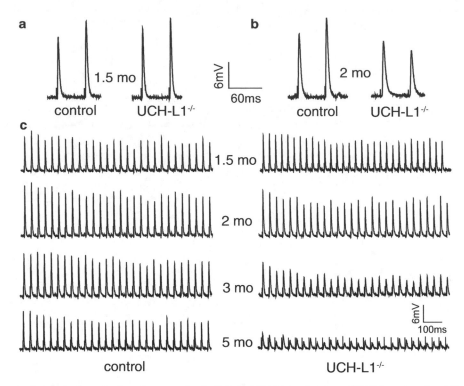

Fig. 3. Impaired short-term neuromuscular synaptic plasticity in *UCH-L1⁻ᐟ⁻* mice. (**a** and **b**): Twin-pulse stimulation induces paired-pulse facilitation (PPF) in WT mice at both 1.5- and 2 months of age. In *UCH-L1⁻ᐟ⁻* mice, twin-pulse stimulation induces PPF at 1.5 months (**a**), but paired-pulse depression 2 months of age (**b**). (**c**) Sample EPP traces in WT and *UCH-L1⁻ᐟ⁻* mice during repetitive nerve stimulation (30 Hz for 1 s) at 1.5-, 2-, 3-, and 5 months of age. Increased synaptic depression in *UCH-L1⁻ᐟ⁻* mice is observed at 2-, 3-, and 5 months of age, compared with age-matched WT mice (modified from ref. 15).

8. For example, use 2.5 ml Ringer's: add a 25 µl of 100 µM stock solution to make a final concentration of 1 µM.

9. In adult mouse muscles, the final concentration 1 µM of Mu-conotoxin is usually effective to prevent muscle contraction. Occasionally, we double the dose to 2 µM if muscle contraction persists after 30 min of incubation.

10. During the recording, the RP should remain stable.

11. Various stimulation protocols can be programmed in pClamp.

References

1. Stalberg, E., Schwartz, M. S., and Trontelj, J. V. (1975) Single fibre electromyography in various processes affecting the anterior horn cell, *J Neurol Sci 24*, 403–415.

2. Killian, J. M., Wilfong, A. A., Burnett, L., Appel, S. H., and Boland, D. (1994) Decremental motor responses to repetitive nerve stimulation in ALS, *Muscle Nerve 17*, 747–754.

3. Similowski, T., Attali, V., Bensimon, G., Salachas, F., Mehiri, S., Arnulf, I., Lacomblez, L., Zelter, M., Meininger, V., and Derenne, J. P. (2000) Diaphragmatic dysfunction and dyspnoea in amyotrophic lateral sclerosis, *Eur Respir J 15*, 332–337.

4. Fischer, L. R., Culver, D. G., Tennant, P., Davis, A. A., Wang, M., Castellano-Sanchez, A., Khan, J., Polak, M. A., and Glass, J. D.

(2004) Amyotrophic lateral sclerosis is a distal axonopathy: evidence in mice and man, *Exp Neurol 185*, 232–240.

5. Frey, D., Schneider, C., Xu, L., Borg, J., Spooren, W., and Caroni, P. (2000) Early and selective loss of neuromuscular synapse subtypes with low sprouting competence in motoneuron diseases, *J Neurosci 20*, 2534–2542.

6. Bruijn, L. I., Becher, M. W., Lee, M. K., Anderson, K. L., Jenkins, N. A., Copeland, N. G., Sisodia, S. S., Rothstein, J. D., Borchelt, D. R., Price, D. L., and Cleveland, D. W. (1997) ALS-linked SOD1 mutant G85R mediates damage to astrocytes and promotes rapidly progressive disease with SOD1-containing inclusions, *Neuron 18*, 327–338.

7. Bruijn, L. I., Houseweart, M. K., Kato, S., Anderson, K. L., Anderson, S. D., Ohama, E., Reaume, A. G., Scott, R. W., and Cleveland, D. W. (1998) Aggregation and motor neuron toxicity of an ALS-linked SOD1 mutant independent from wild type SOD1, *Science 281*, 1851–1854.

8. Kennel, P. F., Finiels, F., Revah, F., and Mallet, J. (1996) Neuromuscular function impairment is not caused by motor neurone loss in FALS mice: an electromyographic study, *Neuroreport 7*, 1427–1431.

9. Williamson, T. L., and Cleveland, D. W. (1999) Slowing of axonal transport is a very early event in the toxicity of ALS-linked SOD1 mutants to motor neurons, *Nat Neurosci 2*, 50–56.

10. Wong, F., Fan, L., Wells, S., Hartley, R., Mackenzie, F. E., Oyebode, O., Brown, R., Thomson, D., Coleman, M. P., Blanco, G., and Ribchester, R. R. (2009) Axonal and neuromuscular synaptic phenotypes in Wld(S), SOD1(G93A) and ostes mutant mice identified by fiber-optic confocal microendoscopy, *Mol Cell Neurosci 42*, 296–307.

11. Coleman, P. D., and Yao, P. J. (2003) Synaptic slaughter in Alzheimer's disease, *Neurobiol Aging 24*, 1023–1027.

12. Nelson, P. G. (2005) Activity-dependent synapse modulation and the pathogenesis of Alzheimer disease, *Curr Alzheimer Res 2*, 497–506.

13. Selkoe, D. J. (2002) Alzheimer's disease is a synaptic failure, *Science 298*, 789–791.

14. Raff, M. C., Whitmore, A. V., and Finn, J. T. (2002) Axonal self-destruction and neurodegeneration, *Science 296*, 868–871.

15. Chen, F., Sugiura, Y., Myers, K. G., Liu, Y., and Lin, W. (2010) Ubiquitin carboxyl-terminal hydrolase L1 is required for maintaining the structure and function of the neuromuscular junction, *Proc Natl Acad Sci USA 107*, 1636–1641.

16. McArdle, J. J., Angaut-Petit, D., Mallart, A., Bournaud, R., Faille, L., and Brigant, J. L. (1981) Advantages of the triangularis sterni muscle of the mouse for investigations of synaptic phenomena, *J Neurosci Methods 4*, 109–115.

17. Angaut-Petit, D., Molgo, J., Connold, A. L., and Faille, L. (1987) The levator auris longus muscle of the mouse: a convenient preparation for studies of short- and long-term presynaptic effects of drugs or toxins, *Neurosci Lett 82*, 83–88.

18. Carlsen, R. C., Larson, D. B., and Walsh, D. A. (1985) A fast-twitch oxidative-glycolytic muscle with a robust inward calcium current, *Can J Physiol Pharmacol 63*, 958–965.

19. Lomo, T., and Rosenthal, J. (1972) Control of ACh sensitivity by muscle activity in the rat, *J Physiol 221*, 493–513.

20. Balice-Gordon, R. J., and Thompson, W. J. (1988) The organization and development of compartmentalized innervation in rat extensor digitorum longus muscle, *J Physiol 398*, 211–231.

21. Balice-Gordon, R. J., and Thompson, W. J. (1988) Synaptic rearrangements and alterations in motor unit properties in neonatal rat extensor digitorum longus muscle, *J Physiol 398*, 191–210.

22. Ribchester, R. R. (2009) Mammalian neuromuscular junctions: modern tools to monitor synaptic form and function, *Curr Opin Pharmacol 9*, 297–305.

23. Aickin, C. C., Betz, W. J., and Harris, G. L. (1989) Intracellular chloride and the mechanism for its accumulation in rat lumbrical muscle, *J Physiol 411*, 437–455.

24. Clark, A. W., Bandyopadhyay, S., and DasGupta, B. R. (1987) The plantar nerves-lumbrical muscles: a useful nerve-muscle preparation for assaying the effects of botulinum neurotoxin, *J Neurosci Methods 19*, 285–295.

25. Harris, G. L., and Betz, W. J. (1987) Evidence for active chloride accumulation in normal and denervated rat lumbrical muscle, *J Gen Physiol 90*, 127–144.

26. Ross, J. J., Duxson, M. J., and Harris, A. J. (1987) Neural determination of muscle fibre numbers in embryonic rat lumbrical muscles, *Development 100*, 395–409.

27. Liley, A. W. (1956) An investigation of spontaneous activity at the neuromuscular junction of the rat, *J Physiol 132*, 650–666.

28. Cruz, L. J., Gray, W. R., Olivera, B. M., Zeikus, R. D., Kerr, L., Yoshikami, D., and Moczydlowski, E. (1985) Conus geographus toxins that discriminate between neuronal and muscle sodium channels, *J Biol Chem 260*, 9280–9288.

29. Hong, S. J., and Chang, C. C. (1989) Use of geographutoxin II (mu-conotoxin) for the study of neuromuscular transmission in mouse, *Br J Pharmacol 97*, 934–940.

Chapter 27

Determination of Neurotransmitter Levels in Models of Parkinson's Disease by HPLC-ECD

Lichuan Yang and M. Flint Beal

Abstract

Parkinson's disease (PD) is a neurological disorder caused by progressive degeneration of dopaminergic neurons in the nigrostriatal area of the brain. The decrease in dopamine (DA) neurotransmitter levels in the striatum and substantia nigra pars compacta is a neurochemistry hallmark of PD. Therefore, determination of dopamine and its metabolites levels in biological samples provides an important key to understanding the neurochemistry profile of PD. This chapter describes the use of reversed-phase HPLC with electrochemical detection (ECD) for simultaneously measuring monoamine neurotransmitters, including dopamine and its metabolites, norepinephrine as well as serotonin and its metabolite. ECD provides an ultrasensitive measurement, which detects at the picogram level. One run for each sample finishes within 18 min, shows clear chromatographic peaks and a complete separation, and produces excellent precision and reproducibility. Once set up, HPLC-ECD is economic and efficient for analyzing a large number of samples. This method has been broadly used for analyzing a variety of biological samples, such as cerebrospinal fluids, plasma, microdialysis elutes, tissues, and cultured cells. In recent days, it has been reported to be able to detect the dopamine level in a single drosophila head.

Key words: Parkinson's disease, Dopamine, 3,4-Dihydroxyphenylacetic acid, Homovanillic acid, Norepinephrine, 5-Hydroxytryptamine, 5-Hydroxyindole-3-acetic acid, HPLC, Electrochemical detection

---

1. Introduction

Parkinson's disease (PD) is a progressive neurodegenerative movement disorder clinically characterized by bradykinesia, rest tremor, and rigidity. The key pathological characteristic of PD is the loss of nigrostriatal dopaminergic neurons. The cell bodies of nigrostriatal neurons are in the substantia nigra pars compacta (SNpc), and project primarily to the dorsolateral putamen. In the progress of nigrostriatal degeneration, the terminal loss in the putamen appears to occur earlier and be more pronounced than the cell loss in SNpc,

Giovanni Manfredi and Hibiki Kawamata (eds.), *Neurodegeneration: Methods and Protocols,*
Methods in Molecular Biology, vol. 793, DOI 10.1007/978-1-61779-328-8_27, © Springer Science+Business Media, LLC 2011

resulting in a significant depletion of the dopamine (DA) neurotransmitter level in the putamen of PD brains. The onset of movement symptoms occurs when the loss of SNpc dopaminergic neurons reaches ~60% and the putamental dopamine is depleted ~80% (1). Current therapeutic strategies fail to block the death of dopaminergic neurons and only provide symptomatic relief by a dopamine replenishment treatment to restore putamental dopaminergic function. Considered as a neurochemistry hallmark of PD, dopamine neurotransmitter has been primarily targeted in studies for the degeneration progress in PD and also the development of therapeutic strategies for PD.

Dopamine is a catecholamine, a subclass of the monoamine neurotransmitters, which derives from amino acid tyrosine. In neurons, dopamine is packaged after synthesis into vesicles, which are then released into the synapse in response to a presynaptic action potential. Two major degradation pathways for inactivating dopamine exist. Mostly, the released dopamine can be taken up by the presynaptic dopamine transporter (DAT), back into the cytoplasm, where it is either repacked into storage vesicles or broken down by the enzyme monoamine oxidase (MAO) to 3,4-dihydroxyphenylacetic acid (DOPAC). The extracellular dopamine left in the synaptic gap can be methylated by the enzyme catecholamine-O-methyltransferase (COMT) to 3-methoxytyramine (3-MT), which can be taken back into the cytoplasm and further oxidized by MAO to homovanillic acid (HVA). DOPAC in the cytoplasm may leak out to the extracellular space and be methylated to HVA by COMT (2). Determination of dopamine and its metabolites DOPAC and HVA in biological samples provides an important key to understanding the PD neurochemical profile resulting from the pathogenic degeneration of dopaminergic neurons in PD patients or in a variety of PD research animal models made by means of either neurotoxins or manipulation of PD genes. The levels of the mediate metabolite 3-MT sometimes are considered to be a reliable marker of dopamine release.

A number of methodologies have been developed to measure monoamines. The early-used radioenzymatic assay is sensitive but very tedious with procedures of manipulating radioactive reagents and now has been less used (3). Gas chromatography/mass spectrometry (GCMS) has been applied to the measurement of monoamines, but the native monoamines need to be derivatized during the assay (4). High-performance liquid chromatography (HPLC) with fluorescence detection can simultaneously measure native catecholamines but provides relatively low sensitivity; to gain a higher sensitivity it will need to add a two-step derivatization procedure in the assay (5). Currently, the most frequently used method for catecholamine measurement is HPLC with electrochemical detection (ECD). Development of the EC detector provides an ultrasensitive analytic measure for detecting compounds which possess electroactive properties, reaching a sensitivity of picogram (pg) levels (6). HPLC-ECD offers a simple and economic analytic system, which is broadly used for analyzing monoamines and their metabolites in a variety of biological samples, including cerebrospinal fluids, plasma, microdialysis elutes, tissues, and cultured dopaminergic cells (7–11). Recently, we also applied this method to measuring the dopamine level in a single drosophila head (12).

This chapter provides a comprehensive protocol for measuring monoamine neurotransmitters in biological samples by using ion-paired reversed-phase HPLC with EC detection. Isocratic elution mode instead of a gradient mode has been applied to this analytic system because of the limitation of EC detector which is extremely sensitive to the changes in the composition of the mobile phase. Ion-pair reagent is used in this system for eliminating the positive charges of the amine groups in an acidic mobile phase to obtain a sufficient separation of monoamines on a reversed-phase octadecyl-silica (ODS) column. By following this protocol, we can simultaneously measure norepinephrine (NE), dopamine and its metabolites DOPAC and HVA, as well as serotonin (5-hydroxytryptamine, 5-HT) and its metabolite 5-hydroxyindole-3-acetic acid (5-HIAA) in less than 18 min from one injection of a tiny amount (15 μL) of the sample extract.

2. Materials

2.1. Analytical HPLC System

1. HPLC pump – Waters 515 HPLC pump (Waters, Milford, MA). To deliver a high-pressure, pulse-free solvent flow to accommodate the highly sensitive EC detector.

2. Autosampler – Waters 717 plus autosampler with sample cooling chamber. To make highly precise and reproducible injections of samples into the analytic system.

3. Coulchem II – ESA model 5200A (ESA, Inc. Chelmsford, MA). Two analytic channels plus a conditioning channel.

4. Analytical cell (dual coulometric electrodes, ESA) – Model 5011.

5. Conditioning cell (single coulometric electrode, ESA) – Model 5021.

6. Column heater (Eppendorf, CH-30) coupled with temperature controller (Eppendorf, TC-50).

7. Pulse damper – ESA PEEK pulse damper (ESA). It smoothes pulsations and maintains constant flow at a high pressure system for reducing baseline noise.

8. Software – ESA 501chromatography data system.

9. Column – HR-80 (RP – C18), 4.6×80 mm; 3 μm; 120A (ESA, Inc.).

10. Guard cartridge – C18, ODS (Phenomenex, Torrance, CA).

11. Filters – post pump and precolumn filters, 0.5-μm PEEK filter element (ESA, Inc.).

2.2. Chemicals and Reagents

1. Chemicals for HPLC standards: Nonbiological compound 2,3-dihydroxybenzoic acid (2,3-DHBA, Sigma-Aldrich, St. Louis, MO) which shows similar electrochemical response with monoamines in the EC detection is chosen as the internal standard. Synthesized monoamines DA, NE, 5-HT, and their metabolites DOPAC, HVA, and 5-HIAA (Sigma-Aldrich, St. Louis, MO) are used for external standards. For standards preparation: (a) A stock solution of internal standard 2,3-DHBA is prepared at a concentration of 1 mg/mL with 0.1 M perchloric acid (PCA, J. T. Baker, Phillipsburg NJ), and then divided to small aliquots in 1.5-mL microcentrifuge tubes, and kept in –80°C freezer for later use. It will be stable at –80°C for 12 months. (b) A diluent of 50 ng/mL 2,3-DHBA in 0.1 M PCA is prepared freshly from its 1 mg/mL stock each time before making the standards of monoamines and their metabolites. (c) Use the 50 ng/mL 2,3-DHBA diluent to make standard solutions of DA, DOPAC, HVA, NE, 3-MT, 5-HT, and 5-HIAA in a series of concentrations in a range from 10 ng/mL to 400 ng/mL so that each standard solution contains the same amount of internal standard 2,3-DHBA (50 ng/mL).

2. 4 M lithium dihydrogen-phosphate (LiH_2PO_4) acidic stock solution: (a) Lithium hydroxide (LiOH, SigmaUtra) and phosphoric acid (H_3PO_4, 85 wt.%, ACS grade) are purchased from Sigma-Aldrich. Chemical reaction is: $LiOH + H_3PO_4 \rightarrow LiH_2PO_4 + H_2O$. (b) Weigh 4 mol LiOH (MW: 42, 4 mol × 42 = 168 g) and put the 168 g LiOH in a 2-L glass beaker sitting on crushed ice. (c) Measure 5 mol H_3PO_4 (MW: 98; 85% w/w; d = 1.685, (5 mol × 98)/85%/1.685 = 342 mL) with a glass graduated cylinder. (d) Carefully pour the 342 mL H_3PO_4 little by little into the beaker containing 168 g LiOH and stir rigorously with a glass stirring stick. A lot of heat will be produced during the chemical reaction, perform the procedure in a lab hood (see Note 1). (e) After the completion of the chemical reaction, build the total volume to 1 L by adding deionized water and stir till the salt is completely dissolved. (f) Keep the stock solution in a glass bottle at room temperature.

3. HPLC mobile phase: 100 mM LiH_2PO_4, 1.5 mM 1-octanesulfonic acid (OSA, ion-pair reagent, SigmaUtra, Sigma-Aldrich), 10% (v/v) methanol (HPLC grade, J.T.Baker, Phillipsburg, NJ). To make 4 L mobile phase, take 100 mL of 4 M LiH_2PO_4 stock solution, 1.298 g OSA (FW 216.3) and 400 mL methanol, add

deionized water to a final volume of 4 L and stir thoroughly for 10 min. Before using for HPLC assay, mobile phase must be filtered and degassed by vacuuming through a 0.2-μm nylon membrane filter (Whatman, Maidstone, England).

4. Deionized water used for HPLC: Resistivity 18.2 MΩ-cm, prepared by using Milli-Q Direct Water Purification System (Millipore, Lincoln Park, NJ).

5. Bovine serum albumen (BSA): 98%, lyophilized powder (Sigma-Aldrich).

6. Bio-Rad Dc protein assay reagents: Reagent A, B, and S (Bio-Rad Lab, Hercules, CA).

7. Bicinchoninic acid (BCA) Protein Assay Reagent: Reagent A and B (Thermo Scientific, Rockford, IL).

2.3. Miscellaneous Equipments

1. Microcentrifuge – Microfuge 22R, refrigerated (Beckman Coulter, Brea, CA).

2. Sonicator – Sonifier ultrasonic cell disruptor/homogenizer (Branson Inc. Danbury, CT).

3. Bioassay Plate Reader – HTS 7000 Plus (Perkin Elmer, Waltham, MA).

3. Methods

3.1. Preparation of Samples

Biological samples for monoamine analysis have to be freshly harvested, swiftly frozen, and kept at temperature of –80°C to avoid enzymatic breakdown before the sample preparation.

Biological samples have to be thoroughly deproteinized before being analyzed with the HPLC assay. Among a variety of deproteinization protocols, PCA precipitation has been extensively used in many different sample preparation procedures, since not only does it have high efficiency in the removal of proteins, but it also functions to stabilize many small-molecule analytes, giving a high analytic recovery (13, 14). Furthermore, by using acidic phosphate buffer for the HPLC mobile phase, there is no need to remove the PCA from the extracted supernatants, which simplifies the deproteinizing procedure (14).

Do not subject samples to additional freeze–thaw cycles once they have been prepared readily for the HPLC assay (see Note 2).

3.1.1. Body Fluids

1. 100 μL of cerebrospinal fluid or plasma sample is placed in a 1.5-mL microcentrifuge tube on crushed ice.

2. Add 1.5 μL of ice-chilled 7 M PCA, making a final PCA concentration of 0.1 M, which is an ideal concentration to obtain a thorough precipitation of protein in the sample.

3. Cap and vortex immediately after the addition of PCA to avoid gelling of the protein in the sample (see Note 3).

4. Centrifuge at 14,000 rpm ($18,000 \times g$) for 15 min, 4°C.

5. Transfer supernatant into a clean 1.5-mL microcentrifuge tube and centrifuge again as above.

6. Take the supernatant for HPLC assay.

3.1.2. Biological Tissues

1. Freshly frozen tissue is placed in a 1.5-mL microcentrifuge tube on dry ice.

2. Add ice-chilled 0.1 M PCA containing internal standard into the tissue tube (~10 mg tissue in 300 μL PCA). Be sure that tissue is submerged in PCA and keep them frozen on dry ice (see Note 4).

3. Put the tubes on crushed ice or at 4°C, until the PCA thaws.

4. Sonicate briefly with a microprobe fitting in the sample tube (6–7 s, duty cycle 80%, output control 3) until the tissue is completely homogenized.

5. Keep the tubes on crushed ice or at 4°C for 10 min.

6. Vortex briefly and take 30 μL of homogenate to save for the protein assay.

7. Centrifuge at 14,000 rpm ($18,000 \times g$) for 15 min, 4°C.

8. Transfer the supernatant into another clean 1.5-mL microcentrifuge tube and centrifuge again as above.

9. Take the supernatant for HPLC assay.

3.1.3. Microdialysis Elutes

1. Microdialysis elutes are collected into sampling vials containing 2 μL of 0.5 M PCA to prevent dopamine degradation.

2. Collected dialysates can be directly applied to HPLC system for analysis without the need of a centrifuging procedure.

3.1.4. Drosophila Heads

1. Drosophilas are collected and frozen on dry ice or liquid nitrogen, and kept at −80°C.

2. Separate the drosophila head from its body on a cryostat plate and put each head in a microcentrifuge tube containing 60 μL of ice-chilled 0.1 M PCA.

3. Sonicate and centrifuge following the same procedures described in Subheading 3.1.2 steps 4–9, except just taking only 5 μL of homogenate and saving it for the protein assay.

3.2. HPLC Analysis

1. The system, in sequence, consists of: pump, pulse damper, autosampler, PEEK in-line filter, column, conditioning cell, analytical cells (see Fig. 1).

2. Prior to HPLC analysis, pass the fresh mobile phase through the system overnight in a recycling mode.

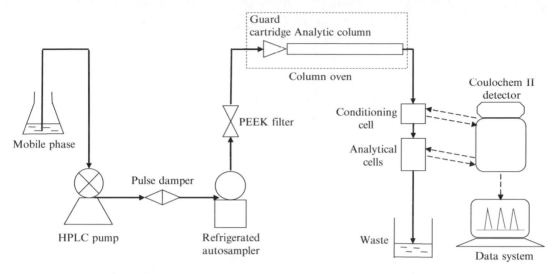

Fig. 1. Schematic diagram of an HPLC-ECD system setup. *Black lines* indicate the mobile phase route; *dotted lines* indicate electric wire connections.

3. Flow rate: 1.0 mL/min.

4. Column oven temperature: 30°C.

5. Injection volume: 15 µL.

6. Autosampler sample chamber temperature: 4°C.

7. Applied potentials (mV): Conditioning cell = +10; analytical cell, E1 = +50; E2 = +340 (see Note 5).

8. Once the system is set up and equilibrated, run the concentration series of the standard mixture of monoamines and their metabolites to be sure of obtaining a clear chromatographic separation (see Fig. 2a). Also, run the individual standard respectively to identify each peak in the mixture by its retention time and voltammetric response.

9. Readings of peak height and area are obtained by using the software of data analytical system.

10. Sensitivity should be determined by cutting off at the ratio of signal peak:noise peak >5.

11. Peak-area (or peak-height) ratios of monoamine or the metabolite: internal standard are used for plotting the standard curve of each monoamine and metabolite; and their least-squares linear regression equations are obtained by using the Microsoft Office Excel (see Note 6).

12. Concentrations of monoamines and their metabolites in the sample are calculated by using their standard equations.

3.3. Assay Validation

3.3.1. Method Performance

All the monoamines and their metabolites in samples are well eluted out in <18 min; and afterward the baseline is clear with no more interfering peaks eluted out (see Fig. 2b, c). The order and

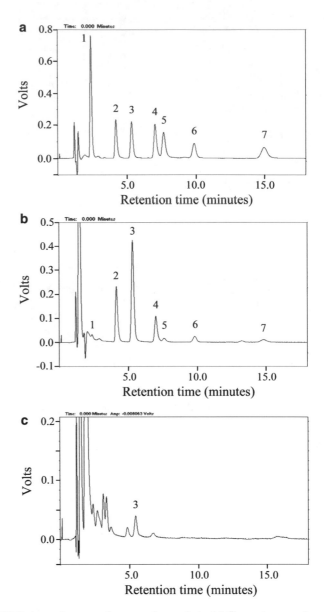

Fig. 2. HPLC chromatograms of monoamine analysis. (**a**) Chromatogram of monoamine standards, concentrations of each compound were 50 ng/mL and the injection volume was 15 µL. (**b**) Chromatogram of a sample extracted from mouse striatum, injection volume was 15 µL. (**c**) Chromatogram of a sample extracted from a single drosophila head, injection volume was 15 µL. Figure notes: 1, NE; 2, DOPAC; 3, DA; 4, 2,3-DHBA; 5, 5-HIAA; 6, HVA; 7, 5-HT.

retention time of the eluted monoamines and their metabolites in the chromatogram are: NE (2.38 min), DOPAC (4.17 min), DA (5.29 min), 2,3-DHBA (6.99 min), 5-HIAA (7.61 min), HVA (9.83 min), and 5-HT (14.9 min).

3.3.2. Recovery and Precision

1. The recovery test was carried out by adding known amounts of DOPAC, DA, 5-HIAA, HVA, and 5-HT to the homogenate from normal mouse striatal tissues.

2. Striatal tissues from nine to ten mice were sonicated briefly in chilled 0.1 M PCA containing 50 ng/mL 2,3-DHBA.

3. Pool sonicates together and vortex, then make aliquots by putting 90 μL to each microcentrifuge tube.

4. Prepare the 90 μL sample aliquots in four groups as the following:

   (a) Unspiked ($n=6$): Add 10 μL of 0.1 M PCA containing 50 ng/mL 2,3-DHBA.

   (b) Diluted ($n=6$): Dilute the unspiked sample to half of its concentration by adding 110 μL of 0.1 M PCA containing 50 ng/mL 2,3-DHBA.

   (c) Spiked ($n=6$): Fortify the sample with DOPAC (20 ng/mL), DA (100 ng/mL), 5-HIAA (20 ng/mL), HVA (20 ng/mL), and 5-HT (20 ng/mL) by adding 10 μL mixture of DOPAC (200 ng/mL), DA (1 μg/mL), 5-HIAA (200 ng/mL), HVA (200 ng/mL), and 5-HT (200 ng/mL).

   (d) Spiked ($n=6$): Fortify the sample with DOPAC (40 ng/mL), DA (200 ng/mL), 5-HIAA (40 ng/mL), HVA (40 ng/mL), and 5-HT (40 ng/mL) by adding 10 μL mixture of DOPAC (400 ng/mL), DA (2 μg/mL), 5-HIAA (400 ng/mL), HVA (400 ng/mL), and 5-HT (400 ng/mL).

5. Process the samples for HPLC assay by following the same procedures described in Subheading 3.1.2 steps 4–9.

6. Recoveries of monoamines and their metabolites were between the range of 96–105% and the precision (six repeats) was between the range of 2.0–3.6% (see Table 1 and Note 7).

3.3.3. Reproducibility and Accuracy

1. Unspiked and spiked samples prepared as described above were kept in a refrigerator at 4°C.

2. Samples were processed and measured by HPLC assay repeatedly in three successive days.

3. Inter-day (over 3 days) precisions of monoamines and their metabolites by this assay were 1.4–3.4% and their accuracies were 95–102% (see Table 2).

3.4. Protein Assay

Serial dilution for protein standard curve is prepared by using BSA.

3.4.1. Bio-Rad Protein Assay

1. This assay is used for measuring biological samples containing high protein levels, such as brain tissue homogenates.

Table 1
Recovery and precision

	Diluted samples (ng/mL, n=6) Mean±SD (CV %)	Unspiked samples (ng/mL, n=6) Mean±SD (CV %)	Spiked samples (n=6)			Spiked samples (n=6)		
			Fortified (ng/mL)	Mean±SD (CV %) (ng/mL)	Recovery (%)	Fortified (ng/mL)	Mean±SD (CV %) (ng/mL)	Recovery (%)
DA	95.4±1.96 (2.1)	188.3±6.9 (3.6)	100	290.4±8.98 (3.1)	102	200	390.4±9.69 (2.5)	98
DOPAC	9.5±0.21 (2.3)	18.4±0.43 (2.3)	20	39.1±1.03 (2.6)	103	40	60.4±1.43 (2.4)	105
HVA	13.4±0.42 (3.1)	28.6±0.71 (2.5)	20	47.9±1.20 (2.5)	97	40	68.6±1.53 (2.2)	100
5-HT	9.1±0.24 (2.7)	19.3±0.43 (2.2)	20	38.8±1.03 (2.7)	97	40	58.0±1.14 (2.0)	97
5-HIAA	6.2±0.15 (2.5)	12.6±0.43 (3.4)	20	31.8±1.02 (3.1)	96	40	50.9±1.17 (2.3)	96

Table 2
Reproducibility and accuracy

	Unspiked samples (n=18; 3 days) Mean±SD (CV %)	Spiked samples (n=18; 3 days)			Spiked samples (n=18; 3 days)		
		Fortified (ng/mL)	Mean±SD (CV %)	Accuracy (%)	Fortified (ng/mL)	Mean±SD (CV %)	Accuracy (%)
DA	187.5±5.92 (3.2)	100	287.2±7.59 (2.6)	99	200	383.9±7.50 (2.0)	98
DOPAC	18.2±0.57 (3.1)	20	38.4±0.98 (2.5)	100	40	59.2±1.66 (2.8)	102
HVA	28.0±0.73 (2.6)	20	47.5±1.33 (2.8)	97	40	68.0±1.50 (2.2)	99
5-HT	19.1±0.53 (2.8)	20	38.5±0.94 (2.4)	97	40	57.9±0.81 (1.4)	97
5-HIAA	12.3±0.42 (3.4)	20	31.7±0.93 (2.9)	97	40	50.4±1.48 (2.9)	95

2. Make the series of BSA standard solutions: 0.625, 1.25, 2.5, 5, and 10 µg/µL, dissolved in 0.1 M PCA.

3. Make the mixture of reagent A and S freshly before use, per 1,000 µL reagent A by adding 20 µL reagent S, gently shake to mix, do not vortex it (see Note 8).

4. Add 10 µL of standards and samples in the 96-well plate in duplicates.

5. Add 50 µL of reagent A and S mixture to each well.

6. Add 200 µL of reagent B and put the plates on the shaker for 15 min while keeping it covered from the light.

7. Prick air bubbles in the well by using a small injection needle (see Note 8).

8. Measure absorbance at a wavelength of 690 nm by using the plate reader.

3.4.2. BCA Protein Assay

1. This assay is used for measuring biological samples containing low protein levels, like drosophila heads and cell culture homogenates.

2. Make the series of BCA standard solutions: 0.05, 0.1, 0.2, 0.4, 0.8, and 1.6 µg/µL.

3. Take reagent A and B and make a mixture of A:B = 50:1.

4. Add 25 µL of standards and samples in the 96-well plate in duplicates.

5. Add 200 µL of A and B mixture in each well and mix very well.

6. Cover the plate and seal it with parafilm, then incubate it at 37°C for 30 min followed by keeping it at room temperature for 10 min.

7. Measure absorbance at a wavelength of 570 nm by using the plate reader.

3.5. Results

Monoamine neurotransmitters and their metabolites in the striatum of normal C57 black mice are measured by using this method (see Table 3).

4. Notes

1. Be especially careful when making the 4 M LiH_2PO_4 stock solution. LiOH is a caustic base and packed as granulated crystal powder; wear goggles and a dust mask during weighing it. Phosphoric acid is a concentrated acid which is very corrosive, be careful for not spilling it outside the container when measuring it.

Table 3
Bioamine levels in the mouse striatum (ng/mg protein)

	Dopamine	DOPAC	HVA	5-HT	5-HIAA
Mouse 1	109.2	8.2	15.5	32.0	10.1
Mouse 2	127.4	10.0	18.5	40.3	12.9
Mouse 3	110.1	7.6	12.7	36.5	8.6
Mouse 4	108.3	8.1	13.9	32.5	8.3
Mouse 5	115.0	8.9	14.2	33.0	8.1
Mouse 6	105.2	7.9	13.2	34.3	8.9
Mouse 7	125.3	11.2	17.0	36.9	9.2
Mouse 8	125.6	10.8	17.0	35.3	11.4
Mouse 9	122.7	10.2	16.7	34.4	10.5

The heat from the reaction of H_3PO_4 and LiOH produces bubbling and hot steam, be sure that this procedure is carried out in a hood and that the beaker is seated on crushed ice. Use glassware for all the procedures. Wear gloves and goggles.

2. Dopamine and its metabolites DOPAC and HVA are stable in 0.1 M PCA for about 1 week when kept at 4°C. Therefore, save the prepared samples and standards left from HPLC loading temporally in refrigerator just in case that if any interruptions occur during the HPLC assay these saved samples can be reloaded. Do not freeze and rethaw these samples.

3. After the addition of concentrated PCA, proteins, lipids, and sugars contained in the plasma sample easily form a chunk of gel-like clot, which is difficult to be spun down. An immediate vortex following the addition of PCA will avoid the formation of gelling.

4. Sample preparation in PCA is carried out at 4°C or on crushed ice. To avoid enzymatic breakdown, biological tissues should always be kept frozen till they are homogenized in PCA, which inactivates enzyme activities.

5. Chromatography recorded in channel 2 (340 mV) is used for quantification. NE, DA, and DOPAC also show lower voltammetric responses in channel 1 (50 mV). Therefore, if a sample contains high concentrations of these compounds and these peaks are cut off in channel 2 by the limit of the recording scale, these peaks recorded in channel 1 can be used for quantification.

6. Generally, peak area is chosen for data calculation. Peak height can also be used for calculation when the chromatographic

peak is significantly high and sharp; under this circumstance, results from peak height and peak area are almost identical.

7. Due to the simplicity of extraction procedures, the recoveries of monoamines and their metabolites reach close to 100% in our validation studies. In this case, the use of internal standard is not a necessity. The calculation can be done by a direct calibration comparing to concentrations of external standards.

8. When using Bio-Rad protein assay, make the mixture of reagent A and S freshly, just before the assay; it turns cloudy when staying long at room temperature. Handshake gently after adding reagent S into reagent A; vortex is not recommended since it creates foams floating on top of the mixture solution. Make sure that there are no air bubbles in the wells before putting the plate in the reader; otherwise, bubbles will interfere in the absorbance, giving incorrect high readings.

Acknowledgments

We would like to acknowledge the support of NIA NS39258 and the Parkinson's Disease Foundation.

References

1. Dauer, W. & Przedborski, S. (2003) Parkinson's disease: mechanisms and models. *Neuron* **39**, 889–909.

2. O'Neill, R. D. (2005) Long-term monitoring of brain dopamine metabolism in vivo with carbon paste electrodes. *Sensors* **5**, 317–342.

3. Peuler, J. D. & Johnson, G. A. (1977) Simultaneous single isotope radioenzymatic assay of plasma norepinephrine, epinephrine and dopamine. *Life Sci* **21**, 625–36.

4. Commissiong, J. W. (1983) Mass fragmentographic analysis of monoamine metabolites in the spinal cord of rat after the administration of morphine. *J Neurochem* **41**, 1313–8.

5. Yoshitake, T., Kehr, J., Todoroki, K., Nohta, H. & Yamaguchi, M. (2006) Derivatization chemistries for determination of serotonin, norepinephrine and dopamine in brain microdialysis samples by liquid chromatography with fluorescence detection. *Biomed Chromatogr* **20**, 267–81.

6. Wei, D., Bailey, M. J., Andrew, P. & Ryhanen, T. (2009) Electrochemical biosensors at the nanoscale. *Lab Chip* **9**, 2123–31.

7. Kalita, J., Kumar, S., Vijaykumar, K., Palit, G. & Misra, U. K. (2007) A study of CSF catecholamine and its metabolites in acute and convalescent period of encephalitis. *J Neurol Sci* **252**, 62–6.

8. Mefford, I. N., Ward, M. M., Miles, L., Taylor, B., Chesney, M. A., Keegan, D. L. & Barchas, J. D. (1981) Determination of plasma catecholamines and free 3,4-dihydroxyphenylacetic acid in continuously collected human plasma by high performance liquid chromatography with electrochemical detection. *Life Sci* **28**, 477–83.

9. Middlemiss, D. N. & Hutson, P. H. (1990) Measurement of the in vitro release of endogenous monoamine neurotransmitters as a means of identification of prejunctional receptors. *J Neurosci Methods* **34**, 23–8.

10. Yang, L., Calingasan, N. Y., Chen, J., Ley, J. J., Becker, D. A. & Beal, M. F. (2005) A novel azulenyl nitrone antioxidant protects against MPTP and 3-nitropropionic acid neurotoxicities. *Exp Neurol* **191**, 86–93.

11. Roy, N. S., Cleren, C., Singh, S. K., Yang, L., Beal, M. F. & Goldman, S. A. (2006) Functional engraftment of human ES cell-derived dopaminergic neurons enriched by coculture with telomerase-immortalized midbrain astrocytes. *Nat Med* **12**, 1259–68.

12. Yang, Y., Ouyang, Y., Yang, L., Beal, M. F., McQuibban, A., Vogel, H. & Lu, B. (2008) Pink1 regulates mitochondrial dynamics through interaction with the fission/fusion machinery. *Proc Natl Acad Sci USA* **105**, 7070–5.

13. Koshiishi, I., Mamura, Y., Liu, J. & Imanari, T. (1998) Evaluation of an acidic deproteinization for the measurement of ascorbate and dehydroascorbate in plasma samples. *Clin Chem* **44**, 863–8.

14. Sakuma, R., Nishina, T. & Kitamura, M. (1987) Deproteinizing methods evaluated for determination of uric acid in serum by reversed-phase liquid chromatography with ultraviolet detection. *Clin Chem* **33**, 1427–30.

Chapter 28

Designing, Performing, and Interpreting a Microarray-Based Gene Expression Study

Giovanni Coppola

Abstract

Microarray-based assays have significantly expanded their scope and range of applications over the last 10 years, and – at least for gene expression – can be considered mainstream applications. High-throughput, microarray-based gene expression studies have proven particularly useful in the study of neurodegenerative diseases, for which they have provided key insights in understanding disease pathogenesis, regional and cellular specificity, and identification of therapeutic targets. Even though many experimental steps are currently performed in specialized core facilities, the key steps of a microarray study – experimental design, and data analysis and interpretation – are performed by the primary investigator. Knowledge of the issues related to these key steps is essential to properly perform and interpret a microarray experiment and constitutes the main focus of the present chapter. The basic analytical steps are covered, and annotated R code for the analysis of a published dataset is provided.

Key words: Microarray, Gene expression, Data analysis, Review, Bioconductor, Normalization, Differential expression, Pathway, Gene ontology

1. Introduction

Only a few years ago, a microarray study was a major undertaking, requiring considerable time and resources to be performed and analyzed, and often presented alone in scientific publications. Currently, 15 years after their introduction to the larger scientific community (1), improved technology and relative standardization of data analysis pipelines (2, 3) allow microarrays to be part of the experimental routine, and many published papers now include a microarray-based section, usually either (1) as a starting screen, performed to identify targets that are then analyzed in detail with other methods; (2) as experimental evidence offered to support

Giovanni Manfredi and Hibiki Kawamata (eds.), *Neurodegeneration: Methods and Protocols*,
Methods in Molecular Biology, vol. 793, DOI 10.1007/978-1-61779-328-8_28, © Springer Science+Business Media, LLC 2011

the scientific hypothesis presented in the paper; or (3) as a way to address the biological question at hand as a whole entity, from a systems biology perspective. In particular, the study of neurode-generative diseases has employed microarray-based methods to address (both in patients' tissues and animal and cellular models) the most challenging aspects of these disorders, including striking regional susceptibility, genetic and clinical heterogeneity, and lack of therapeutic targets.

The Gene Expression Omnibus (GEO, http://www.ncbi.nlm.nih.gov/geo/ (4)) is one of the public repositories collecting microarray data and making them available to the scientific community. GEO currently includes ~550,000 samples (see Fig. 1, accessed August 2010), with more than 10,000 samples submitted per month, further indicating that microarrays have become part of the standard experimental practice.

Most microarray-based gene expression studies are aimed at comparing two or more conditions, functional states, or treatments, in order to improve the understanding of a functional process, disease pathogenesis, or drug treatment. The analysis of these studies usually involves the assessment of differential expression by comparing samples across conditions. More computationally advanced algorithms (i.e., biomarker discovery, time-series analysis,

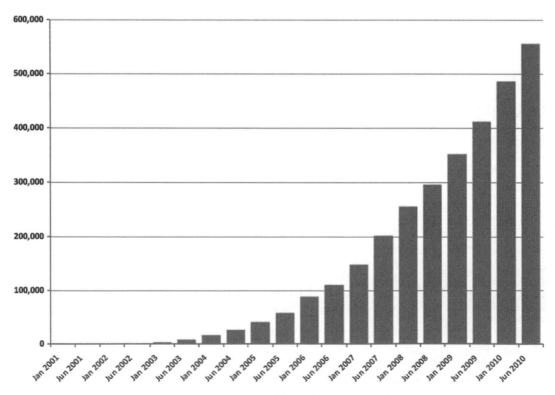

Fig. 1. Number of datasets submitted to the Gene Expression Omnibus (GEO), 1999–2010.

and classification algorithms), are used to identify biomarkers associated with disease status in peripheral, easily accessible tissues, or to identify biomarkers of drug effect, in order to track them in clinical trials.

Although in many cases core facilities perform most of a microarray experiment, the three major experimental steps of a gene expression study – experimental design, data analysis, and interpretation – are essentially always performed by the primary investigator and are the focus of this chapter. After reading this chapter, the reader should have a general understanding of the procedures involved in a microarray experiment, the analytical framework, and the main issues that are encountered in the everyday practice. Supplemental material includes annotated R code from a real-life experiment, designed to characterize the expression changes occurring in cultured neurons after oxidative stress, and the effect of the deletion of the *ATF4* gene (5). By reviewing and possibly rerunning this code, the reader will appreciate some of the practical issues and technical challenges of current microarray data analysis.

Although a number of platforms and methods are available, we chose to focus on a specific platform and RNA amplification method, in order to present a detailed and reproducible protocol. Although the protocol presented here is specific for the Illumina platform, the basic steps are also represented in similar protocols for other platforms. Similarly, an ever-growing number of applications are available for data mining, gene annotation, and pathway analysis. We focus on a few and present a more extended list in a table, leaving to the reader the choice of the most appropriate tool for his/her needs.

Major experimental steps of a microarray-based gene expression analysis experiment include:

1. Experimental design
2. RNA extraction and quality check
3. RNA amplification and labeling
4. Microarray hybridization and scanning
5. Data analysis
   (a) Quality control and data filtering
   (b) Normalization
   (c) Differential expression analysis
   (d) Advanced analysis (e.g., Weighted Gene Co-expression Network Analysis, WGCNA)
6. Data mining and pathway analysis
7. Confirmation of results

In this chapter, we go over the general principles of all these steps, and provide a detailed description of the analytical process, including computer code and sample dataset. Finally, we list the most common sources of problems and propose possible trouble-shooting based on the literature in the field and our own experience.

2. Materials

2.1. Experimental Design

Animals, tissues, cell lines, and treatments (see Note 1).

2.2. RNA Extraction and Quality Check

1. RNA extraction: Multiple kits are available. One point to keep in mind in the choice of an RNA extraction kit is whether small RNAs and microRNAs (miRNAs) are retained, as many kits exclude them from the total RNA fraction during the cleaning process. Therefore, consider the use of miRNA-retaining kits if a downstream miRNA analysis is a possibility.

2. RNA Quality check:

   (a) Agilent Bioanalyzer (http://www.chem.agilent.com/en-US/products/instruments/lab-on-a-chip/2100bioanalyzer/pages/default.aspx).

   (b) RNA 6000 Nano chips for Agilent Bioanalyzer (Agilent). RNA 6000 Pico chips are used for small (in the picogram range) RNA amounts.

2.3. RNA Quantification, Amplification, and Labeling

1. RNA quantification: Thermo Scientific NanoDrop 2000 (http://www.nanodrop.com/Productnd2000overview.aspx).

2. Amplification and labeling:
   Illumina TotalPrep RNA Amplification and biotin labeling kit (Ambion). 50–500 ng total input RNA are recommended for this kit.

   For small starting amounts, a stronger RNA amplification is needed. Two possible choices are:

   (a) NuGEN Ovation RNA Amplification System V2 for total RNA starting amounts between 5 and 50 ng (http://www.nugeninc.com/nugen/index.cfm/products/amplification-systems/ovation-amp-v2/).

   (b) Epicentre TargetAmp™ 2-Round Biotin-aRNA Amplification Kit 3.0 for total RNA starting amounts between 5 and 500 pg (http://www.epibio.com/item.asp?id=517).

2.4. Microarray Hybridization and Scanning

Illumina hybridization and scanning system: iScan (http://www.illumina.com/systems/iscan.ilmn, usually part of core facilities).

2.5. Data Analysis

1. *Computer.* Most average-sized experiments can be analyzed on a modern desktop computer, although advanced analyses requiring computational power are often performed on central servers to handle large datasets with speed.

2. *Software*

   (a) R installation (http://cran.r-project.org/ (6)).

   (b) Bioconductor packages (http://www.bioconductor.org/ (7)).

   (c) Example analysis: Download the supplementary zipped file at http://149.142.212.78/giovanni/MMB/.

   (d) Software and R packages for WGCNA (http://www.genetics.ucla.edu/labs/horvath/CoexpressionNetwork).

2.6. Data Mining Tools (Table 1)

1. Single-gene analysis (Gene Cards).

2. Gene ontology analysis (DAVID, WebGestalt).

3. Pathway analysis (Ingenuity, BioCarta, KEGG).

4. Literature analysis (ChiliBot, PubMatrix, iHOP).

5. Transcription factor analysis (TRANSFAC, MEME).

6. Other tools & databases (BLAT, GEO, ArrayExpress, Celsius, GeneNetwork, GATHER).

2.7. Data Confirmation

Data confirmation can be performed with a variety of methods (real-time PCR, comparison with other datasets, *in situ* hybridization, western blot, functional studies) which are not described in detail in the present chapter.

3. Methods

3.1. Experimental Design

A good experimental design is key for the success of any laboratory experiment, including a microarray study (see Note 1). The availability of probes querying the expression of 20–40,000 transcripts allows microarrays to effectively provide a snapshot of the transcriptional landscape of a cell, tissue, or organ at a given time. In most cases, a high-throughput gene expression experiment is performed to identify differences in gene expression among different conditions, which can include different genotypes, treatments, tissues, or cell types. Other applications include classification and identification of biomarkers related to functional or disease status. The goal of this step is to plan a balanced study and maximize the chances of getting reliable differential expression data. A few issues arise frequently during the planning phase and are addressed below.

Should I use a single-color or a two-color design? In single-color designs, each sample is labeled with one dye and hybridized alone (single color) onto one array. After normalization of the entire set of arrays (inter-array normalization), samples from different conditions are compared. In two-color designs, two samples (labeled with different dyes) are hybridized on the same slide, and therefore directly compared after being normalized to each other (intra-array normalization). This allows the best possible control for technical factors related to array hybridization, washing and scanning, but limits the comparisons that can be reliably performed across samples on different arrays. Typically, the only technical factor that is different between two samples hybridized on the same slide – i.e. the labeling dye – is corrected for by hybridizing two-color arrays in duplicate with dye swap. Many platforms (e.g., Affymetrix, Illumina) only allow single-color designs, whereas others (e.g., Agilent) allow both single- and two-color designs. In summary, if the experimental design includes multiple comparisons across samples, a single-color design is recommended.

Which platform should I use? Many platforms are available and they are for the most part functionally equivalent, but sometimes the experimental design can dictate the platform to choose. For example, (1) if there is the need to compare the current study to previous studies performed by the same group and on the same experimental model, it is better to use the same microarray platform used previously; (2) if the goal is to compare paired conditions, a dual-color platform is recommended. Sometimes, for less common experimental models (e.g., *S. cerevisiae*, *D. melanogaster*, *C. elegans*), the platform choice is limited to fewer options.

How many replicates are needed? Thanks to remarkable improvements in array technology, technical replicates (i.e., the same amplified RNA hybridized multiple times on distinct arrays) are no longer needed; however, biological replicates (i.e., separate biological replicates of the same condition, obtained and processed independently) are still strongly recommended, in order to estimate measurement variability and increase confidence in the findings. As for any other experiment, the magnitude of the effect one is trying to measure determines the number of biological replicates that are needed (8), and a minimum number of three replicates is generally recommended for most experiments with good-quality RNA and standard amplification/labeling methods. Double-amplified samples pose additional problems since heavily amplified RNA can show a skewed representation of the original transcript distribution (9, 10), therefore some researchers consider the possibility of double-amplifying the same RNA sample multiple times, in an attempt to estimate the bias introduced by the amplification process. This is conceptually an intermediate level between technical and biological replication. A higher number of replicates (5 or more) is needed for noisier conditions, such as low-quality samples, double-amplified RNA, and highly variable conditions (e.g., FACS-sorted cells, etc.). In general, the number of replicates should be higher if the difference is predicted to be small, or the noise level across samples high. All the conditions in the experimental design should be represented and replicated. For example, in a time-series experiment it is better to compare experimental and control conditions at each time point, rather than comparing all the conditions to the same baseline measurement (see below).

Can I pool multiple RNA samples and run them on a single array? Unless RNA quantity limitations make this the only option available, it is generally not recommended because variability in sample quality can influence the overall picture in a way that cannot be controlled for. In other words, while it is possible to exclude technical outliers if they are hybridized on individual arrays, it is challenging to tease apart technical and biological variability if a poor-quality sample is pooled with others and hybridized on the same array.

3.2. RNA Extraction and Quality Check

The success of a microarray experiment is heavily based on the quality of the starting RNA (see Note 2). Standard procedures for RNA handling are recommended to avoid degradation. Quality and quantity of starting RNA need to be carefully examined before proceeding to RNA amplification and labeling. The Agilent Bioanalyzer is a standard method for assessing RNA quality, although a simple standard gel electrophoresis can be used when large amounts of RNA are available. When total RNA is run on a gel, two peaks corresponding to 18S and 28S ribosomal RNAs are evident (see Fig. 2), as well as a signal in the small molecular weight area, corresponding to small RNAs. Based on this and other parameters, the Agilent software computes an RNA integrity number (RIN) which is used as an indicator of RNA quality. RINs > 9 are considered excellent and >7 acceptable quality for microarray studies (see Fig. 2), although we have observed exceptions.

3.3. RNA Quantification, Amplification, and Labeling

An accurate RNA quantification is crucial in order to select the appropriate amplification/labeling kit (see Note 3). A good approximation for concentrations in the ng/μl range and higher is obtained using the Nanodrop spectrophotometer (NanoDrop Technologies, Rockland, DE) although accurate RNA quantification is usually performed using the RiboGreen RNA quantitation assay and kit ((11) Molecular Probes Inc.). All RNA preparation methods for microarray hybridization (including reverse transcription) involve some degree of amplification. A total RNA amount of 50–500 ng in a maximum of 11 μl is needed for the Ambion TotalPrep kit, which is used in the present protocol. Lower amounts require alternative protocols, such as the Epicentre and the NuGEN kits (see Subheading 2), both constituting viable alternatives, as both have protocols compatible with Illumina arrays and can

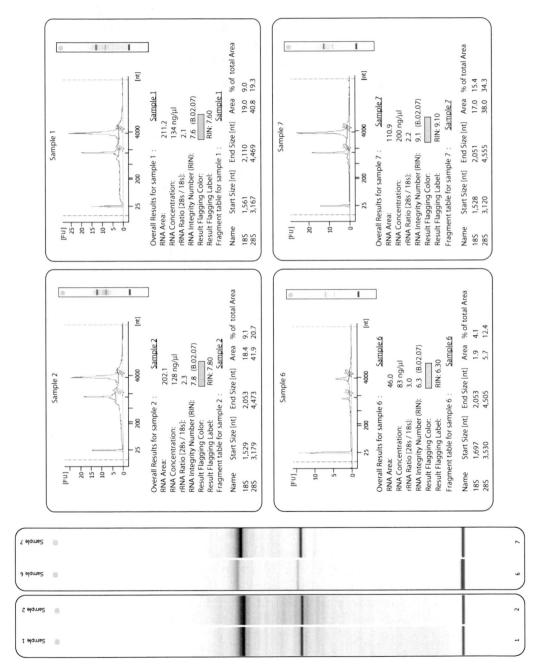

Fig. 2. Agilent Bioanalyzer output. Poor quality (RNA Integrity Number, RIN = 6.3, Sample 6); good quality (RIN = 9.1, Sample 7) and intermediate quality (RIN = 7.6–7.8, Samples 1–2).

be used with starting RNA amounts in the pg range. Optionally, it is possible to run amplified RNA on an Agilent Bioanalyzer for quality check, and this is recommended after double-amplification protocols in order to check for consistency of fragment length and amplification across samples.

3.4. Microarray Hybridization and Scanning

These steps are usually performed in specialized core facilities, and their detailed description is beyond the scope of the present protocol. Briefly, in the case of an Illumina experiment, amplified RNA is biotin labeled and hybridized onto bead arrays (see below). Nonhybridized biotinylated RNA is then removed in the washing step. After hybridization and washing, a laser scanner is used to detect the signal from biotinylated RNA bound to arrays. The raw output of a microarray scanner usually consists of a list of intensity values, in addition to some quality measures based on control probes (see Note 4).

3.5. Data Analysis

Upon scanning, it is possible to perform initial data transformation using the Illumina GenomeStudio software. For the purposes of the protocol presented here, the raw data is not background-subtracted or normalized by the GenomeStudio software.

Illumina bead arrays query each transcript using a set of beads (50 on average) randomly distributed across the slide and decoded using a unique address (12). Each bead has covalently linked multiple copies of the same 50-mer probe, which usually maps toward the 3′ untranslated region (UTR) of the target transcript. GenomeStudio summarizes the data across a bead-set and outputs an average signal, a signal standard deviation, and the number of beads comprising each bead-set.

Typically, the GenomeStudio output file has a 7-line header describing the hybridization data and the normalization and background subtraction methods used (if any). Starting with line 8, all the probe signals are reported (one per line) in the same format, in a .csv or .txt delimited format. Samples are reported in columns. Therefore, if all the arrays were hybridized and scanned at the same time, a single .txt file will contain the complete experimental dataset. The column format can vary based on the GenomeStudio settings, but these four basic fields are present:

(1) probe name; (2) average bead-set probe intensity; (3) bead-set detection score; and (4) bead-set standard deviation.

Goal of the analysis is (1) to use these parameters to assess the overall array quality; and (2) to use the probe intensities to compute average expression levels for each transcript and compare them across conditions. Several commercial data analysis software packages are available, many of them with intuitive graphical interfaces and good technical support. Another option, with a steeper learning curve but with potential for more flexibility, is to use scripts in R, Matlab or other programming language. A large number of scripts are available for the R interpreter and are regularly maintained (Bioconductor project, http://www.bioconductor. org/ (7)). In this protocol, R and Bioconductor scripts are used and provided in the accompanying code. The numbers in parentheses (c**) are referred to the corresponding page within the code document (see also Note 5).

3.5.1. Data Quality Check and Filtering

A data quality control is usually the first step in any comprehensive analytical strategy. After reading data within the R environment and log-2 transformation (which is typically performed in order to obtain normally distributed intensity values), the first analytical step is usually data quality check. Most of the quality indicators are relative to the experiment being analyzed (i.e., each array is considered relative to the other arrays within the current experiment). We estimate the overall data quality by looking at the following parameters: (1) log2 signal distribution (c4), (2) detection score distribution (c5); and (3) array clustering and outlier detection based on average Pearson correlation, median absolute deviation (MAD, (13), c9).

There is no perfect rule on how to exclude outlier or poorly performing arrays, as the signal variability is dependent on RNA variability, and ultimately on the experimental design. Often, simple data visualization provides enough information to exclude outliers, but sometimes it is necessary to perform the analysis with and without dubious arrays to estimate the effect of keeping or excluding one or more samples from the analysis. One possible rule of thumb is to consider for exclusion arrays for which the average Pearson correlation with all the other arrays (when measured over all probe sets) is two or more SD from the average. Clustering arrays using all the probes on the array is usually helpful in identifying technical issues (batch effect, outliers, etc.). Selecting probes before clustering by variance or differential expression can lead to clustering based more on biological effects.

3.5.2. Normalization

The normalization step is aimed at removing systematic biases across arrays, therefore allowing comparisons between samples and conditions. A basic assumption in microarray analysis is that most of the genes present on the array and expressed in a given tissue are not differentially expressed in the conditions under which they are being compared. While this assumption is maintained in most experimental designs, there are some cases for which it needs to be verified: for example, when comparing two very different tissues, or when comparing arrays with a small number of probes (bouquet arrays) which might be selected based on particular pathways. After considering these exceptions, we can assume that most genes are not differentially expressed and therefore normalize our data. A number of normalization methods have been developed and published, some of them platform-specific. Consideration of the relative merits of different normalization methods is beyond the scope of this chapter and reviewed elsewhere (14). We typically apply a quantile normalization (15) to Illumina data (c8).

3.5.3. Analysis of Differential Expression

In this step, we compare conditions according to our experimental design. Some researchers prefer to filter the genes below a certain threshold in order to consider only the "present" genes in the analysis of differential expression (16). Although this is a reasonable strategy – which also reduces the number of multiple comparisons – it is sometimes difficult to establish a threshold to call a gene "present," and all the probes on the array will be considered for the purposes of the present analysis. Analysis of differential expression is performed here by fitting a linear model and then applying a Bayesian measure of differentially expression (LIMMA package, (17)). Briefly, the difference between this method and a standard t-test lies in the fact that this method takes into account – for each probe – the overall signal variability for probes of similar expression level. One needs to take into account this "neighborhood" signal variability because low-signal probes are typically more variable than high-signal probes, and therefore different confidence measures are needed for different signal neighborhoods.

Correction for multiple comparisons. This is a common issue in high-throughput data analysis, and is related to the concept itself of statistical analysis. A p-value is estimating the probability that the observed result is part of the normal variability (e.g., a p-value of 0.001 estimates this possibility in 1 out of 1,000 comparisons). Since modern microarray platforms contain 20–40,000 features, under the conservative assumption that they are independent, if we perform a statistical test 40,000 times, at $p < 0.001$ we expect to see 40 instances where the result is significant but not true because of the level of significance we have set. The correction for multiple comparison aims at taking into account the number of tests that are being performed. A popular method is to estimate the false discovery rate (FDR, (18)). A Bayesian version of this test (17) takes into account the fact that gene expression probes are not truly independent. The q-value obtained by this test is functionally analogous to the uncorrected p-value, and estimates the falsely discovered genes. For example, a gene list selected for having an FDR q-value of 0.05 or lower means that, in that list, 5% of the genes are expected to be falsely discovered. As mentioned, reducing the initial set of probes considered in the differential expression analysis is another strategy to allow for a lower number of statistical tests performed, and less stringent corrected p-values. Finally, another approach is to predetermine the list of genes to be examined (by pathway or gene ontology) and estimate the overall significance of a particular gene-set (19). In general, there is no absolute p-value threshold for differential expression analysis, and other factors, such as the level of confirmation with alternative techniques, consistency across replicates, magnitude of change, etc., need also to be considered. Often, less stringent thresholds (e.g., higher p-value cutoff, or filtering based on fold change rather than on p-value) are used for exploratory analyses, pathway analyses, or comparisons with other datasets.

Data visualization. Data visualization is an important step of the analysis. A heatmap with a cluster dendrogram is a standard way to represent differentially expressed genes. Samples and genes are clustered by similarity and color-coded in two main ways: (1) by scaling across absolute values and representing each sample separately; and (2) by representing ratios deriving from comparisons between samples (see Fig. 3). In the former, shades of a single color (e.g., blue) are used to represent absolute expression levels; in the latter, shades of red and green are used to represent upregulated and downregulated probes, respectively (c18). Expression patterns of single genes across conditions can be visualized using conventional scatterplots or smaller heatmaps.

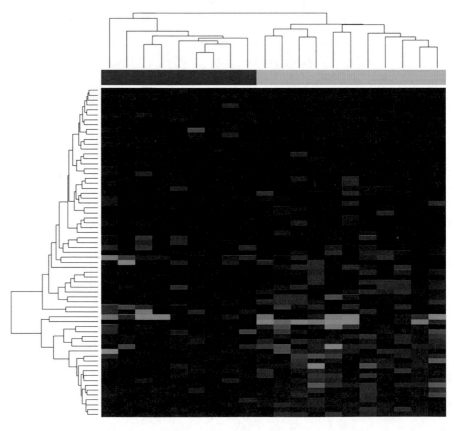

Fig. 3. Representative heatmap with cluster dendrogram. *Genes* are in rows and *samples* are in columns. *Shades of red* indicate upregulation compared to the control condition, *shades of green* indicate downregulation. Samples and genes are clustered by similarity (represented by the dendrogram).

3.5.4. Advanced Analytical Methods (WGCNA)

Conventional gene expression methods only capture part of the complexity of the transcriptome, by looking for absolute changes in expression levels. However, genes are usually coexpressed in functional families and this transcriptional organization can be lost or altered in the absence of a difference in absolute gene expression. In other words, two genes can differ in coexpression across conditions, without being differentially expressed. Gene coexpression methods aim at identifying the transcriptional correlations between families of genes (modules) and their alterations in experimental conditions. WGCNA has proven effective in identifying modules of coexpressed genes that, based on the complexity of the underlying transcriptome, have been linked to evolutionary species differences (20, 21), brain regions (22), cellular and neuronal subtypes (23, 24), and biological pathways (25). An extensive description of WGCNA methods (including tutorial, code, and R packages) is available on the web (http://www.genetics.ucla.edu/labs/horvath/CoexpressionNetwork/).

3.6. Data Mining

Although it is considered by many the final result of a microarray experiment, a list of differentially expressed probes is actually the beginning of a data mining process, which can be laborious and time consuming, but also very rewarding (see Note 6). This step aims at putting in a biological context the individual genes identified in the differential expression analysis. As mentioned, a large number of web sites and commercial packages are available to perform these analyses and we focus here on a few (see Table 1). In general, basic steps of data mining include a gene ontology (GO) analysis (e.g., DAVID) to identify overrepresented gene categories in our list of differentially expressed genes, and a pathway analysis (KEGG pathway database or the commercial software Ingenuity). Additional analyses include assessing overrepresentation of transcription factor-binding sites, microRNA targets, or overrepresentation of specific epigenetic targets. Natural language-based search web sites allow to query literature databases for co-occurrence of gene names and specific keywords (e.g., Pubmatrix, Chilibot, iHOP). With the ever-growing number of datasets contributed to shared repositories of gene expression, the comparison with published datasets is becoming routine, further supporting high-throughput data sharing in public repositories. Some servers (e.g., Celsius, (26)) host a large number of datasets that have been reanalyzed and can be queried for correlation analyses. Finally, GeneNetwork is a unique repository of genetic, genomic, and phenotypic information on a large number of recombinant inbred mouse strains (27). The GeneNetwork web site allows, among others, coexpression and quantitative trait loci analyses in a large number of datasets.

Table 1
Resources for data mining

	URL	Application	References
GeneCards	http://www.genecards.org/	Gene info	(39)
DAVID	http://david.abcc.ncifcrf.gov/	GO analysis	(40)
WebGestalt	http://bioinfo.vanderbilt.edu/webgestalt/	GO analysis	(41)
KEGG pathway database	http://www.genome.jp/kegg/pathway.html	Pathway analysis	(42)
Ingenuity	http://www.ingenuity.com/	Pathway analysis (commercial)	Ingenuity® Systems, http://www.ingenuity.com
BioCarta	http://www.biocarta.com/	Pathway analysis	http://www.biocarta.com/
Chilibot	http://www.chilibot.net/	Literature mining	(43)
PubMatrix	http://pubmatrix.grc.nia.nih.gov/	Literature mining	(44)
iHOP	http://www.ihop-net.org/	Literature mining	(45)
GATHER	http://gather.genome.duke.edu/	Gene relationship annotation	(46)
TRANSFAC	http://www.biobase-international.com/pages/index.php?id=transfac	Transcription factor analysis	(47)
MEME	http://meme.sdsc.edu/meme/	Motif analysis tools	(48)
BLAT	http://genome.ucsc.edu/cgi-bin/hgBlat	Probe mapping	(49)
GEO	http://www.ncbi.nlm.nih.gov/geo/	Data repository	(4)
ArrayExpress	http://www.ebi.ac.uk/arrayexpress/	Data repository	(50)
Celsius	http://celsius.genome.ucla.edu/	Warehouse of publicly available Affymetrix microarrays	(26)
GeneNetwork	http://www.genenetwork.org/	Multidimensonal data in recombinant imbred mouse strains	(51)
COEXPRESdb	http://coxpresdb.jp/	Gene co-regulation	(52)

3.7. Data Confirmation

The data quality produced with the current microarray platforms is such that a technical confirmation on the same RNA samples is rarely needed; however, a biological confirmation using independent samples is usually recommended (see Note 7). Real-time quantitative PCR using Taqman or SYBR green is the most popular method. Additional validation methods include immunohistochemistry, *in situ* hybridization, and functional confirmation. As mentioned, *in silico* validation using publicly available datasets is increasingly possible.

4. Notes

4.1. Experimental Design

Partial assessment of the transcriptome: A microarray-based study provides an extensive, but not complete characterization of the transcriptome. In fact, most alternatively spliced mRNA transcripts, microRNA, short RNAs, and mitochondrial RNAs are usually not assayed. Finally, mRNA-based expression studies assume that mRNA levels are a good proxy for protein levels, and this may not always be the case (28). In addition, post-translational modifications (e.g., glycosylation, phosphorylation), which can greatly alter the functional status of a protein as well as its degradation rate, are not captured by an mRNA-based gene expression study.

Other strategies to characterize the transcriptome: RNA-sequencing. Novel high-throughput methods based on next-generation sequencing are revolutionizing not only the sequencing applications (assembly of new genomes (29), exome and targeted resequencing (30), and whole-genome resequencing (31)), but also gene expression analysis (32): in fact, it is now possible and cost-effective to sequence the transcriptome, and use quantitative information from the sequencing run in order to estimate the relative abundance of individual transcripts (33, 34). In addition, RNA sequencing (RNA seq) is much more powerful for detecting novel transcripts and isoforms (34). This method provides a much more detailed picture of the transcriptome, and eventually might replace microarrays, but the cost per sample of an RNA-seq experiment (at the depth needed to identify novel isoforms, etc.) is still a factor of five to ten higher than a conventional microarray experiment, although it is expected to decrease in the near future. Basic procedural steps in data analysis for RNA-seq data are similar (i.e., quality analysis, exclusion of outlier samples, analysis of differential expression, etc.), but this distinct method requires a different analytical approach, and more computational power.

Insufficient power. Finally, flaws in the experimental design can be responsible for batch effects or for not having an adequate number of samples to assess differential expression with confidence.

4.2. RNA Extraction and Quality Check

Poor RNA quality is the single most common cause of failure for a microarray experiment. RNA quality should always be carefully checked before amplification and labeling. Sometimes, for limited or unique samples it is not possible to obtain RNA of acceptable quality, and in these cases additional caution is needed in drawing conclusions and additional effort is needed at the confirmation and validation steps.

4.3. RNA Quantification, Amplification, and Labeling

Occasionally, the amplification and the labeling step fail. Quantity and quality of the amplified product can be checked as described previously, although there is not a reliable way to check for a successful biotin labeling.

4.4. Array Hybridization and Scanning

Hybridization and scanning rarely present problems. Slides can be scanned again, but this can introduce bias (dye bleaching). If one or more slides need to be rescanned, it is recommended to rescan the whole array set.

4.5. Data Analysis

Data analysis is arguably the most challenging aspect of a microarray experiment. Numerous are the possible issues at this step:

Loss of samples after quality control. Sometimes, the outlier exclusion process limits the amount of comparisons that can be performed. For example, if one condition is represented by three replicates, and two of them are excluded at the quality control step, it is impossible to reliably compare that condition to the others. One can attempt a mathematical comparison, but the results will have $n = 1$ and will be unreliable.

Batch effects. Batch effect is still an insidious confounder in microarray studies (35). We define it as a technical artifact derived from factors that are inherently related to technical (RNA extraction process, amplification, hybridization, experimental conditions, etc.) rather than biological aspects. We can identify at least three sources of batch effect: (1) experimental design; (2) inter-array batch effect; (3) in some cases (e.g., Illumina), intra-array batch effect. Each one of these effects can be strong enough to alter or mask biological effects, and therefore it is critical to have a balanced design and technical execution in order to avoid such confounders.

(a) The experimental design can be unbalanced and introduce technical artifacts, especially when the effects we are trying to detect are weak and variable. For example, comparing multiple conditions to the same, flawed control will introduce a technical artifact.

(b) The inter-array batch effect is related to the simultaneous handling (and therefore exposure to shared conditions, including extraction, amplification, hybridization, etc.) of multiple samples in batches. For example, in an experiment that includes 36 arrays, one will hybridize arrays 1–12 in the first batch, 13–24 in the second batch, and 25–36 in the third batch. While it would be optimal to reduce differences in treatment and hybridize all the samples at the same time, this is often not possible. In these cases, it is recommended to balance the conditions hybridized in each batch (for example, make sure that all the conditions are represented at every hybridization batch so that potential confounding factors will be diluted across multiple batches). In addition, recent technological advances have allowed remarkable feature miniaturization so that it is now possible to fit multiple arrays on a single-microarray slide. For example, ink-jet printed Agilent microarrays can fit up to 16 samples on the same slide, and Illumina bead-based arrays up to 12 samples. As for the inter-array batch, samples on the same slide are exposed to the same microenvironment, and this can drive differences in expression data.

(c) Finally, we (unpublished data) and others (36) have noticed a position effect within Illumina slides, that is, array in position #1 will have higher signal on average than array in position #2, etc. One possible way to deal with these potential confounders is to completely randomize the samples across slides and within slides, in order to minimize the risk of position- or batch-related artifacts. Alternative normalization methods specific for the Illumina system and taking into account this slide bias have been proposed (37).

Algorithms have been developed to correct batch effects (38). However, when the batch and a biological condition match exactly (for example, all the controls were extracted and hybridized on one day and all the experimental samples on another day), teasing apart the technical artifacts from the biological effect can be challenging or virtually impossible. This is another reason why careful experimental design and randomization of samples across arrays are important when planning the experiment and gathering samples. One can also run a few replicate samples in each batch to provide a reference for cross-batch normalization.

Too3 few or too many differentially expressed genes. The number of differentially expressed genes is strictly dependent on the statistical threshold that is being applied. Too stringent thresholds can reduce to 0 the number of genes, even though some changes are expected (or proven with other methods) to occur. In general, correction for multiple comparisons and the use of a 5 or 10% FDR provides solid results and high confirmation rate. When these standard thresholds yield very little changes, more permissive thresholds can be applied, perhaps associated with a fold-change filter, with the caveat that the resulting lists may include a high number of false positives.

Similarly, more stringent thresholds can be applied when the number of genes is too overwhelming. A caveat is that in cases involving the comparison of very different conditions (e.g., two different tissue or cell types, extreme stress conditions, or drug treatments) the number of differentially expressed genes can truly reach the thousands, therefore an overly stringent threshold can lead to underestimating the results.

4.6. Data Mining

Caveats in GO analysis. Sometimes, the GO results are driven by the type of sample that is being analyzed. For example, a neuronal mRNA sample will be enriched by definition in neuronal or synaptic genes, etc., and the likelihood of finding an overrepresentation of neuronal genes in the list of differentially expressed genes is high. To get around this, one can use brain-expressed or neuronal genes as the input/denominator list for comparison.

Lack of confirmation of previous data. In many cases, previous studies provide a set of gene changes that are expected to be found. For example, a study comparing a KO for gene X to a wild-type animal is expected to yield gene X as strongly downregulated in the KO sample, or (2) previous studies have characterized an experimental system, developmental stage, etc., and therefore some changes (e.g., upregulation of gene Y) are well characterized and expected to occur. In most cases, the array results confirm the expectations, providing additional confidence in the data. Sometimes however, gene X is only slightly downregulated, or gene Y is not upregulated at all. A few points need to be considered in these cases:

(a) The gene may be low or not expressed in the experimental sample. As mentioned earlier, changes in low or not expressed genes are more difficult to detect.

(b) The probe is mapping to a gene region not affected. For example, some KO transgenic strains are produced by excision of a single exon, which leaves intact the rest of the transcript. Also, the probe may map to a different isoform. Probe sequences are available for most platforms, and mapping on a gene transcript using tools like BLAST or BLAT can help to solve these issues.

(c) The expected changes occur at the protein level. As mentioned, the correspondence between mRNA and protein levels or activity is not complete, and therefore mRNA levels of – for example – an enzyme can remain high in presence of reduced enzymatic activity.

Overall comparisons with previous studies can be made difficult by different gene names across species. Most genes have the same gene symbol in mouse and human, but in other organisms (including rat) considerable differences in gene names can affect data mining.

Poorly annotated probes. Unknown genes often show up at the top of the list of differentially expressed genes. Although sometimes frustrating, this is an unique opportunity for discovery. The first step in the assessment of unknown genes is to check the most updated annotation, as the probe annotation provided with the array may be obsolete. One possible method involves mapping the probe sequence against the genome and looking for annotated transcripts. The second level of investigation is to check homologous regions in other organisms (for example, for an unknown mouse gene, check the paralogous human region): sometimes, the paralogous gene has been annotated. The third level is the investigation of the protein sequence for functional domains or homologies with known proteins. GeneCards (*see* Table 1) is a very good starting point for gene-centric data mining.

4.7. Data Confirmation

Sometimes, the results of qPCR quantification of gene levels does not confirm the array data. As for lack of confirmation of previous or expected changes (*see* above), this can be due to (1) low probe expression; (2) probe targeting a different region or transcript. Specifically, for qPCR additional sources of variability should be considered: (1) the PCR primers may not be specific and amplify other genes; (2) DNA contamination is present and the primers amplify a pseudogene; (3) lastly, one obvious possibility to take into account is that the biological replicates do not present the changes in gene expression observed in the microarray experiment.

In conclusion, when properly planned, designed, and interpreted, microarray-based gene expression studies can provide valuable insights about physiological and pathological phenomena. A decade of microarray experiments has led to substantial protocol and analysis standardization, and to significant advances in our knowledge of the transcriptome.

Acknowledgments

The author would like to thank Fuying Gao and Jeremy Davis-Turak for technical assistance, and Drs. Michael Oldham and Daniel Geschwind for critically reading the manuscript.

References

1. Schena, M., Shalon, D., Davis, R. W., and Brown, P. O. (1995) Quantitative monitoring of gene expression patterns with a complementary DNA microarray. *Science.* **270**, 467–470.

2. Shi, L., Reid, L. H., Jones, W. D., Shippy, R., Warrington, J. A., Baker, S. C., Collins, P. J., de Longueville, F., Kawasaki, E. S., Lee, K. Y., Luo, Y., Sun, Y. A., Willey, J. C., Setterquist, R. A., Fischer, G. M., Tong, W., Dragan, Y. P., Dix, D. J., Frueh, F. W., Goodsaid, F. M., Herman, D., Jensen, R. V., Johnson, C. D., Lobenhofer, E. K., Puri, R. K., Schrf, U., Thierry-Mieg, J., Wang, C., Wilson, M., Wolber, P. K., Zhang, L., Amur, S., Bao, W., Barbacioru, C. C., Lucas, A. B., Bertholet, V., Boysen, C., Bromley, B., Brown, D., Brunner, A., Canales, R., Cao, X. M., Cebula, T. A., Chen, J. J., Cheng, J., Chu, T. M., Chudin, E., Corson, J., Corton, J. C., Croner, L. J., Davies, C., Davison, T. S., Delenstarr, G., Deng, X., Dorris, D., Eklund, A. C., Fan, X. H., Fang, H., Fulmer-Smentek, S., Fuscoe, J. C., Gallagher, K., Ge, W., Guo, L., Guo, X., Hager, J., Haje, P. K., Han, J., Han, T., Harbottle, H. C., Harris, S. C., Hatchwell, E., Hauser, C. A., Hester, S., Hong, H., Hurban, P., Jackson, S. A., Ji, H., Knight, C. R., Kuo, W. P., LeClerc, J. E., Levy, S., Li, Q. Z., Liu, C., Liu, Y., Lombardi, M. J., Ma, Y., Magnuson, S. R., Maqsodi, B., McDaniel, T., Mei, N., Myklebost, O., Ning, B., Novoradovskaya, N., Orr, M. S., Osborn, T. W., Papallo, A., Patterson, T. A., Perkins, R. G., Peters, E. H., Peterson, R., Philips, K. L., Pine, P. S., Pusztai, L., Qian, F., Ren, H., Rosen, M., Rosenzweig, B. A., Samaha, R. R., Schena, M., Schroth, G. P., Shchegrova, S., Smith, D. D., Staedtler, F., Su, Z., Sun, H., Szallasi, Z., Tezak, Z., Thierry-Mieg, D., Thompson, K. L., Tikhonova, I., Turpaz, Y., Vallanat, B., Van, C., Walker, S. J., Wang, S. J., Wang, Y., Wolfinger, R., Wong, A., Wu, J., Xiao, C., Xie, Q., Xu, J., Yang, W., Zhong, S., Zong, Y., and Slikker, W., Jr. (2006) The MicroArray Quality Control (MAQC) project shows inter- and intraplatform reproducibility of gene expression measurements. *Nat Biotechnol.* **24**, 1151–1161.

3. Shi, L., Campbell, G., Jones, W., Campagne, F., Wen, Z., Walker, S., Su, Z., Chu, T.-M., Goodsaid, F., Pusztai, L., Shaughnessy, J., Oberthuer, A., Thomas, R., Paules, R., Fielden, M., Barlogie, B., Chen, W., Du, P., Fischer, M., Furlanello, C., Gallas, B., Ge, X., Megherbi, D., Symmans, F., Wang, M., Zhang, J., Bitter, H., Brors, B., Bushel, P., Bylesjo, M., Chen, M., Cheng, J., Cheng, J., Chou, J., Davison, T., Delorenzi, M., Deng, Y., Devanarayan, V., Dix, D., Dopazo, J., Dorff, K., Elloumi, F., Fan, J., Fan, S., Fan, X., Fang, H., Gonzaludo, N., Hess, K., Hong, H., Huan, J., Irizarry, R., Judson, R., Juraeva, D., Lababidi, S., Lambert, C., Li, L., Li, Y., Li, Z., Lin, S., Liu, G., Lobenhofer, E., Luo, J., Luo, W., McCall, M., Nikolsky, Y., Pennello, G., Perkins, R., Philip, R., Popovici, V., Price, N., Qian, F., Scherer, A., Shi, T., Shi, W., Sung, J., Thierry-Mieg, D., Thierry-Mieg, J., Thodima, V., Trygg, J., Vishnuvajjala, L., Wang, S. J., Wu, J., Wu, Y., Xie, Q., Yousef, W., Zhang, L., Zhang, X., Zhong, S., Zhou, Y., Zhu, S., Arasappan, D., Bao, W., Lucas, A. B., Berthold, F., Brennan, R., Buness, A., Catalano, J., Chang, C., Chen, R., Cheng, Y., Cui, J., Czika, W., Demichelis, F., Deng, X., Dosymbekov, D., Eils, R., Feng, Y., Fostel, J., Fulmer-Smentek, S., Fuscoe, J., Gatto, L., Ge, W., Goldstein, D., Guo, L., Halbert, D., Han, J., Harris, S., Hatzis, C., Herman, D., Huang, J., Jensen, R., Jiang, R., Johnson, C., Jurman, G., Kahlert, Y., Khuder, S., Kohl, M., Li, J., Li, M., Li, Q.-Z., Li, S., Li, Z., Liu, J., Liu, Y., Liu, Z., Meng, L., Madera, M., Martinez-Murillo, F., Medina, I., Meehan, J., Miclaus, K., Moffitt, R., Montaner, D., Mukherjee, P., Mulligan, G., Neville, P., Nikolskaya, T., Ning, B., Page, G., Parker, J., Parry, M., Peng, X., Peterson, R., Phan, J., Quanz, B., Ren, Y., Riccadonna, S., Roter, A., Samuelson, F., Schumacher, M., Shambaugh, J., Shi, Q., Shippy, R., Si, S., Smalter, A., Sotiriou, C., Soukup, M., Staedtler, F., Steiner, G., Stokes, T., Sun, Q., Tan, P.-Y., Tang, R., Tezak, Z., Thorn, B., Tsyganova, M., Turpaz, Y., Vega, S., Visintainer, R., von Frese, J., Wang, C., Wang, E., Wang, J., Wang, W., Westermann, F., Willey, J., Woods, M., Wu, S., Xiao, N., Xu, J., Xu, L., Yang, L., Zeng, X., Zhang, J., Zhang, L., Zhang, M., Zhao, C., Puri, R., Scherf, U., Tong, W., and Wolfinger, R. (2010) The MicroArray Quality Control (MAQC)-II study of common practices for the development and validation of microarray-based predictive models. *Nat Biotechnol.* **28**, 827–838.

4. Edgar, R., Domrachev, M., and Lash, A. E. (2002) Gene Expression Omnibus: NCBI gene expression and hybridization array data repository. *Nucleic Acids Res.* **30**, 207–210.

5. Lange, P. S., Chavez, J. C., Pinto, J. T., Coppola, G., Sun, C. W., Townes, T. M., Geschwind, D. H., and Ratan, R. R. (2008) ATF4 is an oxidative stress-inducible, prodeath transcription factor in neurons in vitro and in vivo. *J Exp Med.* **205**, 1227–1242.

6. R Development Core Team (2008) R: A language and environment for statistical computing. R Foundation for Statistical Computing, Vienna, Austria. ISBN 3-900051-07-0, URL http://www.R-project.org.

7. Gentleman, R., Carey, V., Bates, D., Bolstad, B., Dettling, M., Dudoit, S., Ellis, B., Gautier, L., Ge, Y., Gentry, J., Hornik, K., Hothorn, T., Huber, W., Iacus, S., Irizarry, R., Leisch, F., Li, C., Maechler, M., Rossini, A., Sawitzki, G., Smith, C., Smyth, G., Tierney, L., Yang, J., and Zhang, J. (2004) Bioconductor: open software development for computational biology and bioinformatics. *Genome Biol.* **5**, R80.

8. Lee, M. L., and Whitmore, G. A. (2002) Power and sample size for DNA microarray studies. *Stat Med.* **21**, 3543–3570.

9. Klur, S., Toy, K., Williams, M., and Certa, U. (2004) Evaluation of procedures for amplification of small-size samples for hybridization on microarrays. *Genomics.* **83**, 508–517.

10. Wilson, C. L., Pepper, S. D., Hey, Y., and Miller, C. J. (2004) Amplification protocols introduce systematic but reproducible errors into gene expression studies. *Biotechniques.* **36**, 498–506.

11. Jones, L., Yue, S., Cheung, C.-Y., and Singer, V. (1998) RNA Quantitation by Fluorescence-Based Solution Assay: RiboGreen Reagent Characterization. *Analytical Biochemistry.* **265**, 368–374.

12. Fan, J.-B., Gunderson, K., Bibikova, M., Yeakley, J., Chen, J., Wickham Garcia, E., Lebruska, L., Laurent, M., Shen, R., and Barker, D. (2006) Illumina universal bead arrays. *Methods Enzymol.* **410**, 57–73.

13. Yang, Y. H., Dudoit, S., Luu, P., Lin, D. M., Peng, V., Ngai, J., and Speed, T. P. (2002) Normalization for cDNA microarray data: a robust composite method addressing single and multiple slide systematic variation. *Nucleic Acids Res.* **30**, e15.

14. Stafford, P. (2008) *Methods in Microarray Normalization*, Vol. **10**, Taylor & Francis.

15. Bolstad, B. M., Irizarry, R. A., Astrand, M., and Speed, T. P. (2003) A comparison of normalization methods for high density oligonucleotide array data based on variance and bias. *Bioinformatics.* **19**, 185–193.

16. Bourgon, R., Gentleman, R., and Huber, W. (2010) Independent filtering increases detection power for high-throughput experiments. *Proc Natl Acad Sci USA.* **107**, 9546–9551.

17. Smyth, G. K., Gentleman, R., Carey, V., Dudoit, S., Irizarry, R., and Huber, W. (2005) Limma: linear models for microarray data, In *Bioinformatics and Computational Biology Solutions using R and Bioconductor*, pp 397–420, Springer.

18. Hochberg, Y., and Benjamini, Y. (1990) More powerful procedures for multiple significance testing. *Stat Med.* **9**, 811–818.

19. Mootha, V. K., Lindgren, C. M., Eriksson, K. F., Subramanian, A., Sihag, S., Lehar, J., Puigserver, P., Carlsson, E., Ridderstrale, M., Laurila, E., Houstis, N., Daly, M. J., Patterson, N., Mesirov, J. P., Golub, T. R., Tamayo, P., Spiegelman, B., Lander, E. S., Hirschhorn, J. N., Altshuler, D., and Groop, L. C. (2003) PGC-1alpha-responsive genes involved in oxidative phosphorylation are coordinately downregulated in human diabetes. *Nat Genet.* **34**, 267–273.

20. Oldham, M., Horvath, S., and Geschwind, D. (2006) Conservation and evolution of gene coexpression networks in human and chimpanzee brains. *Proc Natl Acad Sci USA.* **103**, 17973–17978.

21. Miller, J. A., Horvath, S., and Geschwind, D. H. (2010) Divergence of human and mouse brain transcriptome highlights Alzheimer disease pathways. *Proc Natl Acad Sci USA.* **107**, 12698–12703.

22. Johnson, M., Kawasawa, Y., Mason, C., Krsnik, Z., Coppola, G., Bogdanovi, D., Geschwind, D., Mane, S., State, M., and Sestan, N. (2009) Functional and Evolutionary Insights into Human Brain Development through Global Transcriptome Analysis. *Neuron.* **62**, 494–509.

23. Oldham, M. C., Konopka, G., Iwamoto, K., Langfelder, P., Kato, T., Horvath, S., and Geschwind, D. H. (2008) Functional organization of the transcriptome in human brain. *Nat Neurosci.* **11**, 1271–1282.

24. Winden, K. D., Oldham, M. C., Mirnics, K., Ebert, P. J., Swan, C. H., Levitt, P., Rubenstein, J. L., Horvath, S., and Geschwind, D. H. (2009) The organization of the transcriptional network in specific neuronal classes. *Mol Syst Biol.* **5**, 291.

25. Horvath, S., Zhang, B., Carlson, M., Lu, K. V., Zhu, S., Felciano, R. M., Laurance, M. F., Zhao, W., Qi, S., Chen, Z., Lee, Y., Scheck, A. C., Liau, L. M., Wu, H., Geschwind, D. H., Febbo, P. G., Kornblum, H. I., Cloughesy, T. F., Nelson, S. F., and Mischel, P. S. (2006) Analysis of oncogenic signaling networks in glioblastoma identifies ASPM as a molecular target. *Proc Natl Acad Sci USA.* **103**, 17402–17407.

26. Day, A., Carlson, M. R., Dong, J., O'Connor, B. D., and Nelson, S. F. (2007) Celsius: a community resource for Affymetrix microarray data. *Genome Biol.* **8**, R112.

27. Chesler, E. J., Lu, L., Shou, S., Qu, Y., Gu, J., Wang, J., Hsu, H. C., Mountz, J. D., Baldwin, N. E., Langston, M. A., Threadgill, D. W., Manly, K. F., and Williams, R. W. (2005) Complex trait analysis of gene expression uncovers polygenic and pleiotropic networks that modulate nervous system function. *Nat Genet.* **37**, 233–242.

28. Cox, J., and Mann, M. (2007) Is proteomics the new genomics? *Cell.* **130**, 395–398.

29. Metzker, M. L. (2010) Sequencing technologies – the next generation. *Nat Rev Genet.* **11**, 31–46.

30. Ng, S., Turner, E., Robertson, P., Flygare, S., Bigham, A., Lee, C., Shaffer, T., Wong, M., Bhattacharjee, A., Eichler, E., Bamshad, M., Nickerson, D., and Shendure, J. (2009) Targeted capture and massively parallel sequencing of 12 human exomes. *Nature.* **461**, 272–276.

31. Biesecker, L., Mullikin, J., Facio, F., Turner, C., Cherukuri, P., Blakesley, R., Bouffard, G., Chines, P., Cruz, P., Hansen, N., Teer, J., Maskeri, B., Young, A., Manolio, T., Wilson, A., Finkel, T., Hwang, P., Arai, A., Remaley, A., Sachdev, V., Shamburek, R., Cannon, R., and Green, E. (2009) The ClinSeq Project: Piloting large-scale genome sequencing for research in genomic medicine. *Genome Res.* **19**, 1665–1674.

32. Wang, Z., Gerstein, M., and Snyder, M. (2009) RNA-Seq: a revolutionary tool for transcriptomics. *Nat Rev Genet.* **10**, 57–63.

33. Mortazavi, A., Williams, B., McCue, K., Schaeffer, L., and Wold, B. (2008) Mapping and quantifying mammalian transcriptomes by (RNA)-Seq. *Nat Methods.* **5**, 621–628.

34. Pan, Q., Shai, O., Lee, L., Frey, B., and Blencowe, B. (2008) Deep surveying of alternative splicing complexity in the human transcriptome by high-throughput sequencing. *Nat Genet.* **40**, 1413–1415.

35. Leek, J., Scharpf, R., Bravo, H. É. c., Simcha, D., Langmead, B., Johnson, E., Geman, D., Baggerly, K., and Irizarry, R. Tackling the widespread and critical impact of batch effects in high-throughput data. *Nat Rev Genet.* **11**, 733–739.

36. Verdugo, R., Deschepper, C., Munoz, G., Pomp, D., and Churchill, G. (2009) Importance of randomization in microarray experimental designs with Illumina platforms. *Nucleic Acids Res.* **37**, 5610–5618.

37. Shi, W., Banerjee, A., Ritchie, M., Gerondakis, S., and Smyth, G. (2009) Illumina WG-6 BeadChip strips should be normalized separately. *BMC Bioinformatics.* **10**, 372.

38. Johnson, E., Li, C., and Rabinovic, A. (2007) Adjusting batch effects in microarray expression data using empirical Bayes methods. *Biostat.* **8**, 118–127.

39. Rebhan, M., Chalifa-Caspi, V., Prilusky, J., and Lancet, D. (1997) GeneCards: integrating information about genes, proteins and diseases. *Trends Genet.* **13**, 163.

40. Huang, D. W., Sherman, B., and Lempicki, R. (2009) Systematic and integrative analysis of large gene lists using DAVID bioinformatics resources. *Nat Protocols.* **4**, 44–57.

41. Zhang, B., Kirov, S., and Snoddy, J. (2005) WebGestalt: an integrated system for exploring gene sets in various biological contexts. *Nucleic Acids Res.* **33**, W741-W748.

42. Ogata, H., Goto, S., Sato, K., Fujibuchi, W., Bono, H., and Kanehisa, M. (1999) KEGG: Kyoto Encyclopedia of Genes and Genomes. *Nucleic Acids Res.* **27**, 29–34.

43. Chen, H., and Sharp, B. (2004) Content-rich biological network constructed by mining PubMed abstracts. *BMC Bioinformatics.* **5**, 147.

44. Becker, K., Hosack, D., Dennis, G., Lempicki, R., Bright, T., Cheadle, C., and Engel, J. (2003) PubMatrix: a tool for multiplex literature mining. *BMC Bioinformatics.* **4**, 61.

45. Hoffmann, R., and Valencia, A. (2004) A gene network for navigating the literature. *Nat Genet.* **36**.

46. Chang, J., and Nevins, J. (2006) GATHER: a systems approach to interpreting genomic signatures. *Bioinformatics.* **22**, 2926–2933.

47. Matys, V., Fricke, E., Geffers, R., Gössling, E., Haubrock, M., Hehl, R., Hornischer, K., Karas, D., Kel, A. E., Kel-Margoulis, O. V., Kloos, D. U., Land, S., Lewicki-Potapov, B., Michael, H., Münch, R., Reuter, I., Rotert, S., Saxel, H., Scheer, M., Thiele, S., and Wingender, E. (2003) TRANSFAC: transcriptional regulation, from patterns to profiles. *Nucleic Acids Res.* **31**, 374–378.

48. Bailey, T. L., and Elkan, C. (1994) Fitting a mixture model by expectation maximization to discover motifs in biopolymers. *Proceedings/ International Conference on Intelligent Systems for Molecular Biology; ISMB. International Conference on Intelligent Systems for Molecular Biology.* **2**, 28–36.

49. Kent, J. (2002) BLAT – the BLAST-like alignment tool. *Genome Res.* **12**, 656–664.

50. Parkinson, H., Sarkans, U., Kolesnikov, N., Abeygunawardena, N., Burdett, T., Dylag, M., Emam, I., Farne, A., Hastings, E., Holloway, E., Kurbatova, N., Lukk, M., Malone, J., Mani,

R., Pilicheva, E., Rustici, G., Sharma, A., Williams, E., Adamusiak, T., Brandizi, M., Sklyar, N., and Brazma, A. (2010) Array Express update – an archive of microarray and high-throughput sequencing-based functional genomics experiments. *Nucleic Acids Res.* **39**, D1002–4.

51. Wang, J., Williams, R., and Manly, K. (2003) WebQTL: web-based complex trait analysis. *Neuroinformatics.* **1**, 299–308.

52. Obayashi, T., and Kinoshita, K. (2011) COXPRESdb: a database to compare gene coexpression in seven model animals. *Nucleic Acids Res.* in press.

Part VII

Therapeutic Approaches in Neurodegeneration

Chapter 29

Adeno-Associated Viral Gene Delivery in Neurodegenerative Disease

Peter F. Morgenstern, Roberta Marongiu, Sergei A. Musatov, and Michael G. Kaplitt

Abstract

The advent of viral gene therapy technology has contributed greatly to the study of a variety of medical conditions, and there is increasing promise for clinical translation of gene therapy into human treatments. Adeno-associated viral (AAV) vectors provide one of the more promising approaches to gene delivery, and have been used extensively over the last 20 years. Derived from nonpathogenic parvoviruses, these vectors allow for stable and robust expression of desired transgenes in vitro and in vivo. AAV vectors efficiently and stably transduce neurons, with some strains targeting neurons exclusively in the brain. Thus, AAV vectors are particularly useful for neurodegenerative diseases, which have led to numerous preclinical studies and several human trials of gene therapy in patients with Parkinson's disease, Alzheimer's disease, and pediatric neurogenetic disorders. Here, we describe an efficient and reliable method for the production and purification of AAV serotype 2 vectors for both in vitro and in vivo applications.

Key words: Gene therapy, Adeno-associated Virus, Neurodegenerative disease, Purification

1. Introduction

Viral gene therapy technology is a valuable tool for the exploration of molecular biological relationships in cell culture and animal models, and has been investigated in the treatment of human disease. In the face of challenges presented by the complexity and unique biology of the central nervous system, many vector types have been used, including Lentivirus (1, 2), Adenovirus, Herpes-Simplex virus (HSV) and Adeno-Associated Virus (AAV) (3). Of these, AAV vectors have been widely studied, and have been shown to provide safe and stable long-term gene expression in the brain of rodents, primates, and humans (4–6).

Giovanni Manfredi and Hibiki Kawamata (eds.), *Neurodegeneration: Methods and Protocols*,
Methods in Molecular Biology, vol. 793, DOI 10.1007/978-1-61779-328-8_29, © Springer Science+Business Media, LLC 2011

Fig. 1. Schematic representation of AAV plasmid. The main elements in the AAV vector are: AAV-2 inverted terminal repeats (ITR), hybrid CMV enhancer/chicken b-actin (CBA) promoter; chicken b-actin/rabbit b-globin (CBA-RBG) intron; woodchuck hepatitis virus posttranscriptional regulatory element (WPRE); bovine growth hormone polyadenylation signal (BGH poly(A)). The vector has a carrying capacity of 4.7 kb, which is adequate for most applications.

The AAV vector is derived from the single-stranded DNA of nonpathogenic parvoviruses (7). Their expression in the host cell through the lytic cycles requires helper virus functions to package the DNA into effective viral particles. In the absence of helper functions, they are able to integrate into the host-cell genome site-specifically into human chromosome 19; note that this sequence is not present in rodents. The components of the AAV vector are described elsewhere (8, 9) and the AAV expression plasmid is depicted in Fig. 1. Because of the size limitations imposed by viral packaging, the gene of interest must be less than 4.7 kb when using the AAV vector (10). Modern vectors not expressing AAV rep do not integrate with much efficiency, and the majority of genomes appear to reside in the episomal space (11).

A variety of different promoters can be used to drive expression of the inserted transgene depending on the ultimate goal of gene therapy. Early studies demonstrated stable expression for up to 4 months using the CMV promoter (5), and longer lasting expression was produced using a hybrid CMV/human β-globin promoter (12). High expression levels have been achieved using the neuron-specific enolase (NSE) (13) or chicken beta actin promoter (14). We use the hybrid CMV/chicken beta actin promoter, which has been shown to elicit more sustained transgene expression than the CMV promoter alone (15). Additionally, drug responsive promoters and various regulatory elements have been used to accommodate regulatory control of transcription in vitro (16) and in vivo (17–19).

One consistent characteristic of AAV vectors, regardless of promoter selection, is their tropism for neuronal populations. This is particularly true of AAV type 2 (20). Though it is capable of infection in many cell lines of varying origins (21), it infects neurons preferentially when injected into the CNS (5, 20, 22, 23). This demonstrated tropism and the stability of the postmitotic CNS neuron population makes it an ideal target for AAV-mediated gene therapy.

Among the AAV serotypes, AAV2 was the first cloned and has been the most commonly used in brain (5, 24). However, others have been used successfully in the laboratory. AAV9 has been used to deliver human erythropoietin into the rat striatum in the investigation of Parkinson's disease (PD) (25). A recent study has demonstrated the efficacy of AAV2/5 in primate color blindness (26).

AAV serotypes 2, 4, and 5 have been examined for properties useful in customizing vector capsids (27). In addition to these applications, different serotypes have shown varying affinities for neurons and other cell types. AAV5 and AAV1, for example, transduce both neurons and glial cells while AAV4 infects ependymal cells most efficiently (28, 29). A primate AAV strain (rh10) has recently been approved for use in a human trial of gene therapy for Batten disease, based on preclinical data suggesting that it would yield wider gene delivery throughout the brain compared with AAV2 used in the first human trial (30, 31).

Viral gene therapy has been hindered by concerns about viral provocation of an immune response. In the case of AAV, it appears that this is affected by the route of administration, though the response in the brain is typically limited (32). Tissue damage at the injection site is minimal when the appropriate viral titer and dose are used. Pathogenicity has not been seen (22, 33). Furthermore, repeated administration of AAV vectors has been shown to be safe, provided that the injections are spaced by appropriate intervals (34). Very large infusions of AAV into the brain have been associated with induction of immune reactions, which do not cause pathology but which limit future gene expression (35). We can therefore conclude that chronic diseases may be addressed through AAV vector technology, as extended treatment over time is likely to be well tolerated.

AAV vectors, particularly serotype 2, have demonstrated usefulness for research and clinical applications in a number of human illnesses. PD, Huntington's disease (HD), neuropathic pain, amyotrophic lateral sclerosis, and other neurodegenerative diseases have all been considered as targets of AAV-mediated gene therapy (2, 36–42). The small packaging size and the slow onset of expression may limit the use of AAV in settings, where large or multiple genes may be required or very rapid expression is needed; on the other hand, the better spread due to the small size and the stability of expression over years are often beneficial (43). Given the large interest and increasing clinical use of AAV for PD, Alzheimer's disease, and several genetic disorders (4, 30, 36, 44, 45), translation of this technology to clinical settings appears to be imminent, particularly in the wake of new approaches to capsid development (4, 46). Several human clinical trials are underway, having demonstrated the safety of these therapies in humans (4, 6, 45).

Our experience has shown a role for AAV vectors in the study of HD, PD, and more recently in depression. These human disorders have been studied effectively through rodent models (41, 47–52). Furthermore, we have used AAV to transfer the glutamic acid decarboxylase (GAD) gene into the subthalamic nucleus of human patients with PD. A phase I trial demonstrated the safety and potential efficacy of this therapy (4), and results of a randomized, double-blind sham-controlled phase II trial are expected soon.

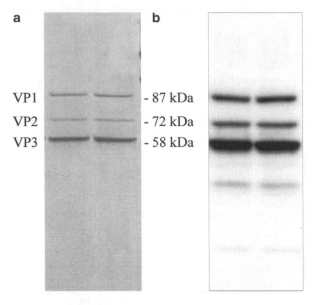

Fig. 2. Representative SDS-PAGE and Western blotting of two AAV2 viral vectors. AAV2 vectors were produced and purified following the protocol described and 15 μL (10^{12} viral particles/mL) were loaded on a gradient 4–12% Bis Tris polyacrylamide gel. (**a**) Coomassie Blue staining of the polyacrylamide gel shows only three bands corresponding to the capsid viral protein 1 (VP1), viral protein 2 (VP2), and viral protein 3 (VP3) indicating the purity of the preparation. (**b**) Western blotting staining with anti-VP1.VP2.VP3 antibody confirms the identity of the VP proteins.

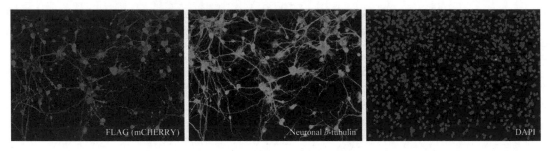

Fig. 3. Expression of recombinant mCherry-Flag in primary VMN neurons. The neurons were isolated from the VMN (VMH minus ARC) of 4-week-old female rats and maintained as a mixed culture in astrocyte-conditioned media. The cells were treated with the AAV2-mCherry vector on the day of plating and immunostained after 7 days in vitro. Note that while the culture consisted of neurons and nonneuronal cells, the vector transduced almost exclusively neurons.

Here, we present our protocol for generating high-titer, highly purified AAV vectors for research use (see Note 1) (see Fig. 2). Clinical-grade vectors are produced with similar methodology, but there are a variety of additional purification and testing steps unique to human applications, which are not outlined here. These AAV vectors can efficiently transduce cells in primary culture, including neurons (see Fig. 3). Furthermore, infusion of AAV vectors into the CNS can be used to overexpress or silence a gene of interest in a discrete brain region (see Fig. 4). This technology has been successfully used to explore the role of genes and neural networks in animal models of human diseases.

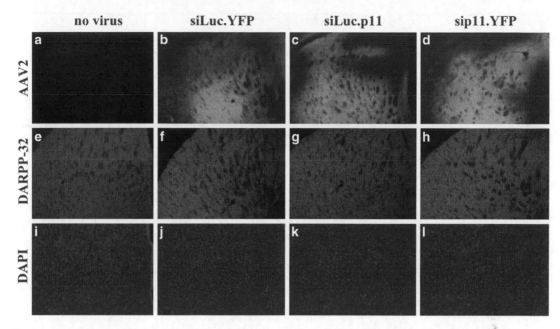

Fig. 4. Assessment of viral transduction in dorsal striatum of C57BL/6 mice. Two microliters of 10^{11} particles/mL AAV2 virus were injected into the mouse dorsal striatum (AP +0.5, ML ±2.1, DV −4.0) using a stereotactic frame. Mice were maintained for 6 weeks to allow maximal viral expression. They were subsequently sacrificed by perfusion and immunohistochemistry demonstrates viral expression in the striatum. (**a**) Striatum from a mouse not injected with any virus used as a negative control. (**b**) and (**d**) AAV-mediated expression of the marker gene green fluorescent protein (GFP) from two vectors also expressing small hairpin RNAs against genes for luciferase (siLuc) or the adaptor protein p11 (sip11). (**c**) Immunostaining for p11 overexpression from an AAV vector expressing p11 along with the luciferase small-hairpin RNA (siLuc). DARPP32 staining was used in all cases as a marker for striatal neurons and showed that AAV2 injections (**f**–**h**) did not cause any neuronal damage when compared to no surgery (no virus) controls (**e**). (**i**–**l**) DAPI staining for nuclei of all live cells further supports the absence of cytotoxicity following AAV-mediated gene expression.

2. Materials

2.1. Transfection

1. CellSTACK Chamber (Corning Inc., Lowell, MA).

2. AAV vector plasmid.

3. AAV2 helper plasmid (or other AAV serotype helper plasmid as needed) pNLrep and adenovirus helper plasmid pFΔ6.

4. HEPES buffer (HBS): 3 g HEPES, 100 mL ddH$_2$O, and 4.1 g NaCl. Dissolve and adjust pH to 7.03–7.05 using 2 M NaOH.

5. 2.5 M CaCl$_2$ solution in double distilled water (ddH$_2$O), filtered after preparation.

6. Dulbecco's modified Eagle's medium with l-glutamine and sodium pyruvate (DMEM, Cellgro®, Mediatech Inc., Manassas, VA) with 5 or 10% fetal bovine serum (FBS) and 1% penicillin/streptamycin (Cellgro®, Mediatech Inc.).

7. Dulbecco's phosphate buffered saline (DPBS, Cellgro®, Mediatech Inc.).

2.2. Harvesting

1. 0.5 M ethylenediaminetetraacetic acid (EDTA).

2. Tris buffered saline (TBS): 150 mM NaCl, 20 mM Tris–HCl pH 8.0, 1 mM $CaCl_2$, and 2 mM $MgCl_2$ in ddH_2O.

3. Benzonase (Sigma Aldrich, Co.).

4. 500 mL plastic centrifuge bottles.

5. 10% sodium deoxycholate in ddH_2O.

6. 2.5 M $CaCl_2$ in ddH_2O.

7. Sonicator probe: Sonifier 450 (Branson Ultrasonics Corporation, Danbury, CT).

2.3. Purification

1. Syringe filters, 0.8 and 0.45 μm.

2. Syringe pump (KD Scientific, Holliston, MA).

3. HiTrap™ Heparin HP Columns (GE Healthcare, Uppsala, Sweden).

4. NaCl solutions: 150 mM, 250 mM, and 550 mM all with 20 mM Tris–HCl pH 8.0.

5. Millex® GV Syringe-Driven Filter Unit, Hydrophilic Durapore® (PVDF), 0.22 μm (Millipore Corporation, Billerica, MA).

6. Amicon® Ultra Centrifugal Filters, Regenerated Cellulose 100,000 MWCO (Millipore).

7. Phosphate buffered saline (PBS).

8. PBS-Mg: 2 mM $MgCl_2$ in PBS.

2.4. Titering

1. Deoxyribonuclease I (DNase I) reaction buffer with $MgCl_2$ 10× (Life Technologies, Carlsbad, CA).

2. DNAse I (Life Technologies).

3. 25 mM EDTA.

4. 2 M NaOH solution.

5. 10 mM Tris–HCl pH 8.0.

6. Standards: Linearized AAV vector plasmid to 10^4, 10^5, 10^6, and 10^7 per 5 μL concentrations.

7. Primers designed to woodchuck posttranscriptional response element (WPRE).

8. Power SYBR® Green PCR Master Mix (Applied Biosystems, Warrington, UK).

3. Methods

3.1. Transfection

1. Plate 293 cells in CellSTACK chamber in 200 mL medium (DMEM with 10% FBS and 1% penicillin/streptomycin) so that at the moment of transfection they reach 80% confluence.

2. When the cells reach 80% confluence, they are ready to be transfected. While preparing the transfection reagents, change the incubator CO_2 level to 3%.

3. Prepare transfection reagents and set aside at room temperature: 2× HBS, sterile ddH_2O and 2.5 M $CaCl_2$.

4. Prepare DNA/$CaCl_2$ solution: Dilute the 2.5 M $CaCl_2$ solution in a volume of 10 mL with sterile ddH_2O (final concentration 0.25 M). Add total of 200 µg (see Note 2) of DNA (1:1:1 molar ratio for plasmids) to the 0.25 M $CaCl_2$ solution.

5. Prepare the transfection mix: Place 10 mL of 2× HBS in a 50-mL tube. Using a 5-mL pipette, bubble from the bottom of the tube while slowly adding the $CaCl_2$/DNA mix. Bubbles should appear at approximately the same rate at which the drops are released into the solution. The final volume at the end of this step will be 20 mL: 1/10 of the total volume of media on the cells in the cell stack. If higher volumes of reagents are needed, the final volume can be adjusted according to this ratio.

6. Place approximately 20 µL of the transfection mix on a tissue culture plate and view under a 10× objective of the inverted microscope. Fine, uniform, and unaggregated particles should be visible in the center of the drop. Incubate transfection mix for 15 min. During incubation, good particles tend to grow in size but not to aggregate.

7. If they look like aggregates (clamps and/or strings), discard the mix and attempt to rebubble the solution. If unsuccessful, prepare new solutions and repeat.

8. Add the transfection mix to 200 mL of fresh 5% DMEM (DMEM with 5% FBS and 1% penicillin/streptomycin).

9. Gently remove media from the cell stack and replace it with the 220 mL 5% fresh medium/transfection mix. Swirl the stack gently and return it to the incubator. Incubate for 20–24 h; do not disturb the cells during this time.

10. Check the cells. The precipitate from the transfection mix should be visible on the cells and mostly in the areas of the stack with no cells attached. Discard the medium and add 50 mL of PBS or medium (see Note 3). Gently rock the stack to wash and discard the solution. Replace the medium with 200 mL of fresh DMEM with 5% FBS and incubate at 5% CO_2 for 48 h.

3.2. Harvesting

1. Pour the medium off the cell stack into a 500-mL centrifuge bottle. Separate three 50 mL aliquots of this medium in conical tubes. Add 0.5 M EDTA to each of the first two tubes, to a final concentration of 10 mM.

2. Add the contents of one tube of medium/EDTA to the cell stack and rock gently, tapping the sides of the stack to detach

the cells. Incubate at room temperature for approximately 5 min and continue to gently rock the stack until the cells are fully detached. Return this medium containing the cells to the centrifuge bottle and repeat with the second tube of medium/EDTA. Wash with the third tube of medium, without EDTA.

3. Return the contents of all three conical tubes to the 500-mL centrifuge bottle and spin down at 4°C for 10 min at $1,000 \times g$. Pour off the medium completely and, if necessary, use a pipette to remove any remaining drops. It is very important that all medium be removed.

4. Resuspend pellet in 10 mL TBS. You gently vortex or use a pipette to carefully accomplish this. Transfer the suspension to a clean 50-mL tube and wash the bottle for any remaining cells with another 10 mL of TBS.

5. Freeze the cells at −80°C to be processed at a later date.

6. Thaw the cells at 37°C (about 10 min) and lyse with a sonicator probe: output control – 4, duty cycle% – 30. Sonicate for ten beats.

7. Add benzonase to a final concentration of 50 U/mL.

8. Mix gently and incubate for 45 min at 37°C. Mix several times during incubation.

9. Put the samples on ice for 10 min. Vortex and spin down at $5,000 \times g$ for 20 min at 4°C. Transfer supernatant, which contains the virus, to a new tube and keep on ice.

10. Resuspend the pellet in 3 mL of 550 mM NaCl solution containing 1 mM $CaCl_2$ and 2 mM $MgCl_2$. Sonicate (output control – 4, duty cycle% – 30, ten beats). Incubate for 5 min at 37°C, vortex, cool on ice, and spin down at $5,000 \times g$ for 20 min at 4°C. Transfer supernatant to two 2-mL Eppendorf tubes. Spin down $5,000 \times g$ 5 min. Combine all the supernatant with the previous one.

11. Add 8 mL of 20 mM Tris–HCl pH 8.0 containing 1 mM $CaCl_2$ and 2 mM $MgCl_2$ to the total volume. Total volume should be approximately 30 mL.

12. Add 1.5 mL of 10% deoxycholic acid to achieve the final concentration of 0.5%. Spin sample down at $5,000 \times g$ for 20 min at 4°C and transfer supernatant to a clean tube.

3.3. Purification

1. Draw up sample in a 30-mL syringe. Filter by hand through a 0.8-μm filter

2. Using the same syringe, filter the volume through a 0.45-μm filter.

3. Set up pump to a flow rate of 1.0 mL/min and equilibrate the heparin column with one full 10-mL syringe of 150 mM NaCl solution. Before starting, clear the syringe adaptor with ethanol

and ddH_2O. Be careful to avoid air bubbles when attaching the column and adaptor to the syringe (see Note 4). Check that column is free of air (white spots) when equilibration is complete. Unscrew the column from syringe and set aside.

4. Draw up the sample in a 30-mL syringe making sure to expel all air. Reattach the column "drop to drop." Pump at 0.5 mL/min to load the column.

5. After loading the sample into the column, draw up 30 mL of 150 mM NaCl in a new syringe and wash the column at 1 mL/min.

6. Repeat with 30 mL 250 mM NaCl at a rate of 1 mL/min.

7. Elute the virus from the column into a 15-mL conical tube with 5 mL of 550 mM NaCl in a 10-mL syringe at a rate of 1 mL/min (see Note 5).

8. Equilibrate the Millipore® membrane in a 15-mL conical tube with 5 mL PBS-Mg: Add solution to the membrane and centrifuge at $2,000 \times g$ until all liquid has passed through (about 5 min).

9. Discard the PBS-Mg from the bottom of the conical tube and load the virus solution. Centrifuge at the same speed until only 1 mL remains above the filter (about 10 min). Add 4 mL PBS-Mg and pipette up and down with a P1000 pipette without touching the membrane. Spin down as before, until 1 mL remains. Repeat five times. During the last wash, spin down and leave only 300–500 μL above the membrane in which to recover the virus (see Note 6).

10. Wash down the sides of the membrane by pipetting this volume. Transfer to a 1.5-mL Eppendorf tube. Wash twice with 100–200 μL PBS-Mg while scratching back and forth over the membrane with the pipette tip.

11. Spin down on a table top centrifuge at maximum speed for 30 s to clear the solution of pieces of membrane and transfer supernatant to fresh tube. Sonicate six bursts (output control 1.5) to break up viral aggregates. Centrifuge at $5,000 \times g$ for 10 min at 4°C.

12. Wash a 0.22-μm filter with 500 μL PBS-Mg. Pass the virus through the filter with a 1-mL syringe free of air, maintaining a secure hold on the filter so that it does not burst under pressure. Collect in a 1.5-mL Eppendorf tube.

3.4. Titering by RT-PCR

1. Prepare DNA probe: 2 μL virus, 1 μL 10× DNAse 1 buffer with $MgCl_2$, 1 μL DNAse 1, and 6 μL ddH_2O to a PCR tube. Incubate at 37°C for 15 min.

2. Inactivate DNAse: Add 2 μL EDTA (25 mM stock solution) to the tube and incubate at 70°C for 10 min.

3. Digest capsid: Add 10 μL 2 M NaOH and incubate at 56°C for 20 min. Then, dilute immediately 1:100 in 10 mM Tris pH 8.0.

4. Standards: Prepare standards as described in Subheading 2.4.

5. Using RT-PCR, analyze sample and standards and calculate titer (see Note 7).

4. Notes

1. This protocol is optimized for AAV2. Other AAV subtypes require an additional helper plasmid and may require minor changes in the protocol. For example, different heparin columns may be needed, as the columns used in this protocol have higher affinity for AAV2 than for AAV1/2 or AAV1.

2. The minimum amount of DNA for transfection is 200 μg. More, up to 400 μg, can be used if transfection is inefficient or final titer is low.

3. Fluorescence can be used to monitor transfection efficiency, or cells can be observed under light microscopy. Well-transfected cells should look rounded, generally unhealthy, and stop growing. If cells look normal and continue to grow, transfection is poor and should be repeated before proceeding.

4. Checking for bubbles in the heparin column frequently will help avoid reduced and inefficient binding of virus to the column and loss of samples.

5. After eluting the virus from the heparin column, titering can be performed to aid in determining the final volume in which to recover the virus from the Millipore® membrane.

6. If there are low titer issues in virus preparation, elution in a lower volume at the conclusion of purification will increase the titer.

7. Amplification of the CMV promoter can also be used for titering by RT-PCR. Alternatively, enzyme-linked immunosorbent assay (ELISA) can be used to determine titer.

References

1. Dreyer, J. L. Lentiviral vector-mediated gene transfer and RNA silencing technology in neuronal dysfunctions *Methods Mol Biol* **614**, 3–35.

2. Jarraya, B., Boulet, S., Ralph, G. S., Jan, C., Bonvento, G., Azzouz, M., Miskin, J. E., Shin, M., Delzescaux, T., Drouot, X., Herard, A. S., Day, D. M., Brouillet, E., Kingsman, S. M., Hantraye, P., Mitrophanous, K. A., Mazarakis, N. D., and Palfi, S. (2009) Dopamine gene therapy for Parkinson's disease in a nonhuman primate without associated dyskinesia *Sci Transl Med* **1**, 2ra4.

3. Manfredsson, F. P., and Mandel, R. J. (2010) Development of gene therapy for neurological disorders *Discov Med* **9**, 204–211.

4. Kaplitt, M. G., Feigin, A., Tang, C., Fitzsimons, H. L., Mattis, P., Lawlor, P. A., Bland, R. J.,

Young, D., Strybing, K., Eidelberg, D., and During, M. J. (2007) Safety and tolerability of gene therapy with an adeno-associated virus (AAV) borne GAD gene for Parkinson's disease: an open label, phase I trial *Lancet* **369**, 2097–2105.

5. Kaplitt, M. G., Leone, P., Samulski, R. J., Xiao, X., Pfaff, D. W., O'Malley, K. L., and During, M. J. (1994) Long-term gene expression and phenotypic correction using adeno-associated virus vectors in the mammalian brain *Nat Genet* **8**, 148–154.

6. Kaplitt, M. G. (2009) Gene therapy clinical trials in the human brain. Protocol development and review of current applications *Front Neurol Neurosci* **25**, 180–188.

7. Buning, H., Perabo, L., Coutelle, O., Quadt-Humme, S., and Hallek, M. (2008) Recent developments in adeno-associated virus vector technology *J Gene Med* **10**, 717–733.

8. During, M. J. (1997) Adeno-associated virus as a gene delivery system *Adv Drug Deliv Rev* **27**, 83–94.

9. Fitzsimons, II. L., Bland, R. J., and During, M. J. (2002) Promoters and regulatory elements that improve adeno-associated virus transgene expression in the brain *Methods* **28**, 227–236.

10. During, M. J., Young, D., Baer, K., Lawlor, P., and Klugmann, M. (2003) Development and optimization of adeno-associated virus vector transfer into the central nervous system *Methods Mol Med* **76**, 221–236.

11. Smith, R. H. (2008) Adeno-associated virus integration: virus versus vector *Gene Ther* **15**, 817–822.

12. Mandel, R. J., Rendahl, K. G., Spratt, S. K., Snyder, R. O., Cohen, L. K., and Leff, S. E. (1998) Characterization of intrastriatal recombinant adeno-associated virus-mediated gene transfer of human tyrosine hydroxylase and human GTP-cyclohydrolase I in a rat model of Parkinson's disease *J Neurosci* **18**, 4271–4284.

13. Klein, R. L., Meyer, E. M., Peel, A. L., Zolotukhin, S., Meyers, C., Muzyczka, N., and King, M. A. (1998) Neuron-specific transduction in the rat septohippocampal or nigrostriatal pathway by recombinant adeno-associated virus vectors *Exp Neurol* **150**, 183–194.

14. Miyazaki, J., Takaki, S., Araki, K., Tashiro, F., Tominaga, A., Takatsu, K., and Yamamura, K. (1989) Expression vector system based on the chicken beta-actin promoter directs efficient production of interleukin-5 *Gene* **79**, 269–277.

15. Klein, R. L., Hamby, M. E., Gong, Y., Hirko, A. C., Wang, S., Hughes, J. A., King, M. A., and Meyer, E. M. (2002) Dose and promoter effects of adeno-associated viral vector for green fluorescent protein expression in the rat brain *Exp Neurol* **176**, 66–74.

16. Sieger, S., Jiang, S., Kleinschmidt, J., Eskerski, H., Schonsiegel, F., Altmann, A., Mier, W., and Haberkorn, U. (2004) Tumor-specific gene expression using regulatory elements of the glucose transporter isoform 1 gene *Cancer Gene Ther* **11**, 41–51.

17. Haberman, R. P., McCown, T. J., and Samulski, R. J. (1998) Inducible long-term gene expression in brain with adeno-associated virus gene transfer *Gene Ther* **5**, 1604–1611.

18. Ye, X., Rivera, V. M., Zoltick, P., Cerasoli, F., Jr., Schnell, M. A., Gao, G., Hughes, J. V., Gilman, M., and Wilson, J. M. (1999) Regulated delivery of therapeutic proteins after in vivo somatic cell gene transfer *Science* **283**, 88–91.

19. Li, X. G., Okada, T., Kodera, M., Nara, Y., Takino, N., Muramatsu, C., Ikeguchi, K., Urano, F., Ichinose, H., Metzger, D., Chambon, P., Nakano, I., Ozawa, K., and Muramatsu, S. (2006) Viral-mediated temporally controlled dopamine production in a rat model of Parkinson disease *Mol Ther* **13**, 160–166.

20. Bartlett, J. S., Samulski, R. J., and McCown, T. J. (1998) Selective and rapid uptake of adeno-associated virus type 2 in brain *Hum Gene Ther* **9**, 1181–1186.

21. Muzyczka, N. (1992) Use of adeno-associated virus as a general transduction vector for mammalian cells *Curr Top Microbiol Immunol* **158**, 97–129.

22. McCown, T. J., Xiao, X., Li, J., Breese, G. R., and Samulski, R. J. (1996) Differential and persistent expression patterns of CNS gene transfer by an adeno-associated virus (AAV) vector *Brain Res* **713**, 99–107.

23. During, M. J., Samulski, R. J., Elsworth, J. D., Kaplitt, M. G., Leone, P., Xiao, X., Li, J., Freese, A., Taylor, J. R., Roth, R. H., Sladek, J. R., Jr., O'Malley, K. L., and Redmond, D. E., Jr. (1998) In vivo expression of therapeutic human genes for dopamine production in the caudates of MPTP-treated monkeys using an AAV vector *Gene Ther* **5**, 820–827.

24. Coura Rdos, S., and Nardi, N. B. (2007) The state of the art of adeno-associated virus-based vectors in gene therapy *Virol J* **4**, 99.

25. Xue, Y. Q., Ma, B. F., Zhao, L. R., Tatom, J. B., Li, B., Jiang, L. X., Klein, R. L., and Duan, W. M. AAV9-mediated erythropoietin gene delivery into the brain protects nigral dopaminergic neurons in a rat model of Parkinson's disease *Gene Ther* **17**, 83–94.

26. Mancuso, K., Hauswirth, W. W., Li, Q., Connor, T. B., Kuchenbecker, J. A., Mauck, M.

454   P.F. Morgenstern et al.

C., Neitz, J., and Neitz, M. (2009) Gene therapy for red-green colour blindness in adult primates *Nature* **461**, 784–787.

27. Van Vliet, K. M., Blouin, V., Brument, N., Agbandje-McKenna, M., and Snyder, R. O. (2008) The role of the adeno-associated virus capsid in gene transfer *Methods Mol Biol* **437**, 51–91.

28. Davidson, B. L., Stein, C. S., Heth, J. A., Martins, I., Kotin, R. M., Derksen, T. A., Zabner, J., Ghodsi, A., and Chiorini, J. A. (2000) Recombinant adeno-associated virus type 2, 4, and 5 vectors: transduction of variant cell types and regions in the mammalian central nervous system *Proc Natl Acad Sci USA* **97**, 3428–3432.

29. Wang, C., Wang, C. M., Clark, K. R., and Sferra, T. J. (2003) Recombinant AAV serotype 1 transduction efficiency and tropism in the murine brain *Gene Ther* **10**, 1528–1534.

30. Worgall, S., Sondhi, D., Hackett, N. R., Kosofsky, B., Kekatpure, M. V., Neyzi, N., Dyke, J. P., Ballon, D., Heier, L., Greenwald, B. M., Christos, P., Mazumdar, M., Souweidane, M. M., Kaplitt, M. G., and Crystal, R. G. (2008) Treatment of late infantile neuronal ceroid lipofuscinosis by CNS administration of a serotype 2 adeno-associated virus expressing CLN2 cDNA *Hum Gene Ther* **19**, 463–474.

31. Sondhi, D., Hackett, N. R., Peterson, D. A., Stratton, J., Baad, M., Travis, K. M., Wilson, J. M., and Crystal, R. G. (2007) Enhanced survival of the LINCL mouse following CLN2 gene transfer using the rh.10 rhesus macaque-derived adeno-associated virus vector *Mol Ther* **15**, 481–491.

32. Brockstedt, D. G., Podsakoff, G. M., Fong, L., Kurtzman, G., Mueller-Ruchholtz, W., and Engleman, E. G. (1999) Induction of immunity to antigens expressed by recombinant adeno-associated virus depends on the route of administration *Clin Immunol* **92**, 67–75.

33. Mastakov, M. Y., Baer, K., Xu, R., Fitzsimons, H., and During, M. J. (2001) Combined injection of rAAV with mannitol enhances gene expression in the rat brain *Mol Ther* **3**, 225–232.

34. Mastakov, M. Y., Baer, K., Symes, C. W., Leichtlein, C. B., Kotin, R. M., and During, M. J. (2002) Immunological aspects of recombinant adeno-associated virus delivery to the mammalian brain *J Virol* **76**, 8446–8454.

35. Lowenstein, P. R., Mandel, R. J., Xiong, W. D., Kroeger, K., and Castro, M. G. (2007) Immune responses to adenovirus and adeno-associated vectors used for gene therapy of brain diseases: the role of immunological synapses in understanding the cell biology of neuroimmune interactions *Curr Gene Ther* **7**, 347–360.

36. Bartus, R. T., Herzog, C. D., Bishop, K., Ostrove, J. M., Tuszynski, M., Kordower, J. H., and Gasmi, M. (2007) Issues regarding gene therapy products for Parkinson's disease: the development of CERE-120 (AAV-NTN) as one reference point *Parkinsonism Relat Disord* **13 Suppl 3**, S469-477.

37. Beutler, A. S., and Reinhardt, M. (2009) AAV for pain: steps towards clinical translation *Gene Ther* **16**, 461–469.

38. Bjorklund, T., and Kirik, D. (2009) Scientific rationale for the development of gene therapy strategies for Parkinson's disease *Biochim Biophys Acta* **1792**, 703–713.

39. Daya, S., and Berns, K. I. (2008) Gene therapy using adeno-associated virus vectors *Clin Microbiol Rev* **21**, 583–593.

40. Hester, M. E., Foust, K. D., Kaspar, R. W., and Kaspar, B. K. (2009) AAV as a gene transfer vector for the treatment of neurological disorders: novel treatment thoughts for ALS *Curr Gene Ther* **9**, 428–433.

41. Kim, J., Yoon, Y. S., Lee, H., and Chang, J. W. (2008) AAV-GAD gene for rat models of neuropathic pain and Parkinson's disease *Acta Neurochir Suppl* **101**, 99–105.

42. Brantly, M. L., Chulay, J. D., Wang, L., Mueller, C., Humphries, M., Spencer, L. T., Rouhani, F., Conlon, T. J., Calcedo, R., Betts, M. R., Spencer, C., Byrne, B. J., Wilson, J. M., and Flotte, T. R. (2009) Sustained transgene expression despite T lymphocyte responses in a clinical trial of rAAV1-AAT gene therapy *Proc Natl Acad Sci USA* **106**, 16363–16368.

43. Hadaczek, P., Eberling, J. L., Pivirotto, P., Bringas, J., Forsayeth, J., and Bankiewicz, K. S. (2010) Eight years of clinical improvement in MPTP-lesioned primates after gene therapy with AAV2-hAADC *Mol Ther* **18**, 1458–1461.

44. Bishop, K. M., Hofer, E. K., Mehta, A., Ramirez, A., Sun, L., Tuszynski, M., and Bartus, R. T. (2008) Therapeutic potential of CERE-110 (AAV2-NGF): targeted, stable, and sustained NGF delivery and trophic activity on rodent basal forebrain cholinergic neurons *Exp Neurol* **211**, 574–584.

45. Christine, C. W., Starr, P. A., Larson, P. S., Eberling, J. L., Jagust, W. J., Hawkins, R. A., VanBrocklin, H. F., Wright, J. F., Bankiewicz, K. S., and Aminoff, M. J. (2009) Safety and tolerability of putaminal AADC gene therapy for Parkinson disease *Neurology* **73**, 1662–1669.

46. Vandenberghe, L. H., Wilson, J. M., and Gao, G. (2009) Tailoring the AAV vector capsid for gene therapy *Gene Ther* **16**, 311–319.

47. Feng, L. R., and Maguire-Zeiss, K. A. Gene therapy in Parkinson's disease: rationale and current status *CNS Drugs* **24**, 177–192.

48. Terzioglu, M., and Galter, D. (2008) Parkinson's disease: genetic versus toxin-induced rodent models *The FEBS Journal* **275**, 1384–1391.

49. Lundblad, M., Picconi, B., Lindgren, H., Cenci, M. A. (2004) A model of L-DOPA-induced dyskinesia in 6-hydroxydopamine lesioned mice: relation to motor and cellular parameters of nigrostriatal function *Neurobiology of Disease* **16**, 110–123.

50. Iancu R, Mohapel, P., Brundin, P., Gesine, P. (2005) Behavioral characterization of a unilateral 6-OHDA-lesion model of Parkinson's disease in mice *Behavioural Brain Research* **162**, 1–10.

51. Brooks, S., Higgs, G., Janghra, N., Jones, L., and Dunnett, S. Longitudinal analysis of the behavioural phenotype in YAC128 (C57BL/6J) Huntington's disease transgenic mice *Brain Res Bull.* 2010 May 10; Epub 2010 May 10.

52. Svenningsson, P., Chergui, K., Rachleff, I., Flajolet, M., Zhang, X., El Yacoubi, M., Vaugeois, J. M., Nomikos, G. G., and Greengard, P. (2006) Alterations in 5-HT1B receptor function by p11 in depression-like states *Science* **311**, 77–80.

Chapter 30

Specific Gene Silencing Using RNAi in Cell Culture

Chunxing Yang, Linghua Qiu, and Zuoshang Xu

Abstract

RNA interference (RNAi) is a conserved cellular mechanism in most eukaryotes that can mediate specific gene silencing. Since its discovery in 1998, rapid progress has been made in understanding its basic mechanism and its application in research and drug discovery. In recent years, the application of RNAi in research, including research in neurodegeneration, has expanded rapidly such that it has become a regular tool for reverse genetics in cultured cells in many labs. However, an incomplete understanding of the RNAi mechanism and worries about its pitfalls still intimidate many others. Here, we present a streamlined and simple protocol for the design and implementation of an RNAi experiment in cultured cells, aiming to enable those who are inexperienced with RNAi to apply this powerful method in their research, particularly in the field of neurodegeneration.

Key words: RNAi, shRNA, siRNA, miRNA, Gene silencing, Knockdown

1. Introduction

RNA interference (RNAi) is a conserved gene silencing mechanism in most eukaryotes (1, 2). Since the discovery and the subsequent revelation of this mechanism, RNAi has evolved into a highly effective means for reverse genetics. In mammalian cells, RNAi can be triggered by introducing (1) short double-stranded RNA (dsRNA) of 21 nucleotides in length (also known as small interfering RNA or siRNA), (2) a gene cassette driven by a promoter that synthesizes short hairpin RNA (shRNA), or (3) a gene cassette driven by a promoter that synthesizes primary microRNA (pri-miRNA), from which a precursor microRNA (pre-miRNA) is cut out by an RNase III complex composed of Drosha and DGCR8 (Figs. 1 and 2). Both shRNA and pre-miRNA are exported from the nucleus (see Fig. 2). In the cytoplasm, the siRNA, shRNA, or

Giovanni Manfredi and Hibiki Kawamata (eds.), *Neurodegeneration: Methods and Protocols*,
Methods in Molecular Biology, vol. 793, DOI 10.1007/978-1-61779-328-8_30, © Springer Science+Business Media, LLC 2011

a Target mRNA: E2K

5' ...UCAAUGA GGUUGACAUGAGUAACAUA CAAGAGA... 3'

|———————— 19 nt ————————|

b siRNA

sense siRNA: 5' GGUUGACAUGAGUAACAUA-dTdT 3'
                       | | | | | | | | | | | | | | | | | | |
antisense siRNA: 3' dTdT-CCAACUGUACUCAUUGUAU 5'
                                    19 nt

c shRNA

UU

5' GGUUGACAUGAGUAACAUA CA^U ^U C_A
   | | | | | | | | | | | | | | | | | | | A
3' UU CCAACUGUACUCAUUGUAU GU_A _G A G
              19 nt

d pre-miRNA

```
      UG      UUGA        A    A       UC              G A A
5'  CUG     CAGUG GCG  AGGUUGACAUG   AGUAACAUA CA G U      G  C
                       | | | | | | | | | | |   | | | | | | | | |      C
    GGC     GUCGU CGU  UCCAACUGUAC - - UCAUUGUAUGU GU A G A C A
3' GGGAACUUCA   UC-C    C    C          19 nt
```

Fig. 1. Examples of small RNAs that can be used to silence specific genes in cells. (**a**) The target sequence of 19nt within the target mRNA. E2k is a subunit of α-ketoglutarate dehydrogenase complex (KGDHC) and is used as an example of silencing target here (for more details, *see* ref. 4). Based on this target sequence, corresponding siRNA (**b**), shRNA (**c**), and pre-miRNA (**d**) are designed. *See* the text for more details.

miRNA is further processed by other RNase III complexes composed of Dicer, TRPB, and AGO2, leading to the formation of RNA-induced silencing complex (RISC), which consists of a single-stranded RNA (called guide strand) and a complex of proteins (see Fig. 2). A key component in RISC is AGO2, which is an RNase III enzyme and has RNA slicer activity. In the RISC, the guide strand recognizes the target RNA by Watson-Crick base-pairing and directs the RISC to cleave the target RNA (see Fig. 2), thus destroying the target RNA and leading to gene silencing (2). Because of the simplicity and specificity of the RNAi mechanism, RNAi is increasingly employed in loss-of-function studies for understanding gene function in cells and in vivo. In almost every research article that investigates the role of a particular gene in neurodegeneration, RNAi is used to knockdown the specific genes under investigation, and the effect on neurodegeneration, either aggravation or alleviation, is observed. In this regard, RNAi has become an indispensible tool modern biomedical research, which includes the field of neurodegeneration. Particularly notable is that RNAi has become a highly useful method for target validation in developing

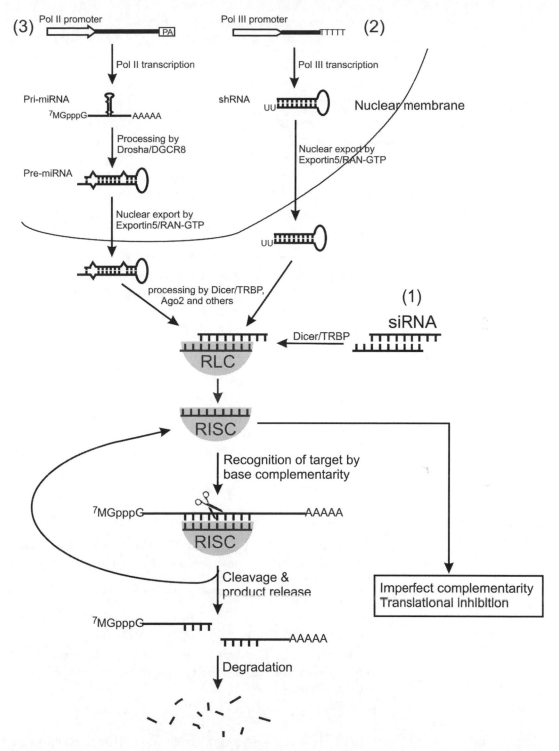

Fig. 2. Three commonly used RNAi expression strategies: transfection of cells with (1) siRNA or constructs that synthesize (2) shRNA or (3) miRNA. RLC stands for RISC loading complex. RISC stands for RNA-induced silencing complex. See the text for more descriptions.

therapeutic strategies ((3); see Note 1). In this chapter, we describe the basics in designing and conducting an RNAi experiment in cultured cells.

2. Materials

2.1. Making shRNA Constructs

1. Doubly distilled water.
2. Synthetic DNA oligonucleotides encoding shRNA (Integrated DNA Technologies, Standard desalting only).
3. 2× DNA annealing buffer: 20 mM Tris–HCl (pH 7.6), 100 mM NaCl, 2 mM EDTA.
4. shRNA expression plasmid pEGFP-U6 ((4); see Note 2).
5. Cloning enzymes (all from New England Biolabs, Ipswich, MA): T4 DNA ligase, T4 polynucleotide kinase, PmeI, EcoRI, calf intestinal alkaline phosphatase.
6. Bacterial culture and transformation reagents: MAX Efficiency DH-5α competent cells (Invitrogen, Carlsbad, CA). S.O.C. medium (Invitrogen, Carlsbad, CA), LB broth (EMD Chemicals, Gibbstown, NJ), LB agar (EMD Chemicals, Gibbstown, NJ), LB plate with 30 μg/mL Kanamycin (Invitrogen).
7. Reagents for colony screen using PCR: forward primer 5′-ATATCCCTTGGAGAAAAGCCTT-3′, reverse primer 5′-GGCCGCGAACCAATGCA-3′, PCR master mix (Promega, Madison, WI).
8. 100 bp DNA Ladder (Promega).

2.2. Making miRNA Constructs

1. pCAG-RFP-miRint-KX plasmid ((4); see Note 2).
2. Three synthetic PCR primers at 10 mM dissolved in distilled water. The sequences are based on the target gene sequence (see Subheadings 3.1 and 3.2 and Fig. 4).
3. PCR reagents: Platinum Pfx DNA Polymerase (Invitrogen, Carlsbad, CA) with 10× PCR buffer, 10 mM dNTP mixture (Promega, Madison, Wisconsin), 50 mM MgSO$_4$.
4. Cloning enzymes: The same enzymes as listed in Subheading 2.1 item 5 replace the two restriction enzymes with two new ones: KpnI and XhoI.
5. PCR product purification kit (Qiagen, Valencia, CA).

2.3. Cell Culture and Transfection

1. Dulbecco's modified Eagle's medium (DMEM) (Invitrogen, Carlsbad, CA) supplemented with 10% fetal bovine serum (FBS, Invitrogen) and antibiotics (100 U of penicillin and 100 μg of streptomycin, Invitrogen).
2. 0.25% Trypsin (Invitrogen) and 1 mM ethylenediaminetetraacetic acid (EDTA).

3. Lipofectamine 2000 (Invitrogen).

4. Opti-MEM I (Invitrogen).

2.4. Western Blot

1. Phosphate-buffered saline (PBS) (10×): 1.37 M NaCl, 27 mM KCl, 100 mM Na_2HPO_4, 18 mM KH_2PO_4 (adjust to pH 7.4 with HCl if necessary), and autoclave. Store at room temperature.

2. RIPA lysis buffer: 150 mM NaCl, 1.0% IGEPAL® CA-630, 0.5% sodium deoxycholate, 0.1% SDS, and 50 mM Tris–HCl, pH 8.0. Store at 4°C.

3. Halt Protease Inhibitor Cocktail (Thermo Scientific, Waltham, MA).

4. BCA Protein Assay Kit (Pierce, Rockford, IL).

5. 4× Laemmli's SDS sample buffer: 250 mM Tris–HCl (pH 6.8), 40% glycerol, 8% SDS, 0.02% bromophenol blue, and 8% beta-mercaptoethanol.

6. BioRad Protean II Mini gel system.

7. Separating buffer (4×): 1.5 M Tris–HCl, pH 8.7, 0.4% SDS. Store at room temperature.

8. Stacking buffer (4×): 0.5 M Tris–HCl, pH 6.8, 0.4% SDS. Store at room temperature.

9. 40% acrylamide/bis solution (29:1) and N,N,N,N'-tetramethylethylenediamine (TEMED).

10. Ammonium persulfate: 10% solution in water. Store at 4°C and make fresh every month.

11. Gel running buffer (10×): 250 mM Tris–HCl, 1.920 M glycine, 1% (w/v) SDS. Store at room temperature.

12. Prestained protein molecular weight markers: Kaleidoscope markers (Bio-Rad, Hercules, CA).

13. Protran nitrocellulose membrane (Whatman, Piscataway, NJ).

14. Transfer buffer (1×): 25 mM Tris–HCl (pH 8.0), 192 mM glycine, and 10–20% methanol.

15. BioRad transblot apparatus.

16. PBST: 0.05% Tween 20 in PBS.

17. Blocking buffer: 5% milk in PBST.

18. Antibodies: primary antibody against the knockdown target protein and secondary antibody conjugated with horseradish peroxidase (HRP) (GE Healthcare Biosciences, Piscataway, NJ).

19. Enhanced chemiluminescent (ECL) reagents (Amersham, Piscataway, NJ).

20. Blot imaging apparatus: Fuji LAS 4000.

21. Stripping buffer (Thermo Fisher Scientific, Waltham, MA).

2.5. Northern Blot

1. Trizol Reagent (Invitrogen).

2. 20× MOPS running buffer: 100 mM sodium acetate, 0.4 M MOPS (pH 7.0), 20 mM EDTA (pH 7.0). Store in dark at room temperature.

3. 20× SSC buffer: 3 M NaCl, 0.3 M sodium citrate (pH 7.0). Store at room temperature.

4. 1.5× loading buffer: 75% formamide, 7.5% formaldehyde (37%), 15% 10× MOPS running buffer, 20 μg/mL ethidium bromide. Best to make fresh each time.

5. Loading dye: Bromophenol blue in 35% glycerol and 1× running buffer.

6. Positively charged Nylon membrane (Hybond N+) (Amersham, Piscataway, NJ).

7. DIG probe kit (Roche, Indianapolis, IN): PCR DIG Probe Synthesis Kit, DIG Easy hyb, DIG block and wash buffer, CDP star.

8. UV crosslinker Stratalinker (Stratagene, LaJolla, CA).

9. HB-1000 Hybridizer (UVP LLC).

10. Blot imaging apparatus: Fuji LAS 4000.

2.6. Reverse Transcription and Quantitative PCR

1. SuperScript® III First-Strand Synthesis System (Invitrogen).

2. DNase I (Promega).

3. Sybr Green mix (Bio-Rad).

3. Methods

3.1. Design of siRNA, shRNA, and miRNA

3.1.1. Background

The three types of RNAi triggers, siRNA, shRNA, and miRNA, are all effective in silencing target genes and they often can be used interchangeably. siRNA is a simple small RNA duplex. The two strands are chemically synthesized separately and then annealed to form the siRNA. It can be efficiently delivered into many types of cells by standard transfection protocols. shRNA is synthesized from a gene construct as a single-stranded RNA, which folds to form shRNA (Figs. 1 and 2). Likewise, miRNA is also produced from a gene construct as a long, single-stranded RNA called pri-miRNA. Therefore, both shRNA and miRNA require the construction of plasmids. These plasmids also can be delivered into cells by standard transfection protocols. In general, the transfection efficiency of an siRNA is higher than constructs for shRNA and miRNA. siRNA can reach its highest concentration in cells immediately after transfection and does not require time to accumulate as shRNA and miRNA constructs do. Therefore, siRNA can trigger gene silencing more rapidly than shRNA and miRNA constructs.

However, the use of shRNA and miRNA also has some advantages. These gene cassettes can be incorporated into viral vectors and used for making cell lines that have the gene silencing construct stably integrated. Therefore, these constructs may be used in experiments where long term, sustained gene silencing is required.

The purpose of siRNA design is to select siRNAs with potent knockdown efficiency and minimal off-target effects. This can be achieved based on predefined siRNA sequence features and experiments testing the knockdown efficacy. It is important to keep in mind that at present no rational design guarantees a perfect siRNA. Therefore, designing multiple siRNAs and empirically testing them are essential. A time-saving approach is to purchase siRNAs and constructs that produce shRNA or miRNA targeting specific genes from commercial sources, some of which offer warranty for a certain degree of knockdown of the target. For targets that one cannot find a commercial source, or if the investigator prefers to design and make their own siRNA, we offer the following procedure.

3.1.2. Designing siRNA

The first step in designing siRNA is to identify the target gene sequence. It is preferred to use the RefSeq of the target gene from NCBI, since the RefSeq collection provides nonredundant, curated, and most reliable sequences. The targeted region of the gene should be located within the open reading frame (ORF) between 50 and 100 nt downstream of the start codon and the stop codon. The UTRs and the beginning of the ORF often bind to regulatory factors and may be less accessible by the RISC. However, this should not be regarded as an absolute limit since highly efficient knockdown has been reported with siRNAs directed against the other regions, including 3'UTRs. Once the target region is identified, target sequences for RNAi can be selected using one or several siRNA design algorithms that are accessible online (see http://i.cs.hku.hk/~sirna/software/sirna.php; ref. 5).

Each siRNA duplex contains a sense strand and an antisense strand. The antisense strand complements the sequence in the target cDNA or mRNA (see Fig. 1). Each strand of an siRNA duplex is 21 nt long, of which the 5' 19 nucleotides complement each other and match the target gene sequence. The 3' end two nucleotides of each strand are TT, which are not required to match the target sequence and constitute the overhang of siRNAs (see Fig. 1). To select the 19nt target sequence, follow some general rules (5, 6). Select sequences with a GC content between 30 and 60%. Avoid sequence repeats, sequences with low complexity, single nucleotide polymorphism (SNP) sites, sequence stretches of four or more bases, such as AAAA, CCCC, strings of six or more G and C (e.g. GGCGCC) or G and T (e.g. GGTTGT) and sequences containing palindromes, which may form unwanted hairpin structures. Use RNA folding software, such as RNAfold (http://rna.tbi.univie.

ac.at/cgi-bin/RNAfold.cgi) to explore possible secondary structures and exclude sequences with undesirable folds. Exclude sequences containing GTCCTTCAA, TGTGT, and CTGAATT, known to induce immune response (7–9). Select sequences with asymmetric thermostability in a manner that the three nucleotides at the 3' end of the sense strand has less G and C than its 5' end. Select two to six target sequences for each target gene. Design a negative control siRNA by scrambling targeted siRNA sequence so that it has the same length and nucleotide composition as the target sequence but contains at least four to five mismatches spread throughout the siRNA. Perform BLAST homology search to avoid sequences that share a certain degree of homology with nontargeted genes to minimize off-target silencing effects. It is particularly important to avoid a perfect or near perfect homology of the seed region (the six to eight bases at the 5' end of the two siRNA strands) with any nontarget genes.

The online algorithms adopt the general rules stated above and some incorporate additional rules derived from their experimentally tested siRNAs. One should check what rules are incorporated in the algorithm and check the siRNA against the rules that are not applied. Once the siRNAs are designed, they can be ordered from many commercial sources and used directly in experiments (see Subheading 3.3). The siRNA sequence can also be incorporated into shRNA or miRNA expression constructs (see below). However, it is not guaranteed that a highly efficient siRNA remains as efficient when its sequence is transplanted into an shRNA or miRNA. Therefore, we advise making multiple constructs and then selecting the most efficient ones to use by measuring the target knockdown.

3.1.3. Designing shRNA

Each shRNA has a stem and a loop. The stem contains the 19nt sense and antisense strands of an siRNA. The antisense strand complements the target mRNA sequence (see Fig. 1a). Pol III promoters, such as U6 and H1, are commonly used to drive the shRNA synthesis. Both promoters have defined initiation nucleotide, a G for U6 and an A for H1. The transcription terminates at a string of four or more Ts. To design an shRNA, insert the 19nt target siRNA sequence after the initiation nucleotide, connect the sense and antisense strands with a loop sequence of CAUUCAAGAGAUG and add six Ts at the 3' end of the antisense strand (see Fig. 3c). Synthesize the two strands of DNA encoding the shRNA with appropriate restriction enzyme sites at both ends (see Fig. 3c) and clone the sequence behind the Pol III promoter in a DNA plasmid (see Subheading 3.2).

3.1.4. Designing miRNA

The expression of miRNAs is usually driven by a Pol II promoter, although Pol III promoters can also work (10). One advantage of using Pol II promoter is that a marker protein can be expressed

a shRNA

b Single-stranded RNA

c Synthetic DNA

Fig. 3. Design of an shRNA construct. (**a**) Plug in the 19nt siRNA core sequence into the shRNA structure between the *two dotted lines*. (**b**) Unfold the hairpin sequence to a linear sequence. (**c**) Convert the linear sequence to a double-stranded DNA and add 6 Ts at the 3′ of the sense strand and 6 As plus TTAA at the 5′ of the antisense strand. This double-stranded DNA can be ligated into the plasmid pEGFP-U6. *See* text for more description. Note that the 5′ of the sense strand must be a "G". If an siRNA does not have a "G" at this 5′ end, this base can be replaced with a G.

under the control of the same promoter that drives miRNA expression (4, 11). The miR30a structural features are commonly employed in designing miRNAs (see Fig. 1d). To express an miRNA, a pre-miRNA has to be designed and inserted into the expression plasmid (see Fig. 4). To design your pre-miRNA, replace the 19nt sequence between the two vertical lines in Fig. 4a with your 19nt core siRNA sequence. Keep the UC bulge in the sense strand in place. Maintain the rest of the miRNA sequence the same as in Fig. 4a. Transform this whole miRNA sequence to a linear DNA sequence and use this as the sense strand (see Fig. 4b). Add the antisense strand and cloning restriction sites (see Fig. 4b, c). The exact steps of making this construct are described below in Subheading 3.2.2.

3.2. Making shRNA and miRNA Constructs

3.2.1. Making shRNA Constructs

1. Order two strands of synthetic DNA oligonucleotides encompassing the entire shRNA and the restriction sites PmeI and EcoRI (Fig. 3c).

2. Anneal the two shRNA oligonucleotide strands: To a 1.5-mL microtube, add 100 μL 2× DNA annealing buffer, 90 μL water, and 10 μL of each oligo (250 μM). Heat the tube with boiling water in a beaker for 5 min. Cool slowly to room temperature by leaving the beaker on bench top for 1 h. The annealed oligonucleotides will have appropriate overhangs for ligation.

3. Add phosphate to the shRNA oligo: To a new 1.5-mL microtube, add 10 μL of the annealed oligo pair (from **step 2**), 5 μL of 10× T4 DNA ligase buffer, and 2 μL of T4 polynucleotide

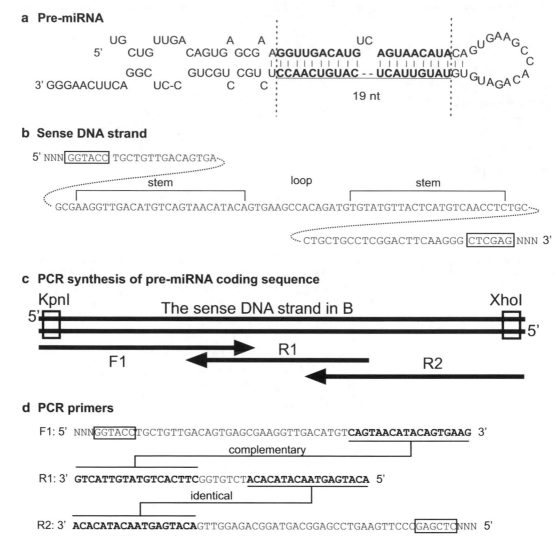

Fig. 4. Design of an miRNA construct. (**a**) Plug in the 19nt siRNA core sequence into the pre-miRNA structure between the *two dotted lines* and keep the UC bulge in the sense strand in place. (**b**) Unfold the pre-miRNA sequence to a linear sequence. (**c** and **d**) Design three PCR primers covering the entire pre-miRNA sequence. Synthesize the double-stranded DNA by PCR and clone the sequence into the plasmid pCAG-RFP-miRint-KX. *See* the text for more descriptions.

kinase. Adjust the volume to 50 μL using H₂O. Incubated at 37°C for half an hour.

4. Digest the expression vector pEGFP-U6 with PmeI and EcoRI: To a 1.5-mL microtube, add 15 μL of 10× NEBuffer 4, 1.5 μL of 100 μg/mL bovine serum albumin (BSA), 4 μg of the vector DNA, 2 μL of 10,000 units/mL PmeI, 2 μL of 20,000 units/mL EcoRI, and H₂O to a final volume of 150 μL. Incubate the reaction at 37°C for 1 h. Dephosphorylate the digested vector as follows: To the above digestion reaction tube, add 2 μL of calf intestinal alkaline phosphatase, incubate

at 37°C for 1 h. The vector DNA can be purified using the PCR purification kit with a final volume of 80 μL. Determine the concentration of the purified vector by OD measurement.

5. Ligate the shRNA oligo into the digested vector as follows: To a 1.5-mL microtube add 2 μL of 10× T4 DNA ligase buffer, 100 ng of vector DNA (from **step 4**), 3 μL of the hairpin insert (from **step 3**), and 2 μL of T4 DNA ligase. Adjust the volume to 20 μL using H₂O and incubate at room temperature for 15 min. Move the tube on ice.

6. Transformation: Add 50 μL of thawed MAX Efficiency DH-5α competent cells to each prechilled 17×100 mm polypropylene tube on ice and 2 μL of the ligation mix (from **step 5**). Mix gently by tapping the tube. Incubate on ice for 30 min. Heat-shock the cells in a 42°C water bath for 45 s. Put the cells on ice for 2 min. Add 0.9 mL of SOC medium that has been warmed to room temperature to the tube and shake the tube at 225 rpm and 37°C for 1 h. Spread 20–200 μL of cells on LB plate with 30 μg/mL kanamycin and incubate the plates overnight at 37°C.

7. PCR screening of colonies: Pick five to ten colonies for each shRNA construct. Each PCR reaction contains 1 μL of 10 μM forward and reverse primers and 10 μL PCR master mix. Adjust the volume to 20 μL by adding H₂O. Use a micropipette tip to pick up cells from single colonies and dip the tip in the PCR tube. PCR reaction is programmed as follows: 5 min at 95°C, followed by 35 cycles of 95°C for 30 s, 56°C for 30 s and 72°C for 1 min. Check the PCR products on a 2% agarose gel. You can see three PCR products: 121 bp from the blank vector control, 91 bp from the self-ligation products and 148 bp from the vector with shRNA insert (91 bp + 57 bp insert). Pick two to four positive colonies for each construct for plasmid DNA preparation and sequence verification.

3.2.2. Making miRNA Constructs

The miRNA constructs can be made using the same steps as the shRNA constructs as described above. The only difference is to synthesize the sense and the corresponding antisense DNA strands based on the pre-miRNA sequence as shown in Fig. 4a–c, with the addition of overhangs matching the appropriate restriction enzyme cleavage sites at the two ends (see Fig. 4c). Order the two synthetic DNA strands and anneal them as described above. You can clone the annealed DNA sequence behind the Pol III promoter, in which case you will need overhangs matching PmeI and EcoRI restriction sites, or behind a Pol II promoter in the plasmid pCAG-RFP-miR-int-KX (see Note 2), in which case you will need the overhangs matching KpnI and XhoI restriction sites. The cloning steps are the same as described in Subheading 3.2.1. We prefer a PCR method to synthesize the DNA encoding the pre-miRNA and clone it behind a Pol II promoter in pCAG-RFP-miRint-KX plasmid, as described below:

1. Design two strands of synthetic DNA oligonucleotides encompassing the entire pre-miRNA and the restriction sites KpnI and XhoI (Fig. 4a–c).

2. Order three strands of oligonucleotide primers, one forward (F1) and two reverse (R1, R2), as shown in Fig. 4c, d. Primer F1 has a Kpn1 site and primer R2 has an Xho1 site at their 5′ end. Primers F1 and R1 have 18 complementary nucleotides at their 3′ ends. Primers R1 and R2 have 18nt overlapping sequence.

3. Obtain the full-length pre-miRNA sequence by PCR amplification: Add the following components to an autoclaved microcentrifuge tube on ice: 5 μL 10× PCR buffer, 1 μL MgSO$_4$ (50 mM), 3 μL oligo F1 (10 mM), 1 μL oligo R1 (10 mM), 3 μL oligo R2 (10 mM), 1 μL dNTP mixture (10 mM), 0.5 μL Platinum Pfx DNA Polymerase (2.5 U/μL). Adjust the volume to 50 μL by adding H$_2$O. PCR using the program: 2 min at 94°C, followed by 35 cycles of 94°C for 30 s, 55°C for 30 s, 72°C for 30 s. Store the reaction at 4°C after cycling or at −20°C until use.

4. Insert the PCR product into the pCAG-RFP-miRint-KX (see Note 2): Digest the PCR product (from step 3) and the pCAG-RFP-miRint-KX vector with Kpn1 and Xho1 according to the enzyme manufacture's instructions. Ligate the PCR products into the digested vector as follows: In a 1.5-mL microtube, add 1 μL of 10× T4 DNA ligase buffer, the digested PCR product (the insert) and 100 ng of the digested vector DNA, and 1 μL of T4 DNA ligase. The amount of the insert added should be such that the molar ratio of the insert to the vector is 3 to 1. Adjust the volume to 10 μL using H$_2$O and incubate at room temperature for 15 min. Move the tube on ice.

5. Transformation: Add 50 μL of thawed MAX Efficient DH-5α competent cells (Invitrogen) to the 10 μL ligation mixture (from step 4). Mix gently by tapping the tube. Incubation on ice for 5 min. Heat-shock the cells in a 37°C water bath for 45 s. Put the cells on ice for 5 min. Spread all cells on LB plate with 100 μg/mL ampicillin and incubate the plates overnight at 37°C.

6. PCR screening of colonies: Pick five to ten colonies for each miRNA construct. Each PCR reaction contains 2 μL 10× PCR buffer, 0.8 μL MgSO$_4$ (50 mM), 0.8 μL oligo F1 (10 μM), 0.8 μL reverse primer (10 μM, binding on the vector), 0.5 μL dNTP mixture (10 mM), and 0.25 μL Taq DNA polymerase (2.5 U/μL). Adjust the volume to 20 μL by adding H$_2$O. Use a micropipette tip to pick up cells from single colonies and dip the tip in the PCR tube. PCR using the program as follows: 2 min at

94°C, followed by 35 cycles of 94°C for 30 s, 55°C for 30 s, 72°C for 30 s. Check the PCR products on a 2% agarose gel. Pick two to four positive colonies for each construct for plasmid DNA preparation and sequence verification.

3.3. Testing siRNA, shRNA, and miRNA Constructs

Once the constructs are obtained, they have to be tested to determine which construct works the best and whether it meets your knockdown requirement. The most direct way of achieving this is to transfect your constructs into cultured cells that express your target gene and measure the target gene expression in the transfected cells. You can determine the levels of the target gene expression by Western blot for the protein expression and Northern blot or reverse transcription PCR (RT-PCR) for the mRNA expression. It is important to use cells with high transfection efficiency, which determines the upper limit for the detection of knockdown. We prefer cells that can be transfected with 70% or higher transfection efficiency. A good construct is expected to knock down the target by 80% or more. If the transfection efficiency is 70%, an 80% knockdown will allow you to detect maximally a ~60% reduction in your target expression. For siRNAs, transfection efficiency is usually not a problem since it can be transfected into most cell types with high efficiency. For shRNA and miRNA constructs, the transfection efficiency could be a problem. However, this problem can easily be monitored by employing plasmids that express a marker, such as green fluorescent protein (GFP) or red fluorescent protein (RFP). If cells that express the target gene are not available or if they are available but transfected too poorly to be useful, alternative approaches are available (see Notes 3 and 4).

3.3.1. Transfecting Cells

1. Choice of cells: The commonly used cells that are easily transfectable are the best to use. We routinely use human cell lines, such as HEK293 and HeLa, and mouse cell lines, such as NF-1, N2a and NSC-34. Below we describe our protocol using NSC-34 cells because these cells are frequently used in neurodegeneration research. This is a cell line derived by fusion of mouse primary motor neurons and a mouse neuroblastoma cell line (12). The transfection efficiency is lower than the other cell types listed above. Therefore, the protocols described below can easily be applied to the other cell types.

2. Prepare cells as follows: Warm the culture medium (DMEM plus 10% FBS, 100 U penicillin and 100 μg of streptomycin) in an incubator to 37°C. Spray and wipe all tubes and bottles with 70% EtOH. Quickly thaw one vial frozen cells in a 37°C water bath. Add 1 mL medium to the tube of cells. Gently mix and transfer the cells from the vial with a pipette to a 15-mL tube. Add 5-mL pre-warmed medium, mix and centrifuge at $400 \times g$ for 3 min. Remove the medium. Add 1 mL fresh

medium to the cell pellet and resuspend the cells by pipetting up and down until no clumps are visible. Add an additional 10 mL prewarmed medium, mix and transfer the cells to a T25 flask. Incubate at 37°C till confluence, which usually takes approximately 3 days.

3. Splitting and plating cells: Aspirate old medium from confluent cell flask and wash cells once with 10 mL warm Dulbecco's Phosphate-Buffered Saline (DPBS) without calcium. Aspirate the DPBS, add 1 mL 0.5% trypsin to cells and swirl over cells for about 3 min. Add 5 mL of fresh medium to the flask and resuspend cells in the medium by pipetting up and down until no clumps are visible. Pipette the cells into a 15-mL tube. Apply desired amount of cells and proper amount of medium to a new flask (splitting ratio at 1:10) or 6-well plates ($1–2 \times 10^6$ cell per well).

4. Transfect the cells with siRNA, shRNA, or miRNA constructs using Lipofectamine 2000 (Invitrogen) according to manufacture's instructions (see Note 5). We routinely use the following procedure: Plate $1–2 \times 10^6$ cells in 2 mL medium in each of the wells in a 6-well plate. The cells will grow to 90–95% confluency by the next day. Before the transfection, change medium to DMEM plus 5% FBS without antibiotics and incubate for 1 h. Aspirate the medium and apply to the cells 500 μL transfection medium containing 100 pmol μL siRNA or 4 μg plasmids encoding miRNA or shRNA, and 5 μL (for siRNA) or 10 μL (for plasmids) Lipofectamine 2000 in opti-MEM I. Incubate at 37°C in a CO_2 incubator for 48–72 h. The cells are harvested for protein or RNA extraction to determine the levels of knockdown.

3.3.2. Determining Protein Knockdown Using Western Blot

1. To extract the protein, remove media from the culture plates. Wash cells twice with cold PBS. Carefully soak up any extra PBS with a piece of filter paper. Add 50 μL of RIPA lysis buffer with 1× protease inhibitor to each well. Use a cell scraper to lift the cells from the bottom of the well. Pass cell lysate through a pipette 20 times to form homogeneous lysate. Transfer lysate to 1.5-mL microcentrifuge tube. Allow samples to stand for 5 min on ice. Centrifuge the resulting mixture at $14,000 \times g$ for 10 min at 4°C to separate cell debris from protein. Transfer the supernatant to a new tube for use or store at −80°C.

2. Take an aliquot of the lysate and measure the protein concentration with BCA Protein Assay Kit (Pierce) based on the manufacturer's instructions.

3. Depending on the abundance of the protein being detected, mix 5–100 μg protein with 4× Laemmli's SDS sample buffer and boil for 5 min. Clear the mixture by a brief (~10 s) centrifugation at the top speed of a microcentrifuge.

4. We routinely use the Protean II Mini gel apparatus for SDS-PAGE. Assemble the gel plates according to manufacturer's instructions. Make the gel mix (for a single 10% separation gel) by mixing 2 mL 40% acrylamide/bisacrylamide (29:1 mixture), 2 mL 4× separation gel buffer and 3.87 mL water. Just before pouring the mix, add 12 μL TEMED and 120 μL 10% ammonium persulfate. Mix and pour immediately to ~1 cm below the anticipated bottom of the wells. Gently add a ~2 mm thick layer of 75% ethanol on top of the gel to block exposure to air and maintain a smooth top edge of the separation gel. Let the gel polymerize for ~15–30 min.

5. Make the 5% stacking gel mix by mixing 0.5 mL 40% acrylamide/bisacrylamide (29:1 mixture), 1 mL 4× stacking gel buffer and 2.5 mL water. After the separation gel is solidified, pour off the ethanol layer and wash 2× with water. Add 8 μL TEMED and 50 μL 10% ammonium persulfate to the stacking gel mix, immediately mix and pour on top of the separation gel using a pasture pipette. Insert the comb gently and leave it to polymerize 10 min. Assemble the electrophoresis unit and immerse it in SDS gel running buffer.

6. Flush the wells clear with gel running buffer. Load the protein samples (from step 3) and 10 μL kaleidoscope prestained standards to the wells with long gel-loading tips. Run through the stacking gel with constant voltage at 60 V and increase the voltage to 80 V after the bromophenol blue enters the separation gel. Run the gel about 2 h until the dye has migrated to the bottom of the gel.

7. While the gel is running, cut a piece of nitrocellulose membrane and two pieces of Waterman filter paper to the same size as the gel (5.5 cm×8.5 cm). Soak the membrane and sponges in 1× gel transfer buffer for 1 min.

8. Disassemble the gel from the gel running apparatus and soak the gel and the Waterman filter papers briefly in 1× gel transfer buffer.

9. Assemble "sandwich" for the transblot apparatus as follows (from the negative to the positive pole): sponge, filter paper, gel, membrane, filter paper, sponge. Transfer with a cold pack at 100 V and 4°C for 1 h.

10. Remove the membrane from the sandwich and wash briefly in PBS. Remove any small pieces of the gel that might stick to the membrane.

11. Incubate the membrane in blocking buffer for 1 h at room temperature. Change to the blocking buffer with the primary antibody and incubate at 4°C overnight. Wash membrane three times with PBST for 10 min each. Incubate with secondary antibody conjugated to HRP in blocking buffer for 1 h at

room temperature. Wash membrane three times with PBST for 10 min each. All the incubation and washing steps should be carried out with gentle agitation.

12. Detection of the knockdown target protein: Remove the ECL detection reagents from storage at 4°C. Mix 1 mL detection solutions A and 1 mL solutions B. Drain the excess wash buffer from the washed membranes and place protein side up on a sheet of SaranWrap. Pipette the mixed detection reagent on to the membrane. Incubate for 2 min at room temperature. Drain off excess detection reagent by holding the membrane gently with forceps and touching the edge against a piece of filter paper. Place the blots protein side down on to a fresh piece of SaranWrap, wrap up the blots and gently smooth out any air bubbles. Image the blot with fuji LAS 4000 or another imaging machine.

13. To confirm the equal loading, strip the blot if necessary with stripping buffer for 10 min with agitation. Wash the membrane with PBST and blot for tubulin or another housekeeping protein as described in steps 11 and 12.

14. Quantify the levels of the target protein using ImageJ or other imaging software and normalize it to the levels of the house keeping protein.

3.3.3. Determining RNA Knockdown Using Northern Blot

1. Make probes: Select a unique segment between 200 and 1,000 bp in length from the target RNA sequence, design a pair of PCR primers to make the DIG labeled-DNA probe by PCR using a PCR DIG Probe Synthesis Kit (Roche). Add to an autoclaved microcentrifuge tube on ice: 5 μL 10× PCR buffer with $MgCl_2$ (vial 3), 2 μL forward primer (10 μM), 2 μL reverse primer (10 μM), 5 μL PCR DIG Probe Synthesis Mix (vial 2), 0.75 μL Enzyme mix (vial 1), and template DNA (1–50 ng genomic DNA or 10–100 pg plasmid DNA). Adjust the volume to 50 μL by adding water. PCR using the program: 2 min at 94°C, followed by 35 cycles of 94°C for 30 s, 60°C (or the annealing temperature of the primers) for 30 s, 72°C for 40 s. Use 5 μL of the reaction mixture to check the results of the reaction on an agarose gel. Store the reminder of reaction at –20°C.

2. Extract RNA from the cells using the Trizol reagent as follows: Remove media from 6-well plates and wash cells once with ice-chilled PBS. Carefully soak up any extra PBS with a filter paper. Add 1 mL of TRIZOL Reagent to each well and scrape with cell scraper. Pass the cell lysate several times through a pipette. Transfer lysate to 1.5 mL microcentrifuge tube. Leave the samples at room temperature for 5 min. Add 0.2 mL of chloroform per 1 mL of TRIZOL reagent. Cap sample tubes securely. Vortex samples vigorously for 15 s and incubate them at room temperature for 2–3 min. Centrifuge the samples at $12,000 \times g$

at 4°C for 15 min. Transfer upper aqueous phase carefully without disturbing the interphase into new 1.5-mL tube. Add 0.5 mL of isopropyl alcohol per sample. Incubate samples at room temperature for 10 min and centrifuge at 12,000×g at 4°C for 10 min. Remove the supernatant completely. Wash the RNA pellet with 1 mL 75% ethanol. Mix the samples by vortexing and centrifuge at no more than 7,500×g at 4°C for 5 min. Repeat above washing procedure once. Remove all leftover ethanol. Air-dry RNA pellet for 5–10 min and dissolve RNA with 15 μL nuclease free water. Store the RNA at −80°C.

3. Measure RNA concentration: Dilute 1 μL of RNA with 99 μL of nuclease free water (1:100 dilution). Measure OD at 260 nm and 280 nm to determine sample concentration and purity. The A260/A280 ratio should be above 1.6. Apply the convention that 1 OD at 260 equals 40 μg/mL RNA.

4. Prepare and run a denaturing RNA gel: To prepare a 200 mL, 0.9% agarose formaldehyde gel, dissolve 1.8 g agarose in 170 mL water in a 65°C water bath, add 10 mL formaldehyde (37% stock), and 20 mL 10× MOPS running buffer. Pour the gel to about 0.6 cm thick in a fume hood and let it polymerize for ~15 min. Mix 10 μg RNA with 1.5× loading buffer and incubate at 65°C for 5 min. Cool on ice. Then, add 2 μL loading dye. Prerun gel at 5 V/cm for in 1× MOPS running buffer 5 min. Load samples and run with the same voltage until the bromophenol blue dye reaches the bottom of the gel (2.5–3.5 h for a 10 cm long gel). Soak the gel three times for 5 min each in distilled water to remove the formaldehyde. Photograph gel on a UV box.

5. Transfer to nitrocellulose membrane: Cut nitrocellulose membrane and eight pieces of Waterman filter paper to the same size as the gel and soak in 5× SSC. Assemble transfer stack as follows: Fill a glass tray with 500 mL 5× SSC. Place a glass plate across the top of the glass tray. Place a wet filter paper bridge on top of the glass plate, and let the two sides drop into the 5× SSC below. Put the following layers on top of the paper bridge on the glass plate: three pieces of wet filter paper, the gel, the wet nitrocellulose membrane, two pieces of wet filter paper, three pieces of dry filter paper and ~15 pieces of dry paper towel. Place another glass plate and weight (150–200 g) on top of this stack. Let transfer for ~3 h or longer.

6. Cross-link the RNA to the nitrocellulose membrane by place it on a new piece of filter paper wet with 5× SSC in a UV cross-linker (Stratalinker) and use auto-cross-link with 1200 J. Remove the membrane and soak for about 10 s in 1× MOPS buffer to remove excess salt and agarose.

7. Hybridization and detection of the target RNA: Place the membrane in a hybridization tube. Prehybridize with 4.5 mL DIG

Easy hybridization solution (Roche) for about 1 h at 50°C with gentle rotation in a hybridization oven (HB-1000 Hybridizer from UVP LLC). Boil 250 ng of DIG labeled-DNA probe (from step 1) in 0.5 mL of DIG Easy hybridization solution for 5 min. Chill quickly in ice-cold water. Add the probe to the hybridization solution in the hybridization tube and let hybridize at 50°C overnight. Wash the membrane twice with 125 mL 2× SSC plus 0.1% SDS at room temperature for 5 min each. Wash the membrane twice in 125 mL 0.1× SSC plus 0.1% SDS (preheated to 68°C) at 68°C with gentle agitation for 15 min each. Rinse the membrane in 1× DIG washing buffer for 1 min. Block the membrane in 50 mL 1× block buffer (5 mL 10× DIG Maleic acid buffer, 5 mL 10× DIG block buffer and 40 mL H_2O) with rocking for at least 1 h. Add 5 μL Anti-DIG antibody to blocking buffer (1:10,000) and incubate at room temperature with rocking for 30 min. Wash membrane twice in 50 mL DIG washing buffer with rocking for 15 min each. Equilibrate the membrane in 10 mL DIG detection buffer for 5 min. Place the membrane with DNA side facing up on a developing folder (or hybridization bag) and apply 1 mL CDP-star working solution. Immediately cover the membrane with the second sheet of the folder and spread the solution evenly without any bubbles on the membrane. Incubate at room temperature for 5 min. Squeeze out the excess liquid and seal the folder completely. Image the membrane with fuji LAS 4000.

8. Strip the blot if necessary by washing the membrane in boiling 0.1% SDS for 5 min, pour off the washing solution, and then wash with 2× SSC at room temperature.

9. Probe the membrane with an 18S or 7S rRNA or any other housekeeping gene probe as a loading control using the same steps 1–7. Quantify the bands using imageJ.

3.3.4. Determining RNA Knockdown Using Reverse Transcription and Quantitative PCR (RT-qPCR)

1. Treat RNA samples (from Subheading 3.3) with DNase: To a microtube, add 2 μL 10× reaction buffer, ten units of RNase-free DNase, 20 μg of RNA, and H_2O to a final volume of 20 μL. Incubate at 37°C for 60 min. Add 2 μL of DNase stop solution to terminate the reaction followed by incubation at 65°C for 10 min to inactivate the DNase.

2. cDNA synthesis: To a microtube, add 2.0 μL RNA from step 1, 1 μL of Oligo (dT)20 (50 μM), 1 μL of dNTPs (10 mM each) and 6 μL of RNase-free H_2O. Incubate at 65°C for 5 min. Move the tube on ice for at least 1 min. In parallel, set up two negative controls, NTC (no template RNA), and NoRT (no reverse transcriptase).

3. Continue the cDNA synthesis: To the tubes from step 2, add 2 μL 10× RT buffer, 4 μL 25 mM $MgCl_2$, 2 μL 0.1 M DTT, 1 μL RNaseOUT (40 U/μL), and 1 μL of SuperScript III RT

(200 U/μL). Mix well followed by a brief spin. Incubate at 50°C for 50 min followed by 85°C for 5 min to terminate the reaction. Chill on ice.

4. Quantitative PCR: The reactions for the target gene and one to two house-keeping genes (as internal controls) should be carried out in parallel. To a microtube, add 2 μL cDNA from **step 3**, 1 μL forward and reverse primer (10 μM each), 15 μL of Sybr Green mix, and 12 μL of H_2O. PCR using the program: 3 min at 95°C, followed by 40 cycles of 15 s at 95°C, 30 s at 56°C (2–5°C below Tm, see **Note 6**), and 30 s at 72°C.

5. Quantify knockdown: Derive the Ct values from the PCR data with the threshold set near the middle of the linear range on a log scale. For each sample, calculate the ΔCt by subtracting the Ct value for an internal control. Subtract the ΔCt from the knockdown samples from the control sample (from cells transfected with nonspecific siRNA, shRNA, or miRNA) to obtain the ΔΔCt, which can be used to calculate the percent knockdown.

4. Notes

1. RNAi is ideal for validation of therapeutic targets. However, this requires functional assays for the targets. Particular care must be taken to rule out nonspecific effects that often are associated with RNAi. A functional phenotype observed with an siRNA must be verified with the following observations: the phenotype is absent with a nonspecific siRNA but observable with at least two or more different siRNAs that silence the same target, and importantly, the phenotype can be rescued by overexpression of the target gene that is made resistant to the silencing by the specific siRNA by silent mutations in the siRNA targeting sequence.

2. The plasmids for making shRNA construct, pEGFP-U6, and miRNA construct, pCAG-RFP-miRint-KX, can be ordered from Addgene (http://www.addgene.org/pgvec1). Information about the plasmids can be accessed by searching the name Zuoshang Xu on the Web site.

3. If the cells that express the target gene are not readily available, use a cell type that has a high transfection efficiency (e.g., HEK293) and cotransfect the siRNA, shRNA or miRNA vectors with a plasmid that expresses the target gene. For detection, one can use an antibody against the target or a tag that is placed on the target. Another alternative is to make a report

construct, which expresses luciferase or GFP and carries the target sequence in the 3′ untranslated region of the mRNA (11).

4. If the cells express the target gene but are transfected poorly, use a vector that coexpresses the shRNA or miRNA with a marker protein, such as EGFP or RFP expression, sort the fluorescent cells and determine the target knockdown in the sorted cells (11).

5. In our hands, lipofectamine 2000 works well for most cell types but could be too toxic for some cells. Users need to empirically determine what is best for their cells. Besides lipofectamine 2000, we also have used FuGENE 6 (Roche, Indianapolis, IN), NeuroPORTER™ (Sigma, St. Louis, MO) and have attained satisfactory results in some cell types.

6. PCR parameters should be adjusted based on the dye and the real-time PCR equipment used. Many online programs are available to help design primers (e.g., http://bibiserv.techfak. uni-bielefeld.de/genefisher2/). In general, quantitative PCR primer pairs should span an intron and amplify a product ranging from 50 to 250 bases from the mature mRNA. This can minimize the effect of contaminating DNA and increase the PCR efficiency. Tm should be in the range of 58–60°C. Make master mix first and then distribute to all reactions to minimize inconsistency.

Acknowledgments

This work has been supported by RO1 NS048145 and R21 NS062230-01.

References

1. Mello, C. C., and Conte, D. (2004) Revealing the world of RNA interference. *Nature.* **431**, 338–342.

2. Tomari, Y., and Zamore, P. D. (2005) Perspective: machines for RNAi. *Genes Dev.* **19**, 517–529.

3. Xia, X., Zhou, H., Huang, Y., and Xu, Z. (2006) Allele-specific RNAi selectively silences mutant SOD1 and achieves significant therapeutic benefit in vivo. *Neurobiol. Dis.* **23**, 578–586.

4. Qiu, L., Wang, H., Xia, X., Zhou, H., and Xu, Z. (2008) A construct with fluorescent indicators for conditional expression of miRNA. *BMC Biotechnology.* **8**, 77.

5. Pei, Y., and Tuschl, T. (2006) On the art of identifying effective and specific siRNAs. *Nat Meth.* **3**, 670–676.

6. Birmingham, A., Anderson, E., Sullivan, K., Reynolds, A., Boese, Q., Leake, D., Karpilow, J., and Khvorova, A. (2007) A protocol for designing siRNAs with high functionality and specificity. *Nat. Protocols.* **2**, 2068–2078.

7. Hornung, V., Guenthner-Biller, M., Bourquin, C., Ablasser, A., Schlee, M., Uematsu, S., Noronha, A., Manoharan, M., Akira, S., de Fougerolles, A., Endres, S., and Hartmann, G. (2005) Sequence-specific potent induction of IFN-alpha by short interfering RNA in

plasmacytoid dendritic cells through TLR7. *Nat. Med.* **11**, 263–270. Epub 2005 Feb 2020.

8. Judge, A. D., Sood, V., Shaw, J. R., Fang, D., McClintock, K., and MacLachlan, I. (2005) Sequence-dependent stimulation of the mammalian innate immune response by synthetic siRNA. *Nat Biotech.* **23**, 457–462.

9. Lan, T., Putta, M. R., Wang, D., Dai, M., Yu, D., Kandimalla, E. R., and Agrawal, S. (2009) Synthetic oligoribonucleotides-containing secondary structures act as agonists of Toll-like receptors 7 and 8. *Biochemical and Biophysical Research Communications.* **386**, 443–448.

10. Borchert, G. M., Lanier, W., and Davidson, B. L. (2006) RNA polymerase III transcribes human microRNAs. *Nat Struct Mol Biol.* **13**, 1097–1101.

11. Zhou, H., Xia, X. G., and Xu, Z. (2005) An RNA polymerase II construct synthesizes short-hairpin RNA with a quantitative indicator and mediates highly efficient RNAi. *Nucl. Acids Res.* **33**, e62.

12. Cashman, N. R., Durham, H. D., Blusztajn, J. K., Oda, K., Tabira, T., Shaw, I. T., Dahrouge, S., and Antel, J. P. (1992) Neuroblastoma x spinal cord (NSC) hybrid cell lines resemble developing motor neurons. *Dev Dyn.* **194**, 209–221.

Chapter 31

Stem Cell Transplantation for Spinal Cord Neurodegeneration

Angelo C. Lepore and Nicholas J. Maragakis

Abstract

Amyotrophic lateral sclerosis (ALS), a disorder that affects 30,000 individuals in the USA alone, is characterized by relatively rapid degeneration of upper and lower motor neurons, with death normally occurring 2–5 years following diagnosis due to respiratory paralysis. Transplantation of various classes of neural precursor cells (NPCs) is a promising therapeutic strategy for the treatment of traumatic CNS injury and neurodegeneration, including ALS, because of the ability to replace lost or dysfunctional CNS cell types, provide neuroprotection, and deliver gene factors of interest. In order to target cellular therapy to diaphragmatic dysfunction in models of ALS, NPCs can be transplanted specifically into the cervical spinal cord ventral gray matter of both SOD1^{G93A} rats and mice. The SOD1^{G93A} rats and mice are currently the most well-studied animal model of the disease.

Key words: Cell transplantation, Engraftment, Graft, Spinal cord, Stem cells, Precursors, Amyotrophic lateral sclerosis, Motor neuron, Degeneration

1. Introduction

Amyotrophic Lateral Sclerosis (ALS). ALS, a motor neuron disorder that affects approximately 30,000 individuals in the USA alone, is characterized by relatively rapid degeneration of upper and lower motor neurons, with death normally occurring 2–5 years following diagnosis due to respiratory paralysis (1). The vast majority of cases are sporadic while 10% are of the familial form. Approximately 20% of familial cases are linked to various point mutations in the Cu/Zn superoxide dismutase 1 (SOD1) gene on chromosome 21 (2). Transgenic mice (2–5) and rats (6, 7) carrying mutant human SOD1 genes $^{(G93A, G37R, G86R, G85R)}$ have been generated, and, despite the existence of other animal models of motor neuron loss, are currently the most highly used models of the disease. The cause of

Giovanni Manfredi and Hibiki Kawamata (eds.), *Neurodegeneration: Methods and Protocols*,
Methods in Molecular Biology, vol. 793, DOI 10.1007/978-1-61779-328-8_31, © Springer Science+Business Media, LLC 2011

selective death of motor neurons is at present unknown; however, a number of mechanisms have been shown to at least contribute to ALS pathogenesis (reviewed in ref. 1).

Advantages of NPCs for the treatment of ALS. Transplantation of neural precursor cells (NPCs) offers a strategy for the treatment of traumatic and degenerative CNS disorders (8). A number of the benefits of engrafted NPCs coincide nicely with the neuropathological obstacles associated with ALS (reviewed in ref. 9). NPC-derived motor neurons have the potential of replacing motor neurons, reconstructing motor circuits or creating novel connections (10). Dysfunction in the important physiological properties of astrocytes is also associated with ALS (reviewed in ref. 11). For example, graft-derived glial cells may be able to replace dysfunctional host astrocytes, thereby restoring control over levels of extracellular signaling molecules, including glutamate, and in turn restoring some level of normal synaptic communication and preventing further excitotoxic motor neuron cell death (12). Despite the attention placed on motor neurons, the reconstitution of CNS astrocytes may be the simplest and most attractive approach to ALS therapies. In addition to replacement, NPC transplants may provide trophic support (13), thereby reducing endogenous motor neuron loss. NPCs can also be engineered to deliver therapeutic factors (14) (some of which they express without engineering) aimed at promoting enhanced host plasticity, decreasing and/or modulating immune cell infiltration and subsequent cell death (15), and delivering trophic signals.

Diaphragm function and ALS. In human ALS, patients ultimately succumb to disease because of respiratory compromise due to loss of phrenic motor neuron innervation of the diaphragm (16–18). Llado et al. (19) have shown cervical motor neuron loss, phrenic nerve axonal loss, diaphragm atrophy, and progressive reduction of phrenic nerve compound muscle action potential amplitudes in the $SOD1^{G93A}$ rat. Despite the obvious importance of using NPC transplantation for diaphragm function preservation, few studies have been undertaken to assess this therapeutic approach (12).

Astrocyte replacement for ALS via transplantation. By specifically targeting astrocyte replacement, transplantation of glial progenitors (20, 21) into the cervical spinal cord ventral horn extends animal survival and disease duration, attenuates motor neuron loss, and slows declines in forelimb motor and respiratory physiological function in the $SOD1^{G93A}$ rat model of ALS (12). These findings demonstrate the feasibility and efficacy of transplantation-based astrocyte replacement, and show that targeted multisegmental cell delivery to cervical spinal cord is a promising therapeutic strategy for slowing focal motor neuron loss associated with the disease.

With appropriate modifications, this method can also be applied to other cell types of interest, disease models, and anatomical targets of cell injection (22–24). Outcome measures include detailed analysis of cell transplant fate, biochemical, histological, and functional states of the spinal cord, and behavioral changes relevant to the location of transplantation. In this chapter, detailed methods for this procedure are provided.

2. Materials

2.1. Cell Preparation

1. 1.5-mL Eppendorf tubes.
2. Cell culture medium specific to cell type being used.
3. Hank's buffered salt solution (HBSS).
4. 0.05% trypsin or other enzyme used to gently remove the cell type from monolayer culture.
5. Soybean trypsin inhibitor optional to deactivate trypsin/ decrease cell death.
6. 37°C incubator.
7. Wet ice.
8. Ice bucket.
9. Centrifuge.
10. Hemocytometer.
11. Trypan blue.
12. Light microscope.
13. Pipettes and tips.
14. 20.0-µL pipette-man.
15. 20.0-µL pipette tips.
16. 15.0-mL and/or 50.0-mL conical/Falcon tubes.

2.2. Animal Preparation for Surgery

1. Scale for weighing animals: Digital Mettler Toledo PG802-S.
2. Chloral hydrate solid.
3. Optional: Ketamine, xylazine, and acepromazine injectable cocktail for anesthesia (acepromazine maleate (0.7 mg/kg; Fermenta Animal Health), ketamine (95 mg/kg; Fort Dodge Animal Health), and xylazine (10 mg/kg; Bayer)).
4. Optional: Isofluorane inhalation anesthesia.
5. 1.0 mL "Sub-Q" syringes with 26-gauge needles.
6. 0.9% sterile saline.
7. Electric shaver: Oster Golden A5.
8. Gauze: Kendall (from Tyco).
9. Povidine-iodine antiseptic solution: Applicare.

2.3. Surgery

1. Surgical microscope: We use a Nikon SMZ645.

2. Light source: Volpi Intralux 4000-1.

3. Ring-light.

4. Surgery board: Thermo Scientific brand.

5. Surgical gloves.

6. Rolled gauze pad for surgery.

7. Four retractors: Paperclips (see inset of Fig. 1b).

8. Blades: #11 feather surgical blade for #3 holder.

9. Scalpel blade holder.

10. Cotton-tipped applicators: Fisher brand 6-in.

11. Rat-toothed forceps: Fine Science Tools (rat: FST #11023-15; mouse: FST #11042-08).

12. Medium-sized spring scissors: Fine Science Tools (FST #15012-12).

13. Mini spring scissors: Fine Science Tools (FST #15000-10).

14. Rongeurs: Fine Science Tools (rat: FST #16121-14; mouse: FST #16221-14).

15. #5 Forceps: Fine Science Tools (Dumont #5).

16. Microknife: Fine Science Tools (FST #10056-12).

17. Needle holders: Fine Science Tools (FST #12502-14).

18. Suture: 4-0 Sofsilk #S-183.

19. Staples: 9 mm (Autoclip #427631).

20. Stapler: 9 mm (Reflex #203-1000: World Precision Instruments #5000344).

21. Glass bead sterilizer: Steri250 from Inotech.

22. Water pump: Gaymar T/Pump #P/N 07999-000.

23. Blanket for water pump.

2.4. Cell Injection

1. Cyclosporin A (CSA): Novartis/Sandimmune – 250.0 mg/5.0 mL ampules: immunesuppression for rodent-derived transplanted cells.

2. FK-506: LC Laboratories: Immunesuppression for human-derived transplanted cells.

3. Rapamycin: LC Laboratories: Immunesuppression for human-derived transplanted cells.

4. Injector: World Precision Instruments #UMP2.

5. Micro 4 microsyringe pump controller: World Precision Instruments #UMC4.

6. Micromanipulator: World Precision Instruments #Kite-R.

7. Stand: World Precision Instruments.

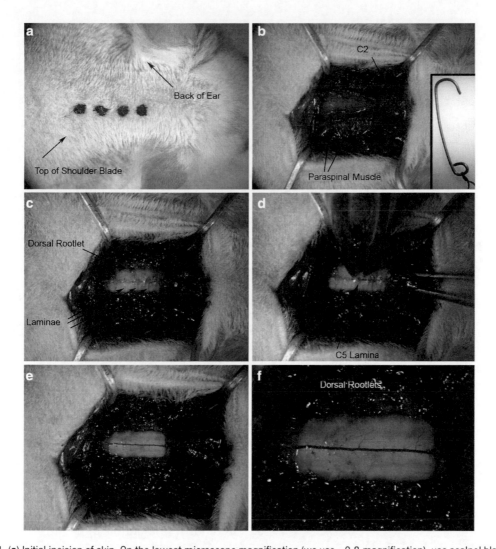

Fig. 1. (**a**) Initial incision of skin. On the lowest microscope magnification (we use ×0.8 magnification), use scalpel blade to make midline incision. Stretch skin laterally with other hand to make skin taut (which makes the skin easier to cut), and make incision from base of skull to shoulder blade (denoted by *dotted line*). (**b**) Exposure of surgical field. The surgical exposure should be square/rectangle-shaped. This shape can be achieved by pulling the surrounding muscle toward four corners using the four retractors. Tape string that is attached to retractors to surgical board in order to secure retractors and properly keep field open. (*inset*) Retractors for exposure of surgical field. The retractors are used to pull back muscle in order to create a surgical field with both clear visibility of the spinal cord and adequate space to perform surgery. Retractors can be made by shaping sturdy paperclips into the desired shape and size. Autoclave the retractors before for surgery. Tie string to retractor. (**c**) Vertebrae. Following removal of paraspinal muscle, thoroughly clean dorsal surface of vertebrae with rongeurs. Individual lamina can be seen, as well as dorsal rootlets entering from lateral aspect of spinal column. (**d**) Laminectomy. Start laminectomy at spinal cord level C5. Secure spinal column by holding muscle overlying level C2 with rat toothed forceps. Grab entire lamina (see diagram: grab near midline) with rongeurs. Position rongeurs so that tool is completely perpendicular to axis of spinal column. Slowly crush lamina. Do not push down into spinal cord, as this will cause damage to spinal tissue. Crush and gently pull broken piece of bone upward. Rongeurs should crush piece so that one can easily pull away to remove. If piece of bone is still attached to rest of laminae, do not tug as this will cause hemorrhage and possible injury to spinal cord. Make sure that rongeurs are clean and sharp. (**e**) Exposure of spinal cord tissue following laminectomy. Extend laminectomy to expose all of C4–C6 spinal cord. Make one continuous opening in the bone over three spinal levels. Do not extend laminectomy too far laterally because this will cause hemorrhage. In order to target ventral horn, the injection site is relatively medial, so it is unnecessary to extend laminectomy to the complete lateral extent of the vertebral bone. (**f**) High magnification of dorsal surface of spinal cord. The prominent dorsal blood vessel can be seen running down the midline of the spinal cord. This blood vessel pattern is observed in most cases; however, some animals display a nonmidline trajectory of the blood vessel. The dorsal rootlets can be seen at the lateral aspects of the spinal cord dorsal surface. Relative to the dura overlying the spinal cord, these nerves have a hazy appearance.

8. Metal base: World Precision Instruments.

9. 10.0-μL Hamilton syringe.

10. Hamilton needles: 33-gauge, 45° bevel, 1 in.

11. Timer.

12. Optional: 26-gauge blunt Hamilton metal needle.

13. Optional: Glass 20.0-μL microcapillary pipettes.

14. Optional: Epoxy: purchase from hardware store.

3. Methods

3.1. Cell Preparation

As an example, we describe the procedure for preparing glial progenitor cells (20, 21) for transplantation because of our experience with this cell type. However, the specifics of the protocol, including medium and use of trypsin for example, depend on the particular cell type being used for transplantation.

1. Prewarm all solutions to 37.0°C in water bath.

2. Rinse flask 2× with HBSS.

3. Add 5.0 mL/T-75 flask of 0.05% trypsin/EDTA.

4. Incubate flask for 3 min in 37.0°C incubator.

5. Triturate 3× gently in flask with 5 or 10-mL pipette.

6. Optional: Add 5.0 mL/T-75 flask of 1.0 mg/ml soybean trypsin inhibitor in DMEM/F12.

7. Rinse each flask 2× in 5.0 mL of medium.

8. Pool cells and rinses.

9. Keep cells on ice for all subsequent steps.

10. Spin at 200–300×g for 5 min in conical tube: Preferably in centrifuge cooled to 4°C.

11. Discard (and save) supernatant. Resuspend cells in 1.0 mL of medium, and transfer to 1.5-mL Eppendorf tube.

12. Count cells with hemocytometer using trypan Blue to determine viability.

13. Spin again in 1.5-mL tube (800 RPM for 10 min: preferably in centrifuge cooled to 4°C).

14. Discard (and save) supernatant.

15. Resuspend cells in desired final volume of medium to achieve appropriate cell density.

16. See notes on suggestions for choosing an appropriate cell dilution.

17. Keep cells on wet ice until transplantation.

18. Distribute cell suspensions to multiple 1.5-mL Eppendorf tubes (see Note 1).

3.2. Preparation Prior to Surgery

1. Age- and sex-match the animals within a group, and distribute animals within the same litter to different groups (see Note 2).

2. Consider the age/point of disease prior to the transplantation of cells into animal (see Note 3).

3. Conduct relevant baseline behavior assessments prior to surgery.

4. Immune suppression: Subcutaneous cyclosporine A (CSA: 10.0 mg/kg of body weight) or intraperitoneal (IP) FK-506/ rapamycin (1.0 mg/kg of body weight) should be started at least 3 days prior to transplantation, and should be given daily until sacrifice. Administration in the drinking water (instead of daily injections) is not advised. CSA is our immunosuppressant of choice when transplanting rodent-derived cells while FK-506/rapamycin is our choice when transplanting human-derived cells.

5. Autoclave surgical tools prior to surgery. Prepare a neat and clean place to do surgery, as well as an uncluttered place to put tools. A glass-bead sterilizer is also useful. Keep all of tools clean (especially rongeurs) during surgery, as this will make everything much easier.

3.3. Surgery: Animal Preparation and Surgery

Animal Prep:

1. Weigh animal.

2. Anesthesia: 4.0% chloral hydrate (alternative anesthesia: Isofluorane inhalation or cocktail of acepromazine maleate (0.7 mg/kg), ketamine (95 mg/kg), and xylazine (10 mg/kg)).

3. Inject chloral hydrate via IP: 1.0 mL of 4.0% chloral hydrate per 100.0 g of body weight.

4. Pinch toe and/or touch eye ball with cotton-tipped applicator to determine if animal is properly anesthetized.

5. Shave back hair with electric razor (no attachment needed). Shave from slightly rostral to ears to the middle of the animal's back. Shave well, and remove cut hair from skin to prevent hair getting into incision.

6. Soak piece of gauze with povidine-iodine antiseptic solution, and apply to skin over shaved area.

7. Place animal on rolled gauze pad (see Note 4).

Surgery:

8. On the lowest microscope magnification (we use 0.8× magnification), use scalpel blade to make midline incision. Stretch skin laterally with other hand to make skin taut (which makes

the skin easier to incise), and make incision from base of skull to scapula (see Fig. 1a).

9. Use scalpel blade to make incision through three muscle layers over the spinal column (see Note 5).

10. With two cotton-tipped applicators, use twisting motion to separate overlying muscle from paraspinal muscle (see Note 6).

11. Retract muscle with four homemade retractors (see Note 7, Fig. 1b, inset).

12. Clear paraspinal muscle from dorsal surface of spinal column (see Note 8).

13. Clean surface of vertebral bone well. Use rongeurs (the most important surgical instrument) to pull away muscle from levels C4, C5, and C6 (see Note 9) (see Fig. 1c).

14. Start with laminectomy of C5. Again, secure by holding C2 with rat-toothed forceps. Grab entire lamina (see diagram: grab near midline) with rongeurs. Position rongeurs so that tool is completely perpendicular to axis of spinal column. Slowly crush lamina. Do not push down into spinal cord, as this will cause damage to spinal tissue. Crush and gently pull broken piece of bone upward. Rongeurs should crush piece so that one can easily pull away to remove. If a piece of bone is still attached to the rest of the lamina, do not tug as this will cause hemorrhage and possible injury to spinal cord. Make sure that rongeurs are clean and sharp. (see Fig. 1d)

15. Extend laminectomy to all of C4–C6 laminae. Make one continuous opening in the bone over three spinal levels.

16. Do not extend laminectomy too far laterally because this will cause hemorrhage. In order to target ventral horn, the injection site is relatively medial, so it is unnecessary to extend laminectomy far laterally (see Fig. 1e).

17. Increase magnification to ~1.5×.

18. Clean off connective tissue (and possible dried blood) on top of dura with sharp/straight #5 forceps (see Note 10).

19. Incise dura, a very tough meningeal layer, with either mini-spring scissors or microknife. Make incision parallel to axis of spinal column just medial to the entry zone of the dorsal rootlets. This will allow one to target the ventral horn (see Fig. 1f) (Figure with diagrams: sagittal from top; transverse section of cord).

20. Use sharp/straight #5 forceps to grab and slightly lift dura off of spinal cord. Do not pinch/injure spinal cord. This takes some practice. Make incisions with microknife or very small spring scissors. Do not injure spinal cord.

21. Extend incision slightly in rostral–caudal axis (see Note 11).

22. Make incisions at all appropriate sites for each intended injection (see Note 12). Inject cells bilaterally at three levels (six sites total) to target large region of cervical enlargement. This, of course, depends on the specific experiment.

23. Cerebrospinal fluid will pour out. Dry surface of cord and entire exposure with gauze or cotton-tipped applicator.

24. For cell injection, use 10.0-μL Hamilton gastight RN syringe (see Note 13).

25. Attach 33-gauge/45-degree beveled metal RN Hamilton needle (see Note 14).

26. Cells will fall out of suspension easily inside the Eppendorf tube on ice. Immediately prior to loading the injection syringe, gently tap tube until cells go back into suspension. Do not flick too intensely as this will injure cells and/or cause bubbles. Alternatively, use 20.0-μL pipette-man to gently mix cells: pipette cell suspension only one or two times up and down.

27. Load enough cell suspension volume for only one injection site. Cells will fall out of suspension inside injection syringe if one plans to do multiple injections. Slowly take up cells into syringe to avoid bubbles and/or damage to cells (see Note 15).

28. Lower tip toward the spinal cord dorsal surface using microscope. Line up syringe/needle with axis of spinal column to properly target desired anatomical region (see Fig. 2a).

29. Aim needle just medial to entry zone of dorsal rootlets (see Fig. 2b). Gently touch surface of spinal cord with tip of needle. Slightly depress spinal cord with needle. Retract needle until spinal cord is back to normal. Record this position as the $z = 0.0$ using the ruler on the micromanipulator.

30. Slightly lower needle into spinal cord. Make sure to look in microscope while doing this. Lower needle to the depth of 1.5 mm to target ventral horn in adult rats (the age and sex of the animal does not make much of difference on depth). Lower needle to the depth of 0.75 mm to target ventral horn in adult mice (see Fig. 2c). Of course, depth and lateral position depend on the specific region of interest (see Note 16).

Wait 2 min (longer is even better) after lowering needle to desired depth before injection.

32. Inject 2.0 μL over 2–5 min at constant rate using pump controller.

33. Wait 2 min (longer is better) after injection before slowly removing needle from spinal cord.

34. Clean syringe with dH$_2$O after each injection in order to prevent clogging. Slowly draw up and expel 3–5 times. Do not draw air into syringe.

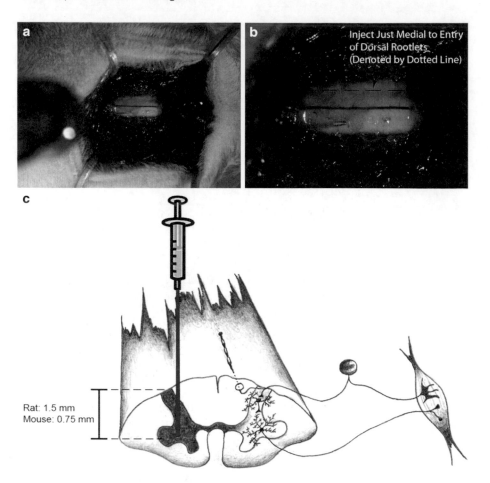

Fig. 2. (**a**) Injection setup. Lower injection tip toward the spinal cord dorsal surface using the microscope. Line up syringe/ needle with axis of spinal column to properly target desired anatomical region. Gently touch surface of spinal cord with tip of needle. Slightly depress cord with needle. Retract needle until spinal cord is back to normal flat state. Record this position as $z = 0.0$ using the ruler on the micromanipulator. (**b**) High magnification of spinal cord injection. Aim needle just medial to the entry zone of the dorsal rootlets (denoted by *dotted line*). (**c**) Diagram of spinal cord: how to target anatomical region of interest. When attempting to target the ventral horn, incise dura parallel to axis of spinal column just medial to the entry zone of the dorsal rootlets. This will allow one to target the ventral horn. Lower needle to depth of 1.5 mm to target ventral horn in adult rats (the age and sex of the animal does not make much of difference on depth). Lower needle to depth of 0.75 mm to target ventral horn in adult mice. Of course, depth and lateral position depend on the specific region of interest.

35. Return to lowest magnification (same as used at the beginning of surgery: ~0.8×) (see Note 17).

36. Optional: Suture paraspinal muscle closed with 4–0 suture.

37. Suture closed three overlying muscle layers at one time with 4–0 suture. Suture muscles at three locations in the rostral–caudal axis (see Fig. 3a).

38. Staple skin closed with 9.0-mm staples (see Fig. 3b). Tighten staples with needle holders to prevent animal from pulling off staples prior to full wound healing. Space staples approximately 0.5 mm apart.

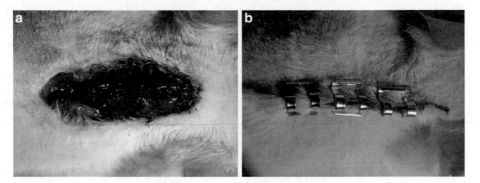

Fig. 3. (**a**) Closure of surgical site. Suture closed three overlying muscle layers at one time with 4-0 suture. Suture muscles at three locations in rostral-caudal axis. (**b**) Stapling of skin. Staple skin closed with 9.0-mm staples. Tighten staples with needle holders to prevent the animal from pulling off staples prior to full wound healing. Space staples approximately 0.5 mm apart.

39. Apply povidine-iodine antiseptic solution to wound area with soaked gauze.

40. Allow animal to recover on circulating warm water pad.

41. It should take several hours before animal is walking again.

42. Continue CSA daily until sacrifice.

43. See Notes 18–20.

4. Notes

1. Do not use 0.75-mL tubes, as the Hamilton syringe/needle cannot fit deeply into this tube. Multiple injections will likely be made during the surgery sessions, and one wants to avoid disturbing the same tube of cells many times. Try to prepare at least 50% more volume of cell suspension than is needed. Keep cells on ice throughout the surgery session. Transplant cells within 4–5 h of preparing cell suspensions in order to assure greatest viability of cells posttransplantation.

2. It is preferable to use all animals from the same sex because the disease that is being studied may differ between males and females; however, it may also be useful to have enough animals from both sexes to detect possible sex-specific effects, as this phenomenon has been reported with SOD1^{G93A} rats and SOD1^{G93A} mice.

3. We transplant SOD1^{G93A} rats at approximately 90 days of age. This age is slightly before the onset of hindlimb symptoms and overall weight loss in rats that live on average to ~120 days of age. However, the literature (and our own experiences) report that SOD1^{G93A} rats can lose transgene copy number. In this

case, there is a "drift" to an average endstage of ~155 days of age and, in some cases, over 200 days of age.

4. For an adult rat, the rolled gauze pad should be ~1.0 in. in thickness and ~6.0 in. in length. Tape this pad to surgery board so that it does not move. Place animal on pad at level of chest/shoulder. Stretch out both arms on pad, and tape down arms to surgery board. The idea is to prop the cervical spinal cord up so that it is easier to work throughout the surgery since the cervical cord is found deep below the surface of the skin.

5. Slightly squeeze the muscle (from lateral to medial) on both sides with other hand. Make sure to press firmly enough with scalpel to cut all muscle layers with the least number of cuts as possible so that muscle incisions will not be jagged, which makes suturing at the end of surgery easier. However, do not press too firmly to prevent creating damage in deeper tissue (spinal column).

6. One can be vigorous here. Do not cut or tear, as this will cause hemorrhage. Twisting motion with the applicators will nicely tease the tissue apart without causing damage or hemorrhage. If hemorrhage occurs at any stage of the surgery, it is best to be patient. Do not try to rush ahead with lots of hemorrhage because the blood will obscure vision of the surgical field. Cauterization can be used, but we find that patience is best. With practice, there should be little or no hemorrhage throughout this surgery.

7. These retractors can be made from sturdy paperclips shaped into a retractor (see Fig. 1b inset). Autoclave these before surgery. Tie string to retractor. Exposure should be square/rectangle-shaped. This shape can be achieved by pulling to four corners using the four retractors. Tape string to board in order to secure retractors (see Fig. 1b). Commercially available retractors can also be used, but we prefer not to use these. It is crucial to create a good surgical field in order to clearly see throughout the surgery. Do not "blindly" continue along.

8. To deal with paraspinal muscle on the dorsal surface of the spinal column, there are two potential strategies. For the first strategy, use rat-toothed forceps and medium-sized spring scissors to remove paraspinal muscle from dorsal surface of vertebrae. Try to make cuts very close to bone to better expose bone surface. Make cuts parallel to surface of vertebrae, but do not cut down into cord. For the second approach, make a midline incision in paraspinal muscle, and retract laterally with smaller retractors. In the second approach, the muscle can then be sutured back together after surgery.

9. This, of course, depends on the spinal levels that are targeted. While using rongeurs with one hand, secure entire spinal

column by grasping muscle over C2 process with rat-toothed forceps in other hand. C2 is the large rostral process in the surgical field. C3 is slightly under the muscle as well. C4, C5, and C6 are easy to access because they have little muscle overlying them.

10. The difference between connective tissue and dura needs to be learned with experience.

11. Tension of dura will slightly separate dura to create a nice exposure of spinal cord.

12. Dura has a hazy translucent appearance while spinal cord surface has a brighter white-colored appearance. One will need to learn to distinguish these apart. It is advisable not to try to inject through the dura with the injection needle. The dura is tough so that it will be hard to pierce through it. In addition, piercing through the dura can affect the trajectory of the needle's insertion and will therefore affect the anatomically targeted delivery of cells.

13. Having a surgical assistant to anesthetize, shave, "open" and "close" the animals is quite helpful for smoothly getting through the most animals as possible.

14. The needle should be sharp. Use the same needle for only ~20 injections (the needle will be sharper with less injections). One can also use pulled glass capillary tubes for needles. Use epoxy to attach pulled glass tip to 26-gauge blunt Hamilton metal needle. Trim outer tip diameter to ~75.0–100.0 μL (depending on diameter of cells: one does not want to damage cells during the process of injection) using a surgical microscope and micrometer slide. We prefer the 33-gauge metal needles.

15. Specifically for the glial progenitor cells that we work with in our lab, we inject 2.0 μL of cell suspension at a dilution of 50,000–75,000 cells per 1.0 μL over 2–5 min. Do not inject a very large number of cells (particularly if they are larger in size than our glial progenitors: 10.0–20.0 μm in diameter) because this will cause damage to the spinal cord parenchyma. We conducted an extensive cell dosing experiment to determine the optimal numbers of cells to inject at each site. The needs will determine the specific cell type.

16. Do not disturb injection system/surgery board or surgery table while needle is inserted into spinal cord to avoid damage to spinal tissue.

17. Suture dura closed using 9–0 suture, but this is definitely not necessary. It is also challenging. We sutured dura closed in the past, but we find no difference in animal behavior, transplant survival, or spinal cord histology when the dura is not closed with suture.

18. The entire surgery should take less than ~45 min (with six injection sites). Otherwise, the animal will begin to wake up and breathe more heavily, which will adversely affect injection into spinal cord.

19. This same protocol can also be used for both rat and mouse cervical spinal cords, as well as other levels of the rat and mouse spinal cords.

20. The specifics of behavior testing, histological analysis, and other outcome measures analyzed following transplantation should be tailored to the experimental needs/model.

Acknowledgments

We thank: all members of the Maragakis lab for helpful discussion; The Paralyzed Veterans of America (A.C.L.), The Robert Packard Center for ALS Research (N.J.M.), The ALS Association (N.J.M.) and the NIH (F32-NS059155: A.C.L., R01-NS41680: J.D.R. and N.J.M.) for funding.

References

1. Bruijn, L. I., Miller, T. M., and Cleveland, D. W. (2004) Unraveling the mechanisms involved in motor neuron degeneration in ALS. *Annu Rev Neurosci.* **27**, 723–749.

2. Rosen, D. R., Siddique, T., Patterson, D., Figlewicz, D. A., Sapp, P., Hentati, A., Donaldson, D., Goto, J., O'Regan, J. P., Deng, H. X., and et al. (1993) Mutations in Cu/Zn superoxide dismutase gene are associated with familial amyotrophic lateral sclerosis. *Nature.* **362**, 59–62.

3. Bruijn, L. I., Becher, M. W., Lee, M. K., Anderson, K. L., Jenkins, N. A., Copeland, N. G., Sisodia, S. S., Rothstein, J. D., Borchelt, D. R., Price, D. L., and Cleveland, D. W. (1997) ALS-linked SOD1 mutant G85R mediates damage to astrocytes and promotes rapidly progressive disease with SOD1-containing inclusions. *Neuron.* **18**, 327–338.

4. Gurney, M. E., Pu, H., Chiu, A. Y., Dal Canto, M. C., Polchow, C. Y., Alexander, D. D., Caliendo, J., Hentati, A., Kwon, Y. W., Deng, H. X., and et al. (1994) Motor neuron degeneration in mice that express a human Cu,Zn superoxide dismutase mutation. *Science.* **264**, 1772–1775.

5. Wong, P. C., Pardo, C. A., Borchelt, D. R., Lee, M. K., Copeland, N. G., Jenkins, N. A., Sisodia, S. S., Cleveland, D. W., and Price, D. L. (1995) An adverse property of a familial ALS-linked SOD1 mutation causes motor neuron disease characterized by vacuolar degeneration of mitochondria. *Neuron.* **14**, 1105–1116.

6. Howland, D. S., Liu, J., She, Y., Goad, B., Maragakis, N. J., Kim, B., Erickson, J., Kulik, J., DeVito, L., Psaltis, G., DeGennaro, L. J., Cleveland, D. W., and Rothstein, J. D. (2002) Focal loss of the glutamate transporter EAAT2 in a transgenic rat model of SOD1 mutant-mediated amyotrophic lateral sclerosis (ALS). *Proc Natl Acad Sci USA.* **99**, 1604–1609.

7. Nagai, M., Aoki, M., Miyoshi, I., Kato, M., Pasinelli, P., Kasai, N., Brown, R. H., Jr., and Itoyama, Y. (2001) Rats expressing human cytosolic copper-zinc superoxide dismutase transgenes with amyotrophic lateral sclerosis: associated mutations develop motor neuron disease. *J Neurosci.* **21**, 9246–9254.

8. Gage, F. H. (2000) Mammalian neural stem cells. *Science.* **287**, 1433–1438.

9. Lepore, A. C., and Maragakis, N. J. (2007) Targeted stem cell transplantation strategies in ALS. *Neurochemistry International.* **50**, 966–975.

10. Deshpande, D. M., Kim, Y. S., Martinez, T., Carmen, J., Dike, S., Shats, I., Rubin, L. L., Drummond, J., Krishnan, C., Hoke, A.,

Maragakis, N., Shefner, J., Rothstein, J. D., and Kerr, D. A. (2006) Recovery from paralysis in adult rats using embryonic stem cells. *Ann Neurol.* **60**, 32–44.

11. Barbeito, L. H., Pehar, M., Cassina, P., Vargas, M. R., Peluffo, H., Viera, L., Estevez, A. G., and Beckman, J. S. (2004) A role for astrocytes in motor neuron loss in amyotrophic lateral sclerosis. *Brain Res Brain Res Rev.* **47**, 263–274.

12. Lepore, A. C., Rauck, B., Dejea, C., Pardo, A. C., Rao, M. S., Rothstein, J. D., and Maragakis, N. J. (2008) Focal transplantation-based astrocyte replacement is neuroprotective in a model of motor neuron disease. *Nature Neuroscience.* **11**, 1294–1301.

13. Llado, J., Haenggeli, C., Maragakis, N. J., Snyder, E. Y., and Rothstein, J. D. (2004) Neural stem cells protect against glutamate-induced excitotoxicity and promote survival of injured motor neurons through the secretion of neurotrophic factors. *Mol Cell Neurosci.* **27**, 322–331.

14. Cao, Q., Xu, X. M., Devries, W. H., Enzmann, G. U., Ping, P., Tsoulfas, P., Wood, P. M., Bunge, M. B., and Whittemore, S. R. (2005) Functional recovery in traumatic spinal cord injury after transplantation of multineurotrophin-expressing glial-restricted precursor cells. *J Neurosci.* **25**, 6947–6957.

15. Pluchino, S., Zanotti, L., Rossi, B., Brambilla, E., Ottoboni, L., Salani, G., Martinello, M., Cattalini, A., Bergami, A., Furlan, R., Comi, G., Constantin, G., and Martino, G. (2005) Neurosphere-derived multipotent precursors promote neuroprotection by an immunomodulatory mechanism. *Nature.* **436**, 266–271.

16. Kaplan, L. M., and Hollander, D. (1994) Respiratory dysfunction in amyotrophic lateral sclerosis. *Clinics in Chest Medicine.* **15**, 675–681.

17. Miller, R. G., Rosenberg, J. A., Gelinas, D. F., Mitsumoto, H., Newman, D., Sufit, R., Borasio, G. D., Bradley, W. G., Bromberg, M. B., Brooks, B. R., Kasarskis, E. J., Munsat, T. L., and Oppenheimer, E. A. (1999) Practice parameter: the care of the patient with amyotrophic lateral sclerosis (an evidence-based review): report of the Quality Standards Subcommittee of the American Academy of Neurology: ALS Practice Parameters Task Force. *Neurology.* **52**, 1311–1323.

18. Tandan, R., and Bradley, W. G. (1985) Amyotrophic lateral sclerosis: Part 1. Clinical features, pathology, and ethical issues in management. *Ann Neurol.* **18**, 271–280.

19. Llado, J., Haenggeli, C., Pardo, A., Wong, V., Benson, L., Coccia, C., Rothstein, J. D., Shefner, J. M., and Maragakis, N. J. (2006) Degeneration of respiratory motor neurons in the SOD1 G93A transgenic rat model of ALS. *Neurobiol Dis.* **21**, 110–118.

20. Rao, M. S. (1999) Multipotent and restricted precursors in the central nervous system. *Anat Rec.* **257**, 137–148.

21. Rao, M. S., and Mayer-Proschel, M. (1997) Glial-restricted precursors are derived from multipotent neuroepithelial stem cells. *Dev Biol.* **188**, 48–63.

22. Lepore, A. C., Han, S. S., Tyler-Polsz, C., Cai, J., Rao, M. S., and Fischer, I. (2004) Differential fate of multipotent and lineage-restricted neural precursors following transplantation into the adult CNS. *Neuron Glia Biology.* **1**, 113–126.

23. Lepore, A. C., Neuhuber, B., Connors, T. M., Han, S. S., Liu, Y., Daniels, M. P., Rao, M. S., and Fischer, I. (2006) Long-term fate of neural precursor cells following transplantation into developing and adult CNS. *Neuroscience.* **139**, 513–530.

24. Lepore, A. C., Walczak, P., Rao, M. S., Fischer, I., and Bulte, J. W. (2006) MR imaging of lineage-restricted neural precursors following transplantation into the adult spinal cord. *Exp Neurol.* **201**, 49–59.

Chapter 32

Evaluation of Histone Deacetylase Inhibitors as Therapeutics for Neurodegenerative Diseases

Elisabetta Soragni, Chunping Xu, Andrew Cooper, Heather L. Plasterer, James R. Rusche, and Joel M. Gottesfeld

Abstract

Various neurodegenerative diseases are associated with aberrant gene expression. We recently identified a novel class of pimelic *o*-aminobenzamide histone deacetylase (HDAC) inhibitors that show promise as therapeutics in the neurodegenerative diseases Friedreich's ataxia (FRDA) and Huntington's disease (HD). Here, we describe the various techniques used in our laboratories to dissect mechanisms of gene silencing in FRDA and HD, and to test our HDAC inhibitors for their ability to reverse changes in gene expression in cellular models.

Key words: Histone deacetylase, Epigenetics, Friedreich's ataxia, Huntington's disease, Gene activation, Pimelic *o*-aminobenzamide HDAC inhibitor, qRT-PCR, Chromatin immunoprecipitation

1. Introduction

Histone deacetylase (HDAC) inhibitors have received considerable attention as potential therapeutics for cancer (1) and for a variety of neurological and neurodegenerative diseases (2). We recently described a series of pimelic *o*-aminobenzamide HDAC inhibitors (HDACi) that reverse heterochromatin-mediated silencing of the *frataxin* (*FXN*) gene in the neurodegenerative disease Friedreich's ataxia (FRDA) (3–5) and also show efficacy in a mouse model for Huntington's disease (HD) (6). FRDA is caused by the expansion of the simple triplet repeat DNA sequence GAA•TTC within intron 1 of the *FXN* gene, encoding the essential mitochondrial protein frataxin. Repeats over a threshold level of ~70 induce heterochromatin formation (3), and concomitant gene silencing, resulting in decreased

Giovanni Manfredi and Hibiki Kawamata (eds.), *Neurodegeneration: Methods and Protocols*, Methods in Molecular Biology, vol. 793, DOI 10.1007/978-1-61779-328-8_32, © Springer Science+Business Media, LLC 2011

levels of frataxin protein in affected individuals. Importantly for therapeutic development, the pimelic *o*-aminobenzamides cross the blood brain barrier, cause global increases in histone acetylation in cells and in the mouse brain, and show good tolerance in murine models of disease (5, 6). These molecules also directly affect the histone acetylation status of *FXN* gene chromatin in FRDA patient cells and in the mouse brain, increasing acetylation at particular lysine residues on histones H3 and H4, and increase *FXN* gene expression in the brain and heart in a mouse model for FRDA (5). Strikingly, gene expression microarray analysis indicates that most of the differentially expressed genes in FRDA mice revert toward wild-type levels on treatment with the pimelic *o*-aminobenzamide HDAC inhibitor (5). Similar results have been obtained in a mouse model for HD, where one of these compounds ameliorated the disease phenotype and reversed many of the transcriptional abnormalities found in the brain of R6/2 HD mice (6). We find that members of the pimelic *o*-aminobenzamide HDAC inhibitor family only increase *FXN* gene expression in FRDA patient cells and in the FRDA mouse model and are without effect on wild-type human or mouse *FXN* alleles (refs. 3–5 and see below), indicating that these molecule reverse the effects of the GAA•TTC repeat on the transcriptional status of the *FXN* gene. We reported that the pimelic *o*-aminobenzamides are inhibitors of class I HDAC enzymes, with a marked preference for HDAC3 (7), and that molecules with selectivity for HDACs 1 and 2 fail to increase *FXN* gene expression in patient cells (8).

Here, we present the methods used in our laboratories to interrogate the chromatin changes associated with dysregulated genes in cellular models for neurodegenerative diseases, quantitative reverse transcriptase PCR methods for estimation of mRNA levels for these genes and methods for synthesis and characterization of HDAC inhibitors. We focus on Friedreich's ataxia, but the methods employed can be extended to other neurological and neurodegenerative diseases, where transcriptional dysfunction is at the core of the disease etiology.

2. Materials

2.1. Synthesis of Pimelic o-Aminobenzamide HDACi and Activity Assays

1. Detailed methods for the synthesis pimelic *o*-aminobenzamide HDACi have been published (3–5, 8) and two such compounds are commercially available from Calbiochem/EMD Biosciences: HDACi **4b** (N^1-(2-aminophenyl)-N^7-phenylheptanediamide, sold as *Histone Deacetylase Inhibitor IV*) and HDACi **106** (N^1-(2-aminophenyl)-N^7-*p*-tolylheptanediamide, sold as *Histone Deacetylase Inhibitor VII*).

Compound dilution: Test compounds at a starting concentration of 9 mM (in DMSO) are first diluted in DMSO at a 1/3 ratio to give 11 concentration points (9 mM, 3 mM, 1 mM, 0.33 mM, 0.11 mM, 0.037 mM, 0.012 mM, 4.11 μM, 1.37 μM, 0.47 μM, 0.15 μM), and a control with no compound, then the whole series (the 12 concentration points in DMSO) is diluted in HDAC buffer at a 1/20 ratio to give 12 new concentrations. 10 μL inhibitor is used per assay, and the assay volume is 25 μL, giving final compound concentrations of 180 μM, 60 μM, 20 μM, 6.7 μM, 2.2 μM, 0.74 μM, 0.25 μM, 82.3 nM, 27.4 nM, 9.1 nM, 3.0 nM, and 0 nM, respectively.

2. HDAC buffer: 50 mM Tris–Cl, pH 8.0, 137 mM NaCl, 2.7 mM KCl, 1 mM $MgCl_2$.

3. Fluor-de-Lys™ deacetylase substrate: 50 mM in DMSO (Enzo Life Sciences, Plymouth Meeting, PA).

4. BSA: Stock (10 mg/mL), prepared in HDAC buffer.

5. HDAC enzymes: BPS Bioscience, San Diego, CA.

6. Trypsin: Stock (10 mg/mL), prepared in HDAC buffer.

7. 96-well plates: Greiner Bio-one, black, U-shape.

8. Lys-C from EMD: 500 μL HDAC buffer is added to Lys-C vial to provide a stock solution.

9. Equipment: 96-well plate reader, such as Tecan Infinite M200; centrifuge, such as Eppendorf 5810 with rotor adaptor for 96-well plates.

2.2. Growth of Lymphoblastoid Cells in Culture

1. Cell lines GM15850 and GM15851, from an FRDA affected male and from an unaffected male sibling, respectively (Coriell Cell Repositories, Camden, New Jersey).

2. Lymphocyte growth medium: RPMI Medium 1640+ GlutaMAX-I supplemented with 15% fetal bovine serum and antimycotic, antibiotic.

3. Hank's balanced salt solution (HBSS).

2.3. Isolation of Peripheral Blood Mononuclear Cells

1. Lymphocyte growth medium.

2. HBSS.

3. Ficoll-Paque PLUS (GE Healthcare, Uppsala, Sweden).

2.4. Incubation with HDAC Inhibitors

1. DMSO and HDACi of appropriate dilutions in DMSO.

2. CellTiter 96® AQ_ueous Non-Radioactive Cell Proliferation Assay (MTS, Promega, Madison, WI).

2.5. In Cell Deacetylase Assay

1. Phosphate buffered saline (PBS), pH 7.4.

2. HDAC buffer: 50 mM Tris–Cl, pH 8.0, 137 mM NaCl, 2.7 mM KCl, 1 mM $MgCl_2$.

3. Deacetylase lysis buffer: 50 mM Tris–Cl, pH 8.0, 137 mM NaCl, 2.7 mM KCl, 1 mM $MgCl_2$, 1% IGEPAL.

4. BCA Protein Assay Kit (Thermo Scientific).

5. 96-well, flat-bottom, black opaque microplate.

6. Fluor-de-Lys™ deacetylase substrate: 50 mM in DMSO.

7. 10 mg/mL BSA, prepared in HDAC buffer.

8. 10 mg/mL trypsin, prepared in HDAC buffer.

9. Equipment: 96-well plate reader; centrifuge, such as Sorvall RT6000B; Vortex mixer.

2.6. qRT-PCR to Detect FXN mRNA Expression

1. RNeasy Plus Mini Kit (QIAgen Hilden, Germany).

2. qScript™One-Step SYBR® Green qRT-PCR Kit (Quanta Biosciences, Gaithersburg, MD).

3. Primer sequences for the *FXN* gene, 5′-CAGAGGAAACGC-TGGACTCT-3′ and 5′-AGCCAGATTTGCTTGTTTGG-3′.

4. Equipment: MJ Research Chromo4 thermal cycler.

2.7. Western Blotting for Detection of Frataxin and Acetylated Histones

1. RIPA buffer: 50 mM Tris–HCl pH 7.5, 150 mM NaCl, 1 mM EDTA, 1% Triton X-100, 1% sodium deoxycholate, 0.1% SDS, protease inhibitors.

2. Equipment: sonicator, such as Branson sonifier 150.

3. BCA™ Protein Assay Kit (Thermo Scientific, Rockford, IL).

4. 4–12% Bis-tris polyacrylamide gel.

5. MES running buffer.

6. Nitrocellulose membrane.

7. TBS-T: 10 mM Tris–HCl pH 8, 150 mM NaCl, 0.1% Tween.

8. Dry non-fat milk.

9. Antibodies:

   Mouse anti-frataxin antibody (Invitrogen, Carlsbad, CA).

   Rabbit anti-RPL13a antibody (Cell Signaling, Beverly, MA).

   Rabbit anti-histone H3 antibody (Abcam, Cambridge, MA).

   Rabbit anti-acetylated histone H3 antibody (Millipore, Billerica, MA).

   Anti-mouse IgG HRP-linked secondary antibody.

   Anti-rabbit IgG HRP-linked secondary antibody.

10. HRP chemiluminescent substrate.

2.8. Chromatin Immunoprecipitation

1. Equipment: sonicator (Branson sonifier 150); MJ Research Chromo4 thermal cycler.

2. 37% formaldehyde.

3. 2 M glycine.

4. HBSS.

5. Lysis buffer: 1% SDS, 10 mM EDTA pH 8, 50 mM Tris–Cl pH 8, protease inhibitors.

6. Dilution buffer: 1% Triton X-100, 150 mM NaCl, 2 mM EDTA, 20 mM Tris–Cl pH 8, protease inhibitors.

7. Protein A agarose beads.

8. Antibodies:

    Rabbit anti-histone H3 (Abcam).

    Rabbit anti-histone H3AcK9 (Millipore).

    Rabbit anti-histone H3AcK14 (Millipore).

    Rabbit anti-histone H4 (Millipore).

    Rabbit anti-histone H4AcK5 (Millipore).

    Rabbit anti-histone H4AcK8 (Millipore).

    Rabbit anti-histone H4AcK12 (Millipore).

    Rabbit anti-histone H4AcK16 (Abcam).

    Normal rabbit IgG.

9. Low salt wash buffer: 1% Triton X-100, 0.1% SDS, 150 mM NaCl, 2 mM EDTA pH 8, 20 mM Tris–Cl pH 8.

10. High salt wash buffer: 1% Triton X-100, 0.1% SDS, 500 mM NaCl, 2 mM EDTA pH 8, 20 mM Tris–Cl pH 8.

11. Elution buffer: 1% SDS, 100 mM $NaHCO_3$.

12. 20 mg/mL proteinase K.

13. Phenol–chloroform–isoamyl alcohol (25:24:1).

14. 100% ethanol.

15. 3 M sodium acetate pH 5.2.

16. Glycogen.

17. 75% ethanol.

18. SYBR® Green PCR mix.

19. Primer sequences:

    GAPDH coding sequence: 5′-CACCGTCAAGGCTGAGA-ACG-3′ and 5′-ATACCCAAGGGAGCCACACC-3′.

    FXN promoter: 5′-CCCCACATACCCAACTGCTG-3′ and 5′-GCCCGCCGCTTCTAAAATTC-3′.

    FXN upstream of the GAA repeats: 5′-GAAACCCAAAGAA-TGGCTGTG-3′ and 5′-TTCCCTCCTCGTGAAAC-ACC-3′.

    FXN downstream of the GAA repeats: 5′-CTGGAAAAAT-AGGCAAGTGTGG-3′ and 5′-CAGGGGTGGAAGCCC-AATAC-3′.

3. Methods

3.1. In Vitro HDAC Assays

IC_{50} protocol:

1. 5 μL of BSA, at 1 mg/mL, is added to 12 wells in a 96-well plate and 5 μL of enzyme at an appropriate dilution is added to the above 12 wells (given by BPS Biosciences for the individual HDAC enzymes).

2. 10 μL of inhibitor at 12 serial (1/3) dilutions in HDAC buffer are added to each well.

3. The mixture is briefly centrifuged and incubated for 2 h at ambient temperature.

4. 5 μL Fluor-de-Lys™ deacetylase substrate (100 μM) is added to the above 12 wells.

   The solution is centrifuged briefly and incubated for 1 h at ambient temperature.

5. 25 μL trypsin (1 mg/mL) is added to each well, to a final volume of 50 μL.

6. The mixture is centrifuged briefly and transferred to a fluorometric plate reader (excitation wavelength 360 nm, emission wavelength 460 nm). Readings are taken at 1-min intervals. A 20-min data set is used to determine the IC_{50} using GraphPad Prism fitting software.

K_i protocol:

1. 10 μL of BSA (1 mg/mL) is added to eight wells in 96-well plate.

2. 10 μL Fluor-de-Lys™ deacetylase substrate (31.25 μM, 1/1,600 dilution from 50 mM stock solution) is added to each plate.

3. 10 μL of inhibitor at eight serial (1/2) dilutions (10–0 μM) is added to each well.

4. 10 μL Lys-C (1/20 dilution from stock) is added to each well, the solution is centrifuged briefly and incubated for 10 min at ambient temperature.

5. 10 μL enzyme (1/1,000 dilution from stock) is added, the solution is centrifuged briefly and the plate transferred to a fluorometric plate reader. Data are taken from 0 to 90 min to determine the K_i using the GraphPad Prism fitting software.

3.2. Growth of Lymphoblastoid Cells in Culture

1. Lymphoblastoid cells are grown in suspension culture in lymphocyte growth medium at 37°C in a CO_2 incubator. Cells are split 1:3 every 48 h (see Note 1). To split, spin cells for 5 min at $200 \times g$, and resuspend them in HBSS. Spin again for 5 min at $200 \times g$ and resuspend in lymphocyte growth medium.

3.3. Isolation of Peripheral Blood Mononuclear Cells

1. 50 mL of donor human blood is collected in heparin tubes or EDTA tubes (see Note 2).

2. Distribute blood in four 50 mL tubes and add 12.5 mL of HBSS to each tube.

3. Underlay the diluted blood with 12.5 mL of Ficoll-Paque PLUS. Care should be taken not to mix the two solutions.

4. Spin at $400 \times g$ for 40 min (see Note 3).

5. After centrifugation, three layers should be visible: the upper layer of plasma, the middle layer of ficoll, and the lower layer of granulocytes and erythrocytes. At the interface of ficoll and plasma, a thin white layer is also visible. This is the lymphocyte layer or "buffy coat." Aspirate the lymphocyte layer with a plastic pipette (ensure that the whole layer is removed by aspirating a minimum amount of the ficoll layer) and transfer to a new tube.

6. Add at least three volumes of HBSS and centrifuge at $200 \times g$ for 10 min.

7. Remove the supernatant and resuspend in 20 mL of lymphocyte growth medium.

8. Incubate cells at 37°C in a CO_2 incubator for 4–16 h.

3.4. Incubation with HDAC Inhibitors

1. Dilute cells at a concentration of $1–2 \times 10^6$ cells/mL for primary lymphocytes or $3–4 \times 10^5$ cells/mL for lymphoblastoid cells. Distribute the cells in a 6-well plate, 4 mL per well.

2. Make four dilutions of HDACi in DMSO: 10, 5, 2.5, and 1 mM (see Note 4).

3. Add 4 μL of each HDAC inhibitor dilution to each well and 4 μL of DMSO to the vehicle control sample.

4. Incubate at 37°C in a CO_2 incubator and collect cells for in cell deacetylase assay after 4 h, or RNA extraction, Western blot or chromatin immunoprecipitation (ChIP) after 24–48 h.

5. Use a cytotoxicity assay to determine the EC_{50} of each compound (see Note 5). Remove 100 μL and transfer to a 96-well plate. Use the CellTiter 96® AQ_{ucous} Non-Radioactive Cell Proliferation Assay, following manufacturer's instructions (see Note 6).

3.5. In-Cell Deacetylase Assay

1. Collect 2×10^6 cells by centrifugation at $400 \times g$ for 5 min and aspirate supernatant.

2. Resuspend cell pellet in 5 mL PBS, centrifuge at $200 \times g$ for 5 min, and aspirate supernatant.

3. Add 150 μL deacetylase lysis buffer to the pellet and mix by vortex until lysate appears homogeneous.

4. Centrifuge samples at $200 \times g$ for 5 min to pellet insoluble material.

5. Transfer 120 μL of each supernatant to a 2 mL microcentrifuge tube and place on ice.

6. Remove 5 μL from each cell lysate and determine protein concentration using BCA protein assay kit following the manufacturer's instructions.

7. Dilute all samples to the lowest protein concentration using cell lysis buffer (see Note 7).

8. Transfer 100 μL of each cell lysate to one well of a 96-well black opaque microplate.

9. Dilute Fluor-de-Lys™ Substrate 1:100 in HDAC buffer to make a working concentration of 500 μM.

10. Transfer 50 μL HDAC buffer to one well of the assay plate for a "substrate only" control.

11. Add 10 μL of diluted Fluor-de-Lys™ Substrate to 100 μL of each sample. Add 5 μL of diluted Fluor-de-Lys™ Substrate to the "substrate only" and control well.

12. Transfer 55 μL of each sample from one well of the assay plate to a second well of the plate to generate duplicate assay wells.

13. Incubate the assay plate at room temperature (20–23°C) for 30 min.

14. Dilute trypsin solution to 1 mg/mL with HDAC buffer, add 50 μL to each well of the assay plate.

15. Read plate in fluorometric plate reader (excitation wavelength of 360 nm and emission wavelength of 460 nm) at 2-min intervals for a total of 30 min.

16. Analyze fluorescent intensity at 30 min. Inspect fluorescence vs. time curves for possible anomalies, such as nonlinear behavior.

3.6. qRT-PCR to Detect FXN mRNA Expression

1. Isolate total RNA from $1-2 \times 10^6$ cells for primary lymphocytes or $2.5-5 \times 10^5$ cells for lymphoblastoid cells using RNeasy Plus Mini Kit, following manufacturer's instructions (see Note 8) and measure its concentration with a spectrophotometer (see Note 9).

2. qRT-PCR is performed using qScript™ One-Step SYBR® Green qRT-PCR Kit. Each RNA sample is analyzed in triplicate using FXN primer set (see Note 10).

3. Each 20 μL reaction contains: 20 ng of total RNA, 10 μL of Quanta qScript™ One-Step SYBR® Green mix, 0.1 μL of reverse transcriptase, and 10 pmol of each primer.

4. Use the following conditions for the thermocycler (see Note 11): 55°C for 10 min, 95°C for 5 min, 40 cycles of 95°C for 10 s, 55°C for 10 s, and 72°C for 30 s. After cycling is complete set the PCR machine to calculate dissociation curves from 70 to 96°C (see Note 12). Visualize the amplification

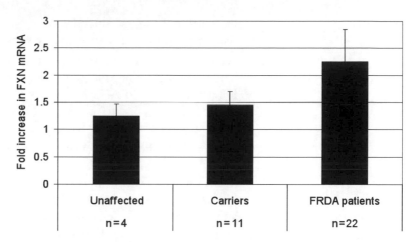

Fig. 1. qRT-PCR determination of *FXN* mRNA levels in primary lymphocytes from unaffected individuals, carriers and Friedreich's ataxia patients, after the treatment with HDACi **4b** at 5 μM, for 24 h. The fold-increase in *FXN* mRNA compared to the DMSO control is shown, and the number of patient samples is shown for each group. HDACi treatment fails to show a statistically significant increase in *FXN* mRNA levels in PBMCs from unaffected individuals, while significant increases in *FXN* mRNA levels in PBMCs from FRDA patients are observed ($p = 0.0036$). *FXN* mRNA levels in this experiment were normalized to GAPDH mRNA, which did not change significantly from the DMSO control. We thank Dr. Ryan Burnett for these data (*see* ref. 3).

curve in a logarithmic scale and set the baseline from cycle 3 to cycle 9. Set the threshold in the linear phase of amplification and record the threshold cycle (C_t) for each well. Average the C_t values for the three replicates and calculate the ΔC_t for each sample ($\Delta C_t = C_t$ (HDACi-treated sample) $- C_t$ (DMSO sample)). The relative amount of *FXN* mRNA compared to the DMSO-treated sample is $2^{-\Delta C_t}$. Figure 1 provides an example of results with compound **4b** (5 μM) in primary lymphocytes from normal donors, nonaffected carriers (individuals who are heterozygous for the expanded GAA repeats in *FXN*) and FRDA patients, after a 24 h incubation of these cells in culture. No effects on *FXN* mRNA levels are noted in peripheral blood mononuclear cells (PBMCs) from healthy donors, an intermediate effect is seen in PBMCs from carriers and a clear ~2- to 2.5-fold increase in *FXN* mRNA is noted in patient cells, showing that this compound only affects transcription of pathogenic *FXN* alleles and has no affect on the wild-type gene.

3.7. Western Blotting for Detection of Frataxin and Acetylated Histones

1. Collect 2×10^6 cells for primary lymphocytes or 1×10^6 cells for lymphoblastoid cells by centrifugation and wash them in HBSS.

2. Resuspend cells in 100 μL of RIPA buffer.

3. Sonicate for 5 s with a Branson sonifier at 4 W output to reduce the viscosity of the solution.

4. Determine protein concentration with the BCA™ Protein Assay Kit, following manufacturer's instructions.

5. Separate 30 μg of total protein extract for frataxin and RPL13a detection or 5 μg of total protein extract for H3 and acetylated H3 detection, onto a 4–12% bis-tris polyacrylamide gel in MES buffer and transfer to a nitrocellulose membrane following manufacturer's instructions.

6. Block membrane with 5% nonfat milk for 1 h at room temperature. For frataxin and RPL13a detection, incubate the blot overnight at 4°C with the antibody in 5% nonfat milk (1:1,000 dilution for frataxin and 1:2,000 dilution for RPL13). For H3 and acetylated H3 detection, incubate the blot 1 h at room temperature with the antibody in 5% nonfat milk (1:10,000 dilution for H3 and 1:5,000 dilution for acetylated H3).

7. Wash the blot four times for 5 min with TBS-T at ambient temperature.

8. Incubate for 1 h at room temperature with HRP-conjugated anti-mouse (for frataxin detection) or anti-rabbit (for RPL13a, H3 and acetylated H3 detection) secondary antibody.

9. Wash with TBS-T four times for 5 min at room temperature.

10. Incubate for 5 min at room temperature with HRP chemiluminescent substrate and expose to X-ray films.

11. Scan the film and quantify band intensity using image analysis software, such as Imagequant or imageJ.

3.8. Chromatin Immunoprecipitation

1. Add formaldehyde to cell media to a final concentration of 1% and incubate 10 min at ambient temperature with shaking.

2. Add glycine to a final concentration of 0.125 M and incubate for 5 min at ambient temperature with shaking.

3. Spin cells for 5 min at $200 \times g$ and wash with HBSS.

4. Resuspend at 1×10^7 cell/mL in lysis buffer and incubate for 15 min on ice.

5. Sonicate four times for 15 s using a Brason sonifier at 9 W output, incubating the samples on ice for 1 min after each round (see Note 13).

6. Spin at $12,000 \times g$ for 10 min at 4°C and collect the supernatant. This is the whole cell extract.

7. Preclear the cell extract by incubating with 60 μL of protein A-agarose for 1 h at 4°C.

8. Spin the sample for 30 s at $800 \times g$ and transfer the supernatant to a new tube.

9. Use 100 μL of whole cell extract per immunoprecipitation (IP, include a sample to be incubated with normal rabbit IgG, as a negative control) and save 10 μL (label this sample as INPUT).

10. Dilute the 100 μL of extract with 0.9 mL of dilution buffer and add 2–4 μg of the desired histone antibody or rabbit normal IgG.

11. Incubate at 4°C overnight on a rotator.

12. The next day add 60 μL of protein A-agarose and incubate at 4°C for 2 h on a rotator.

13. Spin for 30 s at 800×g to pellet the beads and remove the supernatant (see Note 14).

14. Add 1 mL of low salt wash buffer to the beads and rotate for 5 min at 4°C. Spin as above and remove the supernatant. Repeat two times, for a total of three low salt washes.

15. Perform a final wash with 1 mL of high salt wash buffer, spin as above, and remove the supernatant.

16. Add 150 μL of elution buffer and incubate at 65°C for 10 min to elute protein–DNA complexes. Spin as above and collect supernatant.

17. Repeat the elution step with 150 μL of elution buffer and combine the two supernatants.

18. Add 300 μL of elution buffer to the 10 μL of INPUT previously set aside.

19. Add 3 μL of 20 mg/mL proteinase K to the eluates and the INPUT samples and incubate at 37°C for 30 min.

20. Reverse crosslink from 6 h to overnight at 65°C.

21. DNA purification: Add 300 μL of phenol–chloroform–isoamyl alcohol, vortex for 30 s and spin in a microcentrifuge at 12,000×g for 5 min. Collect the upper phase and transfer in a new tube.

22. Precipitate DNA by adding 600 μL of 100% ethanol, 30 μL of 3 M sodium acetate pH 5.2 and 20 μg of glycogen.

23. Place on dry ice for 30 min and centrifuge at 12,000×g at 4°C for 30 min. Remove the supernatant and wash the pellet with 500 μL of 75% ethanol.

24. Spin for 10 min at 4°C, remove the ethanol and air-dry the pellet for 15 min at ambient temperature. Resuspend in 100 μL of water.

25. Prepare the following dilution of the INPUT samples: 1:3, 1:9, 1:27, 1:81, 1:243, and dilute each IP sample 1:4.

26. Each 20 μL qPCR reaction contains: 4 μL of DNA (INPUT or IP), 10 μL of 2× SYBR® Green PCR mix and 10 pmol of each primer. For each primer set, generate a standard curve using the undiluted INPUT sample and its dilutions and analyze each diluted IP sample in triplicate. Use the following conditions for the thermocycler (see Note 11): 95°C for 5 min, 40 cycles of 95°C for 10 s, 55°C for 10 s, and 72°C for 30 s.

27. After cycling is complete, set the PCR machine to calculate dissociation curves from 70 to 96°C (see Note 12).

Visualize the amplification curve in a logarithmic scale and set the baseline from cycle 3 to cycle 9. Set the threshold in the linear phase of amplification and record the C_t value for each well. Average the C_t values for the three replicates of the IP samples and calculate the ΔC_t for each IP sample (normalized to the INPUT sample), using the standard curve generated with the C_t values of INPUT sample. Next, calculate the $\Delta\Delta C_t$ ($\Delta\Delta C_t = \Delta C_t$ (GAPDH) $- \Delta C_t$ (target sequence)). For each IP, the relative recovery of the target sequence normalized to the relative recovery of the GAPDH coding region is $2^{-\Delta\Delta C_t}$.

4. Notes

1. Cells should not be seeded at a density lower than 2×10^5 cells/mL and should not be grown above 10^6 cells/mL. Count cells with a hemocytometer before and after splitting.

2. Extreme care should be taken while handling blood samples. Wear protective equipment (lab coat, goggles, and gloves) throughout the whole procedure and decontaminate hood and pipettes with 25% bleach. Refer to your institution's health and safety manual for practices when handling blood samples.

3. Very slow acceleration and deceleration should be applied to the centrifuge run. At the end of the run carefully remove the tubes from the centrifuge without disturbing the lymphocyte layer.

4. Dilute HDACi in DMSO so that the final concentration of DMSO in cell culture medium does not exceed 0.4%.

5. Use HDACi at a concentration at or below their EC_{50}. Using HDACi at a concentration above their EC_{50} can cause an increase of *FXN* transcript, as a stress response.

6. If an accurate determination of EC_{50} is necessary, more dilutions of HDACi are required to produce a more detailed dose–response curve.

7. The protein concentration should be between 0.1 and 0.5 μg/μL, as 5–25 μg protein is recommended per assay replicate. Samples with protein concentrations under 0.1 μg should be excluded from analysis.

8. When handling RNA samples, always wear gloves, use clean pipettes and nuclease-free tubes, tips, and reagents.

9. Measure absorbance at 230, 260, and 280 nm. A good RNA prep will have a 260/280 and a 260/230 ratio close to 2.

10. Expression of most genes normally used as recovery standards in qRT-PCR changes upon HDACi treatment, hence these genes are not suitable for the quantification of relative *FXN* mRNA. We recommend normalizing to cell number, which is more important for lymphoblastoid cells than for primary lymphocytes, as we observe a dose-dependent decrease in cell number when lymphoblastoid cells are treated with HDACi, likely due to the cytostatic effect of HDACi on actively dividing cells. When we normalized to total RNA with these cells, there is a dramatic dose-dependent decrease in GAPDH expression, with the GAPDH level decreasing approximately 70–80% at the higher HDACi concentrations. When we normalize to cell number at the time of cell collection, the subsequent decrease in GAPDH is much less severe, with at most a 20% decrease in the expression level. In contrast to lymphoblastoid cells, HDACi concentration does not appear to decrease cell numbers of nondividing cells, such as primary lymphocytes. Even though primary lymphocyte cell numbers are not affected by HDACi treatment, when total RNA normalization is compared with cell number normalization an improvement in the stability of GAPDH expression level is observed. We routinely count cells after the treatment with HDACi just before collecting the cells for RNA isolation and still have not observed any dose-dependent changes in cell numbers. However, adjust the volume collected from each treatment to ensure that an equal number of cells are collected for each treatment condition.

11. Cycling conditions depend on the thermocycler used, refer to your instrument's manual for optimal cycling conditions.

12. The dissociation plot for every primer pair should be a single sharp peak.

13. Optimal shearing conditions must be determined for different sonicators. The average fragment length of genomic DNA determines the resolution of the assay. To determine the average size of sheared DNA, remove 20 μL of sheared cell extract and add 300 μL of elution buffer. Add proteinase K to a final concentration of 0.2 mg/mL and incubate at 37°C for 30 min and then from 6 h to overnight at 65°C to reverse crosslinking. Add 300 μL of phenol–chloroform–isoamyl alcohol, vortex for 30 s, and spin in a microcentrifuge at $12,000 \times g$ for 5 min. Collect the upper phase and transfer in a new tube. Precipitate DNA by adding 600 μL of 100% ethanol, 30 μL of 3 M sodium acetate pH 5.2, and 20 μg of glycogen. Place on dry ice for 30 min and centrifuge at $12,000 \times g$ at 4°C for 30 min. Remove the supernatant and wash the pellet with 500 μL of 75% ethanol. Spin for 10 min at 4°C, remove the ethanol and air-dry the pellet for 15 min at room temperature. Resuspend in 10 μL of water and load onto a 1% agarose gel to check the fragment size. An average fragment size of 400–500 bp is recommended.

14. Remove as much supernatant as possible without aspirating the agarose beads. The use of 0.4 mm flat tips helps minimizing the amount of beads aspirated with the supernatant.

References

1. Marks, P., and Breslow, R. (2007) Dimethyl sulfoxide to vorinostat: development of this histone deacetylase inhibitor as an anticancer drug. *Nat. Biotechnol.* **25**, 84–90

2. Kazantsev, A. G., and Thompson, L. M. (2008) Therapeutic application of histone deacetylase inhibitors for central nervous system disorders. *Nat. Rev. Drug Discov.* **7**, 854–868.

3. Herman, D., Jenssen, K., Burnett, R., Soragni, E., Perlman, S. L., and Gottesfeld, J. M. (2006) Histone deacetylase inhibitors reverse gene silencing in Friedreich's ataxia. *Nature Chem. Biol.* **2**, 551–558.

4. Rai, M., Soragni, E., Chou, C. J., Barnes, G., Jones, S., Rusche, J. R., Gottesfeld, J. M., and Pandolfo, M. (2010) Two new pimelic diphenylamide HDAC inhibitors induce sustained frataxin upregulation in cells from Friedreich's ataxia patients and in a mouse model. *PLoS One* **5**, e8825.

5. Rai, M., Soragni, E., Jenssen, K., Burnett, R., Herman, D., Gottesfeld, J. M., and Pandolfo, M. (2008) HDAC inhibitors correct frataxin deficiency in a Friedreich ataxia mouse model. *PLoS ONE* **3**, e1958 doi:1910.1371/journal.pone.0001958.

6. Thomas, E. A., Coppola, G., Desplats, P. A., Tang, B., Soragni, E., Burnett, R., Gao, F., Fitzgerald, K. M., Borok, J. F., Herman, D., Geschwind, D. H., and Gottesfeld, J. M. (2008) The HDAC inhibitor 4b ameliorates the disease phenotype and transcriptional abnormalities in Huntington's disease transgenic mice. *Proc. Natl. Acad. Sci. USA* **105**, 15564–15569.

7. Chou, C. J., Herman, D., and Gottesfeld, J. M. (2008) Pimelic diphenylamide 106 is a slow, tight-binding inhibitor of class I histone deacetylases. *J. Biol. Chem.* **283**, 35402–35409.

8. Xu, C., Soragni, E., Chou, C. J., Herman, D., Plasterer, H. L., Rusche, J. R., and Gottesfeld, J. M. (2009) Chemical probes identify a role for histone deacetylase 3 in Friedreich's ataxia gene silencing. *Chem. Biol.* **16**, 980–989.

INDEX

Giovanni Manfredi and Hibiki Kawamata (eds.), *Neurodegeneration: Methods and Protocols*,
Methods in Molecular Biology, vol. 793, DOI 10.1007/978-1-61779-328-8, © Springer Science+Business Media, LLC 2011